Flight Performance of Fixed and Rotary Wing Aircraft

To my mother

Flight Performance of Fixed and Rotary Wing Aircraft

Antonio Filippone

AMSTERDAM • BOSTON • HEIDELBERG • LONDON • NEW YORK • OXFORD
PARIS • SAN DIEGO • SAN FRANCISCO • SINGAPORE • SYDNEY • TOKYO

Butterworth-Heinemann is an imprint of Elsevier

Butterworth-Heinemann is an imprint of Elsevier
Linacre House, Jordan Hill, Oxford OX2 8DP
30 Corporate Drive, Suite 400, Burlington, MA 01803, USA

First edition 2006

British Library Cataloguing in Publication Data
A catalogue record for this book is available from the British Library

Library of Congress Cataloguing in Publication Data
A catalogue record for this book is available from the Library of Congress

ISBN–13: 978-0-7506-6817-0
ISBN–10: 0-7506-6817-2

For information on all Butterworth-Heinemann publications visit
our web site at http://books.elsevier.com

Printed and bound by CPI Group (UK) Ltd, Croydon, CR0 4YY
Transferred to Digital Print 2012

Contents

Preface

This book originates from a series of lectures taught to students of aerospace engineering at the University of Manchester over the course of five years. These courses include aircraft performance, aerospace flight mechanics and helicopter theory. In the process, I have taken advantage of other material I have taught in supersonic aerodynamics. Over time, the material grew in scope, to include more advanced research issues of interest to the professional engineer. The book is now proposed as a reference on the subject of atmospheric flight mechanics and performance. The material presented covers a broad range of aspects in aircraft flight, relative to fixed- and rotary-wing aircraft. I have excluded from the discussion rockets and re-entry vehicles, which are powered by other means, and operate in transatmospheric flight. These vehicles are the subject of space sciences.

This book is not an introduction to flight, nor a pilot's handbook. A background knowledge of aerodynamics and differential calculus is required. I have assumed that readers are familiar with the aerodynamics of airfoils and wings. My main scopes are:

- To introduce the subject of aircraft performance analysis and mission planning;
- To provide a critical analysis of the aircraft performance parameters;
- To present a unified approach of aircraft performance for fixed and rotary wing;
- To present the principles of aircraft noise performance;
- To present the subject of aircraft performance optimization.

I have found that most textbooks focus on aerodynamics, even at the elementary level, and contain oversimplified presentations of the aircraft performance. I have decided to move on from this approach. Several performance problems requiring numerical solutions are proposed. They represent the modern approach to aerospace engineering problems.

I have included some more advanced topics, such as supersonic acceleration, transient roll, optimal climb of propeller-driven aircraft, propeller performance, long-range flight with en-route stop, zero-gravity flight in the atmosphere, V/STOL operations, ski jump from aircraft carrier, some optimal flight paths at subsonic and supersonic speed, range/payload analysis of fixed- and rotary-wing aircraft, performance of tandem helicopters, lower-bound noise estimation, and sonic boom calculations. I have removed the classical tables of the International Standard Atmosphere. It is now more convenient to calculate the air properties from simple computer programs.

Noise performance is now considered essential for aircraft certification, and for the public's acceptance of air traffic near populated areas. Other environmental effects, such as engine emissions, have been left out, because this subject is better covered by modern gas turbine textbooks. However, there is a clear indication that sustainability is now becoming an issue in aerospace engineering. The International Panel for

Climate Change (IPCC) publishes timely reports on all aspects of aviation and the environment.

The inclusion of rotorcraft in a performance textbook is unusual. This topic is not currently taught in courses on aircraft performance, design and aerodynamics. The performance calculation of even simple cases requires a considerable amount of analysis. Only two examples of books covering both subjects come to my mind: McCormick's[104] classic book on aerodynamics, aeronautics, and flight mechanics (1995) and Seckel's[151] book on aircraft control and stability (1964).

All the results presented are the results of computer programs. There are a large number of algorithms in the book. Some of these algorithms have been included in Appendix C and can be found at http://books.elsevier.com/companions, along with other engineering material (engine charts and aerodynamic data). The calculations presented are realistic, but do not correspond to a real aircraft. The performance data change from aircraft to aircraft and the number of hidden variables in a real-life performance calculation can be intimidating.

In the mathematical formulation of the equations the reader will find an unusual notation. I have taken the habit of collecting all the known parameters in constant factors. This strategy has two advantages: it allows a closer examination of the free parameters of the equations, and is computationally efficient when programmed for a numerical solution.

I have struggled to maintain a list of symbols that is coherent with the engineering practice. Unfortunately, there are many cases of *alias* within the subject that I have presented, therefore in some instances I had to resort to non-standard symbols, while avoiding too many sub- and superscripts, which are annoying. Nevertheless, the list of symbols is complete, and most definitions are cross-referenced.

A word of caution is needed with respect to the aircraft data provided throughout the book. They have been inferred, interpolated, extrapolated, and calculated. In no circumstances do they represent data of any specific airplane. Aircraft operations are heavily regulated. The regulations include virtually everything, therefore I have included a list of organizations where additional relevant information can be found (p. xix).

The book is divided into three parts. Part I deals with performance problems of fixed-wing aircraft. Part II deals with rotary-wing aircraft. Part III deals with V/STOL and noise performance.

Each chapter ends with a set of problems. Some of these problems are from past exams, others have been formulated specifically for this textbook. I have avoided questions such as *prove that* and *show that*, that are exercises in applied mathematics unappealing to young aerospace engineering students. The problems proposed are practical solutions to fixed- and rotary-wing aircraft performance, and sometimes critical analyses requiring additional research. As a part of this, sometimes it is required to fill in missing data, that may be available from the aircraft data in the appendix. Judgement, educated guesses and critical thinking are part of the engineering practice. This strategy partly conforms with some recent trends in teaching, such as problem-based learning or enquiry-based learning.

I have excluded from the references a number of quality publications, available from departments in industry, government and the military. These references are sometimes difficult to find, in limited editions, in electronic form, restricted to a few people, out of print, or even disappeared from the records. The papers in this class that are cited were freely available over the Internet, or were considered of particular interest. Much (but not all) of NASA's, NATO's and Aeronautical Research Council's (ARC) bibliographical material is accessible in electronic format. The bibliographical material is limited to that published in the English language, which may not be representative of all the work published around the world.

Books, I have learned, are open-ended projects. I have not covered all the flight vehicles, nor all the things I wished to cover. I hope the book will be judged for what is included. The specialist will find additional material in the bibliography. I have intentionally left out problems of flight in ground effect, high-altitude flight, flight in adverse weather, human-powered flight, lighter-than-air aircraft, unmanned flight, two- and three-degree of freedom flight mechanics, inertial coupling, trim conditions, some advanced fighter performance, and most optimization problems requiring the use of control theory. I just mention in this context that some unmanned vehicles are capable of performing missions not described in this book. In spite of the usual gloom in the aerospace world, I believe there is still a great deal of technology that we have not seen – including "flying saucers".

Some ideas on parts of this book stemmed from serendipitous events. My ideas on soaring flight were inspired by observations of the soaring flight of turkey vultures in Cuba; the birds' migration problems arose from observations of Berwick swans at Martin Mere, England; the formation flight from pictures I took years ago in Denmark.

A. Filippone
Manchester, United Kingdom

Acknowledgments

A number of individuals and institutions have contributed to this work. Dan S. Sørensen, of Novo Nordisk A/S, Copenhagen, provided the program `gtpoints` to digitize and extrapolate performance data from charts. I have used this program extensively when there was no other means of extracting reference data; `gtpoints` is a powerful tool for the digital age.

Brian Riddle, at *The Royal Aeronautical Society* in London, provided great help with historical research at the Society's Library. The pictures of the Avro-F were found in lost archives by Harry Holmes, Bob Pick and Nick Forder at the Manchester Museum of Science and Technology. I want to acknowledge the fruitful discussions we had on this airplane over the past few years. Mrs Sue Barter at the University of Manchester has always responded quickly to my requests of bibliographical material.

Peter Miller at the University of Wisconsin, Ann Harbor, provided feedback on some mathematical aspects. Peter Lissaman, Da Vinci Enterprises, Santa Fe, New Mexico, contributed critically to the theory of formation flight, and reviewed substantial parts of the book (in particular, Chapter 9). Prof. J. Gordon Leishman, at the Center of Excellence for Rotorcraft Research at the University of Maryland, took a great deal of time to review the chapters on the rotary-wing aircraft, particularly my first rough draft. He also provided me with additional data. His promptness in replying has always surprised me. The rotorcraft material was further reviewed by John F. Perry, formerly of Westland Helicopters, and engineer of the world speed record Lynx helicopter. Mr Perry has provided insightful comments from an industrial point of view, as well as some historical details. His views on Chapters 11 to 13 have been particularly important. An anonymous reviewer has offered useful feedback on how to improve the manuscript at a very late stage of my work.

My students at the University of Manchester, Joseph Lopez-Roygon, Alexandre Villedieu, Nicolas Maymard, and Bassam Rashkani, helped with programming some of the flight performance. In addition, Mr Rashkani has done a great amount of proof-reading (Chapters 5, 6 and 17), spotted some errors, and provided comments for which I am grateful. Any incorrect statement that has been overlooked is my own fault.

Some propeller data were provided by *Hamilton Sundstrand*; data for the Airbus-300 were provided by *Airbus*; and some helicopter data were provided by *Eurocopter* representatives (special thanks to Sebastien Voisin and Guillaume Imbert).

My editor at Elsevier in Oxford, Jonathan Simpson, has been an enthusiast of my ideas from the very start. He followed my work with a great passion, and provided insightful opinions on matters of substance and appearance. I must thank Jonathan for the many useful discussions on book publishing. Miranda Turner, Melissa Read and Alan Everett at Elsevier were helpful in the final production process.

Finally, I want to thank my partner Susan Brindle, who had to put up with my digressions whenever a helicopter was flying over head, or when a simulation program did not function, and who waited patiently when I worked long hours. Last but not least, she proof-read parts of the book, and gave me emotional support when I stopped believing in what I was doing. We could have gone walking in the country, but I remained at my desk instead.

List of Tables

Nomenclature: organizations

Most of the organizations listed below regularly publish documents (technical reports, papers, journals, regulations, etc.) relative to issues discussed in this textbook. In addition, the US Department of Defense and NATO publish a dictionary of acronyms and aviation jargon.

AEA = Association of European Airlines (www.aea.be)

AGARD = Advisory Group, Aerospace Research & Development (www.rta.nato.int)

AHS = American Helicopter Society (www.vtol.org)

AIAA = American Institute of Aeronautics & Astronautics (www.aiaa.org)

ARC = Aeronautical Research Council, United Kingdom

ASTM = American Society for Testing and Materials (www.astm.org)

CAA = Civil Aviation Authority (www.caa.co.uk)

EAA = Experimental Aircraft Association (www.eaa.org)

EASA = European Aviation Safety Agency (www.easa.eu.int)

ESDU = Engineering Data Unit (www.esdu.com)

FAA = Federal Aviation Authority (www.faa.gov)

FAI = Fédération Aéronautique Internationale (www.fai.org)

FSF = Flight Safety Foundation (www.flightsafety.org)

HAI = Helicopter Association International (www.rotor.com)

IATA = International Air Transport Association (www.iata.org)

ICAO = International Civil Aviation Organization (www.icao.int)

IPCC = Intergovernmental Panel for Climate Change (www.ipcc.ch)

Jane's = Jane's Information Systems (www.janes.com)

MIL = Military Standards (www.mil-standards.com)

NASA = National Administration for Space and Aeronautics (www.nasa.gov)

NTSB = National Transportation Safety Board (www.ntsb.gov)

ONERA = Office National d'Etudes et de Recherches Aérospatiales (www.onera.fr)

RAeS = The Royal Aeronautical Society (www.raes.org.uk)

SAE = Society of Automotive Engineers (www.sae.org)

SSA = Soaring Society of America (www.ssa.org)

SAWE = Society of Allied Weight Engineers (www.sawe.org)

Nomenclature: acronyms

AEO = All Engines Operating
AF = Activity Factor
AI = Autorotative Index
ASI = Air Speed Indicator
ASW = Anti-Submarine Warfare
ATC = Air Traffic Control
AUW = All-Up Weight
BLF = Balanced Field Length
BPR = By-pass Ratio
BRGW = Brake-Release Gross Weight
BVI = Blade Vortex Interaction
CAS = Calibrated Air Speed
CG = Center of Gravity
CTOL = Conventional Take-off and Landing
DOC = Direct Operating Costs
DOF = Degree of Freedom
EAS = Equivalent Air Speed
EBF = Externally Blown Flap
EPNdB = Effective Perceived Noise, in dB
ETOPS = Extended Twin-Engine OPerationS
GTOW = Gross Take-off Weight
KTAS = True Air Speed in knots
IGE = In Ground Effect
ILS = Instrument Landing System
ISA = International Standard Atmosphere
FL = Fuselage Line; Flight Level
FoM = Figure of Merit
ODE = Ordinary Differential Equation
OEI = One Engine Inoperative
OEW = Operating Empty Weight
OGE = Out of Ground Effect
OSPL = Overall Sound Pressure Level
MAC = Mean Aerodynamic Chord
MIL = Military Standard (USA)
MLW = Maximum Landing Weight
MR = Main Rotor
MRW = Maximum Ramp Weight
MTOW = Maximum Take-off Weight
MZFW = Maximum Zero-Fuel Weight
NOTAR = NOTAil Rotor
PAY = Payload Weight

PAX = Passengers
SAR = Specific Air Range
SFC = Specific Fuel Consumption
SHP = Shaft Horse Power
SI = International Units System
S/L = Sea Level
SPL = Sound Pressure Level
SST = Supersonic Transport Aircraft
TAF = Total Activity Factor
TAS = True Air Speed
TMA = Terminal Maneuver Area
TODA = Take-off Distance Available
TORA = Take-off Distance Required
TOW = Take-off Weight
TPP = Tip Path Plane
TSFC = Thrust-Specific Fuel Consumption
VNE = Velocity Not to Exceed
VMC = Minimum Control Speed
VMO = Maximum Operating Speed
WAT = Weight-Altitude-Temperature (Charts)

Nomenclature: main symbols

$a =$ speed of sound, Eq. 2.35; acceleration

$a_n =$ normal acceleration

$a_t =$ tangential acceleration

$A =$ wing area; disk area

$A_b =$ cross-sectional area of the aircraft

$A_c =$ cross-sectional area of an airfoil

$A_j =$ engine nozzle area

$AI =$ autorotation index, Eq. 14.12

$A\mathcal{R} =$ wing aspect-ratio

$b =$ wing span

$bt =$ moment arm of thrust asymmetry

$c =$ wing, blade, airfoil chord

$c_i =$ constant coefficients $(i = 0, 1, 2, \cdots)$

$C_D =$ drag coefficient

$C_{D_e} =$ drag coefficient of idle engine in flight, Eq. 10.83

$C_{D_o} =$ zero-lift drag coefficient

$C_{D_v} =$ vertical drag coefficient

$C_l =$ rolling moment coefficient

$C_{l_\beta} =$ derivative of rolling moment coefficient w.r.t. side-slip angle

$C_{l_\xi} =$ derivative of rolling moment coefficient w.r.t. aileron deflection

$C_{l_\theta} =$ derivative of rolling moment coefficient w.r.t. rudder deflection

$C_{l_p} =$ damping-in-roll coefficient, Eq. 10.65

$C_L =$ lift coefficient

$C_{Lmax} =$ maximum lift coefficient

$C_{L_\alpha} =$ lift curve slope

$C_{N_\beta} =$ derivative of yawing moment coefficient w.r.t. side-slip angle

$C_{N_\xi} =$ derivative of yawing moment coefficient w.r.t. aileron deflection

$C_{N_\theta} =$ derivative of yawing moment coefficient w.r.t. rudder deflection

$c_p =$ heat of combustion

$C_p =$ specific heat at constant pressure

$C_P =$ power coefficient, Eqs 11.1

$C_Q =$ torque coefficient, Eqs 11.1

$C_T =$ thrust coefficient, Eqs 11.1

$C_W =$ weight coefficient, Eq. 11.2

$C_Y =$ side-force coefficient, Eq. 10.77

$C_{Y_\beta} =$ derivative of side-force coefficient w.r.t. side-slip angle

$C_{Y_\xi} =$ derivative of side-force coefficient w.r.t. aileron deflection

$C_{Y_\theta} =$ derivative of side-force coefficient w.r.t. rudder deflection

$d =$ diameter (propellers, helicopter rotors)

$d' =$ fineness ratio, Eq. 2.1

$dB =$ decibel

$d_r =$ distance between rotor shafts (tandem helicopters)

$D =$ drag force

$D_{ij} =$ mutually induced drag between wings i and j

$D_L =$ disk loading

$e =$ Oswald efficiency factor

$E =$ aircraft energy; endurance

$E_s =$ specific endurance

$f =$ frequency; generic function

$f_e =$ equivalent flat plate area, Eq. 13.33

$f_j =$ the same as TSFC, used in equations for brevity

$FoM =$ Figure of Merit

$g =$ acceleration of gravity

$h =$ altitude; enthalpy

$h_{cg} =$ distance between center of the x rotor and helicopter's CG

$h_E =$ energy height, Eq. 8.71

$H =$ total pressure (head); Hamiltonian function

$I =$ sound intensity, Eq. 17.1

$I_b =$ polar moment of inertia of a blade, Eq. 14.6

$I_{sp} =$ specific impulse, Eq. 5.9

$I_x =$ moment of inertia w.r.t. x (roll)

$I_y =$ moment of inertia w.r.t. y (pitch)

$I_z =$ moment of inertia w.r.t. z (yaw)

$J =$ propeller's advance ratio, Eq. 5.27

$k =$ lift-induced drag factor; induced-power factor for helicopters

$k_o =$ overlap factor of tandem rotors

$k_f =$ fuselage interference factor of tandem rotors

$k_g =$ ground effect correction factor for helicopter rotor, Eq. 12.43

$k_p =$ coefficient in Polhamus Eq. 4.8

$k_r =$ reflection factor in sonic boom

$k_{tr} =$ induced-power factor for tail rotor

$k_v =$ coefficient in Polhamus, Eq. 4.8

$l =$ linear dimension; aircraft's length

$L =$ lift force

$\mathcal{L} =$ rolling moment

$L_p =$ rolling moment derivative w.r.t. the roll rate, Eq. 10.59

$L_\xi =$ aileron effectiveness, Eq. 10.58

$m =$ mass; blade's mass per unit length

$M =$ Mach number; blade's mass

$\mathcal{M} =$ pitching moment

$M_c =$ critical Mach number

$M_{dd} =$ divergence Mach number

$n =$ normal load factor, Eq. 10.4

$\boldsymbol{n} =$ normal unit vector

$N =$ number of blades

$\mathcal{N} =$ yawing moment

$p =$ pressure; pitch rate

$p^* =$ impact pressure, Eq. 6.13

$p_r =$ pressure ratio

p_o = stagnation pressure

P = engine power

P_c = helicopter climb power

P_e = specific excess power

P_h = hover power

P_o = rotor profile power

P_p = airframe parasite power

PL = power loading, Eq. 12.15

q = dynamic pressure, Eq. 6.6; pitch rate

Q = propeller/rotor torque; heat transfer (Chapter 5)

r = non-dimensional radial coordinate; distance between two points

R = propeller/rotor blade radius (Chapter 5); aircraft range (Chapter 9)

R_{out} = equivalent all-out range, Eq. 9.74

$\mathcal{R}$ = ideal gas constant

s_D = aerodynamic penetration, Eq. 4.35

s_L = aerodynamic radius, Eq. 4.38

t = time

$\boldsymbol{t}$ = tangential unit vector

T = engine thrust; net thrust

T_o = static thrust

$\mathcal{T}$ = absolute temperature

U = aircraft velocity in the flight direction

U_{md} = speed of minimum drag

U_{mp} = speed of minimum power

U_T = velocity component parallel to rotor disk (in-plane)

v_c = climb speed (climb rate)

v_i = rotor's induced velocity

v_h = rotor's induced velocity in hover

v_s = sinking speed, or descent rate

V = aircraft's velocity in the horizontal direction

$\boldsymbol{V}$ = velocity vector

V_a = aircraft's volume

w = downwash velocity; vertical velocity component

W = weight; work done (Chapter 5)

x_L = landing run

x_{to} = take-off run

x_{tr} = distance between main and tail rotor shafts

x, y, z = Cartesian coordinates (Earth axes)

x_b, y_b, z_b = Cartesian coordinates centered at CG (body axes)

y = radial coordinate from the rotor center

z = service ceiling

Nomenclature: Greek symbols

$\alpha =$ angle of attack

$\alpha_{CLmax} =$ angle of attack corresponding to C_{Lmax}

$\alpha_o =$ zero-lift angle of attack

$\alpha_T =$ tilt angle

$\beta =$ side-slip (yaw) angle

$\beta = (M^2 - 1)^{1/2}$

$\gamma =$ angle of climb; ratio between specific heats

$\gamma_r =$ runway or ramp slope

$\Gamma =$ circulation

$\delta =$ relative pressure (Chapter 2); boundary layer thickness

$\delta^* =$ boundary layer displacement thickness

$\epsilon =$ thrust angle; small number

$\zeta =$ block fuel ratio

$\eta =$ lift-induced factor in drag, Eq. 4.12; propeller's efficiency

$\eta_p =$ propulsive efficiency, Eq. 13.71

$\theta =$ relative temperature; blade pitch; rudder deflection

$\theta_o =$ collective pitch

$\vartheta =$ polar coordinate; stagger angle (§ 9.15)

$\lambda =$ induced velocity factor in forward flight; Lagrange multiplier; taper ratio

$\lambda_h =$ induced velocity factor in hover

$\lambda_s =$ induced velocity factor in descent

$\Lambda =$ wing's sweep angle; vector of Lagrange multipliers (Chapter 8)

$\mu =$ dynamic viscosity; rolling coefficient; advance ratio (helicopters)

$\xi =$ parameter defined by Eq. 8.59; track angle; aileron deflection

$\Pi =$ throttle position

$\rho =$ air density

$\rho_b =$ specific mass of a rotor blade

$\sigma =$ relative air density; rotor solidity (Eq. 5.23)

$\sigma_1 =$ relative air density (when confusion with rotor solidity arises)

$\tau =$ response time

$\phi =$ roll or bank angle; inflow angle of attack (helicopters, propellers)

$\varphi =$ dihedral/anhedral angle (positive upwards)

$\chi =$ radius of curvature of flight path

$\omega_o =$ resonant frequency; Doppler frequency

$\omega =$ non-dimensional rotor speed

$\Omega =$ rotational velocity

$\Omega_1 =$ engine rotational speed

$\psi =$ heading angle; blade azimuth angle; constraint equation

$\dot{\psi} =$ turn rate, Eq. 10.39

Nomenclature: subscripts/superscripts

$()_o$ = sea level, standard, stagnation or initial conditions
$()_a$ = air
$()_b$ = body
$()_c$ = climb
$()_e$ = value at end point
$()_f$ = fuel quantity
$()_g$ = in ground effect, ground conditions
$()_h$ = hover conditions
$()_i$ = ideal, induced, or value at initial point
$()_{jet}$ = relative to a jet, or jet engine
$()_{le}$ = value at leading-edge
$()_{lo}$ = value at lift-off
$()_{max}$ = maximum value
$()_{min}$ = minimum value
$()_{mr}$ = main rotor
$()_p$ = parasite
$()_{qc}$ = quarter-chord
$()_{ref}$ = reference quantity
$()_{sl}$ = sea level conditions (when alias with "o" occurs)
$()_t$ = transmission, or turn
$()_{te}$ = value at trailing-edge
$()_{tip}$ = value at the blade tip
$()_{to}$ = value at take-off
$()_{tr}$ = tail rotor
$()_v$ = viscous or profile
$()_w$ = wind
$()$ = mean value
$(\cdot)$ = time derivative
∞ = free stream conditions

Supplements to the text

For the student:

- Additional material can be downloaded from the companion website accompanying this book. To access this please visit http://books.elsevier.com/companions and follow the instructions on screen.

For the instructor:

- Solutions to the exercises in this book are freely available to teachers who adopt or recommend the text for course use. For information on accessing this material visit http://textbooks.elsevier.com and follow the instructions on screen.

Part I
Fixed-Wing Aircraft Performance

Chapter 1
Introduction

... but the fact remains that a couple of bicycle mechanics from Dayton, Ohio, had designed, constructed, and flown for the first time ever a practical airplane.

J. Dos Passos, in *The Big Money*, 1932

After an uncertain start at the beginning of the 20th century, aviation has grown to a size on a global scale. By the year 2000, over 100 million passengers traveled through the airports of large metropolitan areas, such as London. In the same year, there have been 35.14 million commercial departures worldwide, for a total of 18.14 million flight hours[1]. Demand for commercial air travel has grown by an estimated 9% a year since the 1960s. The expansion of the aviation services is set to increase strongly. Today, every million passengers contribute about 3,000 jobs (directly and indirectly) to the economy. Therefore, aircraft performance is a substantial subject.

The calculation and optimization of aircraft performance are required to:

- Design a new aircraft;
- Verify that the aircraft achieves its design targets;
- Efficiently operate an existing aircraft or fleet;
- Select a new aircraft;
- Modify, upgrade and extend the flight envelope;
- Upgrade and extend the mission profile;
- Investigate the causes of aircraft accidents;
- Provide data for the aircraft certification (*Certificate of Airworthiness*).

The engineering methods for the evaluation of aircraft performance are based on theoretical analysis and flight testing. The latter method is made possible by accurate measurement techniques, including navigation instruments. Flight testing is essentially an experimental discipline – albeit an expensive one. Performance flight testing involves the calibration of instruments and static tests on the ground, testing at all the important conditions, gathering of data from computers, data analysis, and calibration with simulation models. Wind tunnel testing is only used for the prediction of the aerodynamic characteristics. Graphical methods, such as finding the intersection between two performance curves, belong to past engineering practice. Analytical and numerical methods, including the equations of motion of the aircraft, are the subject of this textbook. Analytical methods yield closed-form solutions to relatively simple problems. Numerical solutions address more complex problems, and allow the aircraft engineers to explore "virtually" the complete parametric space of the aircraft flight. This practice avoids expensive and risky flight testing. Methods for flight testing and evaluation of the fixed- and rotary-wing aircraft performance are discussed by Kimberlin[2], Olson[3] and Cooke and Fitzpatrick[4], respectively.

Due to the variety of requirements, the subject of aircraft performance intersects several other disciplines, such as aircraft design, scheduling, operational research, systems, stability and controls, navigation, air traffic operations, flight simulation, optimization, in addition to aerodynamics, structures, propulsion systems and integration. Therefore, aircraft performance is essentially a multidisciplinary subject. Among the ones well known to the aerospace engineers there is the flight mechanics approach, the dynamics and aerodynamics of flight (for example, Lan and Roskam[5], Anderson[6]).

Personal interests may be involved in selecting the type of flight vehicles, as these include conventional airplanes, high-performance military aircraft, helicopters, V/STOL aircraft, rockets and vehicles for transatmospheric flight. Old and modern books on the subject deal only with some of these flight vehicles – as convenient.

The basic performance of the fixed- and rotary-wing aircraft can be calculated with little mathematical effort, using the one-degree of freedom model. However, a more accurate prediction of any performance parameter, particularly if the aircraft is maneuvering in unsteady mode, is a challenging subject, because it generally involves a number of free parameters in non-linear differential equations. It will be shown how the question of *how fast can an airplane fly* is difficult to answer. In short, it depends on *how* it flies.

Performance prediction is at the base of any concrete aircraft design methodology. The estimation of weights, range and power plant size requires the calculation of basic aircraft performance from a few input data. In this case the approximation is generally good enough for parameter estimation and design. Input from operational parameters and flight testing is required for detailed analysis.

Performance optimization is at the heart of design and operation of all modern aircraft. From the operational point of view, commercial aviation is driven by fuel prices, and operations at minimum fuel consumption are of great relevance. Performance optimization requires notions of optimal control theory, a subject unfamiliar to aerospace engineers.

Performance efficiency goes beyond the design point, and requires that the aircraft produces the best performance over the widest range of its flight envelope. For this reason, the subject of performance optimization is essential in design. The fighter jets Grumman F-14 and McDonnell-Douglas F-15 (1970s) were the first to be designed with the optimization approach, and all the aircraft of later generations were conceived in the same fashion.

In the past 30 years these optimal conditions have been increasingly challenged by environmental concerns, including noise emission, air quality near airports, global climate change and sustainability. Some aspects of the impact of aviation on climate change are the subject of routine review. Publications of relevance include the ones from the Intergovernmental Panel on Climate Change (IPCC), for example ref. 7.

1.1 PHYSICAL UNITS USED

International units (SI) are used whenever possible. Unfortunately, most data in aviation are still in imperial units. Conversion to international units is not foreseen for the immediate future. In most cases, the flight altitudes will be converted to feet, because of the extensive practice of working out the performance parameters in term of this unit. Speeds will also be given in knots or km/h.

The SI nomenclature notwithstanding, some spurious engineering units have had to be retained in some cases. One of the most confusing units ever devised is the kg. This unit is used for both weight (force) and mass: weight in kgf is equal to mass in kgm. This equivalence can fool any experienced engineer. Unfortunately, there is no way around it, because it is more convenient to denote a weight with kgf, rather than the newton. The mass, instead of the weight, appears in the energy equations, which is the main reason for retaining the kgm. By contrast, the weight appears in the aerodynamic coefficients, and if the other parameters are in international units, then the weight must be converted into newtons. Therefore, the confusion is sometimes overwhelming.

To the student approaching the subject for the first time there is a special word of caution. It is easy to miscalculate an aircraft's performance because of the use of non-conformal units. Some of the most common errors arise from using speeds in km/h instead of m/s, and kN or kW instead of N and W (thrust and power, respectively). The units for specific fuel consumption can also be confusing. With some critical thinking these errors can be avoided. A range of 2,600,000 km, instead of 2,600 km, is achieved by an airplane if one oversees the coherence of units. The former result is a distance from Earth to the Moon and back three times, while the correct result is a medium range flight in many parts of the world.

1.2   PERFORMANCE PARAMETERS

A performance parameter is a quantitative indicator representing how a vehicle operates in a specific flight condition. Typical performance parameters are weights, speeds, aerodynamic loads, engine thrust and power, range and endurance, accelerations, emission indexes (noise, exhaust gases) and many more. At least 60 different parameters can be taken into account in a full aircraft performance analysis.

In accident investigation, the flight parameters considered are the air speed, the Mach number, the dynamic pressure, the altitude, the air temperature, the rate of climb or descent, the flight path angle, the side-slip angle, the angular velocities and accelerations, the load factor, the rudder position, the control surfaces position, the fuel load and the engine's status.

It is not obvious what distinguishes a performance parameter from a purely aerodynamic, propulsion, operational parameter. The drag coefficient of the wing section is not a performance parameter, but the aircraft's drag coefficient is. The wing's aerodynamic characteristics are not a subject of aircraft performance, but the aerodynamic characteristics of the aircraft as a whole are. In performance analyses the drag coefficients are the known part of the problem, while in aircraft design they are part of the problem. The thrust is by itself an engine performance; the same engine mounted and integrated on the airframe becomes an aircraft parameter. The analysis must take into account that the system engine/airframe is not the same thing as the engine alone (airframe/engine integration). The stealth capabilities of an aircraft (radar signature, thermal signature, noise emission) depend more on the design of the aircraft than its operation. Not all the parameters will appear on the instrument panels in the cockpit, and some of them are not relevant to the pilot.

Many performance indexes cannot simply be expressed by a single value, but are presented with charts, because they are dependent on other parameters. The combination of those parameters is essential in defining the operation of the aircraft. Typical charts are the air speed relationships, the weight/altitude/temperature charts, the flight envelopes, and the payload range.

Some performance data are readily available from the manufacturer; other data can be inferred by appropriate analysis; others are clouded by secrecy or confidentiality; and others are difficult to interpret, because the conditions under which the aircraft performs are not given. Among the most common data covered by secrecy are the drag data, the stability characteristics, the excess power diagrams and the engine performance. Other examples are 1) the aircraft range, when the payload is not supplied together with the range; 2) the altitude at which this range is achieved; and 3) the radius of action of a military interceptor – this radius, in fact, may lie in the favourite field of enemy fire.

The maximum take-off weight (MTOW) and the operating empty weight (OEW) are available for most aircraft. However, these data are not sufficient to calculate the maximum payload, because the difference between MTOW and OEW must include the mission fuel. Therefore, some educated guess is needed. A weight advantage compared to heavier rivals translates into significant revenue-earning advantages, which in a competitive market is the most important factor for choosing and operating an aircraft. It is not uncommon to find manufacturers unhappy that their performance data and charts are published in the public domain. Performance charts allow customers and competitors to look at various options, to select the most competitive aircraft and to discover the flaws of the competitors' technology: sharing information makes everybody better players.

The purpose of this book is to take the reader through some simple performance calculations, to look at the performance data, and to give an introduction to aircraft performance optimization. A large number of data are published by Jane's Information Systems[8], and Flight International[9]; other valuable data for lesser known aircraft are available in Gurton[10] and Loftin[11]. The latter reference differs from the other ones because of its critical analysis. Loftin, in his extensive bibliography, also points to additional sources of aircraft performance data. Further data have been taken from official documents of the international authorities (FAA, ICAO), and specialized publications such as AGARD[12], ESDU, and by our own research[13]. ESDU (Engineering Sheets Data Units) provide data and methods on all areas of performance, and are of invaluable value to the practitioner engineer. Of particular interest are landing[14] and take-off performance[15], drag of airframes[16], and range and endurance[17].

Updated data that are not proprietary are published regularly by the magazines *Flight International* and *Flug Revue*. All flight manuals report the essential performance curves of the aircraft and its engines, and include data that may not be available elsewhere. Most flight manuals are now available in electronic form, and represent a great wealth of information for the aircraft performance engineer.

As in any other technology sector, the operator of an aircraft is concerned that the performance parameters quoted by the manufacturer match the actual performance, therefore accuracy of performance prediction methods is essential.

1.3   PERFORMANCE OPTIMIZATION

In the early 1950s, computers made their first appearance in aerospace engineering. Bairstow[18] wrote in 1951 that:

> The use of electric calculators is coming in to reduce manual labor, but there is little hope of doing nearly all that we would like to do.

Computer solutions of aircraft performance are now routine jobs, and have reached a phenomenal level of sophistication, to include the coupling between flight mechanics, aerodynamics, structural dynamics, flight system control and differential game theory. With analog computers first, then with digital computers, the problems solved grew in complexity. See, for example, the 1959 edition of Etkin's book on flight dynamics[19] to gain a perspective. In 1982, Ashley[20] exemplified the problems of optimization in a paper titled "On Making Things the Best". The author argued, among other things, that flight planning ceased to be a matter of hand calculation by the time commercial jet propulsion was introduced (late 1950s).

There are two categories of optimization: optimization of aircraft performance during the design phase, and optimization of operational performance for the given airplane. In the former case, one can investigate the alternative changes in configuration that improve one or more performance parameters. This is more appropriately the subject of aircraft design. We will consider some cases of operational optimization. An excellent source for optimization problems with aircraft applications is the classic book of Bryson and Ho on optimal control[21]. Some of these problems, including multistage rocket trajectories, were also reviewed by Ashley[22].

Today there are programs that plan optimal trajectory routes to minimize DOC (Direct Operating Costs), while complying with several airline constraints. These programs have several types of input data: weather conditions, route, aerodynamics, aircraft performance, and flight-specific information, such as payload, fuel cost, etc. On output they provide the amount of fuel for optimal cruise altitude, climb and descent points, optimal cruise speed, and flight path.

1.4   CERTIFICATE OF AIRWORTHINESS

The Certificate of Airworthiness is a document that grants authorization to operate an aircraft. It specifies the limits of operation of the vehicle in terms of weights, take-off and landing requirements, and a number of other parameters, such as maintenance records, service history and compliance with safety regulations.

The certificate proves that the aircraft conforms to the type specified and it is safe to fly. The certificate is valid as long as the aircraft meets the type specification (commercial, commuter, utility, etc.), it is operated safely and all the airworthiness directives are met. The aircraft may lose its certificate for a number of reasons, including modifications, upgrades, and new directives approved by the international organizations that make the aircraft obsolete, not just unsafe to operate. Other documents are generally required, such as the type certificate data sheet, the certificate of maintenance, and a list of other papers. These documents seldom contain detailed performance data.

Certificates of Airworthiness are issued by the Federal Aviation Administration (FAA) in the USA, by the European Aviation Safety Agency (EASA), by the Civil

Aviation Authority (CAA) in the UK and by other national and international bodies around the world. Certification is a complex legal and technical matter that is beyond the scope of this book.

1.5 UPGRADING OF AIRCRAFT PERFORMANCE

The age of bicycle mechanics has long passed. In the current technology situation, most aircraft are likely to be upgraded and modified to fit the changing market and technological advances. The technology that is fitted over the years can be phenomenally different from the first design. The service time of a single aircraft is of the order of 20 to 25 years, and the life of an aircraft family may exceed 50 years. A lifetime career can be devoted to a single airplane. The famed aircraft engineer Reginald J. Mitchell designed 24 aircraft, including the Spitfire, before dying prematurely, aged 42, in 1937.

To be fair, in the early days of aviation, a new aircraft could roll out of the factory in a few months. Indeed, some aircraft were prototypes that logged a few flights and then were scrapped – if they survived a crash. Figure 1.1 shows the Avro Model F (1912) at Manchester. This airplane was the first to have an enclosed cockpit, but it was capable of flying at only one speed, 65 mph. Only two airplanes were built. The picture to the right shows the airplane after it crash-landed in May 1912 due to an engine failure. The photo appears to have been published as a postcard.

It took only 43 days to build the Ryan NYP that made the transatlantic crossing in 1927, which included a total of 850 engineering hours (including performance and flight testing) and 3,000 man-hours for construction[23]. In 1936, it took just one year for the German aircraft designer Kurt Tank to get from concept to first flight of the Focke Wulf Condor Fw-200, the first long-range passenger (and later reconnaissance and bomber) aircraft to fly from Berlin to New York without stopping en-route (1938).

The wings of the Douglas DC-3 (1935), one of the most successful aircraft ever built, had simple performance improvement devices, a split flap for landing and outer-board ailerons for roll control. A jet aircraft of the first generation, such as the

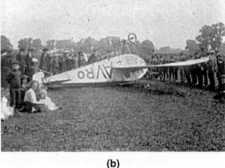

(a)   (b)

Figure 1.1   *The Avro-F, built by A.V. Roe (1912). (a) Photo first published by the magazine* Flight *on 18 June 1942; (b) photo from the AV Roe Archives. Family outing with airplane crash (25 May 1912).*

Boeing 727 (1958), had four outer-board leading-edge slats, three inboard leading-edge Kruger flaps, two banks of triple-slotted trailing-edge flaps, an inboard aileron for high-speed roll control, an outer-board aileron for low-speed flight, and seven spoilers (including five flight spoilers and two ground spoilers), also used as air brakes. This airplane is still flying.

By the 1960s, commercial airplane design and testing required thousands of man-years. The Boeing B-747-100, which first flew in 1969[24,25], required 15,000 hours of wind tunnel testing, 1,500 hours of flight testing with five aircraft over a period of 10 months, and 75,000 technical drawings[1]. The latest version of this aircraft consists of about 6 million parts, 274 km of wiring and 8 km of tubing!

The B-747-400 incorporates major aerodynamic improvements, including a more slender wing with winglets to reduce drag. A weight saving of approximately 2,270 kg was achieved in the wing by using new aluminum alloys. Finally, the version B-747-400ER has an increased take-off weight of 412,770 kg. This allows operators to fly about 410 nautical miles (760 km) further, or carry an additional 6,800 kg payload, for a range up to 14,200 km.

An even older airplane is the Lockheed Hercules C-130A. Its first model was delivered to the US military in 1956. The design of this aircraft actually started several years earlier. By the early 1960s, a V/STOL variant was designed[26]. Since then, the aircraft has progressed through at least 60 different variants. The current C-130J is actually a new airplane. Compared to the earlier popular version C-130E, the maximum speed is increased by 21%, climb time is reduced by 50%, the cruising altitude is 40% higher, the range is 40% longer, and its Rolls-Royce AE-2100DE engines generate 29% more thrust, while increasing fuel efficiency by 15%. With new engines and new propellers, the C130-J can climb to 9,100 m (28,000 feet) in 14 minutes.

Another example is the military utility helicopter CH-47, which has been in service since 1958. The basic performance upgrades for this aircraft (versions A to D) are reported in Table A.19 on page 505. In particular, the MTOW has increased by over 50% and the useful payload has doubled. To the non-expert the aircraft looks the same as it did in the 1960s.

Technological advances in aerodynamics, engines and structures can be applied to existing aircraft to improve their performance. Over time weights grow, power plants become more efficient and are replaced, aerodynamics are improved by optimization, fuselages are stretched to accommodate more payload, and additional fuel tanks are added. This is one of the main reasons why aircraft manufacturers are not challenged to start a brand new design.

The conversion of aircraft for different commercial or military applications, and the development of derivative aircraft from successful aircraft require new performance calculations, and a new certification. For example, the KC-10 tanker aircraft was derived from the commercial jet DC-10 (commercial to military conversion), and the Hercules C-130 was converted to the Lockheed L-100 (military to commercial). The conversion practice is more common with helicopters.

1.6 MISSION PROFILES

A mission profile is a scenario that is required to establish the weight, fuel, payload, range, speed, flight altitude, loiter and any other operations that the aircraft must be able to accomplish. The mission requirements are evidently specific to the type of aircraft. For high-performance aircraft they get fairly complicated, and require some statistical forecasting.

Over the years many commercial aircraft operators have specialized in niche markets, which offer prices and services to selected customers. These niches include the executive jet operators between major business centers, operators flying to particular destinations (oil and gas fields), the all-inclusive tour operators to sunny holiday resorts, and the no-frills airlines flying to minor and underused airports. These operators have different schedules and cost structures.

First, let us start with long range passenger operations, which are serviced for the greatest part by subsonic commercial jets. The basic principle is that the airplane takes off from airport A and flies to airport B along a recognized flight corridor, then returns to A. The main parameters of the mission planning are the distance between the airports, the flight time, the downtime at the airport for getting the airplane ready (also called *time-on-station*), the flight speed, the local air traffic, and the departure times at both ends. Back at the airport of origin, the day is not over for the aircraft, and the operator wishes to utilize the airplane for another flight to the same destination, or to another destination – if possible. The key is the departure time, and the minimization of the curfew. Figure 1.2 shows the typical scheduling profile of such an airplane over a transatlantic route from a major airport in Europe to an airport on the East Coast of the United States.

Due to the time-zone effect, a late morning departure from Europe arrives to the USA in the mid-afternoon. An early evening departure arrives back in Europe in the early hours of the day after. Over the 24-hour period the airplane will have done a return flight and worked about 14 to 16 hours. For a flight arriving late in the evening, a return may not be possible until early morning on the next day. This adds to the operational costs, because of the need of maintaining the crew away from the home port. The time needed to get the aircraft ready for the next intercontinental flight may require up to 3 hours. Boarding of the Boeing-747 requires 50 to 60 minutes.

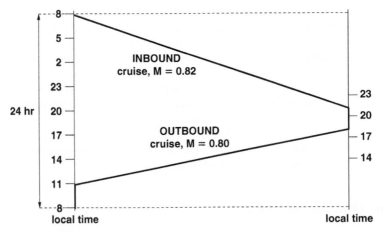

Figure 1.2 *Scheduling of transatlantic flight. The numbers on the left and right side are local times.*

Scheduling of the type shown in Figure 1.2 leads to a block time of the order of 700 to 900 hours per year (depending on aircraft and service route). An airplane flying a day-time shorter route, and returning in the mid-afternoon, should be able to make another return flight, to the same destination or otherwise. For an airline company operating anything above a dozen airplanes, scheduling and optimal operation of the fleet is a complex problem. Events such as bad weather can lead to dozens of airplanes and flight crews out of position for several days. Scheduling and operation of aircraft is a subject for operations research, and is addressed by specialized publications. Gang Yu is a good compendium to start with[27]. It deals with demand forecasting, network design, route planning, airline schedule planning, irregular operations, integrated scheduling, airport traffic simulation and control, and more.

1.6.1 Fighter Aircraft Requirements

The fighter aircraft has evolved from a reconnaissance airplane of the First World War to the most complex aircraft of modern days. Von Kármán[28] reported that fighter aircraft first flew over the battlefields of Europe to spy on enemy lines. Obviously, enemy aircraft wanted to prevent this happening, so their pilots started shooting at enemy aircraft with a pistol. This was the beginning of a dog fight. Toward the end of the war, the Dutchman Anton Fokker, working at the service of the German Army, invented a system that synchronized the shooting of a machine gun through the propeller (*interrupter gear*) – mounted on a single-seater monoplane. With the interrupter pilots had their hands free to maneuver and fight at the same time. This advance was heralded as the birth of the fighter aircraft (see Stevens[29] and Weyl[30] for historical details).

The requirements for fighter aircraft now include multipurpose missions, aircraft with complex flight envelopes, several configurations (changeable in flight), supersonic flight, combat capabilities, delivery of a wide range of weapon systems (all-weather operations), and maneuverability. There are dozens of different mission scenarios, as discussed extensively by Gallagher *et al.*[31] Typical missions are: basic, assault, combat, retrieval, close support, transport, refuel, and reconnaissance. For each of these missions there is a specific take-off weight, mission fuel, payload, range, maximum rate of climb, and service ceiling. This field is now so advanced that engineers use differential game theory and artificial intelligence to study the effectiveness of a given aircraft, and the tactical maneuverability to incoming threats (see, for example, Isaacs[32] for some problems on this subject).

One example of mission profile for this type of aircraft is shown in Figure 1.3. Such a profile must include warm-up, acceleration, take-off, climb, cruise, dash, combat, decelerate/climb, descent, and landing (with allowance for loiter and fuel reserve). In a more detailed breakdown, a typical plan may look like that in Table 1.1. The analysis of the various flight sections is essential in predicting the mission fuel; the mission fuel is essential for returning to base.

An alternative graphic method for indicating a mission profile is shown in Figure 1.4, representing an interdiction operation. The numbers indicate each flight segment. The vertical axis is an arbitrary flight altitude. The graph shows the mission radius (in arbitrary scale) and the point of engagement. Each segment is further specified by requirements such as those on Table 1.1.

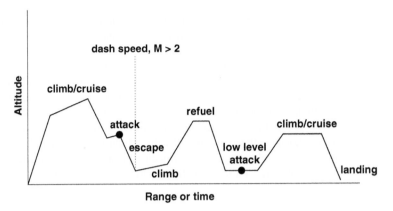

Figure 1.3 *Generic mission profile for fighter aircraft.*

Table 1.1 *Summary of flight segments of a supersonic jet fighter.*

Performance	Description
Warm-up, acceleration, take-off	2 minutes; 0.5 minutes at maximum power
Climb	at maximum power
Dash speed	$M = 1.8$, at 12,000 m (39,370 ft)
Combat	2 minutes, maximum power, dash speed
Subsonic cruise	30 minutes, $M = 0.85$
Decelerate and climb	to $M = 0.8$, at 9,500 m (31,168 ft)
Descent and landing	to sea level; no fuel credit
Loiter	10 minutes at sea level, minimum fuel
Landing	45 s at minimum power
Reserve fuel	5% of total mission fuel

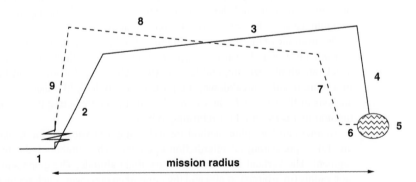

Figure 1.4 *Generic mission profile for interdiction operation.*

The *dash speed* is a supersonic speed that the aircraft can maintain for a limited amount of time, or flight distance. It is usually the maximum speed at the best flight altitude. As indicated in the table, this altitude is around 12,000 m (39,370 feet).

Loiter is the operation around an airport. It usually consists of various turns along prescribed flight corridors, before the aircraft is permitted to land. Delay in landing (and longer loiter) may be due to air traffic control and weather conditions.

A performance index characteristic only of fighter jet aircraft is the *effectiveness*. Effectiveness is defined as the product of ordnance transport rate, availability in war time and kill effectiveness. The ordnance transport weight is the product of the ordnance mass and the number of sorties per day. The availability in war time is the time the aircraft is available for operations (compared to downtime for maintenance, service, loading, etc.). The killing effectiveness is the knocking-out success rate.

1.6.2   Supersonic Commercial Aircraft Requirements

After the Concorde era came to a close, no serious attempts have been made to replace the aircraft and operate a commercial flight at supersonic speeds. Nevertheless, the theoretical analyses regarding the feasibility of such an airplane under modern environmental and financial constraints abound. A new generation of supersonic civil transport aircraft that would replace the Concorde, should be able to fly longer routes, possibly at higher speeds. A Los Angeles to Tokyo route would require a cruise speed of $M = 2.4$ in order to be able to schedule two round trips over a 24-hour period. A replacement for Concorde, operating on the North Atlantic routes, should be able to fly at $M = 2.0$, or possibly lower, if the turn-around time can be reduced. This speed is important in the cycle because it allows the airplane to be serviced at both ends, avoiding long curfews. The operators of Concorde could make a profit (once the mortgage for the acquisition of the aircraft was taken out of the spreadsheets) by having two return flights per day.

PROBLEMS

1. Discuss the possible mission profiles for a V/STOL aircraft, and extract a set of performance criteria that can be applied to all operational conditions.
2. Make a list of all the performance segments of a supersonic jet fighter, and provide a critical discussion. Provide a scenario to deal with a fuel shortage at the end of a scheduled operation, before returning to base.
3. You are asked to plan a flight timetable between London and Berlin. Provide a plan for a subsonic jet transport that maximizes the block time for the operation between the two cities. Analyze the alternative, consisting in operating a turboprop aircraft. Do the necessary research of the data needed for the solution of this problem (flight corridor, distance in nautical miles, estimated flight time, flight speeds, etc.).
4. The Boeing B-52 is one of the oldest aircraft still in service. It has progressed from the first version in 1954, B-52A, to the version B-52G. Do the necessary research to investigate how propulsion, aerodynamics and general performance

parameters have changed from the first to the latest version. List the most important quantities in a spreadsheet, and draw a conclusion.

5. Analyze the ground operations required to get a Boeing B-747-400 ready for an intercontinental flight (refueling, systems checks, food supplies, water, boarding of passengers). Produce a spreadsheet that indicates the time of each operation, and which operations can be performed in parallel. (*Problem-based learning: additional research is required*).

Chapter 2
The Aircraft and Its Environment

Contrary to the British and German military authorities, both of whom believed in nose armament, the Italians took the view that speeds of 250 mph made frontal attacks by fighters unlikely.

J.H. Stevens[29], 1953

This chapter discusses the general aircraft model for the rigid-body approximation, the reference systems, the nomenclature of the aircraft, forces, moments and angles. The aircraft operates in the atmosphere, therefore the standard air model is reviewed, along with relevant approximating functions for performance calculations. Some atmospheric effects in non-standard conditions are briefly reviewed.

2.1 GENERAL AIRCRAFT MODEL

Since the late 1940s[33,34], the accepted aircraft model for performance calculations consists of a point mass concentrated at the center of gravity. The engines are assumed to operate at the aircraft symmetry plane. There have been attempts to improve on the point-movement prediction methods, to include the fact that thrust, aerodynamic center and weight operate at different points. Aircraft flexibility is important at supersonic speed. Although a subsonic aircraft has a nearly "square" dimension, a viable supersonic transport aircraft has a length about double its wing span.

Variable geometry produces changes in the aerodynamic coefficients and in the handling quality. Longitudinal flexibility can be controlled by combined wing, tail-plane and canards. Allowance for wing flexibility is essential in special flight conditions, in order to avoid flutter effects. Wing flexibility at take-off may also contribute to the ground performance. Current research focuses on the aerodynamics, structural dynamics and flight mechanics coupling. However, the point model is quite accurate for most purposes. Therefore, only forces applied to this point and moments calculated around this point must be considered.

The model is shown in Figure 2.1: FL denotes the fuselage longitudinal axis. This is assumed to run through the center of gravity; α is the nominal angle of attack of the aircraft – it is defined as the angle between the FL and the true air speed vector. The angle of attack of the aircraft and the angle of attack of the wing are two different quantities, because the reference line of the wing is a chord-line. The angle of the thrust on the FL is called ϵ. This angle is generally quite small, and for the purposes of this discussion can also be considered zero (e.g. $\alpha + \epsilon \sim \alpha$). The angle of climb γ is the angle between the true air speed vector U and the horizon.

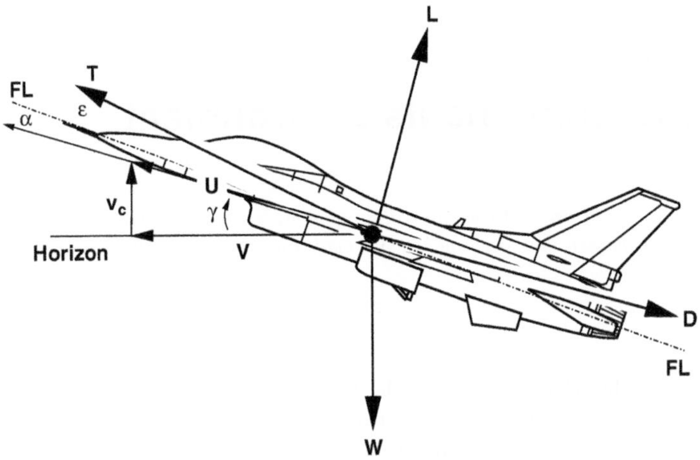

Figure 2.1 *Aircraft model in the vertical plane, with forces concentrated at the CG.*

One shape factor, mostly used in supersonic aerodynamics, is the fineness ratio l/d', where

$$d' = \sqrt{\frac{4A_b}{\pi}}$$ (2.1)

is the equivalent body diameter, A_b is the maximum cross-sectional area of the aircraft, and l is the aircraft's length. The main wing of the aircraft is characterized by a number of essential parameters, as shown in Figure 2.2. The aspect-ratio is

$$\mathcal{AR} = \frac{b^2}{A},$$ (2.2)

where b is the wing span and A is the wing area. Then there is the leading edge and the quarter chord sweep angle (Λ_{le}, Λ_{qc}); the taper ratio (c_{tip}/c_{root}); the root-mean-square thickness ratio

$$t/c = \left[\frac{1}{b/2 - b_r} \int_{b_r}^{b/2} (t/c)^2 \, dy \right]^{1/2},$$ (2.3)

with $b_r =$ spanwise location of the wing root. The mean aerodynamic chord is

$$\text{MAC} = \frac{2}{A} \int_{o}^{b/2} c^2(y) \, dy.$$ (2.4)

The dihedral angle ($\varphi > 0$) is the angle with respect to the ground plane made by the leading edge or quarter-chord line. If $\varphi < 0$, the wing is said to have an anhedral.

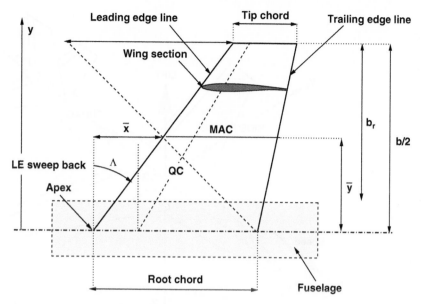

Figure 2.2   *Planform view of wing geometry.*

A number of analytical relationships exist to correlate various wing parameters. These are reported by some reference manuals, such as AIAA[35].

2.2   REFERENCE SYSTEMS

There are three essential reference systems: the Earth system, the body system, and the wind system. Local reference systems may be required for specific reasons (for example, wing aerodynamics).

The aircraft is supposed to fly with respect to a Cartesian system fixed on the ground (*Earth axes*), which for our purposes is considered flat. In fact, most of the performance calculations will be done for relatively short flight times and at relatively low altitudes. The Earth's curvature and rotation are important for inertial navigation systems and to take into account the Coriolis effects (accelerations) over a rotating Earth. The Coriolis acceleration is estimated at less $10^{-3}g$ in atmospheric flight mechanics. The gravitational field is characterized by a constant acceleration of gravity, equal to the standard value of $g = 9.807 \, \text{m/s}^2$. The Earth system has the x axis pointing North, the z axis normal to the ground and pointing downward; the y axis pointing East, and making a right-hand system with x and z.

There are several ways to define reference axes on the airplane; the choice will be limited by the fact that there is always one plane of symmetry. This is not always the case. In the early days of aviation, symmetric wings were relatively unstable, and some aircraft designers used asymmetric concepts to alleviate the rolling problem. For example, the Ansaldo SVA (1917) had unequal wing spans, the Messerschmidt Bf-109/Mf-109 (1935) had an asymmetric fin airfoil, the Republic P-47 (1941) had

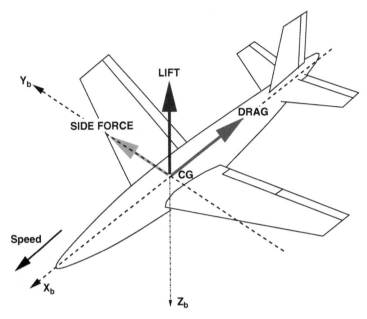

Figure 2.3 *Body reference system and forces on the aircraft.*

an offset fin, and the Spitfire I (1934) had asymmetric radiators under the wings to compensate for the engine-propeller torque*.

Back to the concept of aircraft coordinates, there will be a body-conformal orthogonal reference system, centered at the center of gravity (CG) of the airplane. The subscript "b" will be used to denote body axes. The position of the CG is a known parameter, although its estimation is not straightforward. Also, its position changes with the aircraft loading. This fact is important for the analysis of stability and control. The reference system is shown in the three-dimensional view of Figure 2.3. The longitudinal axis x is oriented in the direction of the forward speed (wind axis); the z axis is vertical (along the acceleration of gravity g) and the y axis makes a right-hand Cartesian system with x and z. The positive y axis is at the starboard side of the airplane.

The main forces on the aircraft (propulsive, aerodynamic, inertial) are applied at the CG. The drag is the opposite direction of the air speed; the lift is at 90 degrees with the drag; the weight is vertical and pointing downward. The thrust generated by the engines on both sides of the aircraft is replaced by a single thrust force.

The correlation between body and Earth reference system is done by three attitude angles. The *pitch attitude* of the aircraft is the angle θ between the longitudinal axis and the horizontal plane (positive with the nose up). The *yaw attitude* is the angle between the aircraft's speed and the North–South direction. It is positive clockwise, e.g. when the aircraft is heading eastwards. This is sometimes called *heading* or

---

* Asymmetric wings can be employed to reduce the wave drag at supersonic speeds, but this is an aerodynamic problem on its own right (Jones[36]).

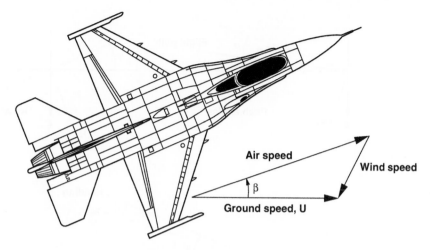

Figure 2.4   *Vector relationship between ground speed and air speed.*

azimuth angle ψ. The *bank attitude* ϕ is the angle between the aircraft spanwise axis y_b and the horizontal plane.

The side force would not normally be present on the aircraft (and its occurrence should be avoided). It is mostly due to atmospheric effects (lateral gusts), asymmetric thrust, and center of mass off the symmetry line (due, for example, to a differential use of the fuel in the wing tanks). The presence of such forces may lead to a yawed flight condition. The yaw angle β is the angle between the longitudinal axis and the true air speed vector.

The velocity (wind) axis reference system indicates the direction of the flight path with respect to the Earth system. At any given point on the trajectory the aircraft will have a track and a gradient. The track is the angle on the horizontal plane between the flight direction and the North–South axis. The gradient is the angle of the velocity on the horizon, which we have called γ in Figure 2.1.

If V is the ground speed, V_w is the wind speed, the air speed is found from

$$V_a = V + V_w. \tag{2.5}$$

The corresponding side-slip angle β is indicated in Figure 2.4.

The transfer of forces between one reference and the other is done through rotation matrices. The order of these rotations is important for the correct development of the flight mechanics equations. The full derivation of these equations may require several pages. A modern presentation is given by Yechout *et al.*[37] A detailed discussion of reference systems and flight paths on a curved surface is available in Miele[38].

2.2.1   Angular Relationships

We consider some simple flight cases to provide a correlation between angles in the different reference systems, Figure 2.5. First, consider the pitch angles. From the

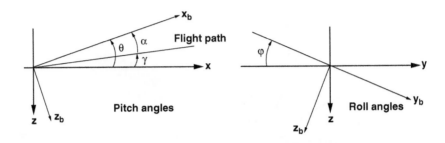

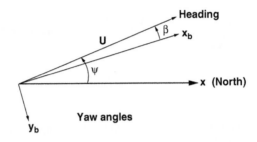

Figure 2.5 *Relationships between angles and reference systems.*

definitions, the pitch attitude θ is related to the angle of attack α and the flight path gradient γ by

$$\theta = \alpha + \gamma. \tag{2.6}$$

Next, consider a yaw. The heading ψ is related to the track ξ and the side-slip by the equation

$$\psi = \xi - \beta. \tag{2.7}$$

If there is no side-slip (normal flight condition) the heading and the track are the same angle. Finally, consider a roll problem. The bank angle ϕ is the inclination of the spanwise body axis on the horizontal plane. The bank attitude is the same as the bank angle on a level flight path. However, if the aircraft climbs or descends the correlation becomes complicated.

2.3 FORCES ON THE AIRCRAFT

For the fixed-wing aircraft the balance of forces on the aircraft, after assuming that these are applied at the CG, is the following

$$F = m\frac{\partial V}{\partial t}, \tag{2.8}$$

where the right-hand side denotes the acceleration of the center of gravity. There will be essentially three types of force: aerodynamic (A), propulsion (T) and gravitational (mg). If we specialize these actions, Eq. 2.8 becomes

$$A + T + mg = m\frac{\partial V}{\partial t}. \tag{2.9}$$

Most of the equations that are solved in the following chapters are one form or another of Eq. 2.9.

An essential concept is the *trim*. An aircraft is said to be trimmed if the sum of all forces is zero, and the sum of all moments is zero. Trimming the aircraft may have the undesirable effect of creating other forces, such as additional drag. Trim is required in most cases, because the resulting forces on the aircraft do not operate at the center of gravity, as assumed in this book.

2.4 MOMENTS OF INERTIA

The moment of inertia is representative of the inertia of a body to rotational acceler-ations, just as the mass is the inertia of a body to a linear acceleration. The moment of inertia about a generic axis is the volume integral

$$I = \int_V r^2 \, dm. \tag{2.10}$$

The term $r^2 \, dm$ is the moment of inertia of the mass dm with respect to the axis at a distance. The units of the moment of inertia are $[\text{kg m}^2]$. For analyses with respect to a Cartesian reference system, there are three principal moments of inertia. For example, the moment of inertia about the x axis is

$$I_x = \int_V x^2 \, dm = \int_V x^2 \rho dV = mr_x^2, \tag{2.11}$$

where r_x is the *radius of gyration* with respect to the principal axis x. The radius of gyration expresses the distance at which the aircraft mass should be concentrated to give the same moment of inertia. Radii of gyration for the aircraft Model C are given in Table A.11.

The calculation of the moment of inertia requires the exact knowledge of the mass distribution around the axis and the geometry of the aircraft. Applications of the moments of inertia to aircraft performance will be shown in Chapter 10 and Chap-ter 14. A first-order approximation for the moments of inertia around the x axis and y axis (roll and pitch) of a generic aircraft is

$$I_x = \frac{1}{4}\frac{b^2 Wr_x^2}{g}, \quad I_y = \frac{1}{4}\frac{l^2 Wr_y^2}{g}, \tag{2.12}$$

where l is the aircraft's length, and r_x and r_y are radii of gyration. These can be derived from the calculation of existing aircraft. The moments of inertia are proportional to the square of the reference length and proportional to the weight.

2.5 FLIGHT DYNAMICS EQUATIONS

The rigid aircraft is defined by six degrees of freedom. These are the parameters needed to identify completely the position and orientation of the aircraft in the Earth axes. The parameters are the coordinates of the center of gravity (x, y, z), and the orientation of the body axes on the Earth axes, (θ, ψ, ϕ).

In the specialized literature dealing with stability and control the angular velocities are called p, q, r, respectively. Unfortunately, these symbols generate confusion with other quantities (pressure p, dynamic pressure q, radius r, etc.). In stability and control the moments are called $\mathcal{L}$, $\mathcal{M}$, $\mathcal{N}$. A summary of symbols is given in Table 2.1.

The rigid-body velocity with respect to a reference system on the ground is

$$V = V_\infty + \mathbf{\Omega} \times r, \tag{2.13}$$

where V_∞ is the velocity of the center of gravity; $\mathbf{\Omega} = (\dot{\theta}, \dot{\phi}; \dot{\psi})$ is the rotation vector centered at the CG; and $r(x, y, z)$ is the vector distance between the reference system on the ground and the CG. The linear acceleration of the aircraft is found from the derivation of Eq. 2.13,

$$a = \frac{\partial V_\infty}{\partial t} + \frac{\partial}{\partial t}(\mathbf{\Omega} \times r). \tag{2.14}$$

These are general equations. However, for practical reasons, in the following chapters we will consider U as the total velocity in the flight path, V the total velocity parallel to the ground and v_c the vertical velocity. If the aircraft accelerates around its axis, the scalar form of Eq. 2.14 is quite elaborate. It is, in fact, of little practical interest in performance calculations, since only particular flight conditions are considered.

However, for a number of problems it is important to calculate the relationship between relative and absolute velocities and accelerations. Consider a reference on the ground $\{O, x, y, z\}$, and a non-inertial reference $\{O_1, x_1, y_1, z_1\}$. The speed of the aircraft in the Earth axes is

$$V = V_\infty + \omega \times r + V_r, \tag{2.15}$$

where V_∞ is the speed of the non-inertial system, ω is its rotational speed, r is the position vector of the CG in this system, and V_r is the relative velocity. The acceleration

Table 2.1 *Nomenclature and symbols in the body-conformal reference system.*

Body-conformal axes	x_b	y_b	z_b
Velocity components	u	v	w
Principal moments of inertia	I_x	I_y	I_z
Rotation angles	θ	ϕ	ψ
Angular velocities	$\dot{\theta}$	$\dot{\phi}$	$\dot{\psi}$
(Alternative symbols)	p	q	r
Moments	M_x	M_y	M_z
(Alternative symbols)	$\mathcal{L}$	$\mathcal{M}$	$\mathcal{N}$
Moment coefficients	C_{Mx}	C_{My}	C_{Mz}

is found from the time derivative of Eq. 2.15. The results of classical mechanics (see, for example, Miele[38]), give

$$a = a_\infty + \omega \times (\omega \times r) + \dot{\omega} \times r + a_r + 2\omega \times V_r. \qquad (2.16)$$

In this equation a_∞ is the acceleration of O_1, and a_r is the relative acceleration of the CG in O_1. The term

$$a_t = a_\infty + \omega \times (\omega \times r) + \dot{\omega} \times r \qquad (2.17)$$

is the transport acceleration, and

$$a_{Cor} = 2\omega \times V_r \qquad (2.18)$$

is the Coriolis acceleration. Consequently, we have

$$a = a_t + a_r + a_{Cor}. \qquad (2.19)$$

In conclusion, the absolute acceleration is the sum of the relative, transport and Coriolis accelerations.

2.6   THE INTERNATIONAL STANDARD ATMOSPHERE

Nearly all the basic calculations of aircraft performance are done in International Standard Atmosphere (ISA) conditions, whose parameters at sea level are given in Table 2.2. The standard humidity is zero, which is far from true. Sometimes the reference to ISA conditions is replaced with a reference to *standard day*. For temperatures above or below the standard day, there is a reference to *hot* or *cold* day.

Observations on the state of the atmosphere at sea level go back hundreds of years, but they have become systematic in the last century with aviation, rocket and satellite data and perfect gas theory. A number of *standard* versions exist: NACA's atmosphere[39] (1955), the ARDC[40] (1959), the US standard[41] (1962, amended in 1976) and the ICAO standard[42]. These tables are basically equivalent to each other up to about 20 km (65,000 ft), that covers most of the atmospheric flight mechanics. We shall be concerned with altitudes below 31 km, or about 100,000 ft.

Table 2.2   *Sea level data of the International Standard Atmosphere.*

Parameter	Symbol	Sea level value
Temperature	T_o	15.15°C
Pressure	p_o	$1.01325 \cdot 10^5$ Pa
Density	ρ_o	1.225 kg/m^3
Viscosity	μ_o	$1.7894 \cdot 10^{-5}$ Ns/m^2
Humidity		0%

Most books of performance and flight mechanics (as well as books of aerodynamics) report these tables. These are made obsolete, from an engineering point of view, by the availability of analytical correlations of great accuracy*.

The atmosphere is divided into a number of layers. These layers are identified for the sole purpose of our discussion into atmospheric aircraft performance.

The atmosphere below 11,000 m (36,089 ft) is called *troposphere*. It is characterized by a decreasing temperature from sea level, and reaches a standard value of −56.2°C. The altitude of 11,000 m is called *tropopause*. The level above is called *lower stratosphere* and covers an altitude up to 20,000 m (65,627 ft), in which the temperature remains constant. The air density keeps decreasing with the increasing altitude. The upper limit of this layer includes most of the atmospheric flight vehicles powered by air-breathing engines. The *middle stratosphere* reaches up to an altitude $h = 32,000$ m (104,987 ft). In this layer the atmospheric temperature increases almost linearly from the value of −56.2°C. The edge of space is generally considered to be at an altitude of 100.5 km, where the gravity is considerably lower than at sea level.

A number of functions are sometimes used to approximate the ICAO data. For the temperature a linear expression is used:

$$T = T_o - 0.0065h, \tag{2.20}$$

where T_o is the standard sea level temperature, and h is the altitude in meters. If we use the equation of ideal gases to describe the atmosphere,

$$\frac{p}{\rho} = \mathcal{R}T, \tag{2.21}$$

then

$$\frac{p}{\rho T} = \frac{p_o}{\rho_o T_o} = \mathcal{R}, \tag{2.22}$$

or

$$\frac{p}{p_o} = \frac{\rho}{\rho_o} \frac{T}{T_o}. \tag{2.23}$$

The value of the gas constant is $\mathcal{R} = 287$ J/kg K. The relative density is called σ, the relative pressure is δ and the relative temperature is θ. Therefore,

$$\delta = \sigma \theta. \tag{2.24}$$

We call the altitude corresponding to a given air density *density altitude*. If, instead, we relate the altitude to the local air pressure, then we have a *pressure altitude*. In order to find the pressure/altitude and the density/altitude relationships we use the buoyancy law for still air along with Eq. 2.20, to find the rate of change of the pressure with altitude. The buoyancy law is

$$\frac{\partial p}{\partial h} = -\rho g. \tag{2.25}$$

---

* Programs that perform ISA calculations can be found from the public domain.

If we insert the differential form of Eq. 2.20, this law becomes

$$\frac{\partial p}{\partial \mathcal{T}} = -\frac{\rho g}{\lambda}. \tag{2.26}$$

The next step is to use Eq. 2.21 to eliminate the density from Eq. 2.26,

$$\frac{\partial p}{\partial \mathcal{T}} = -\frac{pg}{\lambda \mathcal{R} \mathcal{T}}. \tag{2.27}$$

Rearranging of this equation leads to

$$\frac{dp}{p} = -\frac{g}{\lambda \mathcal{R}} \frac{d\mathcal{T}}{\mathcal{T}}. \tag{2.28}$$

Integration of Eq. 2.28 yields

$$\ln p = -\frac{g}{\lambda \mathcal{R}} \ln \mathcal{T} + c, \tag{2.29}$$

where c is a constant of integration that is found from the sea level conditions (Table 2.2). The final result is

$$\ln\left(\frac{p}{p_o}\right) = \frac{g}{\lambda \mathcal{R}} \ln\left(\frac{\mathcal{T}}{\mathcal{T}_o}\right), \tag{2.30}$$

$$\delta = \frac{p}{p_o} = \left(\frac{\mathcal{T}}{\mathcal{T}_o}\right)^{g/\lambda \mathcal{R}}. \tag{2.31}$$

The value of the power coefficient is $g/\lambda \mathcal{R} = 5.25864$. If we insert Eq. 2.20 in Eq. 2.31, we have a *pressure/altitude* correlation:

$$\delta = \frac{p}{p_o} = \left(1 - \frac{0.0065}{\mathcal{T}_o} h\right)^{5.25864} = (1 - 2.2558 \cdot 10^{-5} h)^{5.25864}, \tag{2.32}$$

with h expressed in meters. Equation 2.32 is in good agreement with the ICAO data. A more approximate expression is

$$\delta = \frac{p}{p_o} = (1 - 2.2558 \cdot 10^{-5} h)^{5.25588}, \tag{2.33}$$

The advantage of Eq. 2.33 is that the altitude is related to the pressure ratio, and therefore it can be read directly from the altimeter that is calibrated with the ISA reference value of p_o. To find a *density/altitude* correlation we use Eq. 2.24, with the relative pressure from Eq. 2.33:

$$\sigma = \frac{(1 - 2.2558 \cdot 10^{-5} h)^{5.25588}}{1 - 0.0065 h/\mathcal{T}_o}. \tag{2.34}$$

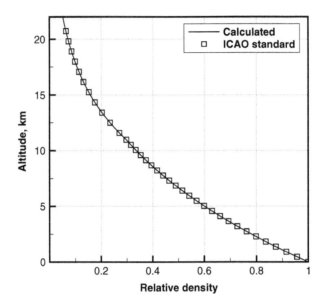

Figure 2.6 *ICAO air density and approximated function.*

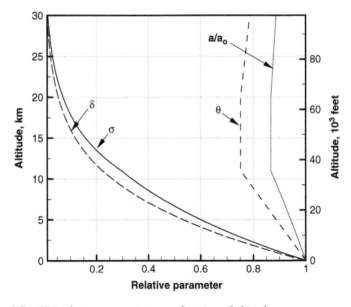

Figure 2.7 *ISA relative parameters as a function of altitude.*

The accuracy of the approximating function is shown on Figure 2.6 for the air density up to an altitude of 22,000 m. This accuracy is good enough for all the type of calculations shown in this book (see Problem 1.).

Figure 2.7 shows the ratios of density, pressure, temperature and speed of sound from sea level to the altitude of 20,000 m. These data are calculated from

Eq. 2.20, Eq. 2.33, and from the definition of speed of sound:

$$a \simeq \sqrt{\frac{p}{\rho}} = \sqrt{\gamma \mathcal{R} T}, \tag{2.35}$$

where γ is the ratio between specific heats ($\gamma = 1.4$).

Finally, the value of the air viscosity from the air temperature can be found from the following relationship

$$\frac{\mu}{\mu_o} = 8.14807 \cdot 10^{-2} \frac{T^{3/2}}{T + 110.4}. \tag{2.36}$$

From a computational point of view, we can construct a routine called

```
atmosphere(h,sigma,delta,theta,asound)
```

that returns the relative air density, pressure, temperature and speed of sound for any input altitude h. This routine will be useful in all the performance calculations presented in later chapters.

Computational Procedure

1. Calculate the local temperature from Eq. 2.20.
2. Calculate the relative pressure from Eq. 2.33.
3. Calculate the relative density from Eq. 2.34.
4. Calculate the speed of sound from Eq. 2.35.

The inverse problem (calculation of the altitude h corresponding to relative density σ) is more elaborate, because it requires to solve a non-linear equation in implicit form. The solution can be found with a bisection method. For the bisection method to work, one has to choose two points at which the function has opposite values. It is safe to choose $\sigma_1 = 0.01$ and $\sigma_2 = 1$, to make sure that the method converges to a solution.

Example

Commercial passenger jets cruise at altitudes between 9,000 and 12,000 m (30,000 to 40,000 ft). The standard temperature at those altitudes is between $-56°$C and $-50°$C. The relative pressure is between 0.29 and 0.19. Without cabin pressurization, air conditioning and heating, it would not be possible to fly passengers. Airplanes such as the Boeing B-737 have two air conditioning systems. If one system does not work, the aircraft can still fly, but it has to maintain a cruise altitude below 7,620 m (25,000 feet, $\delta = 0.3711$). At this altitude it is possible to breathe with some difficulty. In case of failure of the air conditioning, the aircraft must descend to 4,267 m (14,000 feet, $\delta = 0.5875$), for which about 4 minutes are required. In order to maintain air supply to the passengers and the crew, emergency oxygen masks are installed for emergency.

2.7   NON-STANDARD CONDITIONS

The ISA values for the atmospheric parameters are useful to compare aircraft performance over all the range of atmospheric altitudes. In fact, most of the calculations shown in this book are done under ISA conditions. Obviously, this is an idealized case that does not occur in practice. It is not uncommon to encounter temperature inversions, e.g. cooler air at the ground level and warmer air at low altitudes, contrary to the standard model. A comprehensive introduction to weather processes and climatic conditions around the world is available in Barry and Chorley[43].

A detailed performance analysis requires consideration of large deviations from the standard values, to deal with extreme environmental conditions: winters in the northern hemisphere, very hot weather on the ground. In addition, airport altitude, humidity and precipitations, atmospheric winds, lateral gusts, and global air circulation have strong influence on flight and safety. Rain and snow can be so heavy that take-off may have to be aborted. To simplify these matters, the US Department of Defense defines four non-standard atmospheres, referred to as hot, cold, tropical, and arctic (MIL-STD-210A). These profiles are shown in Figure 2.8.

Three important classes of weather-related flight problems are icing, downbursts and atmospheric turbulence. Ice accretion on the lifting surfaces during cruise can give rise to loss of longitudinal control; at take-off it may lead to loss of lift, stall angle reduction, drag penalties and longer ground run. Ice formation on the ground and in flight is discussed by Asselin[44]. Lynch and Khodadoust[45] and Kind et al.[46] are two relevant reviews on the physics and modeling of icing.

The downburst is a weather phenomenon that produces a downward flow of great danger, particularly during take-off and landing. It is created by heavy winds, with

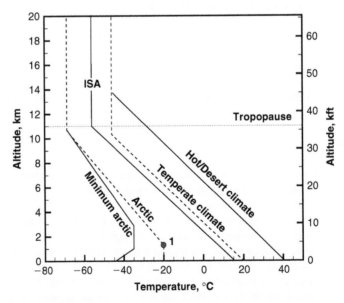

Figure 2.8   *ICAO standard atmospheric temperature and reference atmospheres for flight mechanics calculations.*

speeds up to 30 m/s. These downbursts generally consist of closely spaced small cells (a few km wide) that are separated by relatively calm atmosphere. They act as jets flowing downward and spreading radially from the ground. A basic physical model of the downburst is discussed in detail by Zhu and Etkin[47]. A vortex-ring model has been proposed by Ivan[48]. Take-off and landing simulations with a downburst have been published by Hahn[49], while Zhao and Bryson[50] optimized the flight path of an aircraft under a downburst with a non-linear feedback control program. See also Frost et al.[51] for measurements near the airport and dynamic modeling.

Turbulence is a more familiar weather pattern to the frequent flyer. It includes cases of free air and convective air turbulence, atmospheric boundary layers and mountain ridge waves. No airplane is immune to the powerful gusts of the atmosphere, not even airplanes the size of the Boeing B-747. A thoughtful discussion of turbulence and flight is available in Etkin[52] and Houbolt[53].

Among the global circulation effects there is the jet stream. This is an atmospheric wind with an West–East prevalence, strongest around the troposphere, that affects transatlantic flights between Europe and North America. The jet stream can add or subtract 1 flight hour on such flights. Hale[54] is a good reference for atmospheric wind effects on flight performance.

Temperature variations can be of the order of 80 degrees ($-40°C$ to $+40°C$). Since the temperature does not appear in any of the performance equations, a useful relationship with pressure and density is required.

If the temperature has a constant deviation from the standard value, say a constant $\pm \Delta \mathcal{T}$, the method of §2.6 can still be used, because the temperature gradient is the same (Eq. 2.20). The only difference is that the symbol $\mathcal{T}_o$ denotes the sea level temperature, whatever that may be. The result of such a model is shown in Figure 2.9 for the relative density.

For a temperature profile such as the Arctic temperature in Figure 2.8, a solution method is the following. All the quantities denoted with $(.)_o$ denote non-standard sea level condition, the quantities denoted with $(.)_1$ are relative to point 1 in Figure 2.8. Starting from Eq. 2.27, we have

$$\int_o^h \frac{dp}{p} = -\frac{g}{\mathcal{R}} \int_o^h \frac{dh}{\mathcal{T}}, \tag{2.37}$$

$$[\ln p]_o^h = \ln \left(\frac{p}{p_o} \right) = \ln \delta = -\frac{g}{\mathcal{R}} \int_o^h \frac{dh}{\mathcal{T}}, \tag{2.38}$$

$$\ln \delta = -\frac{g}{\mathcal{R}} \left(\int_o^{h_1} \frac{dh}{\mathcal{T}} + \int_{h_1}^h \frac{dh}{\mathcal{T}} \right) = -\frac{g}{\mathcal{R}} \frac{h_1}{\mathcal{T}_o} - \frac{g}{\mathcal{R}} \int_{h_1}^h \frac{dh}{\mathcal{T}}, \tag{2.39}$$

$$\ln \delta = -\frac{g}{\mathcal{R}} \frac{h_1}{\mathcal{T}_o} + \frac{g}{\mathcal{R}\lambda_1} \ln \left(\frac{\mathcal{T}_1 - \lambda_1 h}{\mathcal{T}_1 - \lambda_1 h_1} \right). \tag{2.40}$$

It is possible to show that if $h_1 \to 0$, then Eq. 2.40 is equivalent to Eq. 2.31 (see also Problem 10.). An amendment to the standard program to calculate deviations from the standard atmosphere is the following.

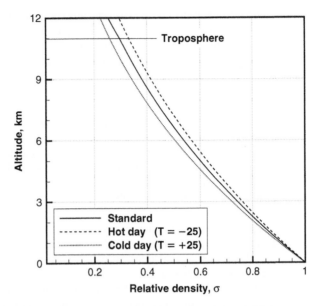

Figure 2.9 *ICAO standard density and deviations due to ±25 degrees around the standard value.*

Computational Procedure

1. Construct a table of actual temperatures $T(h)$.
2. Calculate the relative pressure by solving numerically Eq. 2.38.
3. Calculate the relative density from Eq. 2.24.

PROBLEMS

1. Write a program that solves the equations of the International Standard Atmosphere and on output provide the ratios σ, δ and θ as a function of the altitude h.
2. For a given relative air density $\sigma = 0.2, 0.4, 0.6$, find the corresponding altitude in ISA atmosphere by solving Eq. 2.34 for the unknown h.
3. Plot the air viscosity as a function of altitude, by solving Eq. 2.36. Discuss the origin and the meaning of this equation.
4. Due to improper closing, the door of a certain aircraft is lost in flight. At the time of the incident the aircraft is flying at an altitude $h = 6,000$ m with a speed $U = 400$ km/h. Describe the effects on the aircraft, on the passengers and the cargo. Motivate your answer. Calculate the pressure drop in the cabin by using the data of the International Standard Atmosphere at the flight altitude.
5. Calculate the air mass in the atmosphere from sea level to 5,000 m using the ISA data. Compare this mass with the mass from sea level to 20,000 m. Hint: neglect the changes of surface area with the altitude (this leads to an error less than 0.1%), use the buoyancy law $dp = \rho g dh$; consider a constant value of the gravitational acceleration g; use Eq. 2.34 to evaluate the relative air density.

(About 50% of the air mass above the Earth should be contained within the first 5,000 m from sea level, an indication that the pressure created by the higher layers forces the air closer to the Earth's surface.)

6. From the equation of ideal gases, the relationship between the relative density, pressure and temperature is $\delta = \sigma\theta$. Find the parameter r such that $h = \sigma^r$ and

$$\delta\sqrt{\theta} = \sigma^r$$

which is valid in the troposphere.

7. Which aero-thermodynamic parameters are discontinuous at the tropopause? Can you explain the reasons for this discontinuity?

8. Concorde's cruise performance was found to be quite sensitive to the atmospheric temperature at its normal cruise altitude. Describe how the temperature can affect the cruise conditions and the fuel consumption, and find an engineering solution. (*Problem-based learning: additional research is required.*)

9. Calculate the relationship between the leading-edge and the quarter-chord sweep angle for a wing with a straight leading edge. The taper ratio is λ, the tip and root chords are c_{tip} and c_{root}, respectively; the wing span is b, and the wing area is A.

10. Calculate the relative pressure, density and temperature for the Arctic profile (Figure 2.8), by using the method highlighted in §2.7. The reference point 1 has coordinates: $h_1 = 1.5$ km, $T_1 = -70°$C. The temperature gradient is estimated by $\lambda_1 \simeq -0.0055°$C/m. Stop the calculation at the troposphere.

Chapter 3
Weight Performance

... future growth potential looks unlimited ... one gross weight doubling, possibly two, is predicted by 1985; nuclear power can drive the optimum weight to 5 to 10 million pounds by the year 2000.

F.A. Cleveland[55], 1970

The aircraft's weight influences the flight performance more than any other parameter, including engine power. Calculation of the weight effects will be done in the next chapters. Weight has been of concern since the earliest days of aeronautics; just recall the fact that the first attempts to fly were based on lighter-than-air concepts. Therefore, we devote this chapter to the weight analysis and the relative aspects, such as weight definitions, useful loads, weight reporting formats, and the relative approximations. The discussion will be confined to the operational phase of the aircraft.

3.1 THE AIRCRAFT'S WEIGHT

By 1914, it was believed that the limiting weight of the airplane could not exceed 2,000 lb (about 800 kg). By comparison, Baumann[56] wrote an alarming report in 1920, prompted by the construction of *giant airplanes* in Germany during the war – airplanes weighing as much as 15.5 tons, and powered by as much as 260 hp (195 kW) engines. Interest in large airplanes was sparkled by excitement in the very early days, with F.W. Lanchester and Handley Page[57] expressing their views. Lanchester's opinions on this matter are recorded by Kingsford[58].

Cleveland's forecast in 1970 turned out to be wrong in the opposite direction; nuclear power has never been considered a serious option. Lockheed persevered along these lines for some years, and in 1976 Lange[59] proposed aircraft concepts in the 900 ton (2 million lb) class, including a 275 MW nuclear power plant. By contrast, the design office at Boeing proposed the span-loader concept[60] – a 1,270 ton aircraft without nuclear power (project 759).

In his paper "Quest for a Novel Force", Allen[61] speculated on *antipodal megaliners*, monsters of the future capable of transporting 1,200 passengers from London to Sidney through a transatmospheric flight trajectory. The highly speculative content of Allen's paper is food for thought at a time when aircraft design is a conservative discipline. A further review, focusing on configuration alternatives and economic viability of the big airplanes, is available in McMasters and Kroo[62].

At the start of the 21st century, even the biggest airplanes do not exceed a gross weight of 600 tons (1.3 million lb). The Antonov AN-225, the largest prototype airplane ever built, can lift up to 250 tons of cargo at its design point. Its 88.40-m wing span is a wide as a football field. Its 18.1-m height reaches the top of a six-storey building. The airplane was designed to carry a spaceship as an external load.

The Airbus A-380 commercial liner has a maximum take-off weight of 562 tons and a wing span of 79.8 m. It is powered by four jet engines delivering a static thrust of over 1,000 kN. It can load up to 310,000 liters of fuel, and fly a distance of 15,000 km with over 500 passengers. The volume occupied by the fuel alone is a cube with a 6.8-m side. By way of comparison, the A-380's weight is equivalent to about 15 railway carriages, or five diesel locomotives – by all means a long passenger train.

Cleveland's analysis (recommended reading material to those interested in very large aircraft) contains a discussion of historical growth in size that leads to a *square-cube law*. The argument is that, if technology had not improved, growth would have been halted by the fact that the load stress in airplane structures increases with the linear dimension, when the load is proportional to the weight.

The concept is well illustrated by using as an example a single rectangular beam. The load is proportional to the weight W; the weight is proportional to the cube of the linear dimension l^3; the cross-section of the beam is proportional to the square of the dimension l^2. Therefore, in first approximation load/cross-section $\sim l$. At some point this increase in load reaches the structural limits of the material, and the beam collapses under the effect of its own weight.

While the Airbus A-380 has a wing span six times larger than the Wright Flyer (1903), a wing area about 16.5 times larger, the weight has increased by a factor of 1,650, which appears to defeat the law. The square-cube law would imply that the wing loading on the A-380 would be the same as the Wright Flyer, or about 6.2 kg/m^2. By this rule, the weight of the A-380 should not exceed 5,237 kg! As it turns out, the actual weight is a factor 10^2 times the value calculated at *constant technology*.

If the wing is scaled up while holding wing loading and structural stresses as constant, its weight will grow roughly as $W^{1.4}$. However, when one looks at the details of the components, they do not scale up with the same factor. Cleveland showed that by doubling the gross weight and the payload of the aircraft the wing weight would have to increase by a factor of 2.69; the airframe would grow by a factor of 1.84; and the electrical systems would grow by a factor of 1.40.

Figure 3.1 presents a trend of aircraft MTOW and corresponding wing span b. The analysis shows that the wing span increases slower than the gross take-off weight, according to a function $W \simeq b^n$, with $n < 1$.

From a productivity point of view, the most important factor is not the absolute weight and size of the aircraft, but its useful payload. Historically, this has increased from about 10% of the Wright Brothers Flyer to over 30% of the current generation of airplanes; this ratio has also increased with the increased gross weight. The increase is driven by commercial requirements and by the need to move bulky equipment and machinery. By comparison, Concorde had a payload of less than 1% of its maximum take-off weight, though this is admittedly a completely different design concept.

For aircraft designed to operate in war zones the payload is essentially limited to ordnance. The amount of ordnance that can be carried varies greatly, depending on the type of operation. Figure 3.2 shows some estimated maximum loads for two categories of military aircraft. Figure 3.3 shows the payload ratio of several transport aircraft. The dotted line is a least-square fit of the available data, and although the points are scattered, it indicates that the payload ratio increases with the size of the airplane. Figure 3.4 shows the payload ratio relative to the OEW and a least-square

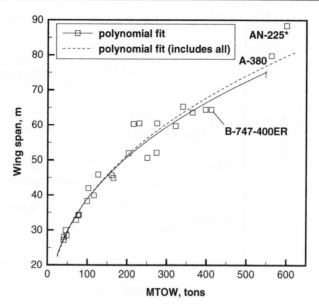

Figure 3.1   *Wing span versus MTOW for large commercial aircraft; AN-225 at its design point.*

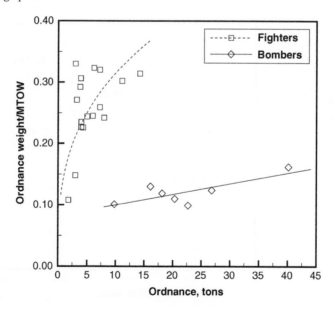

Figure 3.2   *Maximum ordnance versus MTOW for fighter/attack and heavy bomber aircraft.*

fit of the data (dotted line). Again the payload performance increases with the size and weight of the airplane.

Table 3.1, on page 37, summarizes the weight/payload data of the largest cargo airplanes currently in service. The data in the fourth column, R, indicates the maximum range at maximum payload.

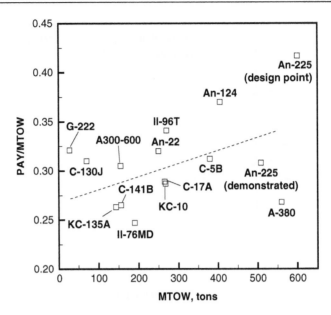

Figure 3.3 *Cargo airplanes maximum payload PAY versus MTOW.*

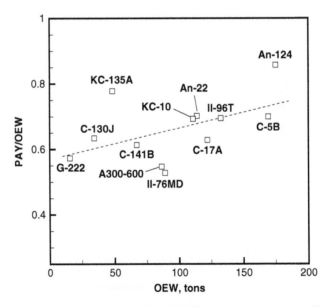

Figure 3.4 *Payload ratio versus the OEW for some transport aircraft.*

To properly understand the sheer size of the vehicles in the table in terms of bulk load that can be carried, a Galaxy C-5B is reported to carry any of the following items: two M1 Abram battle tanks (61,700 kg each); six M2/M3 Bradley infantry vehicles; six Apache helicopters with folded blades; 113,740 kg of relief supplies; a 74,000 kg

Table 3.1 *Cargo aircraft PAY versus MTOW data (estimated); R is the range at maximum payload.*

Aircraft	MTOW	PAY/MTOW	R (km)	Notes
Antonov AN-225, Ruslan	600.0	0.370	4,500	design
Lockheed C-5B, Galaxy	381.0	0.311	5,526	
Lockheed C-130J, Hercules	79.4	0.245	5,400	
Boeing C-17, Globemaster	264.5	0.288	4,700	
Boeing B-747-400F	396.8	0.284	8,240	
Boeing B-747-400ER F	412.8	0.290	9,200	
Airbus A-380-F	560.0	0.268	n.a.	
Satic A-300-600, Beluga	155.0	0.305	1,666	

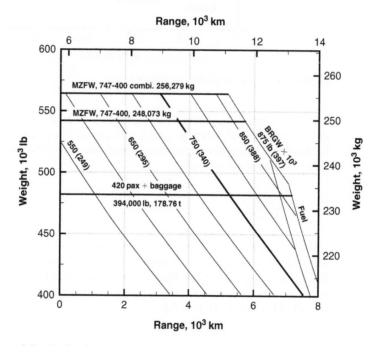

Figure 3.5 *Payload ratio relative to the OEW of some transport aircraft.*

mobile bridge for the US Corps of Engineers, and all the arsenal of the US Army Ordnance.

A passenger Boeing B-747-400 can carry up to 416 paying passengers (or 7.5 coaches), and up to 216,840 liters of fuel. This is enough fuel to fill 4,300 mid-size cars, which would be able to cover a cumulative 3 million km journey. The Boeing B-747-400 at maximum take-off weight would be equivalent to lifting 10 train coaches or 300 mid-size cars at once. The payload/range diagram of this aircraft is shown in Figure 3.5, and was elaborated from Boeing's data[63].

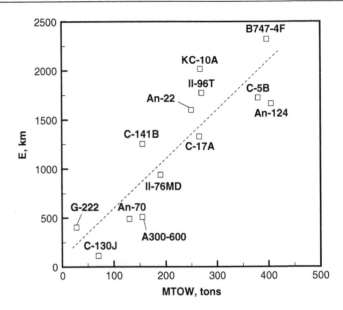

Figure 3.6 *Cargo airplanes maximum range.*

Freight is increasingly transported in standard containers called *pallets*. Loading and unloading of pallets is done quickly with ground-based vehicles with conveyors. The loading of pallets also rationalizes the available space in the cargo hold. A suitable cargo performance parameter is

$$E = \frac{PAY}{MTOW} R.$$ (3.1)

where R is the air range. This parameter is a measure of how much payload can be carried over a given distance. It emphasizes the fact that a certain payload can be flown over a longer or shorter distance; or that for a certain flight distance a larger or lower payload can be carried. This parameter has been estimated for a number of aircraft and is shown in Figure 3.6.

3.1.1 Wing Loading

In the previous discussion the concept of *wing loading* was used without much elaboration. The wing loading is the ratio between the aircraft gross weight and the area of the wing. This definition does not take into account the fact that a (small) portion of the lift may be created by the fuselage or the tail surfaces. Therefore, the wing loading implies that the aircraft's weight rests entirely on the wing. Examples of average wing loadings for different classes of aircraft are given in Table 3.2.

The wing loading of several aircraft has been plotted in Figure 3.7. The MTOW has been considered in the evaluation of W/A. The wing loading is not a constant; it decreases in flight, due to fuel consumption.

Table 3.2 *Average wing loading W/A of different classes of aircraft (conversion between units is approximate).*

Aircraft	kgf/m^2	Pa	lb/ft^2
Light aircraft	150–250	1,500–2,500	31–52
Turboprop transport aircraft	300–450	3,000–4,500	63–95
Business jet	300–450	3,000–4,500	63–95
Subsonic jet aircraft	350–650	3,500–6,500	73–137
Heavy-lift aircraft	450–750	4,500–7,500	94–158
Supersonic fighter	500–950	5,000–9,500	105–200

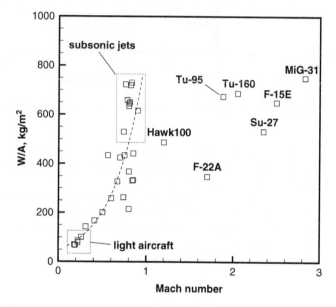

Figure 3.7 *Wing loading of several classes of airplane (estimated).*

A comparison of interest is the one relative to bird flight. Figure 3.8, elaborated from data in Pennycuick[64,65,66] and Greenewalt[67], shows a trend of birds' wing loading versus the birds' gross weight. Data are shown for over 100 birds, averaged between the two sexes; they make no distinction between soaring and flapping flight. The wing loading increases with the weight. The dotted line represents a linear fit of the data within the weight range considered, which can be considered as "equal technology". This is a case where the cube/square law is applicable. Flapping wings require higher wing loading at a given weight, and generally fall above the trend lines. For example, the razor bill (*Alca torda*) is indicated with a high point in the chart, and flies with rapid wing-beats. The same applies to the common loon (*Gavia immer*), a bird that has small wings, beating fast and steady; this bird is unable to soar or glide. The present trend line is compared to von Helmholtz's cube/square law, and shows only

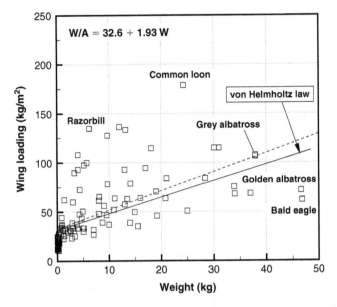

Figure 3.8 *Wing loading of some birds, and trend lines, including von Helmholtz.*

a little deviation. On the other hand, exceptionally good fliers like the bald eagle (*Haliaeetus leucocephalus*) and the golden albatross have a comparatively low wing loading, a sign of aerodynamic efficiency.

The interest in bird flight goes further back in time than aeronautics. For a further discussion of the scale effects on birds and aircraft see Haldane[68], von Kármán[69], Pennycuick[70] (with basic calculation programs), the review of Templin[71], and Tennekes's book[72]. The work of Pennycuick[64] is a fascinating book recommended to those who are interested in the laws of birds in flight.

3.2 DEFINITION OF WEIGHTS

There are several definitions of aircraft weights that are part of the aircraft performance and operations. For some weights, the corresponding acronyms of wide use are reported. A comprehensive discussion of aircraft weights (including historical trends) is given by Staton[73] and Torenbeek[74]. The aircraft's weight has an effect on range, endurance, ceiling, climb rate, take-off and landing distances, maneuverability, to say nothing of direct operating costs and production costs.

The *Manufacturer's Empty Weight* (MEW) is the weight of the aircraft including non-removable items, such as all the items that are an integral part of the aircraft configuration. It is also called *dry* weight.

The *Operational Empty Weight* (OEW) is the aircraft's weight from the manufacturer, plus a number of additional removable items, due to *operational* reasons. These are the engine oil, the unusable fuel, the catering and entertaining equipment, flight and navigation manuals, life vests and emergency equipment.

The *Maximum Take-off Weight* (MTOW) is the maximum aircraft weight at lift-off, e.g. when the front landing gears detach from the ground. The MTOW includes the payload and the fuel. Sometimes there is a reference to a gross take-off weight (TOW); it is intended that this weight is an average value at take-off.

The *Maximum Payload Weight* (PAY) is the allowable weight that can be carried by the aircraft. The sum OEW + PAY is less than the MTOW, the difference being the fuel weight. The payload includes passengers and their baggage, bulk cargo, military weapons, equipment for surveillance and early warning systems. This weight is seldom given for passenger aircraft and high-performance military aircraft, though for different reasons. We need to differentiate between maximum payload based on weight (obviously), volume (space-limited payload, such as number of pallets) or capacity (due to seating limitations). The useful load is the sum between the payload and the mission fuel.

The *Maximum Landing Weight* (MLW) is the weight of the aircraft at the point of touchdown on the runway. It is limited by load constraints on the landing gear, on the descent speed (and hence the shock at touchdown), and sometimes on the strength of the pavement. Permissible loads on the pavement are regulated by the ICAO[75].

The difference between MTOW and MLW increases with aircraft size. For example, the Boeing B-777-200-IGW (Increased Gross Weight) has MTOW = 286,800 kg and MLW = 208,600 kg. This yields a difference of 78,200 kg, corresponding to 27% of MTOW. In extreme cases (landing gear or engine failure), fuel can be jettisoned for unscheduled landing. On the Boeing B-777 fuel is jettisoned through nozzle valves inboard of each aileron. This operation is done more routinely by military aircraft (see also §5.6.1).

The *Maximum Ramp Weight* (MRW) is the aircraft weight before it starts taxiing. The difference between the maximum ramp weight and the maximum take-off weight, *MRW – MTOW*, corresponds to the amount of fuel burned between leaving the air terminal and lift-off. This difference is only relevant for very large aircraft, since it affects the operations on the runway. For example, the Boeing B-747-400 has an MRW = 398,255 kg, and MTOW = 396,800 kg. The difference of 1,455 kg is the fuel that can be burned from the moment the aircraft starts taxiing and the take-off point, about 1,750 liters. This corresponds to the fuel tanks of about 35 mid-size cars!

The *Maximum Brake-Release Weight* (MBRW) is the maximum weight at the point where the aircraft starts its take-off run. The *All Up Weight* (AUW) generally refers to the weight of the aircraft in cruise conditions. Due to variations of weight, as a consequence of fuel burn, there is a change in AUW in flight.

Finally, the *Maximum Zero-Fuel Weight* (MZFW) is the aircraft weight on the ground without usable fuel, and the *Maximum Taxi Weight* (MTW) is the certified aircraft weight for taxiing on the runway. The latter is defined by the structural limits of the landing gear. The unusable fuel is the amount of fuel in the tanks that is unable to reach the engines under critical flight conditions, as specified by the aviation authorities.

There are, of course, the weights of the various aircraft components, such as the engines and the engine installation. For the engines a *dry weight* is sometimes reported, which indicates the weight of the engine without usable lubricant and fuel.

The large take-off weight involved in some airplanes also has consequences on the type of paved runway. For weights above 120 tons, the thickness of the pavement must be around 0.6 m. A thinner pavement would not be capable of sustaining a take-off and landing of such a massive aircraft. Other operational weight restrictions exist, such as the maximum weight for balanced field length, obstacle clearance, noise emission, and available engine power.

3.3 WEIGHT ESTIMATION

One may wonder how the aircraft's weight is measured or estimated, particularly for the very large aircraft. In fact, there is no balance than can hold a Boeing B-747 or an Airbus A-380 at their maximum take-off weight.

There are four classes of methods: conceptual (or component) weight estimation, statistical methods (based on previous aircraft design), quasi analytical methods, and analytical methods. The aircraft's operating empty weight is estimated from the components method, e.g. by summing up the weight of its systems. These are: the wing system (with the control surfaces), fuselage, horizontal and vertical tails, landing gear, power plants, hydraulic and electrical systems, pneumatics, air conditioning and pressurization systems, auxiliary power units, instruments, and furnishings. The latter item is very much dependent on the operator of the aircraft. Therefore, the empty weight tends to change accordingly. These methods are important in aircraft design. What is essential in the present context is the weight management for the operation of the aircraft.

For commercial aviation, the operational weight of the aircraft includes the fuel, payload, and flight crew. For passenger operations, the weight of each passenger is calculated as 95 to 100 kg (including baggage); the volume to be allocated for baggage is 0.015 to 0.018 m^3. Methods for estimating the empty weight are given in a number of textbooks on aircraft design, for example Torenbeek[74], Raymer[76], Staton[73] and more specialized publications[77]. Comparisons of weight breakdown for a number of aircraft, such as in Table 3.3, were published by Beltramo et al.[78]

Table 3.3 shows the weight breakdown from two large aircraft, earlier versions of the B-747 and the C-5. The main difference is in the weight of the furnishings, which is obvious, since the B-747 is a passenger airplane, and the C-5 is a military utility airplane, designed to carry bulky cargo. The weight of the airframe of the C-5 is strengthened for the same reasons. By comparison, the weight of the wing of the -400 version of the B-747 is 43,090 kg, an increase of 9% over the original wing, but its contribution to the total weight is slightly below 11% – an indicator of the technological advances in wing design.

3.4 WEIGHT MANAGEMENT

The weight of almost all aircraft grows over time during their service life. Weight grows due to a number of reasons, namely new performance specifications, re-engineering of the power plant, exploitation of structural design margins, and not least the correction of design flaws, which may come after several years of service.

Weight and balance logbooks are maintained to keep a check on all the modifications done to the aircraft. The manufacturers also provide charts showing the basic weights and position of the center of gravity. Loading of commercial aircraft is done according to the instructions provided. The airline flight management performs basic

Table 3.3 *Weight breakdown for some representative aircraft. All weights in kg.*

System	B-747-100	% weight	C-5A	% weight
Wing	40,200	11.450	37,048	11.234
Tail	5,417	1.543	5,592	1.696
Airframe	31,009	8.833	52,193	15.826
Landing gear	14,596	4.157	17,046	5.169
Nacelle	4,703	1.340	3,838	1.164
Propulsion system	4,352	1.240	3,087	0.936
Flight controls	3,120	0.889	3,143	0.953
Auxiliary power unit	815	0.232	484	0.147
Instruments and navigation	674	0.192	333	0.101
Hydraulics and pneumatics	2,296	0.654	1,956	0.593
Electrical system	2,404	0.685	1,495	0.453
Avionics	1,873	0.534	1,871	0.567
Furnishings	21,748	6.195	3,539	1.073
Air conditioning	1,647	0.469	1,179	0.357
Anti-icing	188	0.054	106	0.032
Load and handling system	104	0.030	124	0.038
Operating empty weight	134,934	38.434	133,028	40.338
Dry engine weight	16,173	4.607	13,137	3.984
Empty weight	151,106	43.041	146,164	44.321
Take-off gross weight	351,076	100.0	329,785	100.0

calculations of passengers and baggage, by assuming a uniform loading of the aircraft. There are models that provide rapid solutions to the aircraft weight and balance as a function of passengers, baggage and fuel.

Generally, fuel consumption must obey special priorities to maintain the balance of forces on the aircraft at different flight regimes. Fuel must be used from the inboard tanks first. Optimal distribution reduces the requirements on aircraft trim, and therefore the drag associated with it, thereby maximizing the profitability of the aircraft.

Concorde was a special case also from the weight management point of view. Its 11 fuel tanks were distributed forward, centrally and aft of the aircraft. The transition from subsonic to supersonic cruise moved the center of pressure about 2 m aft. This movement required trimming of the aircraft, which in subsonic aircraft can be done by operating on the control surfaces. However, at supersonic speeds the trim drag would be an unacceptable penalty, and therefore trimming was done by moving the center of gravity aft by pumping fuel from the forward tanks to the rear tanks. Up to 33,000 kg of fuel can be pumped back and forth to trim the aircraft during subsonic to supersonic transition.

Commercial airlines operate scheduled and unscheduled passenger services, that carry checked-in baggage, cabin luggage and a number of items for passenger comfort. These are primarily food, drinks, magazines, television sets and other entertainment items. The operational weight refers to the weight of the aircraft fully equipped for these operations. This is somewhat higher that the OEW. To reduce this operational weight, some airlines have stopped offering meals altogether; others provide lunch packs to be collected upon boarding.

3.5 RANGE/PAYLOAD DIAGRAM

The distance that can be flown by a given aircraft depends not only on the size of its fuel tanks, but also on the weight of its payload. As anticipated, in most cases the combination of maximum payload and maximum fuel load exceeds the MTOW. If one knows how to calculate the aircraft range (this will be explained in detail in Chapter 9), then it is possible to construct a chart showing how some of the aircraft weights are related to the aircraft range. Since different cruise techniques are available, in principle we could construct different weight/payload diagrams, corresponding to each of the flight programs. Furthermore, there are effects of atmospheric conditions, climb and descent techniques, reserve fuel policy that can change the weight/payload performance considerably. Unless all these conditions are specified, it will not be possible to compare the range/payload performance of two aircraft.

We are interested in range at MTOW and range at maximum PAY, plus some intermediate cases. The range difference is considerable, particularly for cargo aircraft. Consider the Airbus A-300 Beluga, which transports heavy aerospace equipment between industrial plants. Its range at maximum PAY is 1,660 km (about 900 nm); it increases to 2,780 km (1,500 nm) with a 40 ton PAY, and to 4,630 km (2,500 nm) with a "small" 26 ton payload.

Figure 3.9 shows the range/payload diagram for three commercial subsonic jet aircraft of the Airbus family.

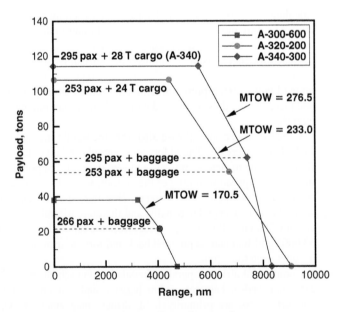

Figure 3.9 *Maximum payload range for Airbus airplanes.*

Start the analysis by considering the following equivalence:

$$\frac{W_e}{W} + \frac{W_p}{W} + \frac{W_f}{W} = 1, \tag{3.2}$$

where W_e is the OEW, W_p is the PAY and W_f is the fuel weight. From Eq. 3.2 the fuel fraction $\xi = W_f/W$ can be written as

$$\xi = 1 - \frac{W_e}{W} - \frac{W_p}{W}. \tag{3.3}$$

Since the empty weight is fixed, Eq. 3.3 is a linear relationship between payload fraction and fuel fraction, with the gross weight W being a parameter. However, the fuel ratio as a function of the aircraft gross weight is different, and is shown in Figure 3.10 for the aircraft model B (in Appendix A), at three values of the payload weight.

Sometimes a fourth term appears in Eq. 3.2 – the engine weight ratio, e.g. an item separated by the operational empty weight. This specification is not useful in discussing the range capability of the aircraft, though it is of interest when considering the efficiency of the propulsion system.

Figure 3.11 shows a complete summary of the weight/payload range. The chart shows the range and weight limits due to payload and fuel.

The longest range is achieved at zero payload and full tanks, which corresponds to a gross weight less than MTOW. This is called *ferry range* and determines the maximum distance that an aircraft can fly non-stop and without in-flight refuelling (range at maximum fuel load). The minimum range is achieved at maximum payload,

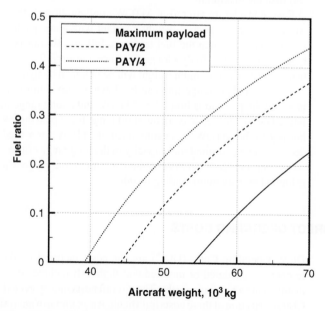

Figure 3.10   *Fuel ratio for aircraft model B.*

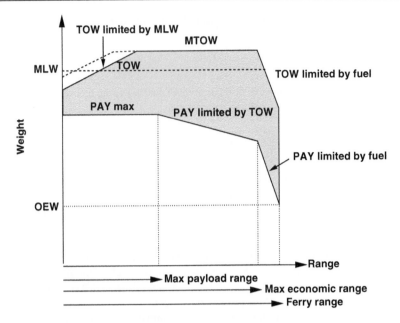

Figure 3.11 *Weight/range chart for a commercial airplane.*

and it is shown how the range at maximum payload is less than half the ferry range (this value is typical of other aircraft). Also of interest is the aircraft range at maximum fuel and maximum take-off weight, which requires the payload to be considerably less than the maximum.

First, consider an aircraft at MTOW configuration. If the mission fuel increases while keeping the aircraft weight constant with some payload, the range increases linearly with it. When the fuel tanks are full, the aircraft is prepared for maximum range at that weight. To fly a longer distance, the aircraft must have a gross weight less than MTOW. The minimum weight will be obtained by flying without useful payload.

There is a short range that can be flown at maximum payload, a medium range in which the payload is limited by MTOW, and a long range in which the payload is limited by the fuel capacity. In the short range flight plan the take-off weight increases linearly to the MTOW, it remains fixed at MTOW for a medium range flight, and decreases due to limited fuel capacity in the long range cruise. Although the diagram starts from a range equal to zero, there is a minimum range where the take-off weight is limited by maximum landing weight.

3.6 DIRECT OPERATING COSTS

Direct Operating Costs (DOC) are the costs incurred by the owner of an aircraft to operate scheduled or unscheduled flights. It includes the cost to fly, insure and maintain the aircraft airworthy. The aircraft costs money even if it stays on the ground. Characterization of these costs is difficult. An example of analysis is given by Beltramo *et al.*[78], who developed cost and weight estimating relationships and weight estimating relationships for commercial and military transport aircraft. Even the parametrization of the fuel costs is aleatory, due to the costs of aviation fuel on the international markets, and to the cost of fuel at different airports around the world.

Kershner[79] has demonstrated that while DOC have consistently decreased over time, the impact of the fuel cost has remained high, and in some historical circumstances has increased to over 50%. The prediction of that part of operational costs due to fuel consumption can be calculated with the methods explained in Chapter 9. Other studies in this area focus on the fuel consumption, see, for example, Isikveren[80] for subsonic flight and Windhorst *et al.*[81] for supersonic transport.

The appearance of *budget airlines* on the major world markets has contributed to a substantial change in the structure of the DOC. It is sometimes exciting to learn that we can buy a ticket to an international destination within Europe or the continental USA at a cost comparable to the cost of this book. Cost items such as ticketing, customer service, seat allocation, baggage handling, ground services, in-flight catering, airport taxes, leasing contracts, etc. have been dissected, reduced or removed altogether from the DOC. However, the cost of the fuel remains essentially the same. This cost is an essential part of the aircraft performance.

PROBLEMS

1. The load on a wing is proportional to the aircraft's gross weight. Describe how and why the stress on the structure increases with the linear dimension of the wing.

2. Discuss the limits to aircraft weight growth, including structural, aerodynamic, propulsion, landing gear, and systems limits; handling qualities, ground services, air traffic control, and costs.

3. You are required to study the feasibility of a military aircraft that has a radius of action of 1,500 km and can carry 5,000 kg of ordnance. The maximum take-off weight cannot exceed 20,000 kg. Assume a reserve fuel ratio equal to 5% of the mission fuel, and a range constant $c = 16,000$ (based on average performance of similar military aircraft). (Methods for calculating the coefficient c will be discussed in Chapter 9.)

4. For aircraft model B (Appendix A) plot the block fuel ratio ξ as a function of the payload fraction W_f/W, where W_f is the fuel weight, W is the aircraft gross weight, for a weight up to MTOW. Discuss the effects on increasing the payload fraction on the mission fuel, and hence on the range.

5. Discuss reasons why the sum of maximum payload weight and maximum fuel weight generally exceeds the maximum take-off weight of an aircraft.

6. Investigate the items of direct operating costs of a modern commercial subsonic jet. Estimate by further research the percentage of each item on the overall costs, and the effects of increasing the fuel cost. (*Problem-based learning: additional research is required.*)

Chapter 4

Aerodynamic Performance

The aircraft [replacement of the SR-71A] has been seen moving at high supersonic speed, with the resultant sonic bangs, over Southern California. It is believed to be powered by a revolutionary new engine which leaves a distinctive "sausage-string" shaped contrail at high altitude, coupled with an unmistakable sound.
The Encyclopedia of World Aircraft, 1997

Aerodynamics is one of the fundamental aspects of aircraft flight. Most books on flight mechanics and performance have prominent chapters on the aerodynamics of the wing and the wing section. In view of the amount of topics to deal with in performance analysis, it will not be possible to review airfoil and wing characteristics. This chapter deals with the aerodynamics of the aircraft as a whole, particularly drag and lift characteristics as a function of the main parameters.

4.1 AERODYNAMIC FORCES

The aerodynamic forces and moments are not directly contributed by the propulsion system. For a symmetric aircraft in level flight, the aerodynamic forces are applied somewhere on the symmetry plane. For first-order calculations, there is a general consensus that these forces are applied at the center of gravity, as discussed in Chapter 2. In reality, the aerodynamic forces are applied at the center of pressure, which is dependent on the flight Mach number. As the aircraft accelerates past the speed of sound, the center of pressure tends to move aft. The opposite happens during deceleration to subsonic speed. The shift of the center of pressure and the consequent stability problems can be solved by pumping fuel aft and fore, as discussed by Leyman[82] for the Concorde (Problem 1.). An important calculation method for the center of pressure of wing/body and wing/body/tail combinations was developed by Pitts *et al.*[83] for bodies at subsonic, transonic, and supersonic speeds.

The aerodynamic force component in the direction of the velocity vector is called *drag*. The force component normal to the drag and pointing upward is the lift. The magnitude of these forces is, respectively,

$$D = \frac{1}{2} C_D \rho A U^2, \tag{4.1}$$

$$L = \frac{1}{2} C_L \rho A U^2, \tag{4.2}$$

where C_D and C_L are dimensionless force coefficients, A is a reference area, and U is the air speed. The reference area is the *wing area*, e.g. the area of the neutral wing projected on the ground, and including the portion inside the fuselage. The wing area is not a constant, because of the use of high-lift devices in take-off and landing operations. However, it is calculated with the control surfaces in the neutral position. The presence of additional aerodynamic surfaces, such as horizontal tail and canards, affects the position of the lift force, the magnitude and the share of lift generated by the aircraft's subsystems.

Two important non-dimensional quantities are the Reynolds number and the Mach number. The Reynolds number is the ratio between the inertial and viscous forces, $Re = \rho U l / \mu$, with l a reference length in the flight direction. The Mach number is the ratio between the air speed and the speed of sound, $M = U/a$.

The force coefficients depend on the Mach number, on the Reynolds number, on the angle of attack, on the geometrical shape of the aircraft, and on whether the flow is steady or transient. For a given aircraft configuration

$$C_L = C_L(\alpha, Re, M, t), \quad C_D = C_D(\alpha, Re, M, t), \tag{4.3}$$

In cases of engineering importance, the influence of the Reynolds number is considerably smaller than the remaining parameters, and is generally neglected. More specifically, this influence can be characterized as a slight decrease in C_D with the increasing speed, due to boundary layer effects, as discussed further in §4.5.1. Unless the aircraft is maneuvering at high angles of attack, the C_L and C_D are taken from steady state operation, hence the time is excluded from the functional parameters. Therefore, we will reduce Eqs 4.3 to

$$C_L = C_L(M, \alpha), \quad C_D = C_D(M, \alpha). \tag{4.4}$$

When the coefficients cannot be expressed in analytical terms, they are tabulated as a function of Mach number and angle of attack.

It is important to understand that the direction of D and L changes with the direction of flight. If the aircraft is flying on a flight path contained in a plane normal to the ground, its position can be characterized by the position of its CG and its attitude γ (three degrees of freedom). The angle between the velocity vector and the longitudinal axis of the aircraft is the nominal angle of attack. The same value of the angle of attack corresponds to an infinite number of attitudes. As shown in Figure 4.1, if the attitude increases, the velocity vector can be adjusted by augmenting the climb velocity by

$$\Delta v_c = V \tan \Delta \theta. \tag{4.5}$$

Under these conditions the angle of attack remains constant. The graph also shows how the direction of the aerodynamic forces is rotated by $\Delta \theta$.

An important aerodynamic parameter is the zero-lift angle of attack, α_o. This is the angle between the velocity vector $V = (V, v_c)$ and the aircraft's longitudinal axis FL at which the lift force vanishes. The zero-lift angle makes an angle, generally small, with the fuselage line.

Assume an aircraft flying in the vertical plane, as in Figure 4.2. The aircraft must rotate around the CG so as to have the fuselage line FL coincident with $V = (V, v_c)$ for the lift to vanish. At attitudes above this, the aircraft will have a positive angle

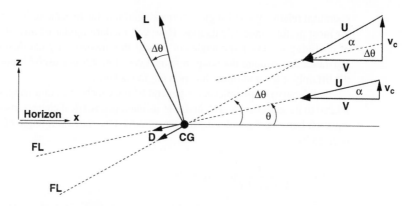

Figure 4.1 *Aerodynamic forces, aircraft attitude and angle of attack in the vertical plane (angles exaggerated for clarity).*

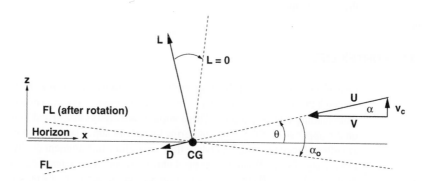

Figure 4.2 *Zero-lift conditions on the aircraft (angles greatly exaggerated).*

of attack. If a perturbation is introduced to α, the direction of the lift and drag force does not change.

4.2 LIFT EQUATION

From aerodynamic theory, the lift coefficient can be written as

$$C_L = C_{L_o} + C_{L_\alpha}(\alpha - \alpha_o),\tag{4.6}$$

where C_{L_o} is the zero-angle of attack lift coefficient and $C_{L_\alpha} = dC_L/d\alpha$ is the lift curve slope. At low Mach numbers and low angles of attack C_{L_α} is constant. As the incidence is increased, a noticeable difference with the linear behavior starts to appear; C_{L_α} may decrease or increase. A case of incipient wing stall is reflected in a decrease of the C_{L_α}. Flow separation starts at the trailing edge and at the outboard regions of the wing; it gradually expands to cover most of the wing surface. When C_{L_α} increases, it is generally a case of vortex lift, discussed further in §4.3. Most airplanes operate at relatively small angles of attack, therefore the linear assumption is adequate for most performance calculations. However, modern fighter jet aircraft are capable of operating at very large angles of attack; this involves a great deal of unsteady separated flow around the wing, in which case Eq. 4.6 is a useful representation of the lift only if the non-linear function $C_{L_\alpha}(\alpha)$ is known.

The lift-curve slope reflects the general lifting capabilities of a wing, because it is related to its aspect-ratio $\mathcal{AR}$, hence to its main geometrical characteristics. A useful equation between C_L and wing aspect-ratio for low-speed low angle of attack flight is given by

$$C_{L_\alpha} = \frac{2\pi}{1 + 2/\mathcal{AR}}(\alpha - \alpha_0), \qquad (4.7)$$

which is valid only for elliptic spanwise loading. Equation 4.7 allows the construction of the three-dimensional wing lift from the airfoil and the geometry of the wing. This equation does not contain the angle of attack, therefore it shows as a straight line in the α–C_L graph.

4.3 VORTEX LIFT

One of the sources of non-linearity in the lift equation is the *vortex lift*. At high angles of attack the wing flow is dominated by two large counter-rotating vortices developing from leading-edge separation. These vortices are always associated to low pressures on the upper side of the wing, and grow linearly or super-linearly downstream. They are often associated with secondary flow separation and augment the lift-curve slope in a variety of ways. Whatever the mechanisms of vortex lift generation, a performance analysis relies on lift curves that can be expressed in a compact form, such as Eq. 4.6. An example of vortex lift on the C_L is shown in Figure 4.3 for a Δ-wing with a 74 degree sweep[84,85,86]. The wing has an aspect-ratio $\mathcal{AR} = 0.5735$. The data show that the leading-edge vortex increases the lift at a super-linear rate. The resulting C_L increasingly diverges from the "conventional" curve, $C_L = 2\pi\alpha$. Also, the stall angle is dramatically increased to about 40 degrees and is more gradual, sometimes yielding a nearly constant lift over a wider range of incidences. This is not uncommon for this type of wing. Therefore, it is an indication that the aircraft can maneuver at angles of attack beyond stall. Aircraft capable of maneuvering in post-stall conditions are called *super maneuverable*.

Angles of attack that can be sustained by modern military aircraft during high-power climb, turning, and aerobatic maneuvers can exceed 50 degrees. This is done to obtain a positional advantage during combat, although the advantage comes at the expense of some energy loss and some stability problems. Calculations under these conditions can be done by considering average values of the lift and drag force, for example from tabulated data like those in Figure 4.3.

A relatively simple method, due to Polhamus[87,88], allows the calculation of the C_L of pointed Δ-wings at incidences below stall, and recovers some of the non-linearities of the vortex lift. The resulting equation is

$$C_L = k_p \sin\alpha \cos\alpha^2 + k_v \cos\alpha \sin\alpha^2, \qquad (4.8)$$

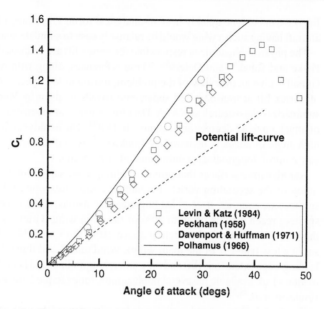

Figure 4.3 *Vortex lift of a slender Δ-wing, sweep angle $\Lambda = 74$ degrees at the apex.*

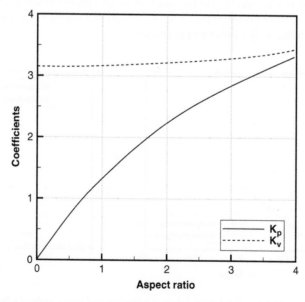

Figure 4.4 *Polhamus coefficients for a pointed Δ-wing.*

where k_p and k_v are coefficients for the potential flow lift and vortex lift components, respectively. These coefficients depend on the aspect-ratio of the wing and their calculation has to be done by other means, for example a vortex lattice method. These coefficients have been calculated by Polhamus, and are shown in Figure 4.4. The value of the coefficients is important, because the calculation of the vortex lift from

Eq. 4.8 can be done quickly in semi-analytical form. Therefore, a high-performance aircraft having such a wing would be relatively easy to simulate in a high-α maneuver.

The physical phenomena responsible for vortex lift are discussed in some detail by Peake and Tobak[89] and Delery[90]. These references, along with Ashley et al.[91] and Gursul[92] give an overview of the problem, mostly at low speeds. A good discussion of vortex lift at transonic and supersonic speeds is given by Wood et al.[93], which summarizes the research at NASA. Data for Δ-wings at supersonic speeds and angles of attack up to 47 degrees are available in Hill[94]. The aerodynamics of the aircraft at high angle of attack is reviewed by Erickson[95], who discussed post-stall maneuver, yaw control, longitudinal stability and other problems.

Aerodynamic surfaces that promote leading-edge separation and control the evolution of the separation vortex are another reason for vortex lift. These additional surfaces are *canards, strakes*, leading-edge extensions, and double Δ-wings. Canard surfaces reach a value of up to 20% of the main wing (Grumman X-29), although they are more likely to be slightly above 10% (the Eurofighter 2000 has a canard surface less than 5% of the main wing). Wind tunnel and flight data on strakes and leading-edge extensions (lift and drag polar, longitudinal and lateral stability at angles of attack) are available in several publications, for example Lamar and Frink[96,97] and Erickson et al.[98,99]

Figure 4.5 shows wind tunnel results of experiments with strakes empirically designed. The C_L curve is smoother; the performance results in lower C_D for a given C_L. The effect of the strakes includes a change in stability characteristics, with a positive pitching moment slope, instead of a negative one. This makes the aircraft unable to fly without proper control.

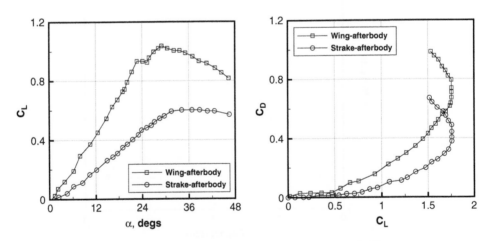

Figure 4.5 *Wind tunnel data for strakes and wing/body combinations at high α, elaborated from Lamar and Frink[96].*

Example

Consider the pointed Δ-wing discussed in Figure 4.3. The aspect-ratio of the wing is $\mathcal{AR} = 2\tan \Lambda/2$, with Λ the apex angle. Therefore, $\mathcal{AR} = 2 \times \tan 37 = 1.507$. The

Polhamus coefficients for this case are estimated as $k_p \simeq 3.20$ and $k_v \simeq 1.82$. The corresponding lift curve is shown as a solid line in Figure 4.3.

4.4 HIGH-LIFT SYSTEMS

The term *high-lift system* denotes all the technical means for increasing lift during the terminal phases of the flight. Low take-off speeds and high climb-out rates are a general requirement for good airfield performance.

Most of the systems are unpowered, and consist of leading-edge slats and trailing-edge flaps in several segments. A single-slotted flap is used when the wing system is not capable of supplying a $C_{L_{max}}$ adequate for approach and landing. If this is still not enough for low-speed control, the landing flap has to be replaced by a double-slotted flap. Calculation of the maximum C_L obtainable from a high-lift system is a complicated matter. Therefore, most of the data available rely on practical experiments and on flight testing. An aircraft operating near its $C_{L_{max}}$ is at risk of wing stall. For take-off and landing operations there are practical rules that avoid operating at too high C_L. For safe operation we have to consider

$$C_{L_{approach}}/C_{L_{to}} \simeq 1.10 - 1.12, \tag{4.9}$$

and

$$C_{L_{to}} \simeq 0.7 - 0.75 C_{L_{max}}, \quad C_{L_{approach}} \simeq 0.6 - 0.65 C_{L_{max}}. \tag{4.10}$$

Therefore,

$$C_{L_{land}} \simeq 1.3 C_{L_{to}} \tag{4.11}$$

A summary of high-lift and control systems on current airplanes is shown in Figure 4.6, where we have plotted the maximum C_L against the mechanical complexity of the systems (arbitrary unit). To the left of the shaded bar is a summary of unpowered systems, which are operated by proper actuators; to the right are some examples of powered systems, which are operated with energy from the main power plant. The maximum C_L achieved by the unpowered systems seldom exceeds 3.0, though it is often below this limit. Some data are the following: the Fokker F-28 has $C_{L_{max}} \simeq 3.35$; the Boeing YC-14 has $C_L \simeq 3.57$. For further lift data the reader is invited to refer to Brune and McMasters[100], Callaghan[101] and Obert[102].

In the top right of Figure 4.6 there are the high-lift performance data of some experimental aircraft, for example the NASA/Boeing QSRA (Quiet Short Haul Research Aircraft), flown in the 1980s (Shovlin *et al.*[103]). This aircraft required only a 300 meter runway. The jet engine exhaust was directed over the wing, which shielded the ground from jet noise during take-off. Figure 4.7 shows a typical C_L polar.

For more details on the mechanics and aerodynamics of these systems we refer to McCormick[104], to the classical paper of Smith[105], the collection of AGARD CP-515[106], Gratzer[107] and the substantial report by Rudolph[108], which is a summary

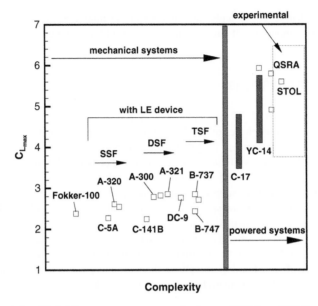

Figure 4.6   *High-lift systems on current airplanes; QSRA and STOL are experimental airplanes. SSF = Single-Slot Flap, DSF = Double-Slot Flap, TSF = Triple-Slot Flap.*

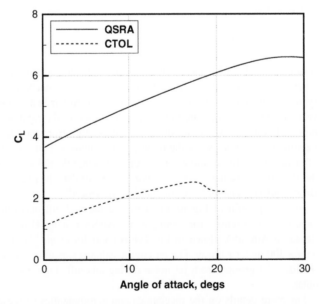

Figure 4.7   *Lift coefficient of NASA QSRA compared with that of a conventional aircraft. Elaborated from Shovlin et al.[103]*

of applications to commercial airplanes. The latter report contains details on the wing systems of nearly all the Boeing, Airbus, and McDonnell-Douglas jet airplanes.

4.5   DRAG EQUATION

Drag is a vast subject in aircraft aerodynamics. It is of particular interest, regarding the matters concerned in this book, to consult McCormick[104] and Stepniewski and Keys[109], where additional specialized references can be found. These authors address the readers to specific drag components and the drag breakdown for the full aircraft. A typical chart of drag components on the aircraft is shown in Figure 4.8.

The drag equation specifies the functional dependence of the C_D on the main state parameters of the aircraft (Eq. 4.4). A general expression to the second order of the angle of attack is

$$C_D = C_{D_o} + \eta C_{L_\alpha} \alpha^2, \tag{4.12}$$

where C_{D_o} is the zero-lift drag coefficient (profile drag), η is an induced drag coefficient, and α is the angle of attack in radians. In Eq. 4.12 all the coefficients are a function of α, the Mach number. They are generally tabulated in the form $C_{D_o}(M)$, $\eta(M)$, $C_{L_\alpha}(M)$. It is assumed that the aircraft operates within the linear range of angles of attack, even at supersonic speeds. An example of drag and lift data is given in Figures A.8 and A.9. The data are reported in Table A.12 for a basic aircraft configuration

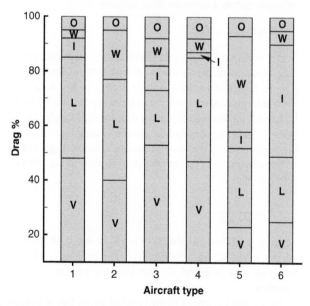

Figure 4.8   *Contributing factors to overall drag. Nomenclature: (1) subsonic transport; (2) SST; (3) business jet; (4) fighter aircraft (subsonic); (5) fighter aircraft (supersonic); (6) helicopter; V = viscous/parasite; L = lift-induced; I = interference drag; W = wave drag; O = other causes.*

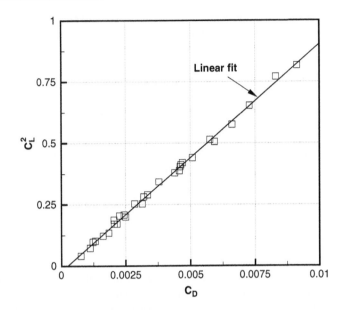

Figure 4.9 *Drag equation for the Douglas DC-10, elaborated from Callaghan*[101].

(Configuration A), and will be used in later chapters for performance calculations. At subsonic speeds a common drag equation is

$$C_D = C_{D_o} + kC_L^2. \tag{4.13}$$

In this equation the angle of attack does not appear explicitly. This is the preferred drag equation for performance calculations of low-speed aircraft and commercial jets. In fact, due to the quadratic term C_L^2, Eq. 4.13 yields several useful closed form solutions. Also, the C_L can be easily associated to the aircraft's gross weight. Figure 4.9 shows some flight test data for the Douglas DC-10 passenger aircraft. The linear fit demonstrates that the parabolic drag Eq. 4.13 is valid over a wide range of lift coefficients, as they can be used in ordinary flight conditions.

For an estimate of the induced drag factor k an equation often used from low-speed aerodynamics is

$$k = \frac{1}{e\pi\mathcal{AR}}, \tag{4.14}$$

where e is a factor variable between 0.74 and 0.88 (Oswald factor), which depends on the spanwise load distribution. For an ideally elliptically loaded wing $e = 1$. Equation 4.14 is useful for checking the validity of the coefficient k in the drag equation.

The C_L and C_D can be plotted against each other in a single graph, to yield the *drag polar*, for a given aircraft speed or Mach number. An example is shown in Figure 4.10, which is relative to the experimental aircraft Lockheed YF-16, as adapted from Webb *et al.*[110] The relationship between C_L^2 and C_D is not linear, therefore the drag (Eq. 4.13) is not suitable for this aircraft.

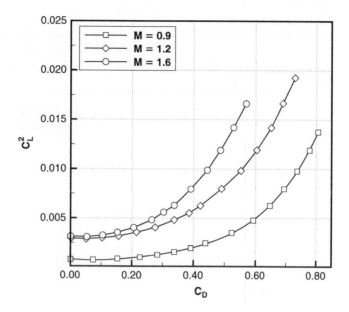

Figure 4.10 *Drag polar of YF-16 at Mach numbers up to M = 1.6 (estimated data).*

4.5.1 Zero-Lift Drag

The zero-lift drag (profile drag) of the aircraft is the resistance due to viscous effects and other causes not directly related to the production of lift, such as the resistance of various subsystems (landing gears, probes, antennas, gaps, etc.). It depends on a number of factors, the Reynolds number and the surface roughness being the most important ones. The percentage of this drag on the total drag count depends on the type of aircraft and its operational configuration, and varies from 25–30% for a supersonic jet fighter to 70–80% in the case of a VTOL aircraft. For a commercial subsonic jet aircraft the contribution of the fuselage to the zero-lift drag is of the order of 30%. The skin friction drag of modern large aircraft is close to the theoretical value from turbulent flat plate theory at the corresponding Reynolds number.

The effect of Reynolds number can be evaluated from a number of semi-empirical relationships (von Kármán-Schoenherr, Prandlt-Schlichting, Schultz-Grünow, and others[111]). At the flight Reynolds numbers these expressions are equivalent to each other.

Figure 4.11 shows the values of the skin friction drag coefficient due to a fully turbulent flow past a flat plate, as a function of the Reynolds number. The estimated C_{D_o} values of modern aircraft are shown in the box for comparison. The Reynolds numbers in these cases are calculated using the average wing chord as a reference length, and the air viscosity at the conditions of the cruise altitude.

The streamlined design of modern aircraft shows that the current values are not too far from the drag obtainable by a corresponding flat plate at the same flight Reynolds number, which leaves little room for improvement, unless a new technology is developed (boundary layer control, flow control, etc.). The current values of the zero-lift drag coefficient for large commercial jets are of the order of 220 to 250 drag counts (1 drag count $= 0.0001$), and sensibly higher for smaller airplanes. By comparison, the Douglas DC-3 of the 1930s had 249 drag counts of skin friction

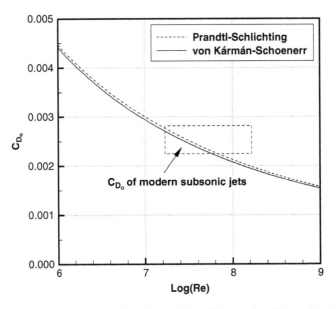

Figure 4.11 *Estimated value of zero-lift coefficient of modern subsonic transport aircraft and comparison with theoretical values from flat plate theory.*

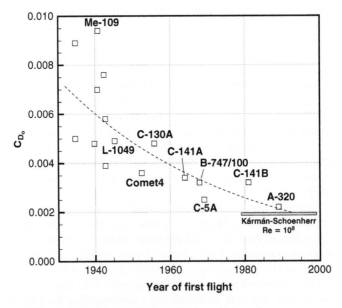

Figure 4.12 *Profile drag coefficient of selected aircraft.*

drag – a very respectable value, which demonstrates that progress in this area has been relatively slow. A historical trend of the profile drag coefficient for selected aircraft is shown in Figure 4.12. The dotted line shows a trend toward the minimum theoretical value of a turbulent flat plate. Additional data of this type can be found in Loftin[11], Cleveland[55] and Anderson[112].

4.6 GLIDE RATIO

This parameter is the ratio between lift and drag; sometimes it is called *glide factor* or *aerodynamic efficiency*. The best glide ratios of current aircraft do not exceed 20, although they are contained within the 14 to 18 range. Some high-performance gliders with very large aspect-ratio wings have glide ratios in excess of 25. By comparison, the best low-drag wing section will have a glide ratio of the order of 150. The L/D values are known to decrease with the increasing Mach number.

The glide ratios of large aircraft at cruise conditions are not that different from those of the most efficient birds. It is estimated that the albatross (*Diomeda exulans*) achieves an $L/D \simeq 20$. The best glide ratio of the Boeing B-52 is about $L/D \simeq 20$, while Concorde struggled to achieve $L/D \simeq 9$; the Wright Flyer had a respectable $L/D \simeq 8.3$. The airfoil Liebeck L-1003 has an optimal $L/D \simeq 220$ at a Reynolds number $Re = 10^6$, e.g. a factor 10 compared to the Boeing B-52.

The glide ratio is a parameter variable with the speed and the weight of the aircraft, and changes during a long-range cruise. Table 4.1 summarizes glide ratio data for some known aircraft in cruise conditions. The product $M(L/D)$ will be discussed later in Chapter 9. Additional data and analyses can be found in Loftin[11], Anderson[112], and for very old airplanes in Ackroyd[113].

In the next chapters we will show that the glide ratio is one of the essential performance parameters, therefore it is of interest to calculate optimal values with respect

Table 4.1 *Average aerodynamic data for selected aircraft (estimated).*

Aircraft	L/D	$M(L/D)$	M	Notes
Boeing B-52G	20.5	16.4	0.80	
Lockheed C-5A	18.6	14.5	0.78	
Boeing B-707/200	18.2	14.4	0.79	
Boeing B-747/100	17.6	14.8	0.84	
Douglas DC-9-30	17.2	13.2	0.77	
Douglas DC-3	14.7	4.1	0.28	propeller
Concorde	9.0	18.0	2.05	
McDonnell-Douglas F-15C	10.0	9.0	0.90	transonic
McDonnell-Douglas F-15C	4.0	6.4	1.60	supersonic
XB-70A	7.55	5.74	0.76	subsonic
XB-70A	5.14	6.22	1.21	
XB-70A	8.72	24.35	2.79	supersonic

to the aircraft's mass (weight), speed and flight altitude. By using the drag Eq. 4.13, the glide ratio becomes

$$\frac{L}{D} = \frac{C_L}{C_D} = \frac{C_L}{C_{D_o} + kC_L^2}, \tag{4.15}$$

and using the definition of C_L,

$$\frac{C_L}{C_D} = \frac{c_1/U^2}{C_{D_o} + c_2/U^4}, \tag{4.16}$$

with $c_1 = (2/\rho)(W/A)$, and $c_2 = kc_1^2$. To find the speed corresponding to maximum L/D, we need to find its derivative with respect to the speed and set it to zero:

$$\frac{\partial}{\partial U}\left(\frac{c_L}{C_D}\right) = \frac{2U(C_{D_o} + c_2 U^4) - 4U^2 c_2 U^3}{(C_{D_o} + c_2 U^4)^2} = 0, \tag{4.17}$$

$$-2c_1 U^{-3}(C_{D_o} + c_2/U^4) - 4(c_1/U^2)c_2 U^{-5} = 0. \tag{4.18}$$

By further simplification, we find

$$U^4 = \frac{k}{C_{D_o}}\left(\frac{2W}{\rho A}\right)^2. \tag{4.19}$$

Another way of expressing Eq. 4.15 is in terms of the mass (or weight) at given flight speed. Again, using the definition of C_L, we find:

$$\frac{C_L}{C_D} = \frac{c_1 m}{C_{D_o} + c_2 m^2}, \tag{4.20}$$

where this time

$$c_1 = \frac{2g}{\rho A U^2}, \quad c_2 = kc_1^2. \tag{4.21}$$

As in the previous case, the glide ratio is optimal at only one mass – for a given flight speed. This can be found from differentiating Eq. 4.20 with respect to the mass and setting the derivative to zero:

$$\frac{\partial}{\partial m}\left(\frac{C_L}{C_D}\right) = \frac{c_1(C_{D_o} + c_2 m^2) - 2c_1 c_2 m^2}{(C_{D_o} + c_2 m^2)^2} = 0, \tag{4.22}$$

which leads to

$$m = \sqrt{\frac{C_{D_o}}{c_2}} = \frac{1}{c_1}\sqrt{\frac{C_{D_o}}{k}} = \frac{\rho A U^2}{2g}\sqrt{\frac{C_{D_o}}{k}}. \tag{4.23}$$

It is straightforward to read the results of Eq. 4.23: the optimal mass (or weight) of the aircraft decreases with the increasing speed and flight altitude – all other parameters

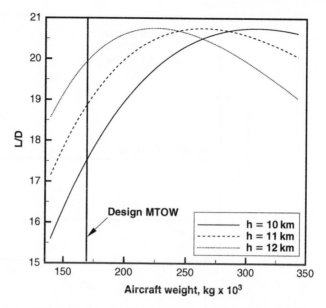

Figure 4.13 *Glide ratio versus aircraft mass at different flight altitudes (as indicated), at a cruise Mach number $M = 0.80$, aircraft model A.*

being constant. The term under square root is an aerodynamic factor that will appear again in other optimal expressions.

A parametric study of Eq. 4.23 is shown in Figure 4.13. The maximum value of L/D does not change with the cruise altitude, but the mass corresponding to its maximum value does. In fact, the mass for maximum L/D decreases as the aircraft climbs. The mass required for the global optima may well exceed the design mass of the aircraft.

Another parametric study, which will turn out useful for the cruise analysis, is shown in Figure 4.14. In this figure we show the estimated L/D of the reference subsonic jet as a function of the cruise altitude, for a fixed mass. Again, the global optimum is achieved at conditions beyond the limits of this aircraft's performance.

4.7 GLIDE RATIO AT TRANSONIC AND SUPERSONIC SPEED

The glide ratio at transonic and supersonic speed cannot be expressed in a simple form. In general, the data are tabulated as a function of the Mach number. We use the lift and drag equation for high-speed flight, Eq. 4.6 and Eq. 4.12, and calculate the glide ratio as

$$\frac{C_L}{C_D} = \frac{C_{L_\alpha}\alpha}{C_{D_o} + \eta C_{L_\alpha}\alpha^2} = \frac{1}{C_{D_o}/C_{L_\alpha}\alpha + \eta\alpha} = f(M,\alpha). \tag{4.24}$$

Exact evaluation of Eq. 4.24 can only be done numerically. The algorithm is the following.

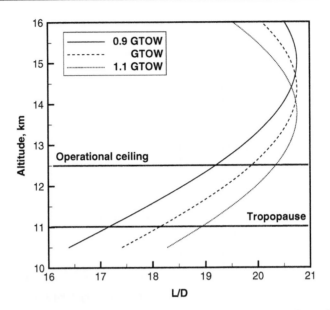

Figure 4.14   *Glide ratio versus altitude at fixed weight, cruise Mach number M = 0.80. The graph shows the effects of gross weight from a nominal value GTOW = 155,000 kg.*

Computational Procedure

- Set the Mach number. Hence, Eq. 4.24 is only a function of α.
- The condition of min/max is found from the derivative of Eq. 4.24 with respect to α:

$$\frac{\partial}{\partial \alpha}\left(\frac{C_L}{C_D}\right) = 0. \tag{4.25}$$

After some algebra, the optimal condition becomes

$$\alpha^2 = \frac{C_{D_o}}{\eta C_{L_\alpha}}, \quad \alpha = \sqrt{\frac{C_{D_o}}{\eta C_{L_\alpha}}}. \tag{4.26}$$

- The value of the root α of Eq. 4.26 is inserted in Eq. 4.24; the corresponding value of C_L/C_D is stored in memory.
- If the value of C_L/C_D has increased, then we set the maximum to the current value.
- Increase the Mach number and repeat the procedure, as above, until the maximum Mach number is reached, or until the C_L/C_D starts decreasing.

The result of this algorithm is presented in Figure 4.15. The maximum value of $C_L/C_D \simeq 16$ at low Mach number ($M < 0.2$), and a cruise angle of attack $\alpha \simeq 3.6$ degrees. Note the transonic dip at low supersonic speeds. The L/D is relatively low, and it reaches a value of about 6 at full supersonic conditions (Concorde managed an

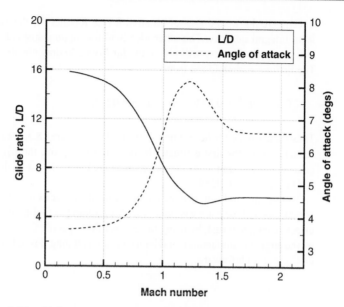

Figure 4.15 *Glide ratio versus Mach number for the supersonic aircraft in baseline configuration.*

L/D around 9). There is a small recovery at high supersonic speeds. This behavior of the glide ratio as a function of the Mach number is characteristic of many high-speed vehicles, as demonstrated by Küchemann[114].

4.8 PRACTICAL ESTIMATION OF THE DRAG COEFFICIENT

The aerodynamic performance of most aircraft is not widely advertised. Nevertheless, there are methods that allow for a practical estimation of the drag characteristics. The values of the lift are not as interesting as the drag. In fact, the lift is mostly depending on the weight. The drag depends on both the weight and the type of aerodynamic surface. We consider as a reference a subsonic jet aircraft, although the following considerations are valid for other fixed-wing aircraft whose drag equation is parabolic.

The glide ratios are known from statistical data. In the present context we will work with $C_L/C_D \simeq 18$, which is not the best nor the worst in its class (see Table 4.1). If the weight of the aircraft is W, then the lift coefficient is found from Eq. 4.2. For example, if the AUW of the aircraft is 160,000 kg, the C_L at cruise conditions ($h = 11,000$ m, $M \simeq 0.80$) is estimated at $\simeq 0.558$.

The aircraft drag will be $C_D = (L/D)C_L \simeq 0.031$, a relatively high figure, although not far off the true value. The next problem is to find the contribution of the zero-lift drag and the lift-induced drag. One possibility is to use the equation

$$C_{D_o} = q\overline{C}_f A_{wet},$$

which requires the knowledge of the total wetted area, A_{wet}, and the average skin friction coefficient, $\overline{C}_f$. These data are not easily found. There are practical methods to estimate the wetted area of an aircraft (see Raymer[76]); the average skin friction is an integral quantity. The other idea is to use again statistical data; for a modern airliner the profile drag will not be too far from 230 drag counts, e.g. $C_{D_o} \simeq 0.0230$. The factor k is then found from

$$C_{D_o} + kC_L^2 \simeq 0.031 \quad \Rightarrow \quad k \simeq 0.0257.$$

These data can be further refined. If the glide ratio were 18.5, then $k \simeq 0.0230$, a 10% reduction over the first estimate. If, instead, we improve the C_{D_o} down to 220 drag counts (a lower limit at the current technology level), then with the initial estimate of L/D we find $k \simeq 0.0289$.

Another method is to use the engine thrust data. For example, an aircraft at cruise conditions will have a measurable fuel flow. From the fuel flow we could extract the effective engine thrust, by using the charts of the specific fuel consumption provided by the engine manufacturer, and hence the aircraft drag. From Eq. 4.1 the calculation of C_D is straightforward.

4.9 COMPRESSIBILITY EFFECTS

As the free stream Mach number is increased beyond $M = 0.4 - 0.5$, some compressibility effects start to appear on the lifting surfaces. These effects are compounded by local accelerations and angles of attack. Systems that are particularly affected are the propellers and rotors. A full compressible flow theory is not necessary until transonic Mach numbers are achieved. Semi-empirical corrections are applied to the aerodynamic characteristics to take into account some of these effects. A number of corrections for lift, drag, pressure and pitching moment coefficients exist to deal with flows at speed range. Prandtl-Glauert, Kármán-Tsien, Chaplygin, Busemann, and others provided relatively simple expressions for the correction of the two-dimensional airfoil characteristics. These correction formulas can be found in most books on applied aerodynamics, for example Bertin[115] and Kuethe and Chow[116]. The Prandtl-Glauert formulas represent a first-order correction, and are given by

$$C_D = \frac{C_D}{\sqrt{1 - M^2}}, \quad C_{L_\alpha} = \frac{C_{L_\alpha}}{\sqrt{1 - M^2}}. \tag{4.27}$$

At full supersonic conditions this correction can be replaced by Ackeret's linearized flow theory. In this case the aerodynamic coefficients of the wing, corrected for three-dimensional effects, are

$$C_L = \frac{4}{\sqrt{M^2 - 1}}\left[1 - \frac{1}{2\mathcal{AR}\sqrt{M^2 - 1}}\right], \quad C_D = C_L\alpha. \tag{4.28}$$

These expressions are valid only when there is no interaction between the Mach cones arising from shocks at the wing tips. The transonic regime, being highly non-linear, is not covered by corrective factors. There exist a number of theories to calculate the lift-curve slope of thin wings at supersonic speeds, for example Hayes[117] (wings with subsonic leading edges) and Harmon and Jeffreys[118] (supersonic leading edges).

4.10 TRANSONIC DRAG RISE

At Mach numbers beyond the validity of the compressibility correction a number of non-linear transonic phenomena start to occur. These phenomena are reflected in Eq. 4.12 by the functional dependence of the coefficients from the Mach number. For reference, the beginning of transonic drag rise is set by the *divergence Mach number*, which is conventionally defined by the point where the derivative

$$\frac{\partial C_D}{\partial M} > 0.1. \tag{4.29}$$

The Mach number at which this occurs is a complex function of the aircraft configuration and operational conditions (angle of attack, lift coefficient, altitude). For subsonic commercial jets it is of relatively limited implication, because both the operating angle of attack and the cruise speed vary over a narrow range. For a high-performance fighter jet the issue is more complex, because it depends strongly on its configuration. An example is reported in Figure 4.16, for the reference supersonic aircraft model C. The graph is indicative of some operational configurations. The drag reaches a maximum somewhere between $M = 0.8$ and 1.2, then it decreases

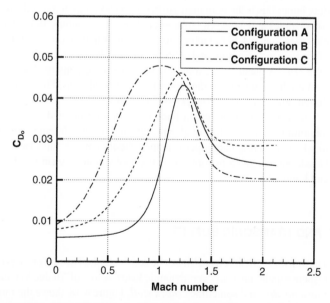

Figure 4.16 *Zero-lift drag coefficient for three different configurations. Configuration A is the baseline.*

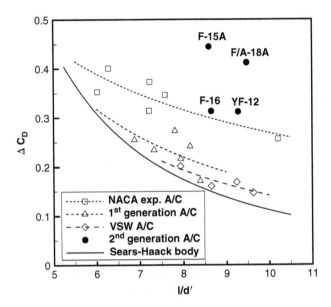

Figure 4.17 *Transonic drag rise for some aircraft.*

and settles to a level higher than the value before the transonic drag rise. Therefore, having an engine powerful enough, the aircraft can go past the *sonic barrier* and fly at supersonic speeds.

The penalty to be paid in supersonic flight is sometimes quantified by the transonic drag jump; this is the drag increase between two reference speeds, which are $M = 1.2$ (supersonic) and $M = 0.8$ (subsonic). The corresponding *transonic drag penalty* is plotted in Figure 4.17 (adapted and extended from Poisson-Quinton[119]), for a number of fighter and experimental aircraft. On the horizontal axis there is the fineness ratio l/d'. A clean configuration has a lower drag penalty. Also, more modern aircraft are less sensitive to the transonic flight conditions. For reference, also the drag discontinuity of a Sears-Haack body is plotted.

The Sears-Haack body is a body of revolution pointed at both ends, having the minimum supersonic drag or given length and volume. It represents an ideal situation, to which an aerodynamic performance should tend. A full discussion of the Sears-Haack body is given in some good textbooks of aerodynamics, for example Ashley and Landhal[120].

4.11 LIFT AND TRANSONIC BUFFET

The transonic effects on the lift can be considerable. The loss of lift is due to aeroelastic response under unsteady aerodynamic loads. A loss of lift due to wing buffet is called *transonic dip*, and sometimes *shock stall*. Figure 4.18 shows the transonic dip at the inception of wing buffeting. One of the first military jet aircraft, such as the North American F-84, suffered badly from loss of lift at transonic speeds. A 10-year older

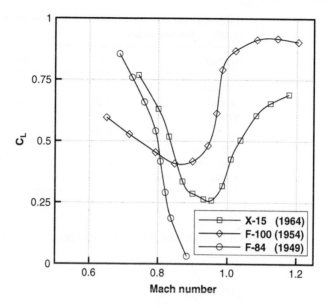

Figure 4.18 *Transonic dip for three reference jet aircraft.*

design, such as the North American F-100 Super Sabre, quickly recovered the lift, although it showed some decrease in the transonic lift.

Another effect of the Mach number is on the maximum lift coefficient of the aircraft, $C_{L_{max}}$. For a subsonic transport aircraft $C_{L_{max}}$ is nearly constant up to the drag divergence point, then it is gradually reduced. The angle of attack corresponding to $C_{L_{max}}$ also decreases. For a supersonic fighter jet aircraft the function $C_{L_{max}}(M)$ is more complicated, and shows a transonic dip with a partial supersonic recovery. Its value at supersonic speeds can be considerably lower than that at subsonic speeds. These effects are discussed in detail by Abbott and von Doenhoff[121] for NACA airfoil sections, and in the NACA publications reported in that book.

The behavior of $C_{L_{max}}(M)$ is important in high-performance maneuver, as it will be described in Chapter 10. A plot of $C_{L_{max}}$ for the supersonic jet aircraft of reference is shown in Figure A.12, on page 493. However, the values of $C_{L_{max}}$ are often approximate, and a satisfactory evaluation of this parameter is difficult.

4.12 AERO-THERMODYNAMIC HEATING

Heating due to the aerodynamics of high-speed flight is due to the conversion of kinetic energy of the flow into thermal energy. The amount and intensity of this exchange is dependent on the free stream Mach number and on the local geometry. It is highest at the stagnation points of the aircraft. This temperature increase is due to the difficulty of transferring momentum to the outside flow domain, therefore the momentum is dissipated in heat. From supersonic aerodynamic theory at moderate Mach numbers

Table 4.2 *Aero-thermodynamic heating at stagnation points (estimated data).*

Aircraft	M	Max temp., K
Concorde	2.0	380
Convair B-58	2.0	420
XB-70	3.0	550
X-15	6.0	900
Waverider	6–8	1,000
Space shuttle	10–12	1,500
ICBM	20–25	6,000

$(M < 3)$, the temperature rise at the stagnation point can be found from the energy equation

$$C_p T + \frac{1}{2} U^2 = C_p T_o, \tag{4.30}$$

which is valid for an iso-entropic transformation; T denotes the absolute temperature, C_p is the specific heat at constant pressure, and "o" denote stagnation conditions. The temperature rise at the stagnation points is found from

$$\Delta T = \frac{U^2}{2C_p} = \frac{1}{2} \frac{a^2 M^2}{C_p} = \frac{1}{2} \frac{\gamma R T M^2}{C_p}, \tag{4.31}$$

where in this case γ is the ratio between specific heats, and is taken equal to 1.4 (approximation valid at temperatures below 900 K). This heating will have consequences on the flight envelope of the aircraft, as discussed in Chapter 6. A summary of estimated stagnation point temperature for some high-speed flight vehicles is reported in Table 4.2

Figure 4.19 shows the map of aero-thermodynamic heating at the stagnation point for supersonic speeds in atmospheric flight, as deduced from Eq. 4.30. The problem of hypervelocity vehicles (waveriders, re-entry vehicles and missiles) is complicated by the fact that the flow does not obey the law of ideal gases, and the aero-thermodynamic heating has to be calculated using real gas relationships. A discussion can be found in specialized studies on hypersonics and missiles (for example, Allen and Eggers[122]) and planetary entry flight mechanics (Vinh *et al.*[123]).

Example

The actual skin temperature will be the value shown minus the difference between altitude and sea level temperature. For example, the estimate for Concorde would be about 173 degrees above the local temperature. Therefore, the sea level temperature would be about $173 - (56 + 15) = 102°C$. The actual temperature found in the

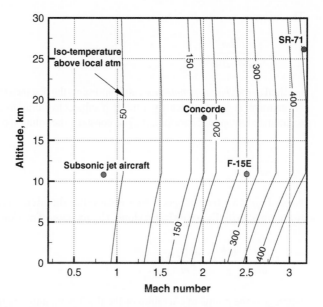

Figure 4.19 *Stagnation point temperatures above local atmospheric temperature.*

technical publications on Concorde report most of the nose above 100°C, the leading edge of the wing at a temperature of about 100°C and most of the fuselage at temperatures of 90 to 97°C. Therefore, the approximation is acceptable.

4.13 AERODYNAMIC PENETRATION AND RADIUS

We define two parameters that have the dimensions of a length. These are both flight mechanics and design quantities. Consider a rigid-body aircraft of mass m at a flight speed U without engine thrust, and subject to a drag force D. The equation of motion in the flight direction is

$$a = -\frac{D}{m},$$
(4.32)

or

$$a = \frac{\partial U}{\partial t} = U\frac{\partial U}{\partial x} = \frac{\rho A C_D U^2}{2m}.$$
(4.33)

Further simplification of this equation leads to

$$\frac{dU}{U} = \frac{\rho A C_D U^2}{2m}\,dx.$$
(4.34)

The factor

$$s_D = \frac{2m}{\rho A C_D U^2} \tag{4.35}$$

is called *aerodynamic penetration*, and represents the distance that the aircraft of mass m can travel at constant density-altitude if subject to a drag force that is proportional to the speed squared. Therefore, if C_D is constant, the solution of the equation above is

$$\frac{U}{U_{ref}} = e^{x/s_D}, \tag{4.36}$$

where U_{ref} is a reference speed. Now assume that the aircraft is accelerated in a direction normal to its flight path by lift alone. The equation of motion along the flight path is

$$U\frac{\partial \gamma}{\partial t} = \frac{U^2}{\chi} = \frac{\rho A C_L}{2m} U^2, \tag{4.37}$$

where χ is the radius of curvature of the flight path, and γ is the local attitude. This equation can also be written as

$$\frac{1}{s_L} = \frac{1}{\chi} = \frac{\rho A C_L}{2m}. \tag{4.38}$$

The *aerodynamic radius* is the radius of curvature of the flight path due to lift. This parameter will be called $s_L = \chi$.

These concepts were first introduced by Larrabee[124] to describe several problems in atmospheric flight and atmospheric re-entry. They both have the dimension of a length. They can be used for the solution of Lanchester's phugoid. Additional applications of these concepts include the wind drift and the lateral maneuver. Wind drift is important in ballistics, parachute drops and unguided atmospheric re-entry. Lateral maneuver is important in air-to-air missiles and combat aircraft.

4.14 AIRCRAFT VORTEX WAKES

When the fixed-wing aircraft generates lift from its lifting surfaces, it also creates a complex vortex system (tip vortices, flap vortices) that travels downwards (downwash). The strength and shape of these vortices depend essentially on the main wing parameters. The downward mass flow depends on the weight of the aircraft. In fact, it is possible to prove that the downwash w creates a downward mass flow rate $\dot{m} \propto W/b$. From low-speed aerodynamic theory, under the approximation of large aspect-ratio and elliptic wing loading downwash, the mass flow rate is

$$\dot{m} = \rho A_{ref}\overline{w} = \frac{2}{\pi}\frac{W}{b}. \tag{4.39}$$

We have assumed that the reference area A_{ref} behind the aircraft is $A_{ref} = \Delta x b = \Delta t \, Ub = Ub$ in the unit of time*. This flow rate is not dependent on the flight altitude. Therefore, for a Boeing B-747 at a cruise speed $M = 0.80$ with an AUW $= 350$ tons, we have $\dot{m} \sim 5,400$ kg. At a density altitude $\sigma = 0.29$ this corresponds to about $15,200$ m^3 of downward flow every second, at an average downwash speed of $\overline{w} = 6$ m/s. If a small airplane of the same class as the Piper 28 flies just below the B-747 at an average speed of 120 kt (222 km/h), the downwash reduces the inflow by about 5.5 degrees – a potentially catastrophic effect. See Rossow and James[125] for a general overview of the problem.

The vortex system creates a hazard for vehicles following at a short distance, or below. This problem is of interest in congested air spaces and around airports with large volumes of traffic. Research in this field has been buoyant, in the attempt to reduce flight hazard and separation times between take-off and landing operations. The literature is too vast to cite; see Spalart[126] for a review, and Rossow[127], and Rossow and Tinling[128,129]. Some research now focuses on two broad areas: (1) the technical means and design solutions to diffuse the vortex system, and (2) on the optimal flight paths that would reduce the runway occupation time. For example, the shortest method to fly away from a trailing vortex is to follow a curved path (by a banked turn). Rossow[130] suggested that with individual flight corridors, created by small changes of operation, the time interval between take-offs can be reduced to the order of seconds.

The air traffic control authorities prescribe separation distances based on the weight of the leading and trailing aircraft. For example, an airplane with a gross weight $W = 100,000$ kg following a larger aircraft with mass of the order of $300,000$ kg must be flying at least 5 nautical miles behind.

The increase in weight of some aircraft brings them up in the separation distance. Obviously, the problem becomes more urgent for the very large aircraft. The Airbus A-380, with a gross weight that is 50% higher than the Boeing B-747, is one of the most recent cases.

There are different separation distances: the *vertical separation* is obtained by assigning different cruising altitudes; the *lateral separation* is obtained by maintaining a distance between the flight paths of two airplanes; the *longitudinal separation* is the distance between the leading and trailing aircraft, if these were to follow the same flight path.

Similar considerations will have to be made for the helicopters, although helicopter wakes are considerably more complicated than the wakes from fixed-wing aircraft. The strongest helicopter wakes occur when the helicopter is operating at low speed (less than 80 km/h).

Another problem, mostly environmental, is due to aircraft contrails. These are condensation trails left behind by the jet engines. As the contrails dissipate, they tend to assume the shape of some clouds, Figure 4.20. The contrails often turn into cirrus clouds and are a sign of changing weather. The environmental conditions required for contrail formation and persistence are discussed by Jensen *et al.*[131] and Schumann[132]. Detwiler and Jackson[133] treated the contrail formation under standard atmosphere,

---

*Note that W is a weight and $\dot{m}$ is a mass flow. The discrepancy in units is only due to the choice of the reference area. To avoid confusion, express all the quantities in SI units.

Figure 4.20 *Vortex trails in the sky: trails intersecting at various degrees of dissipation.*

with focus on the engine cycle. The IPCC[7] estimated that contrails covered about 0.1% of the Earth surface in 1992, and that this figure is set to rise to 0.5% by 2050.

4.15 AERODYNAMICS AND PERFORMANCE

For details on aerodynamics we invite the readers to consult the vast literature on the subject. We recommend, by increasing order of complexity, the books of Bertin[115], Kuethe and Chow[116], Ashley and Landhal[120], and most of the volumes of the Princeton Series in *High Speed Aerodynamics and Jet Propulsion*[134]. The book of R.T. Jones[135] is a classic on wing theory. Katz and Plotkin[136] published a focused book on panel methods and lifting surface methods. These methods are rapid in generating aerodynamic characteristics and stability derivatives of complex wing systems and of the full airplane.

The theoretical foundations in relation to aircraft shape, aerodynamics and performance have been analyzed by Küchemann and Weber[137]. For a more applied approach to the aerodynamics of the airplane, Schlichting and Truckenbrodt[138], Küchemann[114], McCormick[104] and Nickel and Wohlfahrt[139] are important references. They deal with the basic aspects of aerodynamics that are not discussed herein. They have extensive bibliographies, where further information on specific topics is available.

Some of the aircraft flights discussed in this book include non-steady flight. Aircraft maneuvering is simulated with steady state aerodynamics. This approximation is adequate for moderate accelerations on a straight flight path. There is scant research in this area. At the very low end of the speed range (pertinent to birds' flight), experimental results indicate that sudden accelerations increase the lift to values considerably higher than steady state (Minkkinen *et al.*[140]). For accelerations past the speed of sound, the main difference in aerodynamic response is in the transonic area.

The accelerations involved in some maneuvers, such as free roll, turning, and V/STOL operations, occur at high angles of attack and high lift. In these cases accurate unsteady aerodynamics is required (Nelson and Pelletier[141]). Rom's[142] book is one of the few comprehensive publications on aircraft aerodynamics at high angles of attack.

Aerodynamic performance data for airfoils or wing sections are abundant in the specialized literature[121,143,144,145]. More advanced wing sections include the double wedge (Lockheed F-117A), segmented (North American XB-70) and biconvex wings (Lockheed F-22A, Grumman F-104). All the high-speed aircraft have a supercritical wing section. The problem of estimation of the aircraft aerodynamics from flight testing is discussed in detail by Klein[146].

The current technology level requires wing sections and planforms to be designed ad hoc, using appropriate numerical and optimization methods. These methods include constraints on geometry and aerodynamics; they may include multipoint design, and off-design analysis. To get started, see Drela[147] and Selig et al.[148]

Aircraft manufacturers are not likely to venture to the public and show their best aerodynamic data. Nevertheless, some aerodynamic performance data of engineering interest are sometimes found in specialized publications. For example, data on the Douglas DC-10, Lockheed L-1011, and Lockheed YF-16 are found in AGARD CP-242[12]. Data are available for the Boeing B-737-100 (Olason and Norton[149]), B-747-100 (Sutter and Anderson[24], Lynn-Olason[25]), Boeing B-707 (AGARD LS-67), Lockheed C-141B and C-5A in AGARD R-723[150]; data for the Fokker 50 and 100 can be found in AGARD CP-515[106]. Data for the North American F-100, Boeing B-707, Convair B-58, the Bell X-1, and NASA X-15 were published by Seckel[151]. Data for the trainer aircraft Northrop T-38 are found in Brandt et al.[152] Heffley and Jewell[153] provide data for the Northrop F-4C and the North American XB-70.

PROBLEMS

1. A supersonic aircraft is to accelerate from a cruise speed of $M = 0.8$ through the speed of sound. During the acceleration, there can be a change in aerodynamic response from the lifting surfaces, including a loss of lift and a drag rise. Past the speed of sound, at fully developed supersonic flow, another problem appears: the center of pressure moves aft. Investigate how the transonic dip and the drag rise problems are solved. Then investigate the nature of the shift in the center of pressure, and consider solutions to the stability problem that minimize the trim drag. (*Problem-based learning: additional research is required.*)

2. The aerodynamic drag is one the most important performance parameters of an aircraft. Discuss how this force is affected by (1) the speed and the Mach number; (2) lift and gross weight; (3) angle of attack; (4) cruise altitude; and (5) configuration at landing and take-off.

3. Calculate the aero-thermodynamic heating at stagnation points of an aircraft having a Δ-wing with a sweep $\Lambda = 55$ degrees. The aircraft is flying at a supersonic speed $M = 1.8$ at an altitude $h = 9,000$ m. Consider as stagnation points the nose of the aircraft and the leading-edge line of the Δ-wing. (The heating is to be considered as an equilibrium temperature above the atmospheric temperature at the flight altitude.)

4. A generic aircraft has a zero-lift angle of attack α_o inclined over the horizon by $+2$ degrees. The aircraft is on a steady level flight with an angle of attack $\alpha = +3$ degrees. Sketch the direction of the lift and drag forces.
   - If the aircraft rotates around the CG by $\alpha = +1$, how do the lift, drag, normal and axial forces change?
   - Now assume that the aircraft starts a climb, and assumes a constant angle of climb $\gamma = +4$ degrees. Sketch the direction of the aerodynamic forces and indicate the relevant flight angles (angle of attack and climb angle).

5. Estimate the maximum temperature at stagnation points due to aero-thermo dynamic heating for aircraft C (Appendix A) by using Eq. 4.31. Assume that the aircraft is flying level at altitude $h = 12,000$ m and find the corresponding values of the standard atmosphere. Discuss the approximation of the results obtained, and how the heating will be affected by changes in atmospheric conditions.

6. Calculate the mass of airplane model A (data in Appendix A) to cruise at an altitude $h = 11,000$ m with maximum glide ratio L/D. Also, calculate the change in optimal mass if the aircraft drifts down to an altitude $h = 10,000$ m.

7. Prove that the maximum glide ratio L/D for a given jet-driven airplane does not depend on the cruise altitude (refer to the discussion and results of Fig. 4.13).

8. Discuss methods for increasing the lift coefficient of a high-performance fighter jet. Make sketches to explain your ideas.

9. Discuss the reasons why the maximum lift coefficient, $C_{L_{max}}$, of an aircraft is difficult to estimate, and which methods can be used for its evaluation. Also, how do the pilots find out that they are operating the aircraft at (nearly) maximum C_L?

10. Calculate the speed U corresponding to a minimum of the ratio $C_D/C_L^{3/2}$. Consider a parabolic drag equation, with $C_{D_o} = 0.023$, $k = 0.034$.

11. Assume that the subsonic jet transport, whose data are given in Appendix A, in cruise conditions has a zero-lift drag equal to 50% of the total drag. The lift-induced drag is estimated at 35% of the total drag. The remaining drag is due to interference, compressibility effects, excrescences and gaps. Calculate the C_{D_o} and the coefficient k in the parabolic drag equation. Consider an aircraft mass equal to 145,000 kg at cruise conditions. The cruise thrust is $T = 45$ kN.

12. You are required to provide a performance analysis of a certain aircraft. You start from wind tunnel aerodynamic data for the drag. You know that the aircraft model was tested in the wind tunnel as a 1:5 scale, and that the Reynolds number used, $Re = \rho U l/\mu$, matched the actual flight conditions (this does not happen very often). The wind tunnel used was of the variable-density type, with average density at the test section equal to 2 bar. The laboratory conditions were standard atmosphere. The total drag force in the wind tunnel had a value D. Calculate the ratio between the drag force of the actual aircraft and the wind tunnel value.

Chapter 5
Engine Performance

A new impetus was given to aviation by the relatively enormous power for weight of the atomic engine; it was at last possible to add Redmaynes's ingenious helicopter ascent and descent engine to the vertical propeller that had hitherto been the sole driving force of the aeroplane without over-weighting the machine ...

H.G. Wells, in *The World Set Free*, 1914

In this chapter we will look at the engine/airframe as a single system. When an engine is mounted on the airframe, it becomes an integral part of the aircraft. The performance parameters considered are the thrust, power, specific fuel consumption, fuel flow, and the free parameters needed to evaluate them (flight speed, altitude and throttle position). Additional parameters are involved in propeller propulsion. These are the propeller's geometric configuration and its rotational speed. For variable throttle settings, the engine response will be considered instantaneous. Therefore, there will be no time lag between throttle and power or thrust.

We will not provide details on how a gas turbine or a reciprocating engine functions. For reference, some data on jet engine cycles are provided in Table A.3. Data and performance on aero-engines are published regularly by Jane's Information Systems[154]. Chart performance, if not already available in the open literature, is almost impossible to obtain from the manufacturers – unless you buy an engine from them. Some basic concepts will be reviewed in this chapter, with the scope of providing means of calculating some propeller and helicopter performance.

It would not be exhaustive to talk about aircraft engines without emphasizing the extraordinary progress that has been made since the beginning of powered flight, and the technological advances that continue to be made. The words of H.G. Wells quoted above were written in 1914, and imagined the world of aviation in the 1950s. Today, the atomic engine is not a reality, and not even an option. Nevertheless, the progress in aircraft propulsion has been extraordinary – thanks to the jet engine.

5.1 GAS TURBINE ENGINES

The term *gas turbine* is associated with a jet engine consisting of a compressor, a combustion chamber, a turbine and an exhaust nozzle, although the name refers to both jet thrust engines and shaft power engines. The main types of gas turbine engines are the turbojet, the turbofan and the turboprop. The gas turbine is the core of the engine. However, there are other parts whose function is essential (inlet, fuel lines, fuel nozzles, sensors, collectors, thrust reverser).

The turbojet belongs to the first generation of gas turbine engines*. It consists of a single gas flow. The operation of the engine requires a number of aero-thermodynamic

---

* Officially, the first flight of a jet-powered aircraft took place on August 24, 1939. It was the Heinkel He-178, powered by an engine designed by Hans von Ohain.

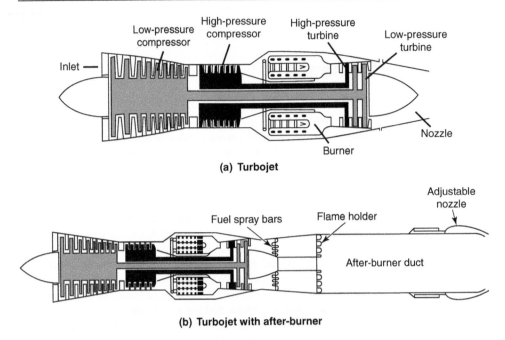

Figure 5.1 *(a) Turbojet; (b) turbojet with after-burner.*

stages: (1) compression of the inlet flow via a number of axial compressor stages; (2) the transfer of the compressed air into the combustion chamber, where it is mixed with fuel; (3) combustion in radially spaced combustion chambers; (4) discharge into a multistage turbine, rotating on the same shaft with the compressor; and (5) ejection of all the exhaust gases as a high-speed hot jet through the nozzle (jet speeds at $M = 2$ and above). The scheme of the engine is shown in Figure 5.1a. The air is captured by an inlet, whose other function is to provide pre-compression by an aero-thermodynamic mechanism called *ram compression*. This is an adiabatic compression in the engine inlet due to flow deceleration.

The main function of the turbine downstream of the combustion chamber is to operate the compressor. A considerable amount of power is generally required by the compressor. The rest of the thermal and kinetic energy associated with the mass flow is transformed into a high-speed, high-noise jet released from the nozzle.

The very first turbojets had centrifugal compressors, but as the engines became better understood, they were replaced by more efficient axial compressors. As the thrust requirements increased, the compressor's architecture became more complicated, with low- and high-pressure units, each with several rotor stages. The exhaust gas leaving the combustion chamber has a high temperature (about 1,000°C). When the compressor and the turbine are connected to the same shaft, their rotational speed is the same. This coupling is referred to as *spool*.

This basic design was applied to engines that powered the early jet airplanes (the Douglas DC-8, the Boeing B-707, the British Comet, and the French Caravelle); it included the JT8D, by most accounts one of the noisiest engines ever to fly. All other

Figure 5.2   *Low by-pass jet engine P&W F-100, with after-burner (McDonnell-Douglas F-15, Lockheed F-16). By-pass ratio 0.36; pressure ratio 32.0; max thrust about 130 kN (sea level, ISA, with after-burner), max diameter: 1.18 m; gross weight: 1695 kg. (Source: Pratt & Whitney.)*

engines, leading to the modern high by-pass ratio engines, are derivatives of this basic idea.

The gas turbine can have an additional combustion stage (reheat, or after-burning). Fuel is injected after the primary combustion for the purpose of increasing the engine thrust. After-burning uses the excess air (about 75%) that does not support the primary combustion.

Gas turbines with this capability operate without reheat most of the time, because the increase in thrust is derived at the expense of considerable fuel consumption. Their application is limited to some military jet aircraft, due to the large fuel consumption by the reheat stage. One exception in the commercial arena is the Rolls-Royce Olympus 593, which powered Concorde.

Figure 5.2 shows a cutaway of the military engine P&W F-100, a low by-pass engine with after-burning thrust. Note the length of the engine and the length of the reheat section.

A turbofan is another derivative of the turbojet. In this engine the excess air that does not support the combustion is channeled through an external annulus, and by-passes the combustor. The *by-pass ratio* (BPR) is the ratio between the by-pass flow rate and the core flow.

This ratio has been increasing over the years, from about 1.1 to values above 5 in modern engines (although the General Electric GE-90 has a BPR = 8.4, and the P&W GP-7000 series has a BPR = 8.7). This results in engines of considerable size, as shown, for example, in Figures 5.3 and 5.4. This is one of the engines that powers the Airbus A-380. The other difference, compared to the basic engine, is a large-diameter fan placed in front of a multistage axial compressor. The function of the fan is to increase the capture area of the inlet, and to channel the by-pass flow through the annulus of the engine. The fan is powered by the engine itself, either on the same shaft as the compressor, or on a separate shaft (dual compressor engine). The advantages of this engine are that the exit flow has lower speed, lower average temperature, and produces far less noise.

A *turboprop* is an aero-engine consisting of a gas turbine unit coupled with a propeller, Figure 5.5. The thrust can be derived both from the jet engine and the propeller, although in practice most of the useful thrust is imparted by the propeller.

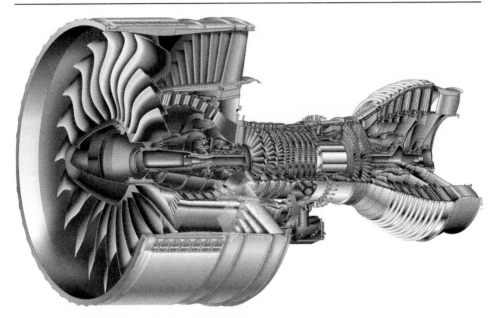

Figure 5.3 *High by-pass turbofan engine P&W GP-7270, that powers the Airbus A-380. By-pass ratio 8.7 (cruise); pressure ratio 45.6 (climb); take-off thrust 311.4 kN (sea level, ISA), flat rated to 30°C; fan tip diameter: 2.95 m; max diameter: 3.16 m. Noise level: 25.6 EPNdB margin to ICAO Stage 3. Combustion emissions: NOx = 0.0509 kg/kN, CO = 0.0227 kg/kN. Other data: single annular combustor. (Source: Pratt & Whitney.)*

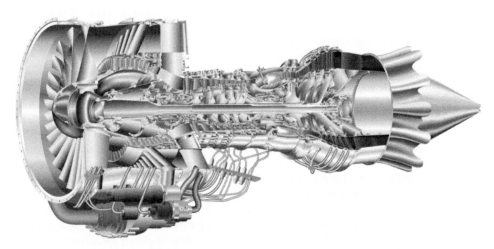

Figure 5.4 *Medium by-pass turbofan engine P&W PW-6000 (Airbus A-318). By-pass ratio 4.9 (cruise); pressure ratio 28.7 (max); max take-off thrust 10.88 kN (sea level, ISA), flat rated to 32.5°C; fan tip diameter: 1.435 m. (Source: Pratt & Whitney.)*

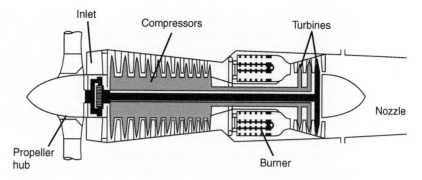

Figure 5.5 *Basic scheme of turboprop engine. In general, the engine has a gear group between the main shaft and the propeller shaft.*

Due to the different speed between the gas turbine and the propeller, these engines have a reduction gear. A gas turbine rotates at speeds of the order of 10,000 rpm, while the propeller's speed is less then one third of this, as limited by the tip Mach number.

Jet engines for helicopter applications are a variant of the turboprop, and consist of one or two gas turbines, a reduction gear of considerable size and a rotor shaft. The reduction in rotational speed is higher than the turboprop, because helicopter rotor speeds rarely exceed 350 rpm.

The subject of aircraft engines is a specialized one, and goes under the field of aerospace propulsion or gas turbines. For a more in-depth presentation the reader is invited to consult the specialized literature, for example Mattingly[155], Oates[156], Archer and Saarlas[157], Kerrebrock[158] and the citations thereof.

5.2 INTERNAL COMBUSTION ENGINES

These engines, also called *reciprocating* or *piston* engines, dominated aviation in its first half century. Today their application is confined to small aircraft and small helicopters. They account for over 60% of the airplanes flying at the present time. At the low end of the propulsion envelope, they are more flexible, lighter and cheaper than gas turbine engines. These engines are always coupled to a propeller; therefore, their limitation is compounded by the limits on propeller propulsion.

Over time, the reciprocating engines grew to deliver shaft power in excess of 3,000 kW. The first engine to be employed for aircraft propulsion was the Wright Brothers six-cylinder, 12 kW, 3.5 liter engine, capable of running continuously at 850 rpm for up to 15 minutes, after which it would melt because of poor cooling capabilities. The engine weighed 82 kg, and therefore it delivered power at a rate of about 0.15 kW/kg. By comparison, this power loading increased to 0.6 kW/kg by the early 1950s, and engines such as the Pratt & Whitney R-4360 Wasp delivered a maximum power of about 3,200 kW, with a total displacement of 71.4 liters from 28 cylinders (four banks of seven). This engine powered a variety of aircraft, from the Boeing B-29 to the Lockheed Constitution.

Reciprocating engines have the advantage of being consistent in terms of efficiency, which does not change much with air temperature or rpm. The most significant factor affecting the efficiency is the throttle setting. A given amount of power can be produced by an infinite number of manifold pressure and rpm combinations, although the engine will be more efficient at higher manifold pressure and lower rpm combinations.

Flight speeds and altitudes can be increased by using turbo-chargers. Turbo-charging can be done by direct coupling of a turbine to the crankshaft (turbo compound), or by supercharging of the intake air. However, the turbo-charger also tends to increase the engine's temperature, because the inlet air is heated by compression. For this reason, above some critical altitude the engine will overheat, unless some preventive action is taken.

For an aspirated reciprocating engine, the brake horsepower decreases with increasing altitude. A turbo-charger maintains the power with altitude until a critical altitude is reached, after which point the power decreases. One of the main differences between automobile and aviation engines is that aviation engines operate at relatively high power most of the time, and at full power at every take-off. Figure 5.6 shows the altitude performance of the Lycoming IO-540, a six-cylinder reciprocating engine for light aircraft. For this engine, a cruise setting of 65% power at 2,400 rpm is recommended.

Generally, curves of this type include lines of constant engine manifold pressure (which are nearly vertical) and correction details for difference in temperature for the inlet air.

Large-diameter propellers are affected by Mach number effects, noise emission, and vibration, hence large step-down gearshafts have to be used, which add to the mechanical complexity and weight of the engine. Useful references on reciprocating engines include Heywood[159].

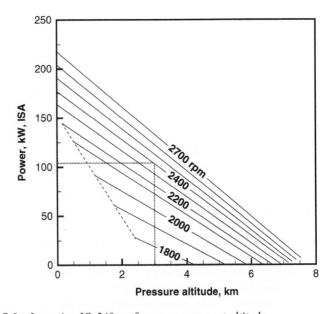

Figure 5.6 *Lycoming IO-540 performance curves at altitude.*

5.3 ENGINE FLIGHT ENVELOPES

Like the aircraft they power, aero-engines have a flight envelope. This is an enclosed area in the altitude-Mach number space where safe and efficient operation is guaranteed. Another way to characterize the flight envelope of the engine is to use the concept of specific impulse.

First, let us consider the comparative flight envelope of various propulsion systems, as shown schematically in Figure 5.7. The lines in the graph show the limits of operation of each engine configuration.

The lower speeds and altitudes are exclusively the domain of the reciprocating engines. Essentially, this type of engine has a low altitude limit, due to the air breathing requirements. Turbo-charged engines provide relatively constant power output up to a critical altitude, $P/P_o = 1$ for $h < h_{crit}$. The power then decreases according to the following semi-empirical relationships

$$\frac{P}{P_o} = \left(\frac{\sigma}{\sigma_{crit}}\right)^{0.765}, \quad h_{crit} < h < 11{,}000\,\text{m}. \tag{5.1}$$

Reciprocating engines provide very limited power above the troposphere.

There is not a single power plant configuration capable of taking the aircraft from take-off to hypersonic speeds. For this reason current research focuses on hybrid engine configurations that include two different power plants. For example, when in the 1960s NASA worked on hypersonic flight with the experimental aircraft X-15, it used the Boeing B-52 as a carrier (under the wing). The X-15's rocket engine was ignited at the cruise speed and altitude of the Boeing B-52.

SpaceShipOne, the first suborbital aircraft to take off from a paved runway (2004) and to land as an airplane, has a hybrid engine system. This consists of a jet engine and a hybrid rocket motor. The carrier aircraft climbs by jet propulsion to about 16,500 m (54,133 ft) with a Mach number $M = 0.85$, then it shifts to a gliding flight and releases

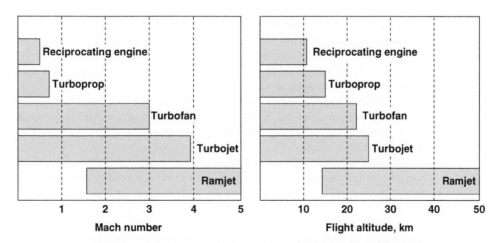

Figure 5.7 *Schematic operation envelopes of different aircraft engines.*

the space ship. The ship fires the rocket motor and accelerates to $M = 3.5$ in a steep climb to about 100 km of altitude.

5.4 POWER AND THRUST DEFINITIONS

Engine and aircraft manufacturers, as well as the aviation regulations, refer to a variety of power and thrust ratings. The term *uninstalled power/thrust* refers to the engine before the integration with the airframe. The installed power and thrust refer to the airframe/power plant combination, and are more pertinent to performance calculations. When doing performance calculations, we call thrust the *net thrust*, i.e. the effective thrust delivered by the engine, which accounts for the intake drag and air bleed.

The *shaft power* usually refers to a propeller engine, and denotes the mechanical power delivered by the engine, not accounting for the efficiency of power conversion and the airframe/engine integration. The installation losses include inlet losses, the exhaust gas losses, and the extracting bleed air.

The *equivalent shaft power* is a parameter applied to turboprops, and refers to the sum of the shaft power delivered to the propeller and the power due to the jet of the exhaust gases from the gas turbine, through the nozzle.

The power output of a gas-turbine engine depends on the thermal stresses it is capable of withstanding, e.g. the maximum temperature at any part of the engine. For this reason the manufacturers quote a *maximum continuous power*. This is the engine power level that can be maintained indefinitely, without compromising the thermal and structural integrity of the engine. The *maximum emergency power* is the engine power that can be delivered by the engine for a short time. This includes events such as One-Engine-Inoperative (OEI) and take-off. The maximum continuous power is lower than the emergency power. *Take-off power* and maximum emergency power are sometimes the same. Otherwise the take-off power rating refers to a time of 5 minutes, or 10 minutes in case of OEI. The *intermediate power* rating is a power that can be maintained over a brief to medium time (less than 30 minutes), and is used for emergency situations.

Since most aircraft require additional power for auxiliary units (cabin pressurization, air conditioning), some air can be extracted from the compressor (*bleed air*) to provide the auxiliary services. This, in turn, reduces the engine's useful power, because the mass flow into the engine and its combustor is reduced. Some engine charts show the effects of bleed air on the power output. These effects appear as discontinuities in the power curve.

For helicopter applications the manufacturers refer to a *maximum transmission power*. This is a power limited by the maximum torque on the transmission system (shaft, gear and hub). Increasing the engine power normally requires re-engineering of the transmission system.

The engine thrust, like the power, is dependent on aircraft speed, flight altitude and rpm. The *static thrust* is the thrust at zero aircraft speed. For a military aircraft flying without the optional after-burning, the maximum thrust is sometimes called *military thrust*.

Sometimes engines must be run at part-throttle, to avoid exceeding a maximum rated thrust. In such cases, they are called *flat-rated* engines. For example, the CFM56-3 engine is flat-rated at ISA +15° (or +30°). This means that the engine is guaranteed to give the rated thrust at full throttle when the ambient temperature is below ISA $+15°$. Above this temperature, the engine will give less thrust because the air has lower density.

For aircraft powered by jet engines, a performance parameter is the *specific thrust* T/W (also called *thrust-ratio*); for the propeller-driven aircraft the corresponding parameter is the *power loading*, P/W (or its inverse). These parameters give an indication of the amount of power or thrust available for a given aircraft gross weight. The values of these parameters vary with the flight altitude and Mach number, one therefore has to be clear as to the conditions at which they are taken: sea level or specified altitude; static $(U = 0)$ or in flight conditions. The data quoted by the manufacturers (when reported at all) generally refer to static thrust at sea level. Useful data, including historical trends, have been published by McCormick[104].

Another performance parameter, more related to the engine itself than the aircraft, is the *engine's specific weight*, i.e. the weight of the engine per unit of power (or thrust). The *engine's specific power* is the power delivered per unit of weight of the engine. However, as engines become bigger, also the installation and integration requirements on the airframe change. This means that the nacelles have to be strengthened to sustain the increased engine weight, and the engine's auxiliary systems have to scale up as well. The weight of engine mounts, pylons, thrust reversers and nozzles are roughly proportional to the engine's thrust. On the other hand, the increase in weight due to accessories (fuel controls, sensors, etc.) grows less rapidly.

The use of the concept of power loading and thrust-ratio for propeller and jet aircraft makes the comparison between the two power plants not completely obvious. The specific thrust is a non-dimensional number (thrust force divided by weight); the power loading is a dimensional number, given in kW/kg or kW/N. A jet engine fixed on the ground does not do any useful work, although it burns fuel and delivers an amount of thrust. However, a useful power can be deduced from a flight condition at speed U, $P = TU$. This expression can be compared to the shaft engine power of the turboprop engine. By this analysis we find that the installed engine power of the Airbus A-380 is a factor 8,500 of the Wright Brothers Flyer.

Extrapolating from a large number of data, we found that the average thrust-ratios of subsonic commercial jets is about $T/W \sim 0.2$ to 0.4; for a fighter jet aircraft these data double, and become even larger if after-burning is used. From the data in Figure 5.8 there is no clear relationship between the thrust-ratios and the MTOW. The complication of the chart also arises from a combination of configurations (external loads, combat mission, use of after-burning).

Example

Let us now take a brief look at the power/weight ratio of some engines. Consider the Wright Flyer's reciprocating engine. Although it was not the best engine of its time, it represented about 20% of the aircraft weight. By comparison, each of the engines on the Airbus A-380 delivers a maximum sea level static thrust of 311.4 kN, or 70,000 pounds*. The weight of the GP-7000 engine is estimated at 4,500 kg, which leads to a total engine weight at MTOW equal to less than 3.5%.

---

* This performance is set to increase to above 80,000 pounds in later versions.

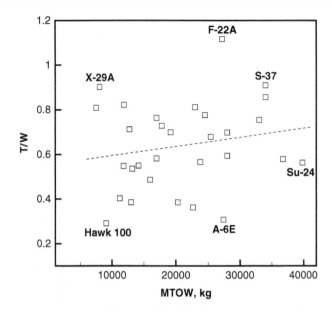

Figure 5.8 *Fighter jet aircraft thrust-ratios versus MTOW. Dotted line is a least-square fit of the available data.*

5.5 GENERALIZED ENGINE PERFORMANCE

The jet engine operation depends on the rotational speed (rpm), the air speed U and the flight altitude h. At ISA conditions the altitude defines the local values for pressure, density and temperature of the inlet flow. The mass flow rate $\dot{m}_f$ and the thrust depend on the inlet diameter d. Therefore, the propulsion parameters are the thrust T, rotational speed rpm, the air speed U (or Mach number M), the temperature T, the pressure p, the fuel flow $\dot{m}_f$, and the diameter d. Since this is a relatively large number of parameters, the engine performance is described by a reduced number of non-dimensional or engineering quantities.

The essential parameters in engine performance are the forward speed, the rotational speed rpm, the thrust T, mass flow rate $\dot{m}_f$, and specific fuel consumption f_j. Therefore, a normalization of these parameters is performed by using the remaining three quantities (T, p, d).

First, the speed U can be replaced by the Mach number (dimensionless) by means of $M = U/a$. The speed of sound is derived from the flight altitude. Second, the rotational speed of the engine is normalized with a/d, to yield the parameter $\mathrm{rpm}\,d/a$. Third, the thrust T is replaced by the parameter T/pd^2 (pressure $\times$ length^2 = force). Finally, the mass flow rate $\dot{m}_f$ is replaced by the dimensionless parameter $\dot{m}_f/(pd^2/a) = \dot{m}_f\,a/pd^2$. A further step is required for some parameters. By using the relative pressure δ from Eq. 2.33

$$\frac{T}{pd^2} = \frac{T}{\delta}\frac{1}{p_o d^2} \simeq \frac{T}{\delta}. \tag{5.2}$$

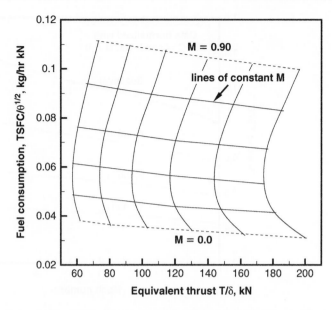

Figure 5.9   *Corrected thrust versus corrected TSFC for the turbofan engine CF6-50 (estimated). Units for corrected TSFC are kgf of fuel flow per hour per kN of engine thrust.*

where p_o is the standard atmospheric pressure at sea level. With the definition of speed of sound, $a = \sqrt{\gamma \mathcal{R} \mathcal{T}}$, we have

$$\frac{\text{rpm}\, d}{a} = \frac{\text{rpm}\, d}{\sqrt{\gamma \mathcal{R} \mathcal{T}}} = \frac{\text{rpm}}{\sqrt{\mathcal{T}}} \frac{d}{\sqrt{\gamma \mathcal{R}}}, \tag{5.3}$$

$$\frac{\dot{m}_f a}{p d^2} = \frac{\dot{m}_f \sqrt{\gamma \mathcal{R} \mathcal{T}}}{p d^2} = \frac{\dot{m}_f \sqrt{\mathcal{T}}}{\delta} \frac{\sqrt{\gamma \mathcal{R}}}{p_o d^2}. \tag{5.4}$$

Instead of using dimensionless parameters, it is a practice in engine performance to use the corrected thrust, rotational speed, and mass flow rate

$$\frac{T}{\delta}, \quad \frac{\text{rpm}}{\sqrt{\mathcal{T}}}, \quad \frac{\dot{m}_f \sqrt{\mathcal{T}}}{\delta}, \tag{5.5}$$

along with the Mach number M. The remaining factors in Eqs 5.2–5.4 are all constant. The parameters in Eq. 5.5 have dimensions. The corrected thrust, the rotational speed and the mass flow rate maintain their physical dimensions. Thrust and mass flow rate are the parameters most widely used in aircraft performance simulation. The rotational speed is generally not given, and calculations are performed at fixed throttle settings, but variable Mach numbers. However, the general dependence of the corrected parameters is

$$\frac{T}{\delta} = f_1 \left(M, \frac{\text{rpm}}{\sqrt{\mathcal{T}}} \right), \quad \frac{\dot{m}_f \sqrt{\mathcal{T}}}{\delta} = f_2 \left(M, \frac{\text{rpm}}{\sqrt{\mathcal{T}}} \right). \tag{5.6}$$

Figure 5.9 shows the CF6 corrected engine performance at constant (full) throttle, at Mach number up to 0.9, and altitudes up to 14,000 meters (about 45,931 ft).

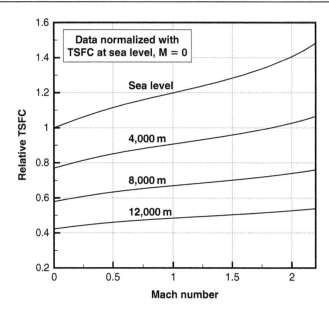

Figure 5.10 *Relative TSFC versus Mach number for the F-100 engine model (military thrust).*

Figure 5.10 shows the estimated TSFC of the F-100 engine (Figure 5.2) versus Mach number at selected altitudes. The TSFC was normalized with the static value at sea level. The results of the analysis show that the TSFC is slightly increasing at cruise altitudes.

5.6 FUEL FLOW

The fuel flow is the rate at which fuel is burned by the engines. It is given in units of mass (or weight or volume) per unit time. For a jet engine the fuel flow is proportional to the thrust. The constant is proportionality is the *thrust-specific fuel consumption* (TSFC)

$$\dot{m}_f = \frac{dm_f}{dt} = TSFC \cdot T. \tag{5.7}$$

Here a word of caution is needed, because generally engine manufacturers, textbooks, flight manuals and technical publications quote the most extravagant units for the TSFC, including kg/kg/h, kg/kN/h, kg/N/h, gr/N/s, mg/N/s, lb/lb/h (where kg and gr are units of force!). Sometimes the units are missing, which leaves the figures open to interpretation. In fact, it is not unusual to find wrong data. If one uses international units, the TSFC will be expressed in N/N/s, e.g. newton of fuel weight per newton of engine thrust per unit of time (seconds). While this seems rigorous, the unit N/N/s is hardly a useful figure. Similar confusion exists on the fuel flow data (lb/h, l/h and gallons/h, where the gallon is sometimes the US gallon and other times the Imperial gallon).

Table 5.1 *Basic jet engine data.*

Engine	SFC, S/L $kgm/s/N$	BPR	$\dot{m}_a$ kgm/s	T_{to} kN
GE-90	$1.051 \cdot 10^{-5}$	8.4	1,350	70.0
CF6-50C2	$1.051 \cdot 10^{-5}$	5.7	590	50.3
CF6-80C2	$9.320 \cdot 10^{-5}$	5.1	800	50.4
RR RB-211-524G/H	$1.595 \cdot 10^{-5}$	4.3	730	52.1
RR Trent 882	$1.566 \cdot 10^{-5}$	4.3	730	72.2
P&W JT-9D-7R4	$1.566 \cdot 10^{-5}$	5.0	690	176.3
P&W JT-8D-17R	$2.337 \cdot 10^{-5}$	6.6	470	30.8
CFM-56-5C2	$1.606 \cdot 10^{-5}$	1.0	150	18.9
V-2500	$1.629 \cdot 10^{-5}$	5.4	355	21.6

Part of the confusion stems from the fact that from a practical point of view one wants to know how much fuel is burned in one hour; and part of the confusion is due to outdated engineering units (pounds, shaft horsepower); and finally there is the confusion between mass, weight and force. One justification for the use of the units lb/h/lb (or N/h/N) is that the SFC is of the order of unity for all jet engines.

It is always a good idea to check the data provided with the gross thrust of the engine. Divide the fuel flow by the engine thrust and make sure that all the units are coherent. Some engine fuel consumption data are provided in Table 5.1. Further fuel consumption data and engine efficiency, with historical trends, are given by Babikian *et al.*[161]

The fuel flow is dependent on the aircraft speed. In fact, from Eq. 5.7, at level flight conditions we have

$$D = T = \frac{\dot{m}_f}{TSFC}. \tag{5.8}$$

If the TSFC is constant, then the speed of minimum drag is also the speed of minimum thrust and speed of minimum fuel flow. The specific fuel consumption is inversely proportional to the thermal efficiency of the engine, and for propeller propulsion it is also inversely proportional to the propulsion efficiency, η.

The specific fuel consumption of gas turbine engines has been decreasing since their first inception. The chart shown in Figure 5.11 was elaborated from data obtained from Rolls-Royce. The datum at the far right of the chart is the 2020 target. The 7–8% improvement on the TSFC is a considerable gain, when one considers the amount of fuel that is burned by an aircraft during one year of operation or during its lifetime in service.

Technical information on aviation fuels is available in Goodger and Vere[162], in addition to some reports on standards. Table 5.2 shows some basic data of aviation fuel useful for performance calculations. The stoichiometric ratios are between 0.0672 and 0.0678.

The thermal efficiency is a measure of the effectiveness of the conversion of the thermal energy of the fuel by combustion to useful power. This efficiency is less than unity, and reflects the inability to use the heat of the exhaust gases, and presence

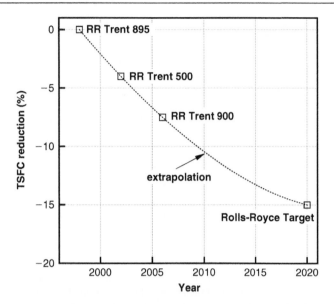

Figure 5.11 *Specific fuel consumption reduction from technological advances on the Trent series of aero engines (data from Rolls-Royce).*

Table 5.2 *Characteristics of aviation fuels at 15°C. Data are averages.*

Fuel	*Wide-cut*	*Kerosene*	*AV gas*
Specific weight	0.762 kg/l	0.810 kg/l	0.715 kg/l
Specific combustion heat	43.54 MJ/kg	43.28 MJ/kg	43.71 MJ/kg

of other losses associated to the engine cycle, such as the losses in compression, combustion, and expansion.

The *specific impulse* is defined as the ratio between the thrust and the specific fuel consumption,

$$I_{sp} = \frac{T}{SFC}. \tag{5.9}$$

It represents the time that a given engine could operate by burning the amount of fuel with a weight equal to the thrust. Sometimes the specific impulse is defined as the amount of time a unit of mass of propellant will last for a unit thrust delivered. This index is useful when comparing different types of propulsion systems (propellers, jets and rockets).

The physical dimensions of the specific impulse are that of speed (force/mass flow rate $= N/(kg\ s^{-1}) = (kg\ ms^{-2})/(kg\ s^{-1}) = m/s$), although it is sometimes given in seconds. Power plants that deliver large amounts of thrust have a large specific impulse. A map of specific impulse for various types of thrust-producing vehicles is shown in Figure 5.12. The specific impulse varies from about 70,000 seconds for a conventional helicopter, to 3,500 seconds (one hour) for a pure jet engine with after-burning.

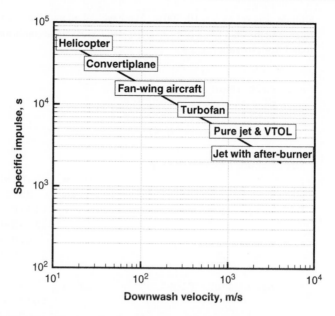

Figure 5.12 *Specific impulse I_{sp} of various propulsive systems.*

Table 5.3 *Selected aircraft data and applications statistical value.*

Aircraft	Type	Engines	TOW, kg	$\dot{m}_f$, kg/h
Boeing B-747/100	wide body	4	340,200	13,770
Boeing B-747/400	wide body	4	394,630	15,208
DC-8-63	narrow body	4	158,900	8,641
MD-11	wide body	3	277,930	9,319
Boeing B-727-200	narrow body	3	95,030	6,980
Airbus A-300/600	wide body	2	161,030	6,351
Boeing B-777	wide body	2	248,570	8,012
Boeing B-737/500	narrow body	2	60,240	2,827
MD-80	narrow body	2	67,810	3,531
Canadair CL-601	regional jet	2	19,550	1,317
Fokker F-100	regional jet	2	44,450	2,748
SAAB 340	regional prop	2	12,700	488
Embraer 120	regional prop	2	11,500	608

Table 5.3 summarizes some fuel flow data for well-known aircraft, taken at their average take-off gross weight. The engines are classified according to the Federal Aviation Administration (FAA), in terms of number of engines and type of airframe.

Example

Below is an example of a calculation of fuel flow and specific fuel consumption at cruise conditions for the airliner whose data are given in Appendix A. The flight altitude was set at $h = 10,500$ m. We have calculated the net thrust from the drag force, and the drag force from an estimated C_D at cruise conditions. The TSFC is then found from solving Eq. 5.8.

```
Aircraft mass at start
     mi          = 150000.0 kg
     fuel mass   =  45000.0 kg
     fuel ratio  =   0.30

Estimated cruise drag
                         CD   =    0.02988
                         Drag =     85.156 kN
Estimated TSFC           TSFC =  0.161E-03 N/s/N
                                 0.164E-04 kg_f/s/N
                                    0.0589 kg_f/h/N
                                    0.5781 lbf/h/lbf


Estimated Fuel Flow          ff =     13.674 N/s
                                     5018.002 kg_f/h
                                    11062.617 lb_f/h


Estimated endurance, full tanks   E     9.76 hours
Estimated endurance w/ actual fuel       8.97 hours

Glide ratio at start,       L/D   =    17.280
Figure of Merit at start M (L/D) =     13.824
```

The program includes a few lines to perform basic unit conversions. The estimate for the endurance (flight time with the available fuel) and glide ratio are also useful to validate our assumptions.

5.6.1 Aspects of Fuel Consumption

In 1998, worldwide consumption of aviation fuel was about 680 million liters per day, an increase of 13% from 1990. This consumption is now expected to be in excess of 700 million liters/day. Every year air travel releases 600 million tons of CO_2 into the atmosphere.

It was previously mentioned that fuel may be jettisoned in emergency situations. The fuel evaporates quickly and does not reach the ground if the altitude of release is 1,500 feet (457 m) or above (see Good and Clevell[162] and Pfeiffer[163]). Although this practice is discouraged, some data indicate that civil aviation jettisons about 0.01% of the fuel consumed. At the current levels of consumption, this corresponds to 17,000 liters/day (5,000 tons/year) – hardly a negligible amount. Clevell[164] also reported that fuel is jettisoned about 1,000 times a year by the US military, for a total of 7,000 tons. Regulations for civil aviation require that fuel is jettisoned at sea, from altitudes above 10,000 ft.

5.7 PROPULSIVE EFFICIENCY

For an engine flying level at speed U, the one-dimensional equation for the mass flow through the engine is found from the energy equation (or first law of thermodynamics) applied to a control volume of the flow into the engine. We consider the control volume limited by the stream tube through the inlet section, as shown in Figure 5.13; we also assume that the flow is in the axial direction.

If $\dot{Q}$ is the heat transferred per unit of time to the air by combustion of the fuel (thermal energy transfer rate), and $\dot{W}$ is the work done on the air, then the energy balance reads:

$$\dot{Q} - \dot{W} = \dot{m}_a \left[(1 + \xi) \left(h_e + \frac{1}{2}(U_e - U)^2 \right) - h_a - \xi \left(h_f + \frac{U^2}{2} \right) \right], \quad (5.10)$$

where the subscript "e" indicates the exit conditions, "a" indicates the air and "f" is relative to the fuel; $\xi = \dot{m}_f / \dot{m}_a$ is the fuel air mixture; h is the enthalpy per unit mass of the flow; and U is the air speed. The work done on the system is

$$\dot{W} = -[TU - (p_e - p_a)A_e], \quad (5.11)$$

where T is the thrust, p_e is the pressure at the exit of the control volume, p_a is the air pressure at the inlet of the control volume and A_e is the cross-sectional area of the exit. The first term in Eq. 5.11 denotes the work done by the thrust; the second term is the work done against the static pressure. By combining Eq. 5.10 and Eq. 5.11, after simplification the thrust equation becomes

$$T = \dot{m}_a \left[(1 + \xi)U_e - U \right] + (p_e - p_a)A_e. \quad (5.12)$$

Equation 5.12 shows that the engine thrust is proportional to the mass flow rate into the engine.

The propulsive efficiency of the jet engine is defined as the ratio of the thrust-power (power generated by the thrust force T at a speed U) and the rate of production of

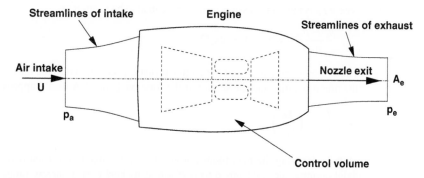

Figure 5.13 *Control volume on the jet engine.*

propellant kinetic energy. This ratio can be written as

$$\eta_p \simeq \frac{TU}{\dot{m}_a[(1 + \xi)U_e^2/2 - U^2/2]}. \tag{5.13}$$

The fuel-to-air mixture ratio is much less than 1 (in fact, $\xi \simeq 0.067$). Often it is appropriate to neglect the pressure term, because $p_e \simeq p_a$, e.g. the exit pressure is nearly equal to the local atmospheric pressure (fully expanded jets). With these two approximations, from Eq. 5.12 and Eq. 5.13, the propulsive efficiency becomes

$$\eta_p = \frac{TU}{TU + \dot{m}_a(U_e^2 - U^2)/2}. \tag{5.14}$$

Equation 5.14 shows that the propulsive efficiency is less than one, and increases as the jet velocity tends to the engine velocity. The difference $(U_e^2 - U^2)/2$ is the residual kinetic energy per unit mass of the jet. It is worth noting that low jet speeds are not only necessary for high propulsive efficiency, but also for engine noise reduction, as will be discussed in Chapter 17. One way to reduce the residual kinetic energy of the jet is to increase the mass flow, and this is in fact achieved by the high by-pass ratio turbofans.

5.8 THRUST CHARACTERISTICS

In order to find an expression for the thrust as a function of the throttle setting, it is important to know how the mass flow rate into the engine is dependent on the pressure ratio and the rotational speed of the engine. The mass flow rate is

$$\dot{m}_a = \rho U A = (\rho U A)_c, \tag{5.15}$$

where the subscript "c" denotes the flow conditions at the exit of the compressor. The cross-sectional area A_c is fixed. If we use the definition of compression ratio and apply the equation of the ideal gas to the compressor flow, we have

$$\dot{m}_a = \rho_c U_c A_c = \frac{\rho_c}{\rho_\infty} U_c A_c \rho_\infty = \left(\frac{p_c}{p_\infty}\right)\left(\frac{T_\infty}{T_c}\right) U_c A_c \rho_\infty$$

$$= p_r \frac{U_c}{T_c} A_c \rho_\infty T_\infty, \tag{5.16}$$

where p_r is the compression ratio. This can also be written as

$$\frac{p_c - p_\infty}{p_\infty} = p_r - 1 = \rho_\infty \Omega^2. \tag{5.17}$$

Therefore, we conclude that the engine thrust is proportional to the mass flow rate, to the pressure ratio, and the square of the rotational speed. A useful approximation is

$$\frac{T}{T_o} = \left(\frac{\Omega}{\Omega_o}\right)^2, \tag{5.18}$$

with T_o denoting the maximum power, and Ω_o the maximum rotational speed. The axial compressor is designed to perform at its best over a narrow range of speeds, usually close to the maximum *rpm*.

As far as the speed is concerned, the mass flow into the engine is obtained through a convergent nozzle. From compressible aerodynamic theory we know that the flow can be accelerated through a convergent nozzle up to the sonic speed, where the mass flow rate is maximum (for a given cross-section and atmospheric conditions). An increase in the speed of the engine will not increase the speed of the inlet mass flow, and the engine becomes chocked. The chocked mass flow rate can be calculated starting from the St Venant equation

$$\frac{\dot{m}_a}{A} = \rho_o \sqrt{\frac{2\gamma}{\gamma - 1} p_o \rho_o \left(\frac{p}{p_o}\right)^{2/\gamma} \left[1 - \left(\frac{p}{p_o}\right)^{\gamma - 1/\gamma}\right]}, \tag{5.19}$$

where the subscript "o" denotes stagnation conditions. Sonic conditions at the throat of the nozzle lead to the chocked mass flow rate

$$\frac{\dot{m}}{A} = 1.8865 \sqrt{p_o \rho_o^3}. \tag{5.20}$$

In fact, to have chocked conditions, the flow must reach a sonic speed at the throat of a convergent-divergent nozzle. In this case, $M = 1$ and $p/p_o = 1.893$, as we find from aerodynamic theory.

In first-order performance calculations the jet engine thrust is assumed to be independent of the speed, and variable with the altitude according to the equation

$$\frac{T}{T_o} = \sigma^r \Pi, \tag{5.21}$$

with the power $r = 0.7 - 0.8$, as found from the statistical analysis of several engines. The movement of the throttle commands the opening and closing of the fuel valve. The valve controls the fuel flow into the engines, and hence the thrust and power delivered. Since excessive fuel flow can create temperature surges in some sections of the engine (combustion chamber, turbine, and nozzle), modern engines have means for overriding excess fuel flows at full-throttle condition.

It is important to mention that most performance calculations are done by assuming either a constant throttle, or an impulsive response of the engine to changes in the throttle settings. More elaborate analysis would require to take into account the time lag between control inputs and engine rotor speeds. Such an analysis may be required in helicopter maneuver, and is discussed in some detail by Newman[165].

5.9   PROPELLER CHARACTERISTICS

Aircraft propulsion by propeller is still the most widespread method of converting engine power into useful thrust, and hence aircraft speed. The mechanism of generating thrust and torque from a propeller is the subject of propeller aerodynamics, therefore we cannot go into details in this context, except for some basic axial momentum theory. The essential theory is available in Glauert[166] and Theodorsen[167], among others.

Data and propeller charts, for the purpose of basic performance calculations, can be found in some old NACA reports, such as Biermann and Hartman[168,169] and

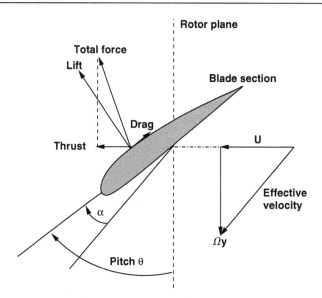

Figure 5.14 *Nomenclature of forces on blade section.*

Theodorsen *et al.*[170]. In these reports the reader can find performance data of propellers with two to six blades. More advanced concepts, broadly related to other problems in aircraft performance, are available in AGARD CP-366[171].

To understand propeller performance, we need to identify the relevant parameters of operation. These parameters are the advancing speed U, the rotational speed rpm, and the tip Mach number M_{tip}. In addition, there are several geometrical quantities that depend on the propeller: the number of blades, the diameter d, the pitch θ, the type of blade section, the chord distribution along the radius, the tip geometry, and the hub geometry.

The pitch is a measure of the orientation of the propeller on a plane normal to the axis of rotation, as shown in Figure 5.14. The reference line for the calculation of the pitch is the chord. Figure 5.14 shows how the local *inflow angle* is calculated. If the blade section is at a radial position y, the total inflow velocity is $\sqrt{U^2 + (\Omega y)^2}$. The direction of the resulting vector velocity is inclined by an angle α on the chord line: this is the local inflow angle, or angle of attack of the blade section.

The thrust generated is the resulting aerodynamic force in the direction of the forward flight. The lift and drag of the blade section are oriented in different directions to the aircraft's global forces. The aircraft's drag is parallel to the thrust element, and has opposite direction.

The *solidity* σ is the ratio between the blade area (projected on the rotor disk) and the rotor disk*

$$\sigma = 2\frac{N\bar{c}}{\pi d}, \tag{5.22}$$

---

* Unfortunately, this is the same symbol as the relative density of the air. We maintain the convention currently used in propeller and helicopter analysis and use σ for the solidity. Therefore, to avoid confusion, either ρ, ρ/ρ_o or σ_1 will be used for the air density.

where N is the number of blades and $\bar{c}$ is the mean chord. The solidity is an important design parameter; in fact, the rotor coefficients are often normalized by σ, to express the notion of effective disk loading.

To make the data more useful from an engineering point of view, we need a measure of performance. This is the efficiency, or the ratio between propulsive power and power at the shaft

$$\eta = \frac{TU}{P}. \tag{5.23}$$

The propulsive efficiency expresses the ability to convert power from the engine (or power at the shaft) into useful power to fly at a speed U. The functional dependence of the efficiency from the other parameters is expressed as

$$\eta = f(U, \text{rpm}, d, \theta, \cdots). \tag{5.24}$$

If the aircraft is stationary on the ground, the conversion efficiency is zero, and all the shaft power generated from burning fuel is lost. The energy E is dissipated at the blades and transferred to the slipstream, which has an axial and rotational velocity. In general terms, the shaft power is

$$P = TU + E. \tag{5.25}$$

By using the dimensional analysis, the forward speed, the rotational speed and the diameter are replaced by another dimensionless group, the *advance ratio*

$$J = \frac{U}{\text{rpm}\, d}. \tag{5.26}$$

Equation 5.26 is just one possibility. Another definition for J is

$$J_1 = \frac{U}{\Omega R}. \tag{5.27}$$

The two values of the advance ratio are proportional to each other. The scaling factor is

$$J_1 = \frac{60}{\pi} J. \tag{5.28}$$

The advance ratio is a measure of the advancement of the propeller in one revolution, measured in number of diameters. The advance ratio is also a scaling parameter, indicating that all the propellers with the same J, and *geometrically similar*, have the same performance index. In other words, a propeller with a diameter of 2 m, operating at 2,000 rpm, and a forward speed 70 m/s ($J = 0.33$), has the same performance as a *scaled-down* propeller of 1.5 m diameter rotating at a speed of 3,000 rpm, and advancing at the same speed.

Finally, a parameter of interest is the tip speed ratio, e.g. the ratio between the propeller's speed and the tip speed:

$$\lambda = \frac{U}{\Omega R}. \tag{5.29}$$

For conventional aeronautic propellers, the tip speed $U_{tip} = \Omega R$ is restricted to subsonic values, in order to keep wave drag losses, noise and vibrations under acceptable limits. If we introduce the Mach number, then

$$M_{tip} = \frac{U_{tip}}{a} = \frac{1}{a}\sqrt{U^2 + \Omega^2 R^2}. \tag{5.30}$$

In principle, to keep the tip Mach number to subsonic values, one can operate on the aircraft speed, on the rpm and on the propeller's radius R.

The blade's Activity Factor (AF) is a non-dimensional parameter that measures the power absorbed by a propeller blade,

$$AF = 10^5 \int_{root}^{tip} \frac{c}{R}\left(\frac{r}{R}\right)^3 d\left(\frac{r}{R}\right). \tag{5.31}$$

The factor 10^5 is used to keep the value of the AF within unit values. The Total Activity Factor (TAF) is defined as the product between the number of blades N, and the single blade's activity factor,

$$TAF = N\,AF. \tag{5.32}$$

Among the parameters that are most effective in setting the value of the efficiency, the propeller pitch deserves special mention. One often finds propeller charts in the form of Figure 5.15, where the efficiency is plotted versus the advance ratio for a fixed value of the pitch. The maximum value of η depends on the advance ratio (a quantity easily measurable). As the pitch decreases, the advance ratio corresponding to maximum propulsion efficiency decreases. At all pitch settings there may be two values of the advance ratio that yield the same value of the propulsion efficiency, one below and one above the optimal advance ratio. This does not mean that the flow condition past the blades is the same. On the contrary, the point of operation past the optimal propulsion efficiency is characterized by unsteady flow separation on parts of the blade sections. This is a condition of blade stall, in a similar fashion to the elementary wing section.

The curves in Figure 5.15 indicate that the efficiency increases up to a maximum, then it decreases sharply; in some cases it almost plunges to zero. This is often an indication that the propeller has stalled.

Propeller charts, where the parameter is the pitch angle of the blades, are to be used in the calculation of the range and the endurance of propeller-driven airplanes. For given flight conditions, it is required to determine the rotational speed that gives the maximum ratio of the propulsive efficiency to the specific fuel consumption.

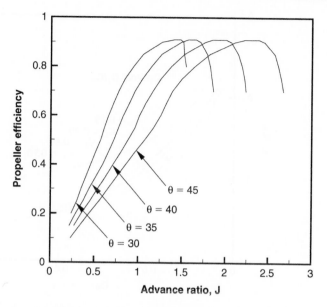

Figure 5.15 *Propeller chart: four-bladed propeller at different pitch angles (as indicated); J versus η (data from Hamilton Sundstrand).*

A number of other dimensionless coefficients are defined, and propeller charts can be found with these parameters, which are the power, thrust and torque coefficients:

$$C_T = \frac{T}{\rho A(\Omega R)^2}, \quad C_P = \frac{P}{\rho A(\Omega R)^3}, \quad C_Q = \frac{Q}{\rho A(\Omega R)^2 R}. \tag{5.33}$$

Large values of these coefficients are an indication of large thrust, power and torque absorption by the propeller. A useful relationship between the efficiency and the advance ratio involves some of the parameters in Eq. 5.33,

$$\eta = J \frac{C_T}{C_P}. \tag{5.34}$$

One type of propeller chart is shown in Figure 5.16. This is the advance ratio J versus the power coefficient C_P, at different values of the average C_L. One curve is the optimum propeller performance. This can be thought of as the envelope of optimal performance in the range of useful advance ratios. For a given advance ratio (e.g. operational conditions), the corresponding power coefficient is read on the vertical axis, if the blade's C_L is known.

Another type of chart is shown in Figure 5.17. The reference axes are the same. These lines of constant propeller efficiency are plotted. From Eq. 5.34 it is found that for a constant η, J and C_T/C_P are inversely proportional. Since C_T is positive, curves of constant efficiency have a positive slope on the plane $J–C_P$. If also C_T is

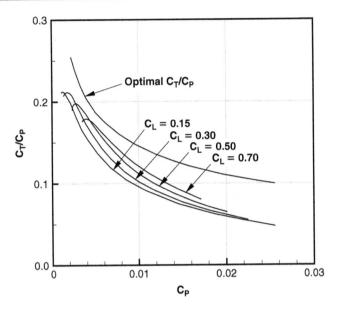

Figure 5.16 *Propeller chart: four-bladed propeller at different design C_L (as indicated); C_P versus C_T/C_P (data from Hamilton Sundstrand[172]).*

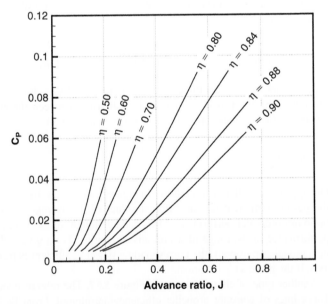

Figure 5.17 *Propeller chart: four-bladed propeller's best efficiency in the plane J–C_P (data from Hamilton Sundstrand[172]).*

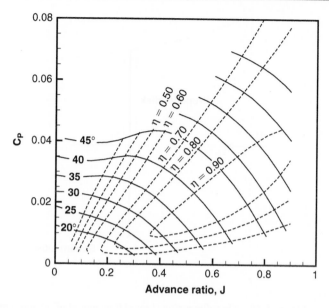

Figure 5.18 *Propeller chart: four-bladed propeller performance. Solid lines are performance indicators at constant pitch angle (as indicated, calculated at 3/4 chord); dotted lines are lines of constant efficiency in the J versus C_P plane (data from Hamilton Sundstrand).*

constant (a reasonable approximation over a reasonable range of advance ratios), the $\eta = $ constant curve is a straight line. Operation at maximum η requires operation over a limited range of advance ratios.

Although of great importance for analyzing the propeller performance, the coefficients given by Eqs 5.33 are not useful for determining the engine power required to fly at a given air speed, altitude and gross weight. In fact, propeller operation is analyzed either at a constant rotational speed, or a constant pitch. Operation with a mix of rotational velocities and pitch angles is also possible (see Figure 5.18).

For this purpose, it is necessary to have propeller data that do not depend on the rotational speed, as in Figure 5.15. If the pitch setting is unique, then the efficiency is a single curve, that can only be changed with the advance ratio, i.e. by a combination of rotational and forward speed.

5.9.1   The Axial Momentum Theory

The most elementary theoretical mechanism for converting the rotating power of a propeller into useful thrust is based on the one-dimensional momentum theory for the flow that passes through the propeller disk (Rankine-Froude momentum theory). This theory is conceptually and practically important. First, it provides a first-order estimate of the power and thrust of an open air screw; second, the method is general and can be applied to the thrust generation from a helicopter rotor (Chapter 11). There

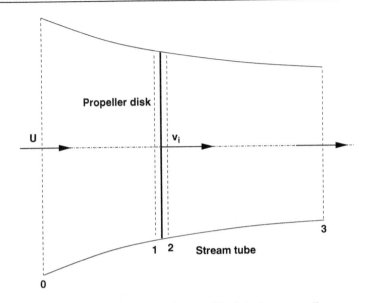

Figure 5.19 *One-dimensional axial flow model of airplane propeller.*

is a unified theory for propellers and helicopter rotors that is useful for a wide range of performance calculations. This method is so powerful that one often forgets its shortcomings, namely the lack of reference to the blades' geometry and the rotational speed.

According to the basic momentum theory, the propeller is reduced to a rotating disk that imparts axial momentum to the air passing through it. For the purpose of this discussion the air is incompressible. For reference, consider sections 0 (far upstream), 1 (just upstream the disk), 2 (just downstream the disk) and 3 (far downstream), as shown in Figure 5.19. The subscripts will refer to quantities at these sections.

The theory assumes that there is no flux along the limiting streamlines and that the velocity is continuous through the propeller disk. The free stream velocity is equal to the propeller's speed $U = U_o$. The continuity equation written for the incompressible mass flow rate within the stream tube is

$$Au = const. \tag{5.35}$$

The propeller thrust is equal to the rate of change of axial momentum at the disk,

$$T = A(p_2 - p_1). \tag{5.36}$$

Now we can apply the Bernoulli equation between sections 0–1 upstream and 2–3 downstream of the propeller, Figure 5.19,

$$p_o + \frac{1}{2}\rho U^2 = p_1 + \frac{1}{2}\rho U_1^2, \quad p_o + \frac{1}{2}\rho u_3^2 = p_2 + \frac{1}{2}\rho U_2^2. \tag{5.37}$$

The difference between the two equations yields the pressure jump through the propeller

$$p_2 - p_1 = \frac{1}{2}\rho(u_3^2 - U^2). \tag{5.38}$$

because from the continuity equation the velocity is continuous at the rotor disk, e.g. $u_1 = u_2$. From Eq. 5.36 the thrust generated by the propeller becomes

$$T = \frac{1}{2}\rho A(u_3^2 - U^2) = \rho A_3 u_3 (u_3 - U). \tag{5.39}$$

The total power is

$$P = T(u_1 + U) = \frac{1}{2}\dot{m}(u_3 + U)^2 - \frac{1}{2}\dot{m}U^2 = \frac{1}{2}\dot{m}u_3(u_3 + 2U), \tag{5.40}$$

that is obtained by substituting Eq. 5.39. The energy imparted to the slipstream is

$$E = \frac{1}{2}A_3\rho u_3(u_3^2 - U^2). \tag{5.41}$$

This energy is minimal when the slipstream velocity is equal to the propeller velocity. The total power is found from summing up Eq. 5.41 and Eq. 5.40. The slipstream velocity far downstream, u_3, can be related to the velocity at the disk u_1 and the free stream velocity U. In fact, the power absorbed by the propeller is

$$T = \dot{m}(u_3 + U) - \dot{m}U = \dot{m}u_3. \tag{5.42}$$

By combination of Eq. 5.40 and Eq. 5.42, we have

$$u_1 = \frac{1}{2}u_3. \tag{5.43}$$

In conclusion, the air speed at the rotor disk is the average between the propeller's speed and the speed far downstream. Also, the far downstream velocity is twice the induced velocity disk. For a static propeller this relationship provides the value of the *induced velocity* at the rotor disk:

$$u_1 = v_i = \frac{1}{2}u_3. \tag{5.44}$$

The corresponding induced power is $P_i = Tv_i$. The thrust becomes

$$T = 2\rho A(U + v_i)v_i, \tag{5.45}$$

where $\rho A(U + v_i)$ is the mass flow rate through the disk, and $2v_i$ is the total increase in velocity. The corresponding power is the product between the thrust and the velocity through the disk

$$P = T(U + v_i) = 2\rho A(U + v_i)^2 v_i. \tag{5.46}$$

This power expression contains two terms: (1) a useful power TU and (2) an induced power Tv_i which is a loss due to the kinetic energy imparted to the flow. The propulsion efficiency is

$$\eta = \frac{TU}{P} = \frac{TU}{T(U + v_i)} = \frac{U}{U + v_i}. \tag{5.47}$$

Therefore, the propulsion efficiency decreases as the induced velocity increases. The induced velocity in terms of the thrust is found from Eq. 5.45, which is quadratic in v_i. The only physical solution of Eq. 5.45 is a positive v_i,

$$v_i = \frac{1}{2}\left[-U + \sqrt{U^2 + \frac{2T}{\rho A}}\right]. \tag{5.48}$$

Under static conditions, $U = 0$, we have

$$v_i = \sqrt{\frac{T}{2\rho A}}, \quad P_i = \sqrt{\frac{T^3}{2\rho A}}. \tag{5.49}$$

These expressions will be useful for rotorcraft calculations (Chapter 11).

An important aspect of this elementary theory is the behavior of the pressure. For points upstream or downstream of the propeller, there is a conservation of total pressure (sum of the static pressure and dynamic pressure, $q = \rho U^2/2$). The total pressure is given by the Bernoulli equation, Eq. 5.37. The engine power is used to increase the kinetic energy of the air passing through the disk. Therefore, we conclude that the theory involves a sudden increase of pressure through the propeller disk, while the velocity of the air in the stream tube is continuous. The flow undergoes an acceleration, therefore the slipstream must contract, according to the continuity Eq. 5.35.

Obviously, various advancements and refinements can be done from the above assumptions. With respect to the non-uniform axial inflow, one needs to integrate the change of axial momentum over the propeller disk, and the result is

$$T = \int_A \rho u_3 (u_3 - U) \, dA, \tag{5.50}$$

A similar expression is obtained for the loss of energy:

$$E = \frac{1}{2} \int_A \rho u_3 (u_3^2 - U^2) \, dA. \tag{5.51}$$

An essential hypothesis is that of axi-symmetric flow through the disk, which is quite reasonable for a propeller in axial flight. This hypothesis leads to a slightly modified theory.

Consider an annulus of width dy, corresponding to an area $dA = 2\pi y dy$. The element of thrust generated by the mass flow is

$$\dot{m} = \rho(U + v_i) dA, \tag{5.52}$$

through this annulus is

$$dT = 2\rho(U + v_i)v_i dA = 4\pi\rho(U + v_i)v_i y dy. \tag{5.53}$$

We define the following induced velocity ratio

$$\lambda = \frac{U + v_i}{\Omega R} = \frac{U + v_i}{\Omega y}\frac{\Omega y}{\Omega R} = \left(\frac{U_n}{U_t}\right)r = \tan\phi\, r, \tag{5.54}$$

where U_n and U_t are the velocity components normal and tangential to the rotor plane, ϕ is the inflow angle. In general, $U_n \ll U_t$, therefore we can make the approximation

$$\tan\phi \simeq \phi = \frac{U_n}{U_t} = \frac{\lambda}{r}. \tag{5.55}$$

If we introduce the inflow velocity ratio, Eq. 5.54, we find

$$dT = 4\pi\rho\left(\frac{U + v_i}{\Omega R}\right)\left(\frac{v_i}{\Omega R}\right)(\Omega R)^2 y dy = 4\pi\rho\lambda\lambda_i(\Omega R)^2\, y dy, \tag{5.56}$$

with $\lambda_i = v_i/\Omega R$. Note that λ_i is the induced velocity ratio in the absence of axial flight velocity. It can also be written as

$$\lambda_i = \frac{v_i}{\Omega R} = \frac{v_i}{\Omega R} + \frac{U}{\Omega R} - \frac{U}{\Omega R} = \lambda - \frac{U}{\Omega R} = \lambda - \lambda_c, \tag{5.57}$$

with $\lambda_c = U/\Omega R$. The resulting thrust is found from integration of Eq. 5.56,

$$T = 4\pi\rho(\Omega R)^2 \int_o^R \lambda\lambda_i\, y dy. \tag{5.58}$$

The element of power is $dP = dTv_i$, therefore

$$P = 4\pi\rho(\Omega R)^3 \int_o^R \lambda\lambda_i^2\, y dy. \tag{5.59}$$

These integrations can only be done if the radial distribution of induced velocity is known, therefore the problem is not closed. A solution can be found by combining the results of the blade momentum theory (next section). Various further advancements have been achieved over the years, and it is now possible to model the case of a propeller/rotor disk inclined by any angle on the free stream, with almost any load distribution on the rotor, see for example Conway[173].

5.9.2 The Blade Element Method

The axial momentum theory provides integral quantities, such as thrust and power. These characteristics do not seem to depend on the propeller's geometry, which is clearly a shortcoming. In the blade element method, these details are yanked back into the theory. In this framework, the blade sections are supposed to operate like two-dimensional sections, with the local inflow conditions derived by appropriate means in the rotating environment, see Figure 5.14. Although the definition of this inflow is not obvious, and the interference between elements on the same blade and between blades is not taken into account, the method is otherwise extremely powerful. It allows the calculation of basic performance from the geometrical details and the two-dimensional blade section aerodynamics. The method described below applies to the helicopter rotor blades as well, though with some small changes.

The relationship between inflow angle, pitch angle and angle of attack is

$$\alpha = \theta - \phi. \tag{5.60}$$

The pitch is a geometrical setting, whilst the angle of attack of the blade section and the inflow velocity are operational free parameters. The lift and drag forces on this section are

$$dL = \frac{1}{2}\rho c C_L U^2 dy, \tag{5.61}$$

$$dD = \frac{1}{2}\rho c C_D U^2 dy. \tag{5.62}$$

These forces, resolved along the direction normal and parallel to the rotor disk give the contributions to the thrust, torque and power for the single blade

$$dT = dL \cos \phi - dD \sin \phi, \tag{5.63}$$

$$dQ = (dL \sin \phi + dD \cos \phi)y, \tag{5.64}$$

$$dP = (dL \sin \phi + dD \cos \phi)\Omega y. \tag{5.65}$$

These elements can be written in non-dimensional form, using the definition of coefficients given by Eq. 5.33. The total thrust, torque and power for N blades will require integration of the above expressions from the inboard cut-off point to the tip.

The integrals are, in fact, not solved directly, but numerically. If we divide the blade section into a number n of elements, each having a radial width dy_j, then

$$T = N \sum_{j=1}^{n} \left(dL_j \cos \phi_j - dD \sin \phi_j \right), \tag{5.66}$$

$$Q = N \sum_{j=1}^{n} \left(dL_j \sin \phi_j + dD \cos \phi_j \right) y_j dy_j, \tag{5.67}$$

$$P = N \sum_{j=1}^{n} \left(dL_j \sin \phi_j + dD \cos \phi_j \right) \Omega y_j dy_j, \tag{5.68}$$

with the forces evaluated from Eq. 5.61 and Eq. 5.62. All the quantities appearing in the aerodynamic forces change with the radial position, including the chord (for complex geometries), the density (for high-speed flows), and the C_L and C_D (depending on local angle of attack α, Reynolds number Re and Mach number M). The problem is to find the inflow angle ϕ, the resultant angle of attack α and the actual inflow velocity U at each blade section.

For the solution of the problem we still have too many unknowns and too few equations. One way of finding the induced velocity ratio is to introduce the axial momentum theory. This combination eliminates the unknowns discussed above, namely the radial distribution of induced velocity v_i.

With the approximation introduced by Eq. 5.55, the element of thrust can be written as $dT \simeq dL$. Since the results of both theories must be the same, we are led to the equivalence between Eq. 5.56 and Eq. 5.61

$$dL = \frac{N}{2}\rho c\, C_L U^2 dy = \frac{N}{2}\rho c\, C_{L_\alpha}\alpha U^2 dy = dT$$
$$= \frac{n}{2}\rho c\, C_{L_\alpha}\alpha U^2 dy = 4\pi\rho\lambda\lambda_i(\Omega R)^2\, ydy. \tag{5.69}$$

Now simplify Eq. 5.69, by using: $U = \Omega y$, Eq. 5.60 (definition of angle of attack), Eq. 5.55 (small inflow angle approximation), and Eq. 5.22 (definition of solidity). After this algebra, the result (Problem 9) is

$$\frac{1}{8}\sigma\, C_{L_\alpha}(\theta r - \lambda) = \lambda(\lambda - \lambda_c). \tag{5.70}$$

Eq. 5.70 must be solved for the unknown inflow velocity ratio λ. It is a quadratic equation with a positive and a negative root. The meaningful solution is

$$\lambda = \left[\left(\frac{\sigma C_{L_\alpha}}{16} - \frac{\lambda_c}{2}\right)^2 - \frac{\sigma C_{L_\alpha}}{8}\theta r\right]^{1/2} + \left(\frac{\sigma C_{L_\alpha}}{16} - \frac{\lambda_c}{2}\right). \tag{5.71}$$

The calculation of the rotor power is done in the same way. All the quantities are known from the previous discussion, and the power can be calculated alongside the thrust, from the equation

$$P = N\sum_{j=1}^{n}\left(dL_j\phi_j + dD\right)\Omega y_j dy_j. \tag{5.72}$$

Computational Procedure

1. Read propeller's geometry (radius, chord, twist distributions).
2. Read operational parameters (rpm, U, density altitude).
3. Read the two-dimensional C_D, and C_L, C_{L_α}, distributions.
4. Divide the radius in a number n of blade elements.
5. Set the radial position (*start loop*).
6. At current r solve Eq. 5.70 to find λ, Eq. 5.71.
7. Solve Eq. 5.55 to find ϕ.
8. Solve Eq. 5.60 to find α.
9. Solve Eq. 5.69 to find the thrust dT.
10. Update the thrust.
11. End loop.

A correction is usually applied to the above procedure, because it implies that the aerodynamic flow around the blade section follows that of the airfoil. However, both the tip and the root are affected by three-dimensional effects, which lead to loss of lift, and hence loss of thrust. One of these corrections is due to Prandtl, and is written as

$$k(r) = \left(\frac{2}{\pi}\right) \cos^{-1} e^{-f(r)}, \tag{5.73}$$

with

$$f(r) = \frac{N}{2} \frac{1-r}{r\phi}. \tag{5.74}$$

The corrected thrust will be $dTk(r)$. A plot of $k(r)$ shows that it is equal to one for most of the span, but it tends rapidly to zero near the tip, with a rate depending on the number of blades and the inflow angle. The alternative to this method is to consider an effective blade radius $R_e < R$, and proceed with the calculations as described above.

The procedure described neglects a number of difficulties. These are related to the format in which the aerodynamic data are available. Ideally one would have data like $C_D = C_D(\alpha, M)$, $C_L = C_L(\alpha, M)$ in a matrix form.

From the performance point of view, it is important to know how to operate the propeller engines, so as to minimize losses and increase fuel efficiency. For a given propeller installation, the parameters that can be changed are the pitch, the rotational speed and the flight speed. Other important aspects of propeller performance are the interaction with the wing and the fuselage[174,175], compressibility effects[176,177], and the propeller noise.

PROBLEMS

1. A turbofan engine has a specific fuel consumption between 0.368 and 0.385 lb/lb/h. Convert these data into units kg/N/s. Is this a conversion into international units? Finally convert the data into kg/kN/s.

2. A certain aircraft has a fuel flow at cruise conditions estimated at 1,400 US gallons per hour. Convert this data into N/s. Also, assume that the aircraft has two engines, and that the total thrust delivered at cruise conditions is $T = 210\,kN$. What is the specific fuel consumption of the engines in $N/N/s$? (For aviation fuel use the data in Table 5.2).

3. The Boeing 747-400 is known to burn fuel at the rate of about 3% of its take-off gross weight every hour. Calculate the amount of fuel in metric tons per hour, in liters per hour, and the number of liters required to fly 1 km at a cruise altitude $h = 11,000\,m$.

4. Find the speed for minimum fuel flow for the jet engine aircraft model A in Appendix A flying at a cruise speed $M = 0.82$ at altitude $h = 10,500$. Calculate the fuel flow at this speed. If the aircraft descends to $h = 10,000\,m$ how would the fuel flow be affected (all other parameters being the same)?

5. Discuss the parameters affecting the mass flow rate into a jet engine. Provide some qualitative plots of $\dot{m}_a$ as a function of flight altitude, Mach number,

engine diameter, and rpm. Use the St Venant equation to calculate the chocked mass flow rate per unit nozzle area as a function of flight altitude.

6. The Boeing B-747/400 has its power plant re-engineered. As a result, its Thrust-Specific Fuel Consumption (TSFC) is decreased by 3%. The other characteristics of the aircraft do not change significantly, therefore its weight can be considered constant. The aircraft is to have an annual block time of 700 hours. Calculate the saving in fuel consumption over the year. Also, considering that the fuel consumption represents 9% of the direct operating costs of the aircraft, calculate the money that can be saved by the improved engines. Consider constant aviation fuel price, equal to 0.70 US dollars/liter. Repeat the analysis for the SAAB 340, and draw a conclusion regarding the effects of fuel efficiency on the operation of the two aircraft. (*Please note that aviation fuel prices change greatly over short times, and depend on the airport and the quantity purchased.*)

7. As an example of pollution created in the high atmosphere by commercial air traffic, we want to calculate the heat released per unit of time by aircraft model A flying at its cruise altitude (10,800 m) at its cruise speed ($M = 0.80$) at ISA conditions. The Mach number of the jets from the nozzle of its two jet engines is $M = 1.3$, and the nozzle cross-sectional area is $A_n = 1.8\,m^2$, and the average jet temperature is estimated at $T_j = 850$ degrees.

8. Consider a four-bladed propeller, whose performance charts are given in Figures 5.14 and 5.15. This propeller has a diameter $d = 1.8\,m$, and is coupled to an engine that delivers a maximum shaft power $P = 300\,kW$. Its maximum tip Mach number at sea level is $M_{tip} = 0.70$. Determine if this propeller can be operated with an efficiency $\eta = 0.90$.

9. Derive Eq. 5.70 from Eq. 5.69 and calculate its only physical solution, Eq. 5.71.

Chapter 6

Flight Envelopes

There is no particular feeling of speed, except that the miles go faster. The sky is not dark and the horizon is not curved.

A. Turcat, Director of Concorde Flight Test, 1969

This chapter deals with atmospheric flight envelopes. The aircraft is to fly level in steady accelerated flight. We describe the shape/speed relationship, give various speed definitions, and the techniques required to measure them. Calculations of absolute ceiling and acceleration problems in supersonic regime are presented.

6.1 GENERAL DEFINITIONS

The *flight corridor* is the speed band at a given altitude where steady state flight is possible. The minimum level speed is called *stalling speed*; the maximum speed, if not called as such, is the dash speed of the aircraft. The aircraft can maintain a steady level flight at any speed within the flight corridor, by adjusting the throttle (engine thrust), attitude (angle of attack) and configuration (control surfaces). The boundaries of the flight corridor are controlled by the aerodynamic, propulsive and structural performance.

The set of all flight corridors from sea level to the absolute ceiling define the *flight envelope*. In other words, the flight envelope is the closed area in the $U - h$ diagram that includes all operating conditions *for a particular aircraft at a given weight*. The flight envelope depends on a large number of factors, including weight, aerodynamics, propulsion system, structural dynamics and atmospheric conditions.

The *absolute ceiling* is the maximum altitude at which an aircraft can keep a steady level flight. Above this altitude, the engine power or thrust are not enough to overcome the aircraft drag. It must be understood that the aircraft can zoom past the absolute ceiling, by exchanging part of its kinetic energy for potential energy (e.g. altitude). This can be done with an inertial climb, even in the absence of sufficient engine power. The ceiling altitude depends on the type of aircraft. It is in the range of 6,000 to 7,000 m (about 19,700 to 23,000 ft) for turboprop aircraft; 10,000 to 12,000 m (about 32,800 to 39,400 ft) for commercial jet aviation; and it increases to 18,000 or 19,000 m (60,000+ ft) for high-performance military aircraft. It is seldom above this altitude. The Lockheed SR-71 had a cruise altitude of 27,000 m (88,500 ft) and an estimated ceiling of 30,700 m (100,000 ft). Experimental high-altitude vehicles aim at the 100,000 ft ceiling.

It is of some interest to report the cruise ceiling of some birds. Various observations indicate that bar-headed geese (*Anser indicus*) migrate through the Himalaya range at altitudes in excess of 8,000 m (26,240 ft) – honking as they fly[178,179]. Whooper swans (*Cygnus cygnus*) have been observed flying over the Hebrides, Scotland, at 8,230 m (27,000 ft). Radar data confirmed their altitude, speed (75 kt) and heading (southbound). Storks, cranes and vultures have been seen soaring at 7,500 m (24,600 ft). Northern pintails (*Anas acuta*) and black-tailed godwits (*Limosa limosa*) have been found at 5,000 m (16,400 ft). Records of bird strikes at altitudes are not uncommon. It is possible that a few dozen bird species fly frequently at altitudes of the commercial air space. Note that at these altitudes even a fit mountain climber has serious difficulty breathing.

6.2 AIRCRAFT SPEED RANGE

The speed of a vehicle is classified on the basis of its Mach number. A low subsonic speed corresponds to $M < 0.3$–0.4; a high subsonic speed requires $M < 0.7$; a transonic speed has a range of Mach numbers below and above the speed of sound, generally $M = 0.7$–1.2. A fully supersonic aircraft will have an $M > 1.2$, although the exact Mach number will depend on the configuration. For speeds above $M = 4$–5 the flow is called hypersonic, and it can be substantially different from the lower-speed regimes. A flight Mach number within the subsonic range is considered a low speed. By contrast, a high-speed aircraft is one that operates at transonic speeds and beyond.

Aircraft cruise speeds have been increasing over the years, as shown in the chart of Figure 6.1. In the first 50 years of powered aviation, cruise speeds and maximum speeds have been increasing roughly exponentially, in part due to more powerful reciprocating engines, and then thanks to the introduction of the jet engine and its derivatives. The aircraft speeds reached a point of diminishing returns around 1970, when most of the commercial long range airplanes were powered by jet engines.

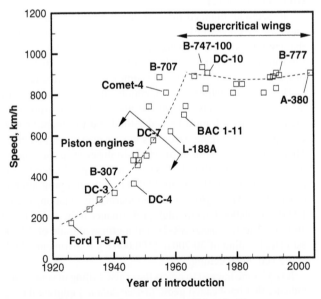

Figure 6.1 *Aircraft cruise speed versus year of introduction.*

Cruise speeds converged toward an average $M = 0.78$ to 0.82. Some advances in aerodynamics have allowed a slight increase in the cruise speed for the latest generation of commercial jet aircraft, which is now estimated at $M = 0.85$.

The concept of high speed, like many other things in life, is not absolute. Gabrielli and von Kármán[180], in a landmark study published in 1950, pointed out the difficulty of measuring the value of speed for a considerable number of airborne and ground systems. One result of their analysis is a chart speed versus power required, in a double-logarithm plot (the Gabrielli-von Kármán chart). It was found that all systems lie above this curve, and at their most efficient point the vehicles nearly touch this line. More recently, Lorentz[181] developed a general power scaling technique for airplanes and helicopters.

6.3 DEFINITION OF SPEEDS

Manufacturers quote a number of different aircraft speeds. The *ground speed* is the aircraft speed measured with respect to a fixed point on the ground. The *Never-to-exceed speed* (VNE) is determined by the structural limits of the aircraft. For a given aircraft and given gross weight, this speed depends on the flight altitude. The *dash speed* is a supersonic speed that can be maintained for a relatively short time (or distance), enough to escape a dangerous theater. After this time, the engines may overheat and fuel consumption cannot be afforded, because of operation with after-burning. A supersonic jet aircraft is said to be *supercritical at all altitudes* if it can maintain supersonic level flight at sea level and above. Some of these speeds are given for the subsonic commercial aircraft in Table A.4 in Appendix A. *Supercruise* is the ability to fly at supersonic speeds without after-burning thrust.

The *block speed* is the air speed in absence of atmospheric winds adjusted in relation to range to compensate for take-off, climb, let-down, instrument approach, and landing. The *air speed* is the aircraft speed relative to the air, and accounts for the presence of atmospheric winds. If an atmospheric wind is aligned with the aircraft speed, the air speed is defined as

$$U = U_g \pm U_w, \tag{6.1}$$

where U_g is the ground speed and U_w is the wind speed, negative for a *tail wind*, positive for a *head wind*. Measuring the air speed is of considerable importance. We know that the atmospheric conditions change with altitude, and that the instruments are unable to read the air density directly. At incompressible speeds, the air speed can be evaluated with the Bernoulli equation,

$$p + \frac{1}{2}\rho U^2 = p_*, \tag{6.2}$$

that gives the *True Air Speed* (TAS)

$$TAS = U = \sqrt{\frac{p_* - p}{2\rho}} = \sqrt{\frac{\Delta p}{2\rho_o}} \frac{1}{\sqrt{\sigma}}. \tag{6.3}$$

In Eq. 6.3 p is the free stream (atmospheric) pressure, p_* is the stagnation pressure, and ρ is the density; the subscript "o" still denotes sea level standard conditions. An instrument that measures difference in pressure between the free stream conditions and the static conditions is the Pitot probe. With the Bernoulli equation the probe converts a pressure difference into air speed, as long as the air density is known. Annoyingly, this quantity depends on the flight altitude, therefore the instrument would not work without additional readings. A partial solution to the problem is to refer the atmospheric conditions to sea level. From Eq. 6.3 the air speed becomes

$$U = \sqrt{\frac{\Delta p}{2\rho_o}}.$$
(6.4)

The velocity in Eq. 6.4 is called *Equivalent Air Speed* (EAS). It is different from the true air speed U due to a density correction,

$$EAS = TAS\sqrt{\sigma}.$$
(6.5)

We note that from Eq. 6.3 and Eq. 6.5 the dynamic pressure is:

$$q = \frac{1}{2}\rho U^2 = \frac{1}{2}\rho_o\, EAS^2.$$
(6.6)

The dynamic pressure can be measured by the probe, therefore

$$EAS = 2\frac{q}{\rho_o}.$$
(6.7)

To find the air speed we need the density, which cannot be measured directly. However, by taking a temperature reading and a static pressure reading at the same time, we can find the value of the relative density σ from the equation of ideal gases,

$$\sigma = \frac{1}{\mathcal{R}\rho_o}\left(\frac{p}{T}\right)_{static}.$$
(6.8)

An *Air Speed Indicator* (ASI) based on this principle is of no use in high-speed flight, where compressibility is important. From compressible aerodynamic theory we know that the ratio between the stagnation pressure p_* and the static atmospheric pressure p corresponding to an isentropic deceleration from a Mach number M is

$$\left(\frac{p_*}{p}\right)^{\gamma-1/\gamma} = 1 + \frac{\gamma-1}{2}M^2.$$
(6.9)

where γ is the ratio between specific heats at constant pressure and constant volume. Its value is constant at moderate temperatures, and equal to 1.4. Solving for the Mach number, we find

$$M^2 = \frac{2}{\gamma-1}\left[\left(\frac{p_*}{p}\right)^{\gamma-1/\gamma} - 1\right].$$
(6.10)

The true air speed will be found from Eq. 6.10, the definition of speed of sound, $a = \sqrt{\gamma \mathcal{R} \mathcal{T}}$ and the equation of ideal gases $p/\rho = \mathcal{R} \mathcal{T}$

$$TAS = \sqrt{\frac{2\gamma}{\gamma - 1} \left(\frac{p}{\rho}\right) \left[\left(\frac{p_*}{p}\right)^{\gamma - 1/\gamma} - 1\right]}$$

$$= \sqrt{\frac{2\gamma}{\gamma - 1} \left(\frac{p}{\rho}\right) \left[\left(\frac{p_* - p}{p} + 1\right)^{\gamma - 1/\gamma} - 1\right]}. \qquad (6.11)$$

If the TAS is given in knots, then it is called *KTAS*. Solving Eq. 6.9 for the stagnation pressure p_*, we find

$$p_* = p \left(1 + \frac{\gamma - 1}{2} M^2\right)^{\gamma/\gamma - 1}, \qquad (6.12)$$

$$p_* - p = p \left[\left(1 + \frac{\gamma - 1}{2} M^2\right)^{\gamma/\gamma - 1} - 1\right]. \qquad (6.13)$$

This expression is the so-called *impact pressure*. At low Mach numbers, when the flow is practically incompressible, the impact pressure is equal to the dynamic pressure, as shown in Figure 6.2. As mentioned, there is no direct way to measure the air density. An indirect way is to take a temperature reading to find p/ρ. An instrument that measures the impact pressure, the local static pressure and the local temperature provides the TAS.

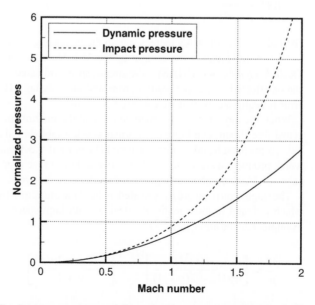

Figure 6.2   *Impact pressure and dynamic pressure versus Mach number.*

Sometimes it is more useful to calibrate the speed to sea level conditions. The local pressure p and density ρ are replaced by the sea level values. This is the same as replacing the local temperature with the sea level temperature. The resulting speed is called *Calibrated Air Speed*. It is abbreviated as *CAS* or *KCAS* (*CAS* in knots),

$$CAS = \sqrt{\frac{2\gamma}{\gamma - 1}\left(\frac{p_o}{\rho_o}\right)\left[\left(\frac{p_* - p_o}{p_o} + 1\right)^{\gamma - 1/\gamma} - 1\right]}. \tag{6.14}$$

It can be verified that the relationship between TAS and CAS is

$$CAS \simeq TAS\sqrt{\frac{T_o}{T}} = \frac{TAS}{\sqrt{\theta}} = TAS\sqrt{\frac{\delta}{\sigma}}. \tag{6.15}$$

In other words, the CAS is equal to the true air speed at standard sea level. If the local static pressure can be measured and the local density is replaced with the sea level ρ_o then we find again the equivalent air speed. EAS is the TAS corrected for changes in atmospheric density,

$$EAS = \sqrt{\frac{2\gamma}{\gamma - 1}\left(\frac{p}{\rho_o}\right)\left[\left(\frac{p_* - p}{p} + 1\right)^{\gamma - 1/\gamma} - 1\right]}, \tag{6.16}$$

a result equivalent to the case of incompressible flow. Therefore,

$$EAS = \frac{TAS}{\sqrt{\sigma}}. \tag{6.17}$$

For a given TAS, the EAS increases with the increasing flight altitude.

This speed is read in the cockpit by the ASI. The ASI may be affected by errors. The *Indicated Air Speed* (IAS) is the aircraft speed indicated by the instrument, which can be affected by errors (position, time and pressure lag). Here we assume that the error is negligible, so that we can call $CAS = IAS$.

When the aircraft flies at supersonic speed the instrument is unable to *sense* the actual free stream conditions. A normal shock establishes ahead of the instrument. Two events must be taken into account: (1) a normal shock wave ahead of the probe, which produces a subsonic Mach number and (2) an isentropic deceleration from a subsonic Mach number to stagnation conditions in the probe.

The stagnation pressure p_* is related to the static pressure p by the Raleigh equation, which is found in most textbooks dealing with high-speed aerodynamics (see, for example, Kuethe and Chow[117]),

$$\frac{p_*}{p} = \left[\frac{(\gamma + 1)^2 M^2}{4\gamma M^2 - 2(\gamma - 1)}\right]^{\gamma/\gamma - 1}\left[\frac{1 - \gamma + 2\gamma M^2}{\gamma + 1}\right]. \tag{6.18}$$

To find the Mach number, this equation must be solved for M.

6.4   STEADY STATE LEVEL FLIGHT

Steady state level flight is obtained when all the forces on the aircraft are balanced. For a case in which the thrust force is aligned with the velocity vector, the balance equations in the direction parallel and normal to the ground are

$$T = D, \quad L = W. \tag{6.19}$$

The weight is balanced by the lift, and the drag is overcome by the engine thrust. The dynamic equations yield

$$C_L = \frac{2W}{\rho A U^2}. \tag{6.20}$$

The level flight speed compatible with a given lift coefficient is found by inverting Eq. 6.20:

$$U = \sqrt{\frac{2}{\rho} \left(\frac{W}{A}\right) \frac{1}{C_L}}. \tag{6.21}$$

The minimum speed compatible with steady level flight is called *stalling speed*. It is also called *zero climb rate speed*, or the lowest speed compatible with OEI or AEO level flight (Pinsker[182]). The stall speed can be obtained from Eq. 6.21 by using the maximum C_L,

$$U_{stall} = \sqrt{\frac{2}{\rho} \left(\frac{W}{A}\right) \frac{1}{C_{L_{max}}}}. \tag{6.22}$$

One should be careful in interpreting this equation, because, as explained earlier, in many cases the correct value of the $C_{L_{max}}$ is not known, therefore at a given altitude and aircraft weight there is a minimum safe speed that prevents the aircraft from stalling. The stalling speed is achieved at one fraction of the nominal $C_{L_{max}}$, as described in §4.4.

6.5   SPEED IN LEVEL FLIGHT

For a jet aircraft at subsonic speeds, whose drag equation is parabolic, the speed is obtained from Eq. 6.19. The solution is found from

$$T - \frac{1}{2}\rho A (C_{D_o} + k C_L^2) U^2 = 0, \tag{6.23}$$

with C_L given by Eq. 6.20. If the thrust is not dependent on the speed, and if the altitude is fixed, the solving equation is

$$c_1 U^4 + c_2 U^2 + c_3 = 0, \tag{6.24}$$

with

$$c_1 = -\frac{1}{2}\rho \frac{A}{W} C_{D_o}, \quad c_2 = \frac{T}{W}, \quad c_3 = -k \frac{2}{\rho} \frac{W}{A}. \tag{6.25}$$

The coefficients contain the thrust ratio and the wing loading, which are propulsion and structural quantities. They also contain the drag coefficients. A parametric study of the speed as a function of engine thrust can be carried out, so as to obtain $U(T)$ at all altitudes. For a given thrust one can study the effects of the aircraft's weight on its speed. The analysis can be repeated for varying flight altitudes, all other parameters being constant (Problem 2). Solution of Eq. 6.24 is straightforward if the unknown is U^2, because it is a parabolic equation

$$U^2 = \frac{-c_2 \pm \sqrt{c_2^2 - 4c_1c_3}}{2c_1}. \tag{6.26}$$

This equation has a physical meaning if $c_2^2 - 4c_1c_3 > 0$. In this case there is a double solution, which is found by replacing the values of the coefficients and simplifying the algebra. The limiting condition is

$$c_2^2 - 4c_1c_3 = \left(\frac{T}{W}\right)^2 - 4C_{D_o}k = 0, \tag{6.27}$$

Therefore, a level speed can be maintained by the aircraft if

$$\frac{T}{W} \geq 2\sqrt{C_{D_o}k}. \tag{6.28}$$

The limiting condition occurs at the absolute ceiling of the aircraft. The minimum speed coincides with the maximum speed. The only speed at the absolute ceiling is

$$U = -\frac{c_2}{2c_1} = \frac{T}{\rho A C_{D_o}}. \tag{6.29}$$

For a propeller-driven aircraft the thrust equation is replaced by the power equation

$$\eta P = DU, \tag{6.30}$$

where P is the power at the shaft. Thus, the level flight condition becomes

$$\eta P = \frac{1}{2}\rho A C_D U^3. \tag{6.31}$$

By replacing the parabolic drag equation, and simplifying the algebra,

$$c_1 U^3 + c_2 \frac{1}{U} + c_3 = 0, \tag{6.32}$$

with coefficients

$$c_1 = -\frac{1}{2}\frac{\rho A C_{D_o}}{W}, \quad c_2 = -2k\left(\frac{W}{\rho A}\right), \quad c_3 = \eta \frac{P}{W}. \tag{6.33}$$

Solution of Eq. 6.32 is more elaborate, because it is non-linear and the engine power is dependent on the speed.

6.6   ABSOLUTE CEILING OF JET AIRCRAFT

Equation 6.29 gives the value of the speed at the absolute ceiling. However, neither the ceiling nor the thrust is known, therefore the equation cannot be used in that form. For a subsonic airplane the ceiling is reached under the condition that the aircraft is unable to climb. If we use the conventional thrust equation, Eq. 5.22, then the equilibrium is obtained with

$$T_o \sigma^r = \frac{1}{2} \rho_o \sigma A C_D U^2. \tag{6.34}$$

Using the same method described above, Eq. 6.20, the condition becomes:

$$c_1 \sigma^2 U^4 + c_2 \sigma^{r+1} U^2 + c_3 = 0, \tag{6.35}$$

with constant coefficients given by Eq. 6.25. Equation 6.35 is an implicit relationship $f(\sigma, U) = 0$, which can be transformed into a function $g(h, U) = 0$. The limiting condition is still given by the discriminant, Eq. 6.27

$$\frac{T_o \sigma^r}{W} = 2\sqrt{C_{D_o} k}. \tag{6.36}$$

Solution of this equation for the unknown relative density σ yields

$$\sigma^r = 2\frac{W}{T_o}\sqrt{C_{D_o} k}, \tag{6.37}$$

$$\sigma = \left[2\frac{W}{T_o}\sqrt{C_{D_o} k}\right]^{1/r}. \tag{6.38}$$

The relationship between σ and the flight altitude is given by Eq. 2.34. Another way to find a solution is by using a numerical technique. The ceiling is found from the *minimal value* of σ that is compatible with a positive root of Eq. 6.35.

Equation 6.38 shows that the absolute ceiling is dependent on the aerodynamic drag characteristics, on the aircraft weight, and the engine thrust at sea level and the throttle setting. From this equation a number of parametric studies can be made to calculate the effects of any parameter on the absolute ceiling (see Problem 5). The numerical solution of this problem for aircraft A, with $r = 0.80$, and gross weight $W = 140,000$ kg, is shown on Figure 6.3.

6.7   ABSOLUTE CEILING OF PROPELLER AIRCRAFT

The calculation of the absolute ceiling of a propeller aircraft is always more difficult, because the engine power is dependent on the speed. A suitable power equation for a gas turbine engine is

$$\frac{P}{P_o} = \delta\sqrt{\theta} = \sigma^r, \tag{6.39}$$

which provides the power as a function of altitude, but not on the speed.

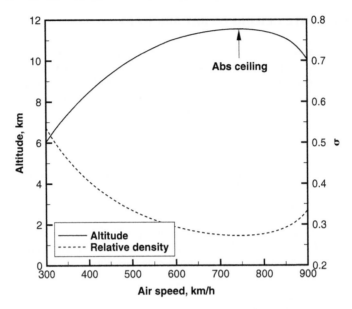

Figure 6.3   *Calculated absolute ceiling for aircraft model A.*

Also, consider that the propeller efficiency is constant. The strategy for finding a solution of this simplified problem is the following

$$\eta P_o \sigma^r = \frac{1}{2} \rho A C_D U^3. \tag{6.40}$$

$$\eta P_o \sigma^r = \frac{1}{2} \rho A (C_{D_o} + k C_L^2) U^3. \tag{6.41}$$

By simplifying the algebra, we are led to

$$\frac{1}{2} \frac{\rho A}{\eta P_o} C_{D_o} U^3 + \frac{kA}{\eta P_o} \left(\frac{2}{\rho}\right) \left(\frac{W}{A}\right)^2 \frac{1}{U} - \sigma^r = 0. \tag{6.42}$$

If we define the coefficients

$$c_1 = \frac{1}{2} \frac{\rho A}{\eta P_o} C_{D_o} = c_1(h), \quad c_2 = \frac{kA}{\eta P_o} \left(\frac{2}{\rho}\right) \left(\frac{W}{A}\right)^2 = c_2(h),$$

$$c_3 = -\sigma^r = c_3(h). \tag{6.43}$$

Equation 6.42 becomes

$$c_1(h) U^3 + c_2(h) \frac{1}{U} + c_3(h) = 0. \tag{6.44}$$

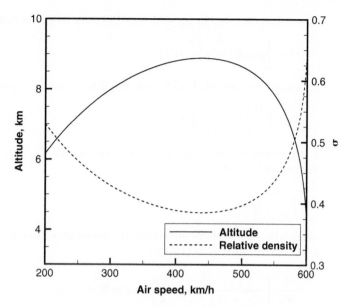

Figure 6.4   *Calculated absolute ceiling for aircraft model B, W = 70,000 kg.*

The solution algorithm is the following:

Computational Procedure

- Assign σ and find the root of Eq. 6.44 for the unknown speed U. This requires the solution of a non-linear equation in which the altitude is incorporated in the coefficients (there are several techniques for root finding of a non-linear equation; please refer to Press *et al.*[183]).
- With the current value of σ find the corresponding altitude from the ISA function, Eq. 2.34. Store this altitude and the corresponding aircraft speed.
- Decrease σ and repeat the procedure, up to a point when there is no solution. The limiting σ that gives a solution of the equation gives the absolute ceiling.

Figure 6.4 shows the predicted absolute ceiling for the turboprop aircraft model B, with a gross weight of 70,000 kg, a constant propeller efficiency $\eta = 0.82$, throttle $\Pi = 1$, and a power coefficient $r = 0.65$. The result is not far off the manufacturer's data. We have used a gross approximation – and perhaps we have been lucky. The predicted ceiling is about 8,800 m at a speed of about 450 km/h (243 kt).

The next step is to use the actual propeller data, e.g. the function $\eta(J)$, the actual engine performance data, $P(h, U)$, and then build a numerical model similar to the one outlined above.

6.8   OPTIMAL SPEEDS FOR LEVEL FLIGHT

We now calculate the speeds relative to minimum drag and minimum power of a generic aircraft, jet- or propeller-driven. Consider an aircraft whose drag equation is parabolic, as described by Eq. 4.12. The drag force on the airplane is

$$D = \frac{1}{2}\rho_o \sigma A U^2 (C_{D_o} + k C_L^2).$$

(6.45)

The speed corresponding to minimum drag is found from the condition that

$$D = \left(\frac{D}{L}\right) L = \left(\frac{D}{L}\right) W$$

(6.46)

is at a minimum. This implies that the glide ratio C_L/C_D is at a maximum. This fraction can be written as

$$\frac{C_L}{C_{D_o} + k C_L^2},$$

(6.47)

which is the function to be minimized. For this purpose, the derivative of Eq. 6.47 is to be derived with respect to the C_L,

$$\frac{\partial}{\partial C_L} \left(\frac{C_L}{C_{D_o} + k C_L^2} \right) = -\frac{C_{D_o}}{C_L^2} + k = 0,$$

(6.48)

$$C_L = \sqrt{\frac{C_{D_o}}{k}}.$$

(6.49)

This is the lift coefficient corresponding to minimum drag, as it can be verified. The corresponding speed will be

$$U_{md} = \sqrt{\frac{2}{\rho_o \sigma} \frac{W}{A} \left(\frac{k}{C_{D_o}} \right)^{1/4}}.$$

(6.50)

An alternative solution is to consider $C_L \sim U^2$, that is: a minimum with respect to C_L will be a minimum with respect to the aircraft speed (and vice versa). The result will be the same, although more elaborate. The conclusion of Eq. 6.50 is that at a given altitude the speed of minimum drag increases with the wing loading and with the profile drag coefficient; it decreases with the increasing lift-induced factor. All other parameters being constant, U_{md} increases with the flight altitude.

The speed corresponding to minimum engine power for the same aircraft is

$$P = TU = DU = \frac{D}{L} WU = \frac{C_D}{C_L} W \sqrt{\frac{2W}{\rho_o \sigma A C_L}} = \frac{C_D}{C_L^{3/2}} \sqrt{\frac{2W^3}{\rho_o \sigma A}}.$$

(6.51)

At a given altitude the terms under square root are constant, therefore the *speed corresponding to minimum engine power is the speed that minimizes the factor* $C_D/C_L^{3/2}$.

Using the same concept as above, the condition of minimum power is found from

$$\frac{\partial}{\partial U} \left(\frac{C_{D_o} + k C_L^2}{C_L^{3/2}} \right) = 0.$$

(6.52)

Using the definition of lift coefficient, with

$$c_1 = \frac{2W}{\rho A}.$$

(6.53)

With some algebra we find

$$\frac{\partial}{\partial U}\left(C_{D_o}c_1^{-3/2}U^3 + kc_1^{1/2}U^{-1}\right) = 0,$$

(6.54)

or

$$3C_{D_o}c_1^{-3/2}U^4 - kc_1^{1/2} = 0.$$

(6.55)

The solution of the latter equation is the speed of minimum power

$$U_{mp} = \sqrt{\frac{2}{\rho_o\sigma}\frac{W}{A}\left(\frac{k}{3C_{D_o}}\right)^{1/4}}.$$

(6.56)

This speed corresponds to a lift coefficient

$$C_L = \sqrt{3\frac{C_{D_o}}{k}}.$$

(6.57)

The value of the parameter $C_D/C_L^{3/2}$ becomes

$$\frac{C_D}{C_L^{3/2}} = \frac{C_{D_o} + kC_L^2}{C_L^{3/2}} = \frac{4C_{D_o}}{C_L^{3/2}} = \frac{4C_{D_o}}{(3C_{D_o}/k)^{3/4}}.$$

(6.58)

The relationship between the C_L of minimum drag and minimum power is simply $\sqrt{3}$, while the ratio between corresponding speeds is

$$\frac{U_{md}}{U_{mp}} = \sqrt[4]{3} \rightarrow U_{md} \sim 1.32U_{mp}.$$

(6.59)

This proves that the speed of minimum drag is about 32% higher than the speed of minimum engine power. Both optimal velocities change with the altitude like the term $1/\sqrt{\sigma}$ (Problem 4).

The variation of drag and power with the aircraft speed is shown in Figure 6.5 for a generic subsonic jet aircraft. The induced component decreases with the speed; the profile drag grows as U^2 and the corresponding power grows as U^3. The sum of the two components has a minimum at an intermediate value of the air speed.

Example

Consider an airplane having wing loading $W/A = 500\,\mathrm{kg/m^2}$, flying at an altitude $h = 6,000\,\mathrm{m}$. The wing area is $A = 180\,\mathrm{m^2}$. Find the speed of minimum drag and minimum power, and plot the drag and power components. In this case, $U_{md} = 102.0\,\mathrm{m/s}$ (Eq. 6.50), $U_{mp} = 78.0\,\mathrm{m/s}$ (Eq. 6.56), and $U_{md}/U_{mp} = 1.31$, not too different from the value given by Eq. 6.59.

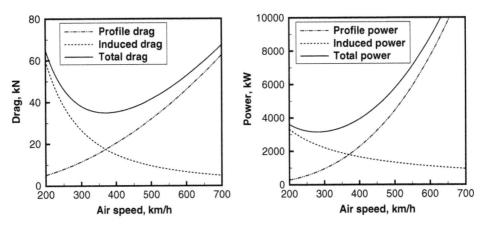

Figure 6.5 *Drag and power characteristics of generic subsonic airplane.*

6.9 GENERAL FLIGHT ENVELOPES

Consider the speed at level flight of a generic subsonic aircraft obtained from the definition of lift coefficient

$$U = \sqrt{\frac{2}{\rho_o} \frac{W}{A} \frac{1}{\sigma C_L}}.$$
(6.60)

The speed limits in level flight are basically set by the wing loading W/A. An analysis of all existing powered aircraft shows that the extreme wing loadings are in the range $W/A = 100 - 1{,}000 \, \text{kgf/m}^2$ ($10^3 - 10^4$ Pa). The lowest W/A causes aircraft stall, and the largest W/A is directly related to the structural limits of the wing.

Figure 6.6 shows a generic flight envelope for a commercial subsonic jet. The graph shows lines of constant CAS and constant Mach numbers. The flight is limited at high speed by the maximum operating speed (VMO), or the maximum operating Mach number, M_{mo}. In general, the aircraft will cruise at a speed slightly lower than this, at the recommended cruise speed (indicated by a thick dashed line). At low speed the envelope is limited by the stalling speed. For safety reasons this speed is higher than that obtainable with fully extended control surfaces.

The ceiling is limited by the cabin pressure. In fact, the cabin pressure is usually maintained to that corresponding to 6,000 m (19,685 ft). The difference in pressure between the cabin and the external atmosphere is

$$\frac{\Delta p}{p_o} = \frac{p_c - p}{p_o} = \frac{p_c}{p} - \delta,$$
(6.61)

where p_o is the sea level standard pressure, p_c is the cabin pressure, and p is the atmospheric pressure at altitude. At 6,000 ft we have $p_c/p_o \simeq 0.80$, therefore

$$\frac{\Delta p}{p_o} = 0.80 - \delta.$$
(6.62)

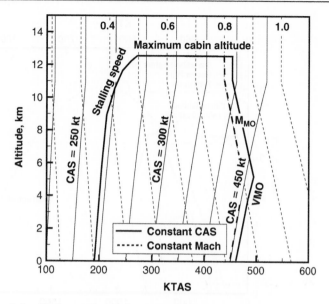

Figure 6.6   *Flight envelope of commercial subsonic jet aircraft.*

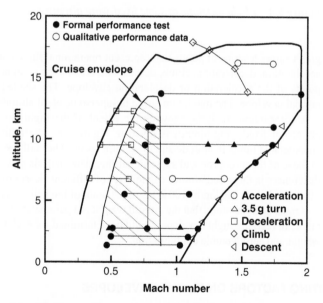

Figure 6.7   *Flight envelope of high-performance jet aircraft.*

This value is important in both design and operation of the aircraft. From the design point of view it will give indication of the loads on the airframe.

Figure 6.7 shows the flight envelope of a high-performance supersonic jet aircraft. The outer envelope is the limit of all operation points from a statistical analysis. The

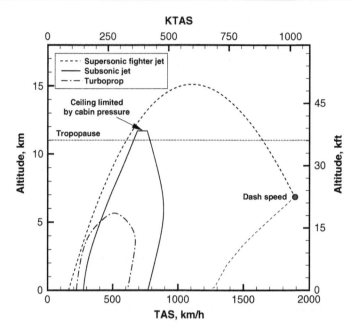

Figure 6.8 *Flight envelope of some fixed-wing aircraft (curves are approximate).*

graph shows a series of points that represent operational flight conditions, including acceleration, deceleration, climb, descent and cruise. The straight lines join extreme points of the acceleration or deceleration envelope. The shaded area indicates the normal envelope. The maximum speed is supersonic at all altitudes.

A comparative analysis, showing the extent of the flight envelopes of different classes of fixed-wing aircraft is given in Figure 6.8. The military fighter jet has considerable limits in altitude and maximum speed. The stall speeds are comparable to those of the turboprop and subsonic airliner. The analysis serves to indicate major differences in the speed/altitude capabilities of different classes of aircraft. The range of speeds below 180 km/h (100 kt) is the domain of the helicopter.

It must be understood that the aircraft weight is a parameter of considerable importance. In fact, the flight envelope should be determined for each representative gross weight and aerodynamic configuration.

6.10 LIMITING FACTORS ON FLIGHT ENVELOPES

The envelopes discussed only take into account the basic aerodynamic and propulsion characteristics, but not other factors, such as structural limits of the aircraft, aero-thermodynamic heating, and other forcing at the extreme flight conditions. Some of the limitations are described in this section, and schematically shown in Figure 6.9. There are four phenomena that need consideration: intake buzz, wing flutter, skin temperature limits and blade stall. The thrust limits are obvious. Other factors, such as cabin pressurization limits, are less so.

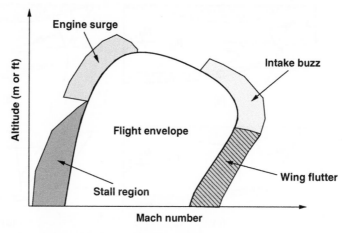

Figure 6.9   *Limiting factors on flight envelopes.*

The *intake buzz* is the interaction between the oblique shock wave at the ramp of the engine inlet and the ramp boundary layer (Seddon and Goldsmith[184]). It creates oscillating conditions inside the intake that assume the shape of forced vibrations. A typical frequency is about 10 Hz. If a shock occurs, it will be formed by two waves: an oblique shock from the external compression flow, and a normal shock. These two shocks meet at a point, and expand outward, as a λ-shock. The cross-sectional area of the engine defined by this shock intersection is called A in Figure 6.10. When buzz starts, the shock intersection point F is inside the cowl lip and a shear flow is established.

If, for a given Mach number, this shear flow is internal to the diffuser, it creates flow separation that reduces the mass captured by the inlet, because the separated flow causes a blockage effect. In the meantime, the compressor runs at constant rpm and tends to suck in all the air at the inlet. The pressure ratio in the engine will decrease. With the pressure p_{i2} decreasing below the stagnation pressure, the shear flow disappears, and so does the blockage effect. Then the process starts again. Some of these problems can be reduced or eliminated by proper intake design (variable geometry).

The *engine surge* is the result of compressor stall in the jet engine. As a result, the complete engine may stall. This is a rare event, appearing first as a loud bang. The air flowing over the compressor blades stalls just as the air over the wing of an airplane. When airfoil stall occurs, the passage of air through the compressor becomes unstable and the compressor can no longer compress the incoming air. The high-pressure air behind the stall further back in the engine escapes forward through the compressor, and out of the inlet. This escape is sudden, like an explosion. Engine surge can be accompanied by visible flames outside the inlet or in the tail pipe. Often the event is so quick that the instruments do not have time to respond. Generally, the instability is self-correcting. In modern engines there are surge valves that pump the disturbed flow out of the engine, and thus limit the instability.

The *wing flutter* is a dynamic coupling between elastic motion of the wing and the unsteady aerodynamic loading. This dynamic response is at first stable, but increases

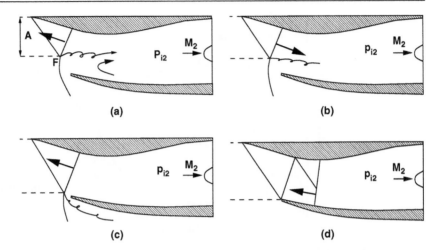

Figure 6.10 *Intake buzz: (a) buzz start; (b) chocking; (c) limit of buzz start; (d) supercritical phase; p_{i2} denotes internal pressure.*

with the Mach number. It depends on the geometry of the wing (aspect-ratio, sweep angle), on its stiffness, on its moment of inertia and some other parameters. This response (frequency and damping) may depend on the acceleration rate of the aircraft. The response can be quantified by the flutter number, which is plotted against the Mach number.

Due to aero-thermodynamic heating created by flight at high speeds (as explained in Chapter 4), high-performance aircraft have a *thermal cost discontinuity*, as shown in Figure 6.11. This means that very high speeds create such as aero-thermodynamic heating that can threaten the structural integrity of the aircraft.

Currently, the $M = 2.5$ speed is considered the practical limit beyond which a change in structural materials is required, from aluminium-based to titanium-based. At this speed, the aero-thermodynamic heating at stagnation point is estimated at about 250°C (see also §4.11).

6.11 DASH SPEED OF SUPERSONIC AIRCRAFT

The next problem deals with the dash speed of an aircraft, whose engine thrust, and transonic drag rise are given in tabulated format from flight testing. These data are shown in Fig. A.8 and Fig. A.9. The flight altitude is fixed.

We will show how the solution of this problem leads to a root finding of non linear algebraic equation. The problem is somewhat complicated, and may lead to non physical and non unique solutions. The conditions that must be satisfied by the aircraft in level flight are:

$$T = D, \quad L = W. \tag{6.63}$$

The solution of the problem is given by the angle of attack that provides the maximum Mach number compatible with Eqs. 6.63. From the equilibrium in the vertical

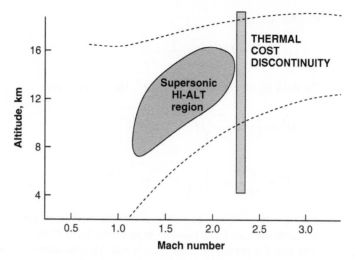

Figure 6.11   *Aero-thermodynamic heating and penetration corridor.*

direction we find the angle of attack and the lift coefficient,

$$\alpha = \alpha_o + \frac{2W}{\rho A a^2} \frac{1}{C_{L_\alpha} M^2}, \tag{6.64}$$

$$C_L = C_{L_\alpha}(\alpha - \alpha_o). \tag{6.65}$$

For the sake of brevity, we will consider $\alpha_o \simeq 0$, but the solution procedure is essentially the same. The equilibrium in the horizontal direction is written as

$$T(M) = \frac{1}{2}\rho A(C_{D_o} + \eta C_{L_\alpha}\alpha^2)U^2 = \frac{1}{2}\rho A a^2(C_{D_o} + \eta C_{L_\alpha}\alpha^2)M^2, \tag{6.66}$$

where $C_{D_o} = C_{D_o}(M)$, $\eta = \eta(M)$, $C_{L_\alpha} = C_{L_\alpha}(M)$. By further simplification, we have

$$T = \frac{1}{2}\rho A a^2 C_{D_o} M^2 + \frac{2W^2}{\rho A a^2} \frac{1}{C_{L_\alpha}} \frac{1}{M^2}. \tag{6.67}$$

Use the coefficients c_1 and c_2 to collect all the constant parameters in the equations

$$c_1 = \frac{1}{2}\rho A a^2, \quad c_2 = \frac{2W^2}{\rho A a^2}. \tag{6.68}$$

Therefore, Eq. 6.66 becomes

$$T = c_1 C_{D_o} M^2 + c_2 \frac{\eta}{C_{L_\alpha}} \frac{1}{M^2}. \tag{6.69}$$

In Eq. 6.69 the only unknown is the flight Mach number. However, the aerodynamic coefficients are also a function of M. Some algebraic simplification from Eq. 6.69 is possible. We have

$$c_1 C_{D_o} M^4 - TM^2 + c_2 \frac{\eta}{C_{L_\alpha}} = 0. \tag{6.70}$$

The latter equation is quadratic in M^2. The solutions are

$$M^2 = \frac{T^2}{2c_1} \pm \frac{1}{2c_1} \sqrt{T^2 - 4c_1 c_2 \frac{\eta C_{D_o}}{C_{L_\alpha}}}, \tag{6.71}$$

$$M^2 = \frac{T^2}{\rho A a^2} \pm \frac{1}{\rho A a^2} \sqrt{T^2 - 4W^2 \frac{\eta C_{D_o}}{C_{L_\alpha}}}, \tag{6.72}$$

and are both positive. The problem with this method is that although the transonic or supersonic speed is in possible in principle, Eq. 6.72 does not given an indication as to how the aircraft can reach that speed. Also, its solution is not straightforward: Eq. 6.72 is implicit in the Mach number, due to the aerodynamic coefficients.

Another suitable method consists in guessing the dash speed and using a Newton-Raphson method around this starting point. This method yields a solution only if it is close enough to the unknown! – In fact, it happens that at the limit Mach number there is still some excess thrust, but the aircraft cannot go past a certain speed, because of heating or buffet limits. Some results are reported in Fig. 6.12, that shows that the residual

$$f(M) = c_1 C_{D_o} M^4 - TM^2 + c_2 \frac{\eta}{C_{L_\alpha}} \tag{6.73}$$

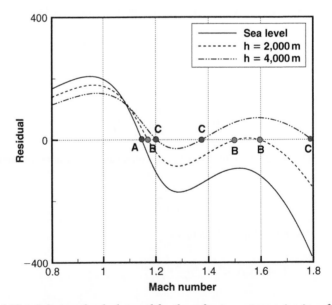

Figure 6.12 *Solutions for dash speed for the reference supersonic aircraft, W = 10,000 kg, at the altitudes indicated.*

is a function of the Mach number. The point A denote the dash speed at sea level (unique); the points B denote the speeds at 2,000 m altitude and the points C are the speeds at 4,000 m altitude. The number of solutions and their value depend on the altitude (as indicated) on the aircraft weight, on the zero-lift angle of attack and on the propulsion characteristics.

In conclusion, the steady-state level flight model is not suitable to calculate the upper limit of the flight envelope of a supersonic aircraft, because of the complexities of the drag equation that lead to multiple solutions, some of which cannot be achieved in level flight.

6.12   ABSOLUTE CEILING OF SUPERSONIC AIRCRAFT

We consider the aircraft in the problem above. The absolute ceiling can be found from the first point at which the drag is tangent to the engine thrust curve. At the absolute ceiling there is only one speed compatible with level flight, therefore from Eq. 6.72 we find

$$T^2 - 4W^2 \frac{\eta C_{D_o}}{C_{L_\alpha}} = 0,$$

(6.74)

or

$$T = 2W \sqrt{\frac{\eta C_{D_o}}{C_{L_\alpha}}}.$$

(6.75)

There is only one positive Mach number at this condition,

$$M = \frac{T}{\sqrt{\rho A a^2}}.$$

(6.76)

By combination of Eq. 6.75 and Eq. 6.76, we find the condition on the Mach number,

$$M = \frac{2W}{\sqrt{\rho A a^2}} \sqrt{\frac{\eta C_{D_o}}{C_{L_\alpha}}} = f(\sigma).$$

(6.77)

We need to solve the non linear Eq. 6.77 for the unknown Mach number (implicit). There are several aspects in the solution of Eq. 6.77. First, the absolute ceiling is given by the minimum density altitude that is compatible with a solution of this equation; second, there are solutions at altitudes below the absolute ceiling.

6.13   SUPERSONIC ACCELERATION

The next problem is to calculate the acceleration of the supersonic jet fighter, model C, from a cruise Mach number (say $M = 0.8$) to supersonic speed. There are different ways to achieve this. One is an acceleration at constant altitude, another is an acceleration at constant angle of attack and constant attitude. Finally, the aircraft can accelerate to phenomenal speeds by doing a zoom-dive. Miele[38] in his book on flight mechanics proposed a method of calculation of accelerated flight at constant engine thrust. Bilimoria and Cliff[185] solved numerically a more general problem of cruise-dash and showed that the trajectories can be a combination of transient and steady state flight. In this context, we will solve only the relatively simple case of acceleration at constant altitude. Any other profile requires a zoom-climb or a zoom-dive, and will be discussed Chapter 8.

6.13.1 Acceleration at Constant Altitude

Examples of constant-altitude accelerations in the flight envelope are shown in Figure 6.9 by lines joining end points of the flight parameters. We now proceed to the calculation of such accelerations. The equation of motion in the flight direction is

$$ m\frac{\partial U}{\partial t} = T - D. \tag{6.78} $$

If we use the definition of speed of sound and rearrange the equation, we find

$$ \frac{\partial M}{\partial t} = \frac{1}{a}\frac{T}{m} - \frac{1}{2}\frac{\rho aA}{m}\left(C_{D_o} + \eta C_{L_\alpha}\alpha^2\right)M^2. \tag{6.79} $$

The angle of attack is established from the equilibrium condition in the vertical direction, $L = W$. This leads to

$$ \alpha = \alpha_o + \frac{2W}{\rho Aa^2}\frac{1}{C_{L_\alpha}M^2}. \tag{6.80} $$

As the aircraft accelerates in the horizontal direction, it must decrease its lift coefficient in order to keep level flight. Solution of the problem requires integration of the ordinary differential. This is efficiently done by using a higher-order integration method, such as Runge-Kutta, discussed previously. Since the acceleration time is relatively small, the aircraft weight can be considered constant.

The computational procedure requires the interpolation of the aerodynamic and propulsive characteristics of the aircraft.

Computational Procedure

- Read aircraft data, aerodynamic and propulsion tables.
- Set initial flight conditions (altitude, Mach number).
- Solve Eq. 6.79 with the condition Eq. 6.80 at current step with Runge-Kutta.
- Calculate other quantities: g-acceleration, specific excess thrust, etc.
- Advance solution to the next step.

For the solution of the differential equation we need to assemble the aircraft forces, as shown in §C.1 in Appendix C.

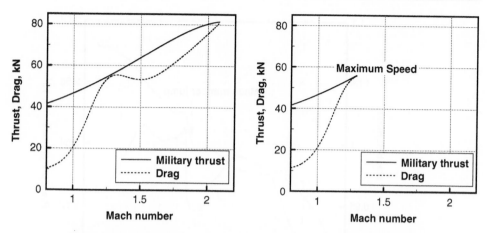

Figure 6.13 *Supersonic acceleration of aircraft model C, constant flight altitude h = 8,000 m, weight W = 10,000 kg (left) and W = 11,000 kg (right). No after-burning thrust, clean configuration.*

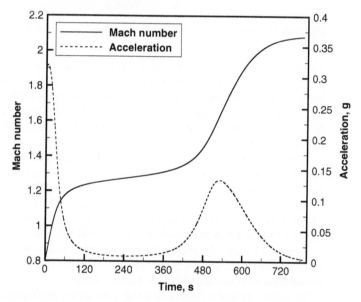

Figure 6.14 *Supersonic acceleration of aircraft model C, as in the case of Figure 6.13 W = 11,000 kg.*

A typical simulation is shown in Figure 6.13 at two different gross weights. It appears evident that, all other conditions being the same, the weight is strategically important in setting the limits on maximum acceleration and supersonic dash. At $W = 10,000$, the drag nearly equals the available thrust, but then the aircraft is capable of going past the $M = 1.25$ limit to reach an $M \simeq 2.1$.

This performance may look extraordinary, but a look at the function $(\partial U/\partial t)/g$, Figure 6.14, shows relatively low accelerations – compared with accelerations in

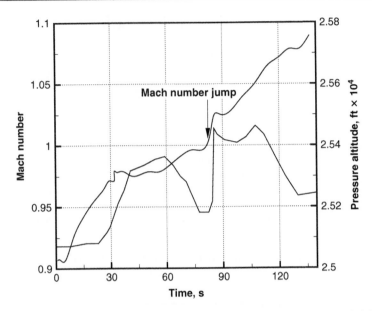

Figure 6.15 *Transonic acceleration of the SR-71, adapted from Moes and Iliff*[186].

high-speed turns and pull-ups. The g-acceleration is about $0.32\,g$ at the start. The acceleration can be greatly improved by using after-burning. A number of different problems can be solved by using the algorithm above: the effects of zero-lift angle α_o, profile drag C_{D_o}, weight W, and altitude h, with and without after-burning. It can be verified that the acceleration time is extremely variable (see Problems at the end of this chapter).

To put these data in perspective, Concorde's best acceleration took the aircraft from $M = 1$ to $M = 2$ in about 7 minutes. After-burning was used in the range $M = 0.96$ to 1.70. In the military arena, the Vought F8U-3 (1958) with after-burning thrust was capable of accelerating from $M = 0.98$ to $M = 2.2$ in 3 minutes and 54 seconds at 11,500 m (about 37,700 ft). The F-15 Strike Eagle is capable of accelerating from $M = 1.1$ to $M = 1.8$ in 56 seconds at about 32,000 ft/9,750 m.

Figure 6.15 shows the transonic acceleration at (nearly constant) altitude of the Lockheed SR-71 (test bed configuration), elaborated from Moes and Iliff[186]. The acceleration was obtained with after-burning thrust.

We finally note that we have used steady state aerodynamic data for the calculations of an accelerated flight performance, and have made no allowance for transient thrust, wing buffet, and other flight dynamic problems.

6.13.2 Other Acceleration Profiles

Another case is the acceleration with dive. The aircraft accelerates to very high speeds while zoom-diving. The speeds that can be achieved are much faster than in level flight.

The final case shown in Figure 6.16 was adapted from Palumbo *et al.*[187] and shows the transonic acceleration of the NASA F15-B aircraft, which was done to study

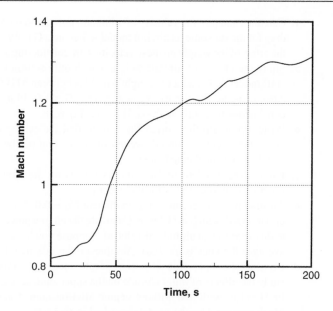

Figure 6.16   *Transonic acceleration at constant dynamic pressure of the NASA F-15B aircraft, adapted from Palumbo et al.[187]*

the transonic capabilities of the aircraft, and to devise experiments for jet engines at transonic speeds. The test was run with a *constant dynamic pressure*, from $M = 0.7$ to $M = 1.5$. Maintaining constant $q = \rho U^2/2$ is difficult, because the pilot does not have a direct reading of q. The acceleration profile is not the fastest, but one with a dynamic pressure constraint (see Problem 13).

PROBLEMS

1. Calculate the impact pressure and the dynamic pressure for a generic static pressure p. Plot these quantities versus the Mach number, and establish the Mach number at which the difference is no longer negligible.

2. Consider the aircraft model A. For this aircraft:
   (a) Study the effects of aircraft weight on the flight corridor, at flight altitude $h = 7{,}000$ m and $h = 10{,}000$ m.
   (b) Study the effects of altitude on the flight corridor with the aircraft weight $W = 0.85$ MTOW.
   (c) Study the effect of throttle position, with $\Pi = 0.5$ to 1.0 on the flight corridor, at $h = 7{,}000$ m, $h = 10{,}000$ m and $W = 0.85$ MTOW.

   For these cases it is useful to work out the solution on a spreadsheet, or on a program with graphical capabilities (MatLab or other).

3. Repeat the analysis of Problem 2 with the turboprop aircraft model B (Appendix A). In this case consider as flight altitudes $h = 4{,}000$ m and $h = 7{,}000$ m.

4. Calculate and plot the velocity of minimum power and the velocity of minimum drag for the subsonic jet aircraft model A, having a GTOW $= 130,000$ kg. Study the effect of the weight on these velocities, by considering a $\pm 10\%$ change on W.

5. Study the effects of aircraft weight and throttle setting on the ceiling of aircraft model A. Consider the eight variables between MTOW and MTOW-PAY. Also, consider the throttle variable $0.75 \leq \Pi \leq 1.0$. Plot the results and draw conclusions from this analysis. Hint: use Eq. 6.38.

6. Write a program that solves Eq. 6.44 to find the ceiling of aircraft model B. Study the effects of aircraft weight on the ceiling performance. Plot the data and draw a critical conclusion.

7. Discuss the problem of *buzz* and its effects on the aircraft speed. Do some research to find methods for its reduction.

8. An airship has an average drag coefficient $C_D = 0.025$ at zero pitch and normal cruise speed, which is 50 km/h. Calculate the engine power required to navigate at this speed, at an altitude of 1,000 m. Suppose you have two propellers, whose average efficiency is $\eta = 0.81$. Assume a reference area $A = 50\ m^2$.

9. Consider the supersonic jet aircraft of reference (Appendix A). For this aircraft study the effects of flight altitude on the supersonic acceleration from $M = 0.8$ to $M = 1.6$, with and without engine after-burning. Calculate the maximum Mach number. Use the model described in §6.13.1.

10. For the supersonic aircraft of reference, study the effects of aircraft gross weight on the supersonic acceleration, at a flight altitude $h = 12,000$ m. Use the model described in §6.13.1.

11. For the supersonic aircraft of reference, study the effects of zero-lift angle α_o on the supersonic acceleration at constant altitude. Consider a gross weight $W = 10,000$ kg, and an altitude $h = 10,000$ m.

12. Consider a modern supersonic jet fighter. Investigate the effects of the supersonic acceleration on intake buzz and wing buffet. (*Problem-based learning: additional research is required*).

13. Consider a transonic acceleration through the speed of sound from $M = 0.8$ and altitude $h = 12,000$ m of the reference supersonic aircraft (Appendix A). Assume that the aircraft has a gross weight $W = 10,000$ kg. Express the conditions required to accelerate at constant dynamic pressure $q = \rho U^2/2$.

14. Find a numerical method to calculate the absolute ceiling for the supersonic aircraft of reference (model C), by starting from Eq. 6.75. Choose the aircraft weight, and use the tabulated data for the aerodynamic and propulsive characteristics of the aircraft, see Tables A.12 and A.13.

Chapter 7

Take-off and Landing

So we'll open the throttle and gather speed,
For the Airman's life is life indeed.
As the engine roars away,
Gone are the cares of the Earth below.
As up through the clouds we go,
With an even-increasing sway.
                    A.C. Kermode[188], 1956

The words of Kermode, in his classical book on flight mechanics, sound eerie to the modern high-flyer. Take-off and landing are nowadays smooth operations servo assisted by on-board computers. Landing still requires good pilot skills, which are best appreciated in bad weather conditions and at night. The first instrumented landing, by James Doolittle in September 1929, is now a forgotten piece of aeronautical history.

Take-off and landing are defined as terminal phases of the aircraft flight, and are accelerated or decelerated flights, respectively. Both operations are affected by prevailing winds. As it will be shown, it is convenient to take off and land in the wind. Side winds can affect the stability of the aircraft.

In this chapter we will solve take-off and landing problems of conventional aircraft on horizontal runways. Cases other than these will be considered in Chapter 16, which deals with V/STOL vehicles. Take-off and landing are heavily regulated by the aviation authorities. Each regulation reports different lift-off C_L, initial gradient, height clearance, etc. For example, in North America the regulations for conventional, utility, acrobatic, and commuter-category airplanes are FAR Title 14, Part 23, Subpart B, of the airworthiness standards.

7.1 DEFINITION OF TERMINAL PHASES

There are five different types of aircraft take-off and landing. The take-off distance and time are the distance and time required to lift off, climb and clear a screen at fixed height.

CTOL denotes Conventional Take-off and Landing. This class includes most civil, commercial and military vehicles, e.g. subsonic jet transport, turboprop airplanes, and cargo airplanes, which take off and land horizontally. These airplanes operate from normal airfields with paved runways, have moderate thrust ratios or power loading, and moderate wing thickness and wing sweep.

The term *RTOL* refers to the Reduced Take-off and Landing vehicles. Compared to the previous class, RTOL vehicles have higher thrust ratios, complex high-lift systems, full span flaps, boundary layer control and thrust reversal systems. They may have parachutes for aerodynamic braking.

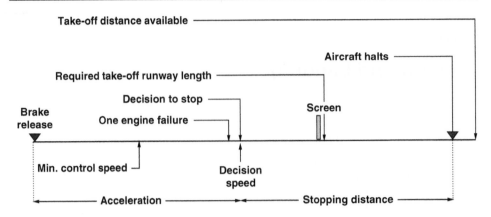

Figure 7.1 *Balanced field length, definitions and nomenclature.*

The term *STOL* refers to the Short Take-off and Landing vehicles. Aircraft in this class may have powered lift systems, vectored thrust, after-burning jet engines, active boundary layer control, wing-lift augmentation (vortex lift) and more advanced aerodynamics. They take off from a ramp, and land on elastic springs, although they are capable of horizontal take-off and landing. V/STOL (or VSTOL) denotes aircraft with both vertical and short take-off and landing.

VTOL refers to the Vertical Take-off and Landing vehicles. Aircraft in this class are supported by direct-lift, either using propellers (helicopters, tilt-rotors) or jet thrust. They do not need runways, and can operate from airfields nearly anywhere. One family of vehicle in this category includes the airship, which will not be possible to cover in this book. Essential performance analyses of airships in all categories of weight are available in Lancaster[189].

STOVL refers to the Short Take-off and Vertical Landing vehicles. Airplanes in this class include hybrid vehicles, such as the BAe Sea Harrier.

The ICAO defines a number of take-off safety factors:

- *TODA* is the Take-off Distance Available. This is the length of the runway available for a take-off run plus a clearway, to take into account such possibilities as aborted take-off.
- *TORA* is Take-off Run Available. This is the length of runway available and suitable for the ground run.
- *BFL* is the Balanced Field Length, e.g. the balance between the distance required to continue the take-off to clear a screen at 35 feet (11.5 m), and the distance required to stop the aircraft on the runway in case of aborted take-off. The basic definitions for the BFL are shown in Figure 7.1.

These lengths, and other relative safety factors are discussed in some detail by Eshelby[190]. Of particular interest for the calculation and requirements of the balanced field length for commercial and civil aircraft is Torenbeek[74]. An estimate of balanced field length for a number of commercial CTOL vehicles is shown in Figure 7.2, calculated at their MTOW.

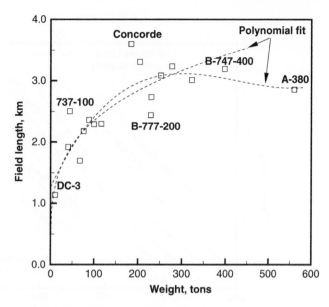

Figure 7.2 *Take-off field length of commercial airplanes. Polynomial fit of the data with and without A-380.*

The field length of the A-380 is actually lower than that of the Boeing 747-400, although there is a leap in take-off gross weight. A detailed chart of the take-off field requirements for the Boeing B-747-400 is shown in Figure 7.3, as elaborated from Boeing's data[63]. The figure shows a set of curves, each corresponding to a pressure altitude. Two limits are indicated: the maximum take-off weight and the maximum tire speed.

Table 7.1 gives some take-off distances for high-performance aircraft. Column A refers to the aircraft in combat configuration, which has a weight below the MTOW. Column B indicates the take-off run at MTOW, for which many data are unknown. The Sukhoi S-34 uses after-burning, and its take-off is quoted from a ramp on an aircraft career.

7.2 CONVENTIONAL TAKE-OFF

A conventional take-off operation starts with the aircraft at rest on the runway, Figure 7.4. The take-off then consists of: (1) an acceleration on the runway; (2) a lift-off of the forward wheels; (3) a rotation of the aircraft; and (4) an initial stage of airborne time with an initial climb gradient to clear an imaginary screen at a conventional height. Therefore, a complete take-off operation consists of a ground run and an airborne phase. The angle of attack is constant during the ground run.

As the aircraft accelerates on the ground, directional control is achieved by steering the forward wheels. When a large enough speed is reached, the control is shifted to the rudder. When the rudder is capable of providing enough yawing moment for lateral control, it is said that the aircraft has rudder effectiveness. The speed at which rudder

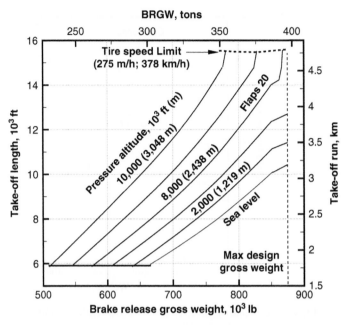

Figure 7.3 *FAR take-off field length requirements for Boeing B-747-400, ISA conditions, CF-80C2B1 engines. No runway gradient, no wind.*

Table 7.1 *Take-off runs of some military aircraft. Configuration A is for combat operations; configuration B is at MTOW.*

Aircraft	Config. A (km)	Config. B (km)
Fairchild A-10	0.44	1.22
Sukhoi S-34	0.12	n.a.
Eurofighter 2000	0.30	n.a.
MiG 29	0.65	n.a.
Lockheed F-16C	0.76	n.a.

effectiveness starts is called *minimum control speed*. If engine failure occurs at a speed below the control speed, the take-off will have to be aborted. Just before reaching the lift-off speed, the aircraft is rotated, e.g. the forward wheels are raised from the ground and the aircraft assumes the attitude of the initial climb. The minimum speed at which the aircraft becomes airborne is called *minimum unstick speed*, and denotes the condition of maximum rotation on the ground, with a safety margin for tail strike. The speed at which rotation of the aircraft is done is called *rotation speed*. At lift-off the aircraft becomes airborne, and must accelerate to a safe climb speed (*minimum control speed, airborne*). This speed depends on whether the aircraft is operating with all engines or with one engine failure. The safe air speed is defined for the most restrictive condition of OEI. It is the minimum air speed that assures a safe climb

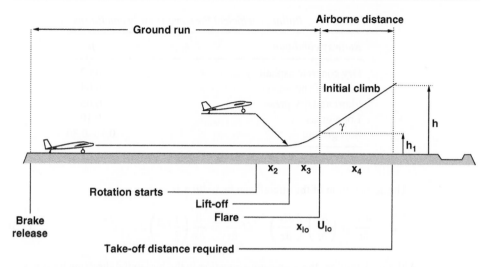

Figure 7.4   *Phases of conventional take-off.*

and directional control of the aircraft with the rudder. In fact, the rudder should be capable of providing the yawing moment required to balance the thrust asymmetry created by OEI.

7.3   GROUND RUN OF JET AIRCRAFT

The ground run is the distance between brake release and lift-off of the forward wheels. The calculations below refer to the condition *All Engines Operating* (AEO), unless otherwise specified.

To calculate the ground run we write the dynamics equations on the center of gravity of the aircraft in the horizontal and vertical direction, for a take-off from a horizontal runway. The engine thrust is aligned with the vector velocity. The equations are, respectively

$$m\frac{\partial U}{\partial t} = T - D - R, \tag{7.1}$$

$$R = \mu(L - W), \tag{7.2}$$

where R is the ground resistance, and μ is a coefficient depending on the runway conditions, tire pressure and tire conditions. In general, it is referred to as a constant depending only on the runway conditions. However, it is possible that this coefficient increases at speeds $U > 200$ km/h (108 kt). Investigations on rolling friction coefficients have been carried out over the years. See, for example, Wetmore[191], Harrin[192], Yager[193] and Agrawal[194]. Kuchinka[195] solved the take-off and landing problem with focus on the dynamic response of the tires. Several sources provide the average data to be used in ground roll calculations. Some of these data are given in Table 7.2.

Table 7.2 *Rolling coefficient for some runway conditions.*

Runway condition	μ
Dry concrete/asphalt	0.02
Hard turf and gravel	0.04
Short and dry grass	0.05
Long grass	0.10
Soft ground	0.10–0.30

The acceleration of the aircraft can be written as

$$a = \frac{\partial U}{\partial t} = \left(\frac{\partial x}{\partial t}\right)\left(\frac{\partial U}{\partial x}\right) = U\frac{\partial U}{\partial x} = \frac{\partial}{\partial x}\left(\frac{1}{2}U^2\right). \tag{7.3}$$

With this definition, the momentum equation in the horizontal direction becomes

$$\frac{1}{2}m\frac{\partial U^2}{\partial x} = T - D - R. \tag{7.4}$$

Integration of the above, performed between brake-release ($x = 0$, $t = 0$) and the lift-off point, yields

$$\frac{1}{2}mU_{lo}^2 = \int_o^x (T - D - R)\,dx. \tag{7.5}$$

Integration of the right-hand side of Eq. 7.5 requires additional information as to how the aircraft forces depend on the speed. For example, the front wheels raise from the ground before the rear wheels. When this occurs, there is a step change in the rolling resistance.

First, consider the lift-off speed. In principle, this speed can be estimated from the stall speed. In practice, it is allowed to have $U_{lo} \simeq 1.1 U_{stall}$; this implies that $C_{L_{lo}} \simeq 0.83 C_{L_{max}}$ – a reasonably safe margin. Therefore, the lift-off speed is

$$U_{lo}^2 = \frac{2}{\rho}\frac{W}{A}\frac{1}{C_{L_{lo}}}. \tag{7.6}$$

The take-off time is obtained by integration of the speed,

$$t_{lo} = \int_o^{x_{lo}} \frac{dx}{U}. \tag{7.7}$$

FAR Part 25 (§25.103) defines the stall speed as

$$U_{stall} \geq \frac{U_{CLmax}}{\sqrt{n}}, \tag{7.8}$$

where n is the load factor in the direction normal to the flight path, U_{CLmax} is the calibrated airspeed obtained when the load factor-corrected lift, nW/qA, is a maximum. The regulations prescribe the methods that must be used to calculate the U_{CLmax}. A stall warning must be clear at all critical flight conditions.

7.4 SOLUTIONS OF THE TAKE-OFF EQUATION

For a first-order solution of Eq. 7.5 we assume that the average thrust is much larger than the other forces, e.g. $\overline{T} >> \overline{D} + \mu \overline{R}$,

$$x_{lo} \sim \frac{1}{2} \frac{m U_{lo}^2}{T_o}. \tag{7.9}$$

By introducing the thrust ratio into Eq. 7.9, with the lift coefficient at take-off (not necessarily equal to maximum value), we find

$$x_{lo} = \frac{1}{\rho_o g} \frac{1}{\sigma} \frac{W}{T_o} \frac{W}{A} \frac{1}{C_{L_{lo}}}. \tag{7.10}$$

Therefore, in first approximation, the lift-off distance is proportional to W/A, and inversely proportional to T/W and $C_{L_{lo}}$. The lift-off run is also affected by the atmospheric conditions, which appear in the factor $1/\sigma$. At ISA conditions this means that take-off from airports at altitude requires a longer field length. For example, the ratio between lift-off lengths at Denver International Airport (altitude $h \simeq 1,600\,\text{m}$) and sea level, for the same airplane at the same take-off weight, would be

$$\frac{x_{lo}}{x_{lo}^{S/L}} = \frac{1}{\sigma} \simeq 1.16. \tag{7.11}$$

Typical thrust ratios to be used in Eq. 7.10 for jet-powered aircraft at $T/W = 0.22$–0.42 for commercial jets and $T/W = 0.40$–0.80 for military fighter aircraft.

A second-order approximation is to take an average value of the thrust, drag and rolling resistance over the ground run. Under these circumstances, the result of the integration is

$$\frac{1}{2} m U_{lo}^2 = (\overline{T} - \overline{D} - \overline{R}) x_{lo}. \tag{7.12}$$

Suitable values of the parameters in Eq. 7.12 are:

$$\overline{T} = T_o, \quad \overline{D} = \frac{1}{2} \rho A C_D \left(\frac{U_{lo}}{2}\right)^2, \quad \overline{R} = \mu (W - \overline{L}) \simeq \frac{1}{2} \mu W. \tag{7.13}$$

By replacing the above equivalences in Eq. 7.12, we find

$$\frac{1}{2} m U_{lo}^2 = \left(T_o - \frac{1}{8} \rho A C_D U_{lo}^2 - \frac{1}{2} \mu W\right) x_{lo}, \tag{7.14}$$

that when solved in terms of x_{to} yields

$$x_{lo} = \frac{4mU_{lo}^2}{8T_o - \rho A C_D U_{lo}^2 - 4\mu W}. \tag{7.15}$$

For the third and more refined solution consider the thrust independent of the speed, and equal to the static thrust. Express the drag force as

$$D = \frac{1}{2}\rho A C_D U^2 \sim \frac{1}{2}\rho A (C_{D_o} + kC_L^2)U^2, \tag{7.16}$$

where the C_L and C_D are relative to the ground configuration of the aircraft. At this point we note that in the aircraft's ground configuration the control surfaces are extended, therefore the C_L can be considerably higher than the corresponding value at cruise conditions. Likewise, the C_D is higher, because of a combination of additional high-drag high-lift devices fully deployed. For reference, Torenbeek[74] provides an empirical relationship to estimate the change in drag coefficient as a function of the flap position, on the wing loading and the aircraft weight during a take-off run for commercial airplanes. This increase can be estimated between 0.013 and 0.024. Haftmann et al.[196] provide data for the Airbus A300-600 (these data are also estimated for aircraft model A, in Appendix A.).

The ground effect may come into play with further changes in the aerodynamic characteristics. These can only be evaluated in the wind tunnel, because of the complexity of the interference between the aircraft and the ground. For a CTOL aircraft, ground effect is negligible at ground clearances above half the wing-chord at the root, and may not be significant at clearances below 1/4 of the chord. For aircraft such as the McDonnell DC-10 the ground effect decreases the effective angle of attack by one degree at $C_L \sim 1.0$, as reported by Callaghan[102].

Back to the analysis of the momentum equation, the rolling resistance will be:

$$R = \mu \left(W - \frac{1}{2}\rho A C_L U^2 \right), \tag{7.17}$$

with W a constant term. The solution is

$$\frac{1}{2}mU_{lo}^2 = \int_o^x \left[T_o - \frac{1}{2}\rho A (C_{D_o} + kC_L^2)U^2 - \mu W + \frac{1}{2}\mu\rho A C_L U^2 \right] dx. \tag{7.18}$$

Simplify Eq. 7.18,

$$\frac{1}{2}mU_{lo}^2 = (T_o - \mu W)x_{lo} - \frac{1}{2}\rho A \int_o^x (C_{D_o} + kC_L^2 + \mu C_L)U^2 \, dx. \tag{7.19}$$

The C_L to be introduced in Eq. 7.19 is a function of the angle of attack of the aircraft (see §4.2), but not the speed. The C_L can be changed by operating on the control surfaces. A suitable approximation is that these surfaces are set to take-off positions, so that the lift coefficient in ground effect is constant. Therefore, Eq. 7.19 becomes

$$\frac{1}{2}mU_{lo}^2 = (T_o - \mu W)x_{lo} - \frac{1}{2}\rho A(C_{D_o} + kC_L^2 + \mu C_L)\int_o^x U^2 \, dx. \tag{7.20}$$

The last step in solving the dynamics equation is the knowledge of the speed $U(x)$.

The acceleration of the aircraft can be considered constant. If an acceleration $a(U) = a_o + c_1 U^2$ is assumed, then it can be proved that if the total acceleration during the ground roll does not exceed 40% of its initial value at lift-off, the "exact" and "constant" acceleration solutions for the ground distance differ by about 2% – by all means an acceptable error from an engineering point of view. By using a constant acceleration, the relationship between speed and time is quadratic. In fact,

$$a = \frac{\partial U}{\partial t} = \frac{\partial U}{\partial x}\frac{\partial x}{\partial t} = \frac{1}{2}\frac{\partial U^2}{\partial x} = \text{const.} \tag{7.21}$$

Integration of the above leads to $U^2 \simeq c\sqrt{x}$, bar a constant of integration. In order to match the lift-off conditions $\{x_{lo}, U_{lo}\}$, the relationship between speed and ground run must be

$$U^2(x) = \frac{U_{lo}^2}{x_{lo}}x. \tag{7.22}$$

With this result, the solution of the integral term in Eq. 7.20 is straightforward,

$$\int_o^x U^2 \, dx = \frac{1}{2}U_{lo}^2 x_{lo}. \tag{7.23}$$

By introducing this expression in the integral of Eq. 7.20, the final result for the ground run is

$$\frac{1}{2}mU_{lo}^2 = (T_o - \mu W)x_{lo} - \frac{1}{4}\rho A(C_{D_o} + kC_L^2 + \mu C_L)U_{lo}^2 x_{lo}, \tag{7.24}$$

$$x_{lo} = \frac{mU_{lo}^2/2}{(T_o - \mu W) - \rho A(C_{D_o} + kC_L^2 + \mu C_L)U_{lo}^2/4}. \tag{7.25}$$

Comparison between the three approximated methods is shown in Figure 7.5, which has been calculated for the subsonic commercial jet of reference (model A, Appendix A). For a given GTOW, the estimates of x_{lo} vary greatly. The solution of Eq. 7.25, supposedly the most accurate, yields intermediate values of the ground run.

The fourth and final method of solution of the take-off equation consists in using the actual thrust and the actual drag. The engine thrust is given in tabulated form in charts. In general, these data can be replaced by polynomial functions of the second or third order, e.g. functions such as

$$T(U) = T_o(1 + c_1 U + c_2 U^2 + c_3 U^3), \tag{7.26}$$

with c_i constant coefficients. Often a polynomial of second order will be good enough for this purpose. We will illustrate the solution method for aircraft model C (supersonic

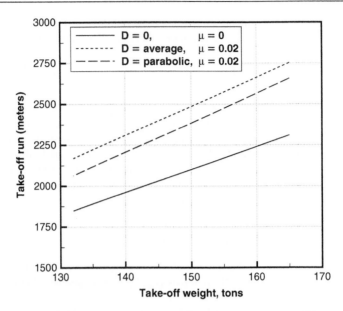

Figure 7.5 *Take-off run estimate for aircraft model A; runway conditions $\mu = 0.02$, take-off thrust $T = 248\,kN$, sea level, ISA.*

jet fighter), because we have the functions $T(U), D(U)$ in tabulated form (Table A.12). The take-off Eq. 7.5 will be written as

$$\frac{\partial U}{\partial t} = \frac{1}{m} \int_o^x (T - D - \mu R)dx. \tag{7.27}$$

The most straightforward method of integration of Eq. 7.27 is to use a Euler method (first-order approximate). More elaborate methods require the use of fifth-order Runge-Kutta methods with variable time stepping, which are stable and accurate. These techniques are now of standard use in engineering, and several software packages offer the integration routines as a black box. A suitable algorithm, shown by Krenkel and Salzman[197], used such an integration scheme to calculate the take-off performance and the balanced field length. Two alternative calculation methods (with constant or variable thrust, with or without head-winds) are given by Powers[198]. Our algorithm is summarized below.

Computational Procedure

- Read all the aircraft data.
- Set the initial conditions, $t = 0$, $U = 0$, $C_L = 0$, etc.
- Set the value of the integration step dt.
- Calculate the current value of the acceleration from

$$a_i = \frac{1}{m}(T_i - D_i - \mu R_i). \tag{7.28}$$

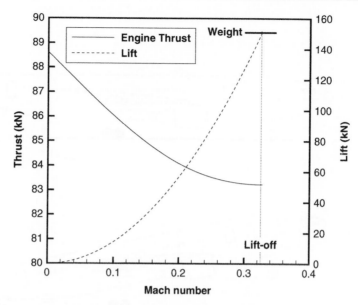

Figure 7.6 *Take-off forces and take-off run of aircraft model C.*

The actual values of the thrust $T_i = T(U_i)$ and the drag force $D_i = D(U_i)$ are found by interpolation of the tables. A cubic spline is generally more accurate than a linear interpolation.

- Calculate current value of the speed and position x_i on the runway

$$U_i \simeq U_{i-1} + a_i dt_i, \quad x_i \simeq x_{i-1} + U_i dt_i. \tag{7.29}$$

- Advance the solution to the next step.

The stopping criterion is the lift-off point, which we assume occurs when $L = W$. The C_L for a ground run can be considerably different than the C_L for climb take-off segment. Since these data are not given anywhere in our tables, we need to make estimates. We use an average $C_L \sim 0.68$ in ground effect (a relatively low value).

The solution procedure requires a few input parameters: the altitude of the airport, the angle of attack during the ground run, the runway conditions, the take-off gross weight, the after-burning option on the engine, the throttle position. We have performed a solution at sea level, with $GTOW = 15{,}500$ kg, $\mu = 0.02$, after-burner off, full throttle. The results are shown in Figure 7.6, which indicates that the take-off run is relatively long for this type of aircraft, about 1.3 km, in about 22.5 seconds, with a speed $U_{lo} \simeq 111.5$ m/s.

In practice, the ground run can be reduced considerably. In fact, aircraft similar to the one under discussion have ground runs of 0.6 to 0.7 km, as shown in Table 7.1 for the Lockheed F-16C. For example, with a $C_L \simeq 1.2$ that take-off run is reduced to 0.81 km, and with $C_L \simeq 1.5$ $x_{lo} \simeq 0.68$ km, corresponding to less than 17 seconds of ground roll.

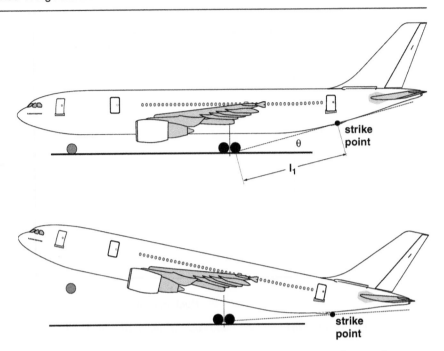

Figure 7.7 *Tail strike at the take-off flare; l_1 is an approximate distance between the point of contact with the ground and the main landing gear.*

There are alternative methods for the solution of the take-off equation, see for example, Vinh[199].

7.5 ROTATION AND INITIAL CLIMB

The lift-off point is reached with at least one landing gear on the ground. The rotation of the aircraft must be done with a small angle, or else there is a risk of a tail strike on the runway. We report that tail strike is not unusual with a wide-body aircraft, with potentially lethal consequences (see Problem 12). Figure 7.7 shows that the tail strike depends on the geometrical configuration of the aircraft, in particular the landing gear height h_g and the distance between the center of the wheels and the strike point, l_1.

Pinsker[200] proposed the solution to a number of problems that occur at lift-off, including conditions for tail strike and minimum ground clearance required for simultaneous banking and pitching. In Pinsker's analysis, the limiting rotation depends on the $C_{L_{lo}}$.

At the start of the lift-off there is an abrupt change in rolling resistance. The point at which the front wheels lift off is more difficult to calculate, because a number of other factors intervene, such as the weight distribution on the landing gear, the position of the aerodynamic center and the center of gravity, see Figure 7.8.

The airborne phase starts with a further rotation of the aircraft, to assume the initial angle of climb. Then the aircraft climbs along a straight path in order to clear

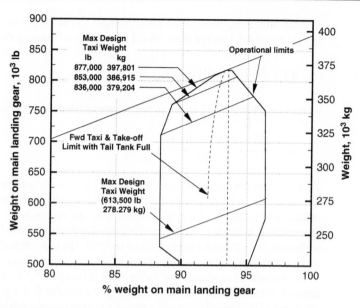

Figure 7.8 *Load on main landing gear of the Boeing B-474-400, elaborated from Boeing's data.*

the reference screen. During flare the aircraft has a centripetal acceleration

$$n = \frac{U^2}{\chi},$$

(7.30)

where χ is the radius of curvature of the flight path. The distance on the ground run during this phase will be $x_3 \simeq \chi\gamma$, and the height reached will be

$$h_1 = \chi(1 - \cos\gamma) \simeq \chi\frac{\gamma^2}{2},$$

(7.31)

where γ is the small climb angle. A first-order estimate for a commercial subsonic jet indicates that $h_1 \simeq 1$ m (3 ft) and the distance covered is of the order of 12 m (40 ft), which is negligible in most cases. The initial climb is done with a constant climb angle, so that

$$x_4 = \frac{h - h_1}{\tan\gamma} \simeq \frac{h}{\tan\gamma}.$$

(7.32)

The total take-off distance will be

$$x_{to} \simeq x_{lo} + x_4.$$

(7.33)

An alternative calculation of the airborne phase can be done with an energy method. The change in total energy from lift-off to the point of clearing the screen is

$$Wh + \frac{1}{2}m\left(U^2 - U_{lo}^2\right) = (\overline{T} - \overline{D})x_4, \tag{7.34}$$

where U is the minimum airborne control speed. Therefore, the airborne distance is

$$x_4 = \frac{Wh + m\left(U^2 - U_{lo}^2\right)/2}{\overline{T} - \overline{D}}. \tag{7.35}$$

There are a number of other cases of interest. The take-off performance of propeller-driven aircraft follows the methods discussed above. Some methods have been discussed by Gasich[201]. Take-off under icing conditions has been studied extensively. One practical example is van Hengst[202], who also provides some lift curves with and without de-icing fluids for the Fokker 50 and 100. The take-off performance with iced surfaces requires at least the knowledge of the correct aerodynamic coefficients.

The take-off of seaplanes and flying boats is complicated by the additional resistance of water and water waves. Essentially, we need to find a good estimate of the resistance of the aircraft. This is the sum of the aerodynamic and hydrodynamic drag. Relevant studies on this subject are those of Perelmuter[203] and Parkinson *et al.*[204] DeRemer[205] showed take-off measurements from shallow lakes of the Cessna 180.

7.6 TAKE-OFF WITH ONE ENGINE INOPERATIVE

Safety requirements dictate conditions with which airplanes have to comply at take-off if one engine fails in the most critical part of the flight (One Engine Inoperative, OEI). Consider a twin-engine airplane that has one engine failure in this phase. To be able to climb to a safe altitude, it must have 100% excess thrust, e.g. it must be able to climb (albeit at a lower climb rate) with the remaining engine. Likewise, a three-engine airplane must be able to climb with two engines (50% excess thrust) and a four-engine airplane must climb with three engines (25% excess thrust). ICAO regulations give the minimum climb angle that these airplanes must retain when one engine fails. Table 7.3 shows these limits. The thrust ratios for each engine configuration are the minimum required for OEI take-off. For example, a four-engine aircraft losing one engine must be able to provide a minimum thrust ratio of 0.18 with the remaining three engines operating at emergency thrust; in this case the aircraft must be able to

Table 7.3 *Thrust ratios and minimum climb OEI gradients.*

	4 engines	*3 engines*	*2 engines*	
T/W	0.24	0.18	0.16	0.12
Gradient	14.1%	8.0%	6.0%	2.0%
Min. gradient	3.0%	3.0%	2.7%	2.4%

climb with a minimum gradient of 3%, although in normal conditions the gradient would be 8%. Due to the lower gradient of the initial climb of an aircraft taking off with OEI, the ground distance required to clear the 35 feet screen is longer.

7.7   CALCULATION OF THE BALANCED FIELD LENGTH

To improve on the results obtained with approximate methods, in this section we calculate the balanced field length of the aircraft model A with All Engines Operating (AEO) and One Engine Inoperative (OEI). In the former case calculation is performed to the point needed for the aircraft to clear a 35 ft (11.5 m) screen.

If the pilot decides to abort the take-off after a critical engine failure, there is a time lag between engine failure and the decision to start braking. The time lag allowance is generally 3 seconds. Also, we need to know the amount of thrust remaining with OEI. From the considerations presented in Chapter 5, we know that the operating engines can augment the thrust they can deliver for a limited amount of time. The worst possible scenario is that failure occurs just before lift-off. At this point two options are possible. First, the pilot decides to take-off with the emergency power and clear the screen at 35 ft (commercial vehicle) or 50 ft (military vehicle); second, the pilot decides to abort the take-off 3 seconds after detecting the failure. In this case, there must be enough field length to take the aircraft safely to a halt. Engine failure at low speed does not provide enough thrust to accelerate the aircraft, lift off and clear the screen. Engine failure close to the point of lift-off is more critical, because the pilot must decide quickly whether to continue the take-off or abort. More details are available in Torenbeek[74]. The computational model for the balanced field length is the following.

Computational Procedure

- Read aircraft data (weight, wing area, etc.).
- Read thrust curves, and aerodynamic characteristics, in- and out-of-ground effect.
- Set environmental parameters (runway conditions, altitude, braking friction, wind speed).
- Solve take-off for AEO,
    1. $U = 0$ to $U = U_r$ (rotation speed);
    2. $U = U_r$ to $U = U_{lo}$ (rotation for specified time);
    3. $U = U_r$ to U_{obs} (climb to obstacle's height).
- Solve take-off with OEI,
    1. Guess critical engine failure velocity;
    2. Calculate both OEI take-off and aborted take-off for the two initial guesses of the failure speed;
    3. The distance to complete stop increases with the speed of engine failure. The distance to take-off decreases with the increasing speed of engine failure. At some point, U_{crit}, these curves intersect, and provide the BFL. We use a bisection method to find the intersection to converge to a solution, so that the BFL definition is satisfied.

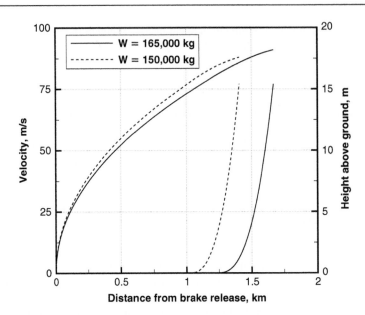

Figure 7.9 *Take-off of Airbus A-300 at the weights indicated; sea level ISA conditions, no wind, dry runway.*

This routine is relatively elaborate, because it integrates the momentum equations for the aircraft after lift-off. This is a climb problem, as discussed in Chapter 8. However, a simplification can be assumed, as indicated in Figure 7.4: the initial climb is straight and at constant speed. A solution to the problem is shown in Figure 7.9, which shows the normal take-off performance. For the case of *GTOW* = 165,000 kg the output is

```
All Engines Operating (AEO) - Normal operation
      Rotation Velocity = 243.83 km/h
      Lift-off Velocity = 268.45 km/h
      Rotation Distance = 881.62 m
      Lift-off Distance = 1094.94 m
   Distance to obstacle = 1308.96 m
Velocity over obstacle = 284.00 km/h
          Rotation Time = 24.94 s
          Lift-off Time = 27.94 s
       Time to obstacle = 30.72 s

      Take-off run     Xto = 1308.95 m
           Take-off time = 30.72 s

One Engine Inoperative (OEI) - Balanced Field Length
      Critical Velocity = 225.97 km/h
      Decision Velocity = 235.42 km/h
      Critical Distance = 744.80 m
      Decision Distance = 937.10 m
```

7.8   GROUND RUN OF PROPELLER AIRCRAFT

The method of calculation follows closely that of the jet aircraft, although there are a number of complications, due to the rotation of the propeller, its pitch and efficiency. The balance of forces along the horizontal direction is

$$m\frac{\partial U}{\partial t} = \eta\frac{P}{U} - D - R, \tag{7.36}$$

or

$$\frac{\partial}{\partial x}\left(\frac{1}{2}U^2\right) = \frac{1}{m}\left(\eta\frac{P}{U} - D - R\right), \tag{7.37}$$

where both the engine power and the propeller efficiency depend on the aircraft speed. We make a number of simplifying assumptions to solve Eq. 7.37. If the power term in the right-hand side of Eq. 7.37 is much larger than the sum of the other two, then

$$\frac{\partial}{\partial x}\left(\frac{1}{2}U^2\right) \simeq \frac{\eta}{m}\frac{P}{U}. \tag{7.38}$$

In addition, if we take an average engine power and propeller efficiency, then we have a first approximated value for the ground run by integrating Eq. 7.38,

$$d\left(\frac{1}{2}U^2\right) = \frac{\eta}{m}\frac{P}{U}dx, \tag{7.39}$$

$$Ud\left(\frac{1}{2}U^2\right) = \frac{\eta}{m}Pdx, \tag{7.40}$$

and finally*

$$x_{lo} \simeq \frac{1}{4}\frac{m}{U_{lo}}\frac{1}{\overline{\eta P}}. \tag{7.41}$$

Let us try to improve on this expression. Retrieve the drag and the rolling resistance, and take average values, as shown by Eqs 7.13

$$\frac{\partial}{\partial x}\left(\frac{1}{2}U^2\right) = \frac{1}{m}\left[\eta\frac{P}{U} - \frac{1}{2}\rho A C_D\left(\frac{U_{lo}}{2}\right)^2 - \frac{1}{2}\mu W\right], \tag{7.42}$$

or

$$\frac{\partial U^2}{\partial x} = \frac{1}{m}\left[2\eta\frac{P}{U} - \frac{1}{4}\rho A C_D U_{lo}^2 - \mu W\right]. \tag{7.43}$$

---

* For the integration we have used the equivalence: $U^2 = z$; $U = z^{1/2}$, which leads to $\int z^{1/2}dz = z^{-1/2}/2$.

Solve this equation in terms of x_{lo}

$$dU^2 = \frac{1}{m}\left[2\eta\frac{P}{U} - \frac{1}{4}\rho A C_D U_{lo}^2 - \mu W\right]dx, \tag{7.44}$$

$$U_{lo}^2 = \frac{1}{m}\left[2\eta\frac{P}{U} - \frac{1}{4}\rho A C_D U_{lo}^2 - \mu W\right]x_{lo}, \tag{7.45}$$

and find

$$x_{lo} = \frac{mU_{lo}^2}{2\eta P/U - \rho A C_D U_{lo}^2/4 - \mu W}. \tag{7.46}$$

As in the case of the jet aircraft, we finally consider the case in which the drag and the rolling resistance change with the speed. If we leave the algebra behind, the governing equation can be written

$$\frac{1}{2}\frac{\partial U^2}{\partial x} = \frac{1}{m}\left[\eta\frac{P}{U} + c_1 U^2 - \mu W\right], \tag{7.47}$$

where the coefficient c_1 is

$$c_1 = -\frac{1}{2}\rho A(C_{D_o} + kC_L^2 + \mu C_L). \tag{7.48}$$

We consider the C_L a constant parameter in the ground run. Integration of Eq. 7.47 yields

$$U_{lo}^2 = \frac{2}{m}\int_o^{x_{lo}}\left(\frac{\eta P}{U} + c_1 U^2 - \mu W\right)dx. \tag{7.49}$$

This equation can only be solved if the propeller's efficiency and the engine power as a function of the speed are known. The efficiency is a function of the advance ratio $J = U/\text{rpm } d$, Eq. 5.27. The problem is to find out if the aircraft is operated with a variable pitch/fixed rpm, or vice versa. Then it is necessary to have a function $\eta(U)$, which, with the engine power curves $P(U)$, allows the solution of Eq. 7.49.

7.9 WAT CHARTS

WAT charts are the Weight-Altitude-Temperature parameters at take-off, and describe the influence of these parameters on the take-off performance of a given aircraft. They may have subcharts that take into account such factors as wind, runway slope, surface conditions, and obstacle height to clear.

For example, for a given altitude and ISA temperature, we can calculate the take-off run or balanced field length for the whole range of gross take-off weights. Then we increase the airport altitude and repeat the calculation. Finally, we can predict the effects of weather conditions by performing the same calculations with temperatures above and below the ISA values at that altitude.

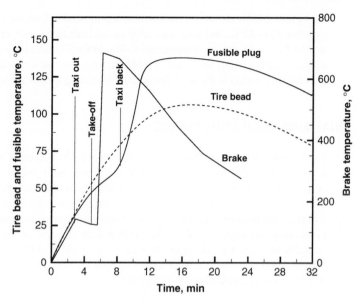

Figure 7.10 *Tire and brake temperatures. Adapted from Clifton and Leonard[206].*

7.10 MISSED TAKE-OFF

The history of aviation is littered with catastrophic failures that in hindsight turned out to be design shortcomings*. In 1963, a French Caravelle III crashed soon after take-off from Durrenasch, Switzerland, due to a tire explosion during landing gear retraction. Official investigations discovered that the tire had exploded due to over-heating during a missed take-off; tire explosion damaged a fuel line, which in turn started an explosion, at the cost of 80 fatalities. The pilot, in fact, decided to clear the morning fog with the aircraft wings, by taxiing the aircraft down to the end of the runway and back. By the time the aircraft left the ground, it had run three times on the runway. What was unknown at the time was the abnormal tire heating that can be created by a heavy aircraft, due to friction effects on the ground, brake operation, and proximity of engine exhausts.

Since then, a missed take-off requires additional performance controls on the tires and landing gear, what are commonly called *load/speed/time charts*. The relevant regulation prescribes the installation of appropriate thermal sensitive devices, and that the wheels do no rotate during retraction. A typical chart showing the heating of the tire bead and the brakes is shown in Figure 7.10.

The chart shows the peak tire temperature at rejected take-off at MTOW, although precise values depend on the tire characteristics, on the aircraft weight, on the atmospheric conditions, and on the runway. For reference, the tire temperature increases by 10 to 15°C per mile (1.6 km) during taxiing; it increases by a further 30 to 35°C

---

* An exhaustive database of aircraft accidents around the world is available at the website of the National Transportation Safety Board, www.ntsb.gov

during take-off. If take-off is rejected, heat is generated to increase the temperature by a further 20 to 25°C, and if the aircraft is taxied back to the start of the runway there will be further increases in temperature. The brakes' temperature can exceed 1,000 degrees. Further to the temperature problem, tires must be stopped before assuming their folded position in the hold. For additional reading on this subject, see Currey[207].

7.11   FINAL APPROACH AND LANDING

The aircraft approaches the runway along a clear path, at decreasing speed and with a low gradient. The final approach consists of three main phases of flight. First, an en-route descent, then a terminal area descent, a final approach, and finally landing. The gradient of the final approach varies from 2 degrees for commercial aircraft to 7 degrees for military aircraft. A low gradient γ is required for a number of reasons, particularly the impact load on the landing gear. To give an idea of this impact, consider an aircraft touching down with an air speed U and a Maximum Landing Weight, MLW. At touchdown the weight rests on the rear wheels. The impact velocity will be $U \sin \gamma$, and the impact energy will be

$$E = \frac{1}{2} \frac{\text{MLW} - \rho A C_{L_g} U^2/2}{g} \sin^2 \gamma. \tag{7.50}$$

The benefit of high lift at landing is evident. This energy decreases rapidly with the gradient of the flight path. A conventional landing starts with the aircraft approaching the runway at a moderate rate of descent, a touchdown with the rear wheels, a rotation of the aircraft, a touchdown of the nose wheels and a ground run. The ground run consists of a deceleration till the aircraft has halted on the runway. During the flare the aircraft is pitched up to increase its effective angle of attack. This is useful to decrease further the impact kinetic energy. This phase is flown with fully extended flaps (landing configuration) that yield reasonably high lift coefficients. A detailed theory of landing flare for large aircraft is available in Pinsker[208], including recovery from lateral gusts. Limits are imposed on roll when the aircraft touches down. The problem is shown in Figure 7.11.

The limiting bank angle ϕ can be estimated from

$$\tan \phi = \tan \varphi - \frac{2h_g}{b - b_t} \tan \theta \tan \Lambda, \tag{7.51}$$

where Λ is the sweep angle of the wing, b_t is the wheel track, h_g is the height of the landing gear, and φ is the wing's dihedral angle. There are cases in which the aircraft strikes the ground with the engine, instead of the wing tip.

The airborne distance in the final approach can be calculated from the gradient of the flight path, assuming that the aircraft clears the screen at 35 or 50 ft. The landing run is the sum of the free roll distance and the distance required to take the aircraft to a halt, although most passenger operations do not stop until they reach the terminal building.

A specific problem of landing operations is the tire spin-up. In fact, at touchdown the tires must assume the rotational speed $\Omega R = U$, where R is the radius of the tires

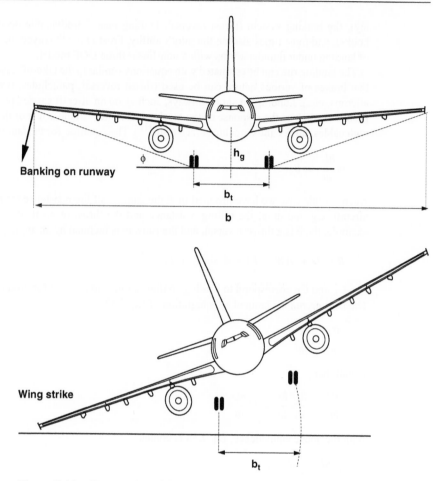

Figure 7.11 *Wing strike at landing.*

and Ω their rotational speed. For lightweight aircraft this is generally not a problem, but for large aircraft a spin prior to landing is required, in order to avoid significant tear and damage. A dynamic model for high-speed landing was developed by Padovan and Kim[209] and was applied to the space shuttle. For reference, the number of cycles (take-off and landing) with a set of tires is of the order of 300 for large commercial jet aircraft, and varies sensibly, depending on the runway conditions, aircraft weight and type of landing. As in take-off, tire burst is a problem that may affect the aircraft at landing. Concorde was particularly unfortunate on this aspect, with a record of one occurrence per 1,500 cycles.

7.12   LANDING RUN

The landing run of an aircraft depends on several external parameters, such as the atmospheric conditions (winds and rain), runway conditions (paved, unpaved, wet, icy), the braking system (thrust reversal, landing gear, aerodynamic decelerators, hooks), and most times also on the pilot's ability. Frost *et al.*[210] solved the problem of landing under thunderstorms with a non-linear three DOF model.

The landing run can be estimated with equations similar to the take-off case, except that brakes of various nature can be used (thrust reversal, parachutes, tire brakes, ailerons, etc.). Landing on short runways, such as on an aircraft carrier, poses other problems, such as the response of the mechanical system to the arrest of the aircraft with cables, hooks and parachutes (see Hsin[211]). The equation for the landing run is

$$m\frac{\partial U}{\partial t} = R, \tag{7.52}$$

where in this case we have grouped in R the sum of all the resistance terms on the aircraft, e.g. the drag, the rolling resistance and the thrust reversal, if any. If, for example, there is a thrust reversal, and the runway is inclined by an angle γ_r, then

$$R = D + \mu(W - L) + W \sin \gamma_r + T. \tag{7.53}$$

The C_L and C_D correspond to landing configuration, and account for ground effects. The landing run is obtained by integration of Eq. 7.52

$$\frac{\partial U}{\partial t} = \frac{R}{m}. \tag{7.54}$$

Recall that

$$\frac{\partial U}{\partial t} = \frac{\partial U}{\partial x}\frac{\partial x}{\partial t} = U\frac{\partial U}{\partial x}. \tag{7.55}$$

With the above equivalence, Eq. 7.54 becomes

$$U\frac{\partial U}{\partial x} = \frac{R}{m}, \tag{7.56}$$

and by separation of variables,

$$\frac{m}{R} U dU = dx. \tag{7.57}$$

Integration of the last equation is performed between the touchdown point (subscript "o") and the point at which the aircraft halts (superscript "1"). The result is the landing distance

$$x_L = \int_o^1 m\frac{U}{R} dU. \tag{7.58}$$

A few solutions are shown below. First, assume that the runway has no inclination ($\gamma_r = 0$), and that there is no thrust reversal ($T = 0$). The solution is found from

$$x_L = m\int_o^1 \frac{1}{D + \mu(W - L)} U dU = m\int_o^1 \frac{1}{\rho A(C_D - \mu C_L)U^2/2 + \mu W} U dU. \tag{7.59}$$

By defining the constant coefficients $c_1 = \rho A(C_D - \mu C_L)/2$ (with C_D and C_L constant) and $c_2 = \mu W$, the integral assumes the form

$$\int_o^1 \frac{1}{c_1 U^2 + c_2} U dU = \frac{1}{2} \int_o^1 \frac{1}{c_1 U^2 + c_2} dU^2. \tag{7.60}$$

In this equation the variable is U^2. The integral has a well-known solution (from textbooks of integral calculus)

$$\frac{1}{2} \int_o^1 \frac{dU^2}{c_1 U^2 + c_2} = \frac{1}{2c_2} \ln(c_1 U_1^2 + c_2). \tag{7.61}$$

Therefore, the landing run is

$$x_L = \frac{m}{2} \frac{1}{c_2} \ln(c_1 U^2 + c_2). \tag{7.62}$$

Equation 7.62 is an oversimplification, because in practice some form of braking is used. The solution is similar to the one outlined if the braking force is either constant or quadratic with the speed.

A constant braking force $-T$ is assumed in the case of thrust reversal; a quadratic expression $F = c_3 U^2$, with c_3 a constant, is assumed in the case of aerodynamic braking (parachute, ground-installed system). Another expression must be found for the actual brakes on the landing gears. First, assume a case of thrust reversal. The integral of Eq. 7.58 becomes

$$x_L = \frac{1}{2} m \int_o^1 \frac{1}{\rho A(C_D - \mu C_L) U^2 / 2 + \mu W + T} dU^2, \tag{7.63}$$

where we have replaced $U\,dU$ with $dU^2/2$. Equation 7.63 can still be solved in closed form. Assuming the constants

$$c_1 = \rho A(C_D - \mu C_L)/2, \qquad c_2 = \mu W + T, \tag{7.64}$$

the solution has the same form as Eq. 7.62. Second, assume a case of aerodynamic braking. The sum of all resistance forces is

$$R = D + \mu(W - L) + T + c_3 U^2. \tag{7.65}$$

The integral of Eq. 7.58 becomes

$$x_L = \frac{1}{2} m \int_o^1 \frac{1}{\rho A(C_D - \mu C_L + c_3) U^2 / 2 + \mu W} dU^2. \tag{7.66}$$

In this case, the coefficients are

$$c_1 = \rho A(C_D - \mu C_L + c_3)/2, \qquad c_2 = \mu W, \tag{7.67}$$

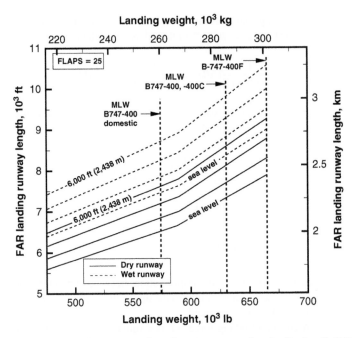

Figure 7.12 *FAR landing runway length requirements for the Boeing B-747, flaps at 25 degrees, no wind, all ground temperature conditions. Both dry and wet runways are considered. Altitudes at 2,000 ft (610 m) intervals from sea level.*

and the solution is like Eq. 7.62. The equations obtained show that the braking force reduces the landing distance as in a logarithm function; the landing run is proportional to the aircraft mass. At this point a parametric study of the effect of different brakes can be performed (Problem 4).

Figure 7.12 shows the FAR runway length requirements for the Boeing B-747 (various versions, as indicated) for a fixed flap configuration (25 degrees), as elaborated from Boeing[63].

Taking the aircraft to a halt on the runway requires finding ways to dissipate the mechanical energy of the vehicle. This energy is roughly proportional to the weight and to the square of the velocity at touchdown. Since all aircraft must take off and land on the available runway length, one can argue that the landing speed is somewhat independent of the gross weight, and therefore the energy that must be dissipated by the braking system is proportional to the aircraft's weight. A detailed analysis with three degrees of freedom of an airplane braking was published by Wahi[212].

Example

Take the Airbus A-380, landing at a speed of 300 km/h (162 kt) with its maximum landing weight certified at 386.0 metric tons. The brake heat sink Q for this aircraft is easily calculated:

$$Q = \frac{1}{2}mU^2 \simeq 1340 \, \text{MJ}. \tag{7.68}$$

This energy is equivalent to the energy released by combustion of 30.8 liters of kerosene (having assumed that the average heat of combustion of aviation fuel is 43.5 MJ/kg). The reader is invited to verify this equivalence.

Landing of sea planes and flying boats requires the modeling of the impact of the aircraft on the water, and cannot be analyzed with the method described. There are fairly elaborate theories on the impact, with flight stability and control (for example, Milwitzky[213] and Smiley[214]) and experimental data (for example, Bell[215]). This type of airplane is mostly used for fire-fighting operations, and includes the De Havilland DHC-6 Twin Otter and the Bombardier CL 415.

7.13 EFFECTS OF THE WIND

As anticipated earlier, take-off and landing are mostly done into the wind, and the aircraft flight to its destination airport may have to change its course to do so. Most commercial runways around the world are designed to align with the prevailing winds in the region. A quantitative explanation is given in this section. Only winds in the direction of the aircraft speed are considered. Hale[54] devotes one chapter of his textbook to the wind effects on aircraft performance.

The true air speed is $U_{TAS} = U \pm U_w$, where U_w is the wind speed, with respect to a reference system on the ground. The sign $+$ denotes head wind, and the sign $-$ denotes tail wind. The cases of practical interest are those with head wind. The solution is formally the same:

$$x_L = \frac{1}{2} m \int_o^1 \frac{1}{\rho A (C_D - \mu C_L) U_{TAS}^2 / 2 + \mu W + T} \, dU_{TAS}^2. \tag{7.69}$$

7.14 GROUND MANEUVERING

On leaving the gate, the aircraft has to maneuver and taxi to the starting point of the runway. Steering is done with the nose wheels, and the aircraft moves forward with the jet thrust. Large aircraft require wide spaces to turn around, and wide paved runways, since the engines must be shielded from ingestion of grass and other foreign objects. The main factors that affect the turn radius of the aircraft are the position of weight, the center of gravity, the engine power settings, the surface conditions, the differential braking, and the ground speed. Maneuvering airplanes the size of the Boeing B-747 requires steering on the nose wheels, as well as the main body wheels (those behind the wing). The steering is hydraulically actuated. Manufacturers provide charts of turning radii as a function of the nose gear steering, with symmetric or unsymmetric thrust. Figure 7.13 shows one case of steering the Boeing B-747 over a 90 degree turn, under the conditions described in the graph[63]. There is a degree of dependence on the pilot's skills.

PROBLEMS

1. Consider an airplane powered by jet engines, which give maximum static thrust at sea level $T_o = 50$ kN. Assume that this thrust does not depend on the speed.

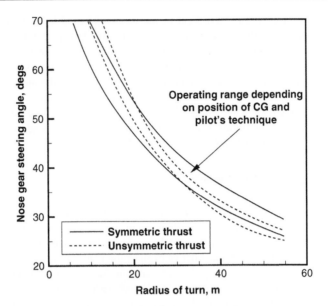

Figure 7.13 *Taxiway turn of 90 degrees for the Boeing B-747, according to ICAO recommendation. Data from Boeing.*

The airplane has a gross take-off mass $m = 25{,}000$ kg, and a wing loading $W/A = 450$ kgf/m^2. The lift coefficient in ground configuration (e.g. with landing gear out, control surfaces in take-off position) is $C_{L_g} = 0.50$. The drag equation is $C_D = C_{D_o} + kC_L^2$, with $C_{D_o} = 0.038$, $k = 0.045$. The airplane is to take off from Mexico City (airport at $h = 2{,}000$ m), with dry conditions on the runway ($\mu = 0.020$). For this aircraft calculate the take-off speed and the take-off run.

2. Consider again aircraft A, and assume ICAO standard conditions both at sea level and at an altitude $h = 1{,}600$ m. Study the effects of altitude on the landing run of the aircraft, all other parameters being the same (weight, touchdown speed, etc.).

3. Calculate the effects of an iced runway on the landing run of aircraft A, by comparing its landing performance to a dry runway (use the data of Table 7.2). Assume that all other parameters (aerodynamic coefficients, landing weight, braking force) are the same. The braking force used is a conventional thrust reversal of magnitude $T = T_o/4$. Devise methods aimed at reducing the landing distance in iced conditions without compromising the stability of the aircraft and the safety of the passengers.

4. Study the effects of landing weight, thrust reversal, and airport altitude on the landing run of aircraft model A. Assume realistic values of the parameters, and produce charts that show the effects of these parameters. For the solution start from Eq. 7.62, and write the appropriate expression for the resistance. The integral is solved using Eq. 7.61.

5. Calculate the energy that must be dissipated by the brakes of the Boeing B-747-400 to bring the aircraft to a halt. Assume that no other braking system (such as

thrust reversal, ailerons, etc.) is to be used. The aircraft is to land at 300 km/h with a weight $W = 285,000$ kg. The aircraft has a braking system on each of the 16 wheels. Describe how the heat generated can be dissipated, and how would the situation change if 60% of the braking could be done with thrust reversal systems.

6. Verify whether the approximation $T >> D + \mu R$ holds in the case of aircraft model A, taking off from a major sea level airport at its maximum take-off weight (rolling data coefficients in Table 7.2). If this is the case, use Eq. 7.9 to estimate the take-off run of the aircraft.

7. Solve the take-off run problem of §7.4, using the actual thrust, drag and lift forces for the same aircraft (Eq. 7.27 and following numerical expressions), and a lift coefficient in take-off configuration on the ground $C_L = 1.25$ (to be considered constant). Plot the functions $x(t)$, $U(t)$, $a(t)$, $C_D(t)$, $C_L(t)$ from brake release to lift-off point. Compare the results with those of Figure 7.6.

8. Calculate the effect of aircraft weight on the lift-off run of aircraft model A. Consider sea level conditions, dry runway conditions (Table 7.2), $GTOW = MTOW$ and $GTOW = 0.9MTOW$. For this calculation use the approximate method of Eq. 7.12, with a constant engine thrust.

9. Calculate the lift-off run x_{lo} for aircraft model A by using different methods of integration. Compare the methods given by Eq. 7.12, Eq. 7.20, and Eq. 7.25 for the same aircraft weight $W = 0.90MTOW$ at sea level ISA conditions. Plot and discuss the results.

10. Study the effect of the runway condition (see data in Table 7.2) on the lift-off run of aircraft model A. For this purpose, solve Eq. 7.12, Eq. 7.20, Eq. 7.25 for two extreme cases, and draw a critical conclusion on the value of the approximations done.

11. Calculate the fuel required to taxi out for aircraft model A (subsonic jet transport), whose data are given in Appendix A. The GTOW is estimated at 150,000 kgf. The taxi out lasts 5 minutes. The runway conditions are: dry, sea level, ISA. For the specific fuel consumption use $TSFC = 9.32 \cdot 10^{-6}$ kg$_m$/s/N.

12. A *tail-strike* problem occurs when at lift-off an aircraft pitches nose up at a high rate. Accidents have been reported with aircraft such as the Boeing B-747, the Airbus A-340, and others. Investigate the possible consequences on the aircraft when the tail scrapes the ground, and do additional research in the current aviation requirements to find out what the procedure to follow is once tail strike has occurred.

Chapter 8
Climb and Gliding

That's it. Not too steep. Give me a little more power, please; after all, the Engine Designer has given you a lot of surplus for occasions like these. Remember that now I have two enemies, Drag and Weight.

A.C. Kermode[188], 1956

This chapter deals with steady and accelerated climb and gliding. Climb is a flight in which the aircraft gains altitude. During a glide, powered or unpowered, the aircraft loses altitude. The aircraft climb, along with the cruise conditions, is among the most common phases of flight.

We discuss separately the problems of steady and accelerated climb. Accelerated climb problems are exclusively the domain of numerical solutions. Yet, the assumption of *quasi-steady flight* is a valid option for many aircraft.

There are essentially two methods for solving climb problems: by solution of the differential equations that govern the motion of the center of gravity, and by the use of energy methods. There is a difference in the climb characteristics of propeller- and jet-driven aircraft. Even within these categories, there is a difference between propellers driven by gas turbines and internal combustion engines.

Although modern flight programs routinely include a turn during climb-out and descent, we will restrict the discussion to a vertical plane. The reason for this is that climb in a three-dimensional space generally requires a detailed knowledge of how the flight controls promote aerodynamic forces that make the aircraft turn.

A number of other climb problems are discussed later in the book: ski jump (Chapter 16) and minimum noise climb (Chapter 17). The helicopter climb profiles will be discussed separately in Chapters 11 and 12.

8.1 GOVERNING EQUATIONS

For a jet aircraft on an arbitrary flight path on the vertical plane, the equations of motions are

$$m\frac{\partial U}{\partial t} = T\cos(\alpha + \epsilon) - D - W\sin\gamma, \tag{8.1}$$

$$mU\frac{\partial \gamma}{\partial t} = T\sin(\alpha + \epsilon) + L - W\cos\gamma. \tag{8.2}$$

The angle of climb γ is the angle between the flight direction and the horizontal plane; the angle of attack α is the angle between the aircraft reference axis (zero-lift axis) and the velocity vector; and the thrust angle ϵ (*vectored thrust*) is the angle between the reference axis and the direction of the engine thrust. While this angle is generally small, and can be neglected, it is not always the case. For example, the

McDonnell–Douglas MD-11 has a central engine mounted above the horizontal tail, whose thrust axis is inclined about 3.5 degrees on the reference axis. This installation provides a small vertical thrust and relieves the horizontal tail.

The flight path will be described by the differential equations

$$\frac{\partial x}{\partial t} = U \cos \gamma, \tag{8.3}$$

$$\frac{\partial h}{\partial t} = U \sin \gamma. \tag{8.4}$$

The fuel flow is also part of the problem, because it affects the aircraft's gross weight. The corresponding equation is

$$\frac{\partial m}{\partial t} = \frac{1}{g} \frac{\partial W}{\partial t} = -\frac{\partial m_f}{\partial t} = -\dot{m}_f. \tag{8.5}$$

The problem is to be closed with a set of initial conditions.

$$t = 0, \quad U = U_o, \quad \gamma = \gamma_o, \quad x = x_o, \quad h = h_o, \quad m = m_o. \tag{8.6}$$

Clearly, the aircraft can climb in an infinite number of ways, but a limited number of climb programs deserve special mention. There are climb programs that include local optimal conditions and fixed starting conditions. These are typically initial-value problems. These programs are: (1) fastest climb; (2) climb at maximum climb angle; and (3) climb at minimum fuel. The maximum angle of climb problem is only important to clear an obstacle in emergency situations, but it is not a normal way of operating the aircraft. The minimum fuel to climb program is also the most economical climb program. There are also special climb programs that require final conditions, such as speed or Mach number at a given altitude. These problems are called two-value boundary problems.

8.2 RATE OF CLIMB

The rate of climb (or *climb rate*) is the aircraft's velocity normal to the ground. The climb velocity is $v_c = U \sin \gamma$. As the aircraft climbs, the power plant delivers less thrust. The aerodynamic drag, on the other hand, is reduced as well, albeit at a lower gradient, because of the air density decreasing. Therefore, the aircraft will reach a point where it can no longer climb – that is the absolute ceiling.

In some operations, the aircraft can zoom past the absolute ceiling, by trading its kinetic energy into potential energy. Past the absolute ceiling, the aircraft might not be able to sustain controllable flight. Zoom-climb altitude maximization of the F-4C and F-15 aircraft was carried out in the 1970s for stratospheric missions reaching 27,000 m (90,000 feet).

Some data of average climb rates are given in Table 8.1 for different types of aircraft. Rates of climb are generally given in the technical and popular literature in meters/minute. This is a misleading unit, because the aircraft cannot maintain its climb rate for any length of time. The instrumentation of the aircraft gives ft/min, and we cannot change that. A more rational unit would be m/s (ft/s) for the maximum rate of climb, including the altitude at which this performance is achieved.

Table 8.1   *Typical climb rates of some fixed-wing aircraft.*

Aircraft type	$v_c(m/min)$
Heavy lift cargo	350–800
Business jets	800–1,300
Business turboprops	500–1,100
Supersonic fighters	9,000–18,000

The maximum known rates of climb are around 18,000 m/min (300 m/s, 984 ft/s), although the MiG-29 claims a maximum rate of climb $v_c \sim 19,800$ m/min (330 m/s, 1,083 ft/s). If this is true, it corresponds to an instantaneous vertical Mach number $M \sim 1$. Also, if the total Mach number of this climb were $M = 2$, the aircraft would fly with a climb angle of the order of 45 to 50 degrees.

8.3   STEADY CLIMB OF PROPELLER AIRPLANE

Imagine a propeller aircraft, whose longitudinal axis is aligned with the speed and the thrust generated by the propellers. The equation of motion in the flight direction for a steady state flight is

$$T - D - W \sin \gamma = 0, \tag{8.7}$$

Multiply this equation by the aircraft speed U

$$TU - DU - WU \sin \gamma = 0. \tag{8.8}$$

Recall that the effective power of the propeller aircraft is

$$TU = \eta P. \tag{8.9}$$

The rate of climb of the aircraft is found by rearranging Eq. 8.8 with Eq. 8.9,

$$v_c = \frac{\eta P - DU}{W} = \frac{\eta P}{W} - \frac{D}{L} U \cos \gamma. \tag{8.10}$$

From the definition of lift coefficient replace the speed U with

$$U = \sqrt{\frac{2W}{\rho_o \sigma A C_L} \cos \gamma}. \tag{8.11}$$

Equation 8.11 neglects the centrifugal acceleration, but in this case it is an acceptable approximation. By replacing Eq. 8.11 with Eq. 8.10, we find

$$v_c = \frac{\eta P - DU}{W} = \frac{\eta P}{W} - \cos^{3/2} \gamma \frac{C_D}{C_L^{3/2}} \sqrt{\frac{2}{\rho_o} \frac{W}{A} \frac{1}{\sigma}}. \tag{8.12}$$

For a small climb angle (say $\gamma < 10$ degrees) assume that $\cos^{3/2} \gamma \sim 1$,

$$v_c = \frac{\eta P - DU}{W} = \frac{\eta P}{W} - \frac{C_D}{C_L^{3/2}} \sqrt{\frac{2}{\rho_o} \frac{W}{A} \frac{1}{\sigma}}. \tag{8.13}$$

The approximation of small climb angle is not required when solving the climb problem with numerical methods, although it makes the algebra more straightforward. We seek a solution of the problem in stages of increasing approximation and computational overhead, and Eq. 8.13 is a useful expression.

8.3.1 Fastest Climb of Propeller Airplane

Fastest climb is a flight program requiring the climb rate to be at a maximum at all altitudes. At a *given altitude* h Eq. 8.13 is only a function of the aircraft speed. The necessary optimal climb condition requires that the derivative of the climb rate v_c with respect to the relevant flight parameters will be zero. Assuming that the throttle is at full position, for a propeller engine the free parameters are the flight speed U and altitude h, the advance ratio J, and the pitch setting θ of the propeller: $v_c = f(h, U, J, \theta)$. The climb rate can be optimized on a point to point basis, therefore the altitude can be taken out of the list.

In the present context, we will reduce the climb rate to a single DOF – the aircraft speed. We consider both the propulsion efficiency and engine power not dependent on the speed. This is equivalent to saying that: as the aircraft speed increases, the increase in engine power is offset by the decrease in propulsive efficiency, so that $\eta P \simeq$ constant. Hence the optimal condition is

$$\frac{\partial v_c}{\partial U} = 0, \tag{8.14}$$

$$\frac{\partial v_c}{\partial U} = -\sqrt{\frac{2}{\rho_o} \frac{W}{A}} \sigma^{-1/2} \frac{\partial}{\partial U}\left(\frac{C_D}{C_L^{3/2}}\right) = 0, \tag{8.15}$$

$$\frac{\partial}{\partial U}\left(\frac{C_D}{C_L^{3/2}}\right) = 0. \tag{8.16}$$

We have seen in §6.8 that the speed that minimizes the ratio $C_D/C_L^{3/2}$ is the speed of minimum power for level flight. The corresponding speed is given by Eq. 6.57. If the propulsive efficiency is constant, the best climb rate at a fixed altitude is given by

$$\frac{\partial}{\partial U} = \left(\frac{C_{D_o} + kC_L^2}{C_L^{3/2}}\right). \tag{8.17}$$

As in §4.16, we find the solution

$$U^2 = \sqrt{\frac{k}{C_{D_o}}}\left(\frac{2W}{\rho A}\right). \tag{8.18}$$

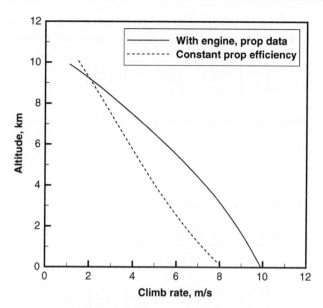

Figure 8.1   *Optimal climb rate of the C-130, model B, GTOW = 55,000 kg.*

The solution of this problem for the aircraft of reference, model B, is shown in Figure 8.1, dashed line.

8.3.2   Optimal Climb with Engine and Propeller Data

Assume now that the aircraft has a fixed-pitch, variable-speed propeller, whose performance charts are given in Figure 5.15. The engine is a gas turbine. If the pitch is chosen, the propulsion efficiency $\eta = \eta(J)$ is a known function (from tabulated performance data). The system parameters are U and J. The engines are run at maximum power. Assuming a fixed gear ratio between the engine and the propeller, the engine rotational speed Ω_1 will decide the propeller's rotational speed. If Ω_{1_o} is the maximum engine speed at sea level, we assume that

$$\left(\frac{\Omega_1}{\Omega_{1_o}}\right)^3 \simeq \left(\frac{\Omega}{\Omega_o}\right)^3 = \frac{P}{P_o}, \tag{8.19}$$

which is solved for Ω

$$\Omega = \frac{\Omega_o}{P_o^{1/3}} P^{1/3}. \tag{8.20}$$

Equation 8.19 is derived from the following considerations. The propeller power is the sum of a number of contributions, as discussed in Chapter 5. For a given propeller, both the profile power and the induced power are proportional to Ω^3. This conclusion can be justified on the grounds of the blade element and momentum theory. Assume a relatively small angle of climb. At a fixed altitude, the optimal climb condition is

$$\frac{\partial v_c}{\partial U} = \frac{1}{W} \frac{\partial(\eta P)}{\partial U} - \sqrt{\frac{2}{\rho} \frac{W}{A}} \frac{\partial}{\partial U}\left(\frac{C_D}{C_L^{3/2}}\right) = 0. \tag{8.21}$$

Assign the following constant coefficient

$$c_1 = -W\sqrt{\frac{2}{\rho} \frac{W}{A}}. \tag{8.22}$$

By substituting this coefficient into Eq. 8.21 we find

$$\frac{\partial(\eta P)}{\partial U} + c_1 \frac{\partial}{\partial U}\left(\frac{C_D}{C_L^{3/2}}\right) = 0, \tag{8.23}$$

$$\left(\frac{\partial \eta}{\partial U}\right)P + \left(\frac{\partial P}{\partial U}\right)\eta + c_1 \frac{\partial}{\partial U}\left(\frac{C_D}{C_L^{3/2}}\right) = 0. \tag{8.24}$$

We need to make another transformation for the derivative of the propulsive efficiency, so that we can calculate it from the tabulated data

$$\frac{\partial \eta}{\partial U} = \left(\frac{\partial \eta}{\partial J}\right)\left(\frac{\partial J}{\partial U}\right) = \left(\frac{\partial \eta}{\partial J}\right)\frac{1}{\Omega R}. \tag{8.25}$$

If we use Eq. 8.25 and Eq. 8.20, then Eq. 8.24 becomes

$$\left(\frac{\partial \eta}{\partial J}\right)\frac{1}{R}\frac{P_o^{1/3}}{\Omega_o}P^{2/3} + \left(\frac{\partial P}{\partial U}\right)\eta + c_1 \frac{\partial}{\partial U}\left(\frac{C_D}{C_L^{3/2}}\right) = 0, \tag{8.26}$$

With the definition of a further constant coefficient

$$c_2 = \frac{1}{R}\frac{P_o^{1/3}}{\Omega_o}, \tag{8.27}$$

the optimal condition finally becomes

$$\left(\frac{\partial \eta}{\partial J}\right)c_2 P^{2/3} + \left(\frac{\partial P}{\partial U}\right)\eta + c_1 \frac{\partial}{\partial U}\left(\frac{C_D}{C_L^{3/2}}\right) = 0. \tag{8.28}$$

There are three derivatives in Eq. 8.28 that need a numerical solution. The derivative $\partial \eta / \partial J$ can be calculated around the current value of J from the efficiency curve. The same holds for the derivative of the engine power. The derivative of the aerodynamic term can be solved in closed form, but the algebra gets quite involved, and adds unwarranted complications to the solution. Therefore, we choose to leave it in the present form, and we will solve it with a numerical method, as the other cases.

Equation 8.26 is a non-linear differential equation. For an arbitrary input value of the aircraft speed the function

$$f(U) = \left(\frac{\partial \eta}{\partial J}\right) c_2 P^{2/3} + \left(\frac{\partial P}{\partial U}\right) \eta + c_1 \frac{\partial}{\partial U} \left(\frac{C_D}{C_L^{3/2}}\right). \tag{8.29}$$

is not zero (it is, in fact, quite a large number). The problem is reduced to finding a value of U that is a root of Eq. 8.29. A suitable algorithm is based on the Newton iterations, as described below.

Computational Procedure

1. Load engine data, propeller data, and aircraft data.
2. Assign initial speed (from take-off, or otherwise) and initial climb angle.
3. Assign propeller rotational speed (throttle).
4. Set the flight altitude h (*outer loop*).
5. Set the value of the aircraft speed U_i, with $i = 1$ (*inner loop*).
6. Calculate numerically all the derivatives in Eq. 8.28 at the current value of U_i. For the calculation we use the method of §C.2, based on small perturbations around the current operation point (second-order accurate).
7. Calculate the value of the residual function Eq. 8.28.
8. Calculate the value of the derivative (gradient) $\partial f/\partial U$ by the same method,

$$\left(\frac{\partial f}{\partial U}\right)_i \simeq \frac{f(U_i + dU) - f(U_i - dU)}{2dU}. \tag{8.30}$$

The use of second-order derivatives reduces the error. The extra calculation of the residual does not add significant time to the procedure.

9. Find a new estimate for the aircraft speed from a Newton step

$$U_{i+1} = U_i - f(U_i)\left(\frac{\partial f}{\partial U}\right)_i^{-1}. \tag{8.31}$$

10. Calculate the value of the residual Eq. 8.28 at U_{i+1}. If

$$f(U_{i+1}) < \epsilon, \tag{8.32}$$

where ϵ is a small tolerance, exit the loop; or else continue iterating (go to point 5).

11. Advance the altitude by dh, set $h = h + dh$, and repeat the above steps. The stopping criterion is the service ceiling condition.

The power curves are given at a few discrete intervals, therefore interpolation of those values is required. The procedure for the interpolation of the data can be quite elaborate. The type of interpolation depends on how the data are arranged. Power curves are best given as a two-dimensional array $P(h, U)$. If the data points are on a grid, then higher-order interpolation can be done in both directions. If not, they may have to be rearranged. The same considerations hold for the propulsive efficiency. In

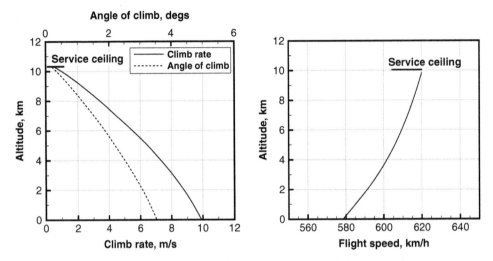

Figure 8.2 *Estimated C-130 climb profile, with maximum rate of climb. Take-off mass m = 55,000 kg.*

the case of a single efficiency curve, higher-order interpolation can be done easily by using a cubic spline. If, however, several η curves are involved, we must seek a different arrangement of the data. This is one of the most time-consuming tasks that occur when doing simulation and optimization based on discrete numerical data.

The problem, as stated above, is a classic example of an initial-value problem: assigned the initial conditions, the solution is advanced in time and space one step at a time, by using information from the previous step. It allows the calculation of most quantities of interest, including the flight path, the cumulative fuel consumption, and the time to climb. For the present purpose, a first-order Euler integration is considered acceptable for finding the flight path. If x is the distance travelled, then

$$dt = \frac{dh}{|v_c|}, \quad V = U \sin \gamma, \quad dx = Vdt, \quad x = x + dx. \tag{8.33}$$

The absolute value of v_c is required, because if the aircraft descends the time increment would be negative. The section of program corresponding to this computational procedure is reported in §C.4. This program can be improved so as to include the effects of the angle of climb, which has been neglected. However, it is fairly general, and can be used for the climb performance calculations of any propeller aircraft, having a known fixed-pitch propeller and engine performance.

A solution for aircraft model B is shown in Figure 8.2, as calculated from the present model. The maximum climb rate is estimated at 10.0 m/s at sea level. The solution includes the effects of changing weight, due to fuel burn. The time to ceiling is estimated at 22.5 minutes.

In the case shown above, the product ηP decreases continuously from sea level to service ceiling, where it reaches a value approximatively half of the initial value. The propeller efficiency is about 0.88 and varies little with the altitude as shown in Figure 8.3.

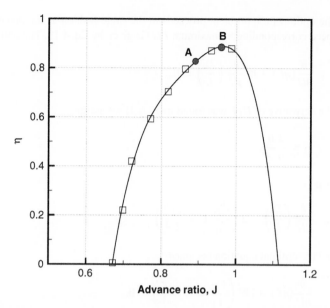

Figure 8.3 *Propeller's reference data for climb optimization. Squares indicate flight data; the solid line is a higher-order interpolation and extrapolation curve. Points A and B denote the limits of operation from sea level to service ceiling of the optimal climb flight shown in Figure 8.2.*

Consider again Figure 8.1. This figure shows the climb profile obtained by solving Eq. 8.13 with the condition Eq. 8.16 and the results of the present method. The maximum climb rate is underestimated by 20%; the ceiling is overestimated. Therefore, we conclude that the derivative of ηP cannot be neglected in a professional simulation environment.

Another climb program is obtained by considering the effects of One Engine Inoperative (OEI). The engine power considered in the previous case consisted of four times the power of the single engine. This was taken as the take-off power. In the present case we can either try to climb with OEI and take-off power on the remaining three engines, or increase the power output of the three functioning engines to the emergency power. These engine data are summarized in Table A.7.

8.3.3 Climb at Maximum Angle of Climb

The angle of climb is given by Eq. 8.48. The condition of maximum for this angle is

$$\frac{\partial}{\partial U}(\sin \gamma) = \frac{\partial}{\partial U}\left(\frac{\eta P - D}{W}\right) = 0. \tag{8.34}$$

If we assume again that the change in weight is negligible then,

$$\frac{\partial}{\partial U}\left(\frac{\eta P}{W} - \frac{D}{L}\right) = 0. \tag{8.35}$$

The optimization problem proceeds in the same fashion as the preceding case. The speed corresponding to maximum L/D is given by Eq. 4.19. The optimal equation is

$$\frac{\partial}{\partial U}(\eta P) - W \frac{\partial}{\partial U}\left(\frac{D}{L}\right) = 0. \tag{8.36}$$

The derivative of D/L was found in §4.6. If we assign

$$c_1 = \frac{2}{\rho} \frac{W}{A}, \tag{8.37}$$

then

$$\frac{\partial}{\partial U}\left(\frac{D}{L}\right) = 2\left(\frac{C_{D_o}}{c_1} U - \frac{c_1 k}{U^3}\right), \tag{8.38}$$

and the optimal equation becomes

$$\frac{\partial}{\partial U}(\eta P) - 2W\left(\frac{C_{D_o}}{c_1} U - \frac{c_1 k}{U^3}\right) = 0, \tag{8.39}$$

$$\left(\frac{\partial \eta}{\partial J}\right)\frac{P}{\Omega R} + \left(\frac{\partial P}{\partial U}\right)\eta - 2W\left(\frac{C_{D_o}}{c_1} U - \frac{c_1 k}{U^3}\right) = 0. \tag{8.40}$$

The algorithm of the preceding section is valid also in this case, although the optimal equation is somewhat different in the powers of U.

8.3.4   Climb Fuel of Propeller Airplane

The instantaneous climb fuel for a propeller-driven aircraft is proportional to the engine power,

$$\dot{m}_f = SFC\, P. \tag{8.41}$$

The climb is characterized by an excess power $\eta P - DU$ and a climb rate

$$v_c = \frac{\eta P - DU}{W} = \frac{\eta}{W}\frac{\dot{m}_f}{SFC} - \frac{DU}{W}. \tag{8.42}$$

From Eq. 8.42, the *instantaneous* fuel flow becomes

$$\dot{m}_f = \frac{SFC}{\eta}(v_c W + DU). \tag{8.43}$$

The first term on the right-hand side denotes the vertical-climb fuel, and the second term denotes the fuel required for forward flight. From Eq. 8.43 the fuel flow depends on the flight program (v_c, U) and requires information as to how the propulsive efficiency changes during the climb. Clearly, it cannot be resolved unless there is a relationship between the climb rate and the aircraft speed. The fuel to climb is found by integration of Eq. 8.43

$$m_f = \int_o^t \dot{m}_f d\tau,$$

(8.44)

which in fact is reduced to a sum in the numerical solution.

8.4   CLIMB OF JET AIRPLANE

The governing equation for the center of gravity of the aircraft in arbitrary flight is given by Eq. 8.2. In the first instance, we will neglect the thrust angle, $\epsilon \simeq 0$. Then we multiply this equation by the aircraft speed U to find

$$TU - DU - Wv_c = mU\frac{\partial U}{\partial t},$$

(8.45)

and hence the climb velocity in general accelerated flight

$$v_c = \frac{T-D}{W}U - \frac{U}{g}\frac{\partial U}{\partial t}.$$

(8.46)

For a steady state flight, the climb rate can be found directly from the specific excess power; the climb angle is found from the specific excess thrust

$$v_c = \frac{T-D}{W}U,$$

(8.47)

$$\sin\gamma = \frac{v_c}{U} = \frac{T-D}{W}.$$

(8.48)

For a high-power maneuver of military aircraft we can define a normal load factor

$$n = \frac{L}{W},$$

(8.49)

which is the ratio between lift and weight. The general expression of the climb angle is

$$\sin\gamma = \frac{T}{W} - \frac{D}{W} = \frac{T}{W} - \frac{n}{L/D}.$$

(8.50)

Equation 8.50 is useful for the calculation of maneuver in the vertical plane. If the centrifugal acceleration is neglected (a reasonable assumption in normal climb programs), then

$$\sin\gamma = \frac{T}{W} - \frac{1}{L/D}.$$

(8.51)

As a further simplification, if the thrust is not dependent on the flight speed, then the maximum angle of climb is obtained with the speed corresponding to maximum glide ratio, $(L/D)_{max}$.

To calculate the maximum v_c we follow the same considerations as in the case of the propeller airplane. The absence of propeller efficiency makes the problem easier. However, we must consider the effects of transonic drag rise.

8.4.1 C_L for Optimal Steady Rate of Climb

We seek the optimal solution of a steady state climb at subsonic speeds, for an airplane whose drag equation is parabolic (Eq. 4.13). We optimize the climb with respect to a single parameter, the C_L. Changes in C_L can be achieved by change of configuration (e.g. by extension of the control surfaces) and by changes in the angle of attack. The optimal flight condition is found from

$$\frac{\partial v_c}{\partial C_L} = 0, \tag{8.52}$$

or

$$\frac{\partial}{\partial C_L}\left[\left(\frac{T}{W} - \frac{C_D}{C_L}\right)\sqrt{\frac{2}{\rho_o \sigma}}\sqrt{\frac{W}{A}}\frac{1}{\sqrt{C_L}}\right] = 0. \tag{8.53}$$

After a few steps of algebra, it becomes

$$\frac{\partial}{\partial C_L}\left[\left(\frac{T}{W} - \frac{C_D}{C_L}\right)\frac{1}{\sqrt{C_L}}\right] = 0. \tag{8.54}$$

Replace the parabolic drag equation and calculate the derivatives:

$$3C_{D_o}C_L^{-5/2} - \frac{T}{W}C_L^{-3/2} - kC_L^{-1/2} = 0, \tag{8.55}$$

or

$$3C_{D_o}C_L^{-2} - \frac{T}{W}C_L^{-1} - k = 0. \tag{8.56}$$

Equation 8.56 is a quadratic equation in the unknown C_L^{-1}. The positive solution is

$$C_L = \frac{6C_{D_o}}{T/W + \sqrt{(T/W)^2 + 12C_{D_o}k}}. \tag{8.57}$$

If the engine thrust can be expressed by Eq. 5.22, we can define a parameter ξ

$$\xi = 1 + \sqrt{\sigma^2 + \frac{C_{D_o}k}{(T_o/W)^2}}. \tag{8.58}$$

Thus, the optimal lift coefficient becomes

$$C_L = \frac{6}{\xi} \frac{C_{D_o}}{(T_o/W)}.$$
(8.59)

A solution algorithm for this problem is the following.

Computational Procedure

1. Set the altitude.
2. Calculate the parameter ξ from Eq. 8.58.
3. Calculate the lift coefficient from Eq. 8.59.
4. Calculate the aircraft speed from $U = \sqrt{2W/\rho A C_L}$.
5. Calculate the climb rate from Eq. 8.47.
6. Calculate the climb angle from $\sin \gamma = v_c/U$.
7. Advance to the next step and repeat the procedure.

The stopping criterion is either $v_c = 0$ (absolute ceiling) or $v_c < 0.5$ m/s (service ceiling). A solution of this climb program for aircraft model A is shown in Figure 8.4. The service ceiling is reached at an estimated 19.2 minutes from take-off. The graph on the left shows the aircraft forces (thrust and drag) as a function of flight altitude. The graph on the right shows the climb rate and Mach number. At the absolute ceiling $T = D$; hence no further climb is possible. From Eq. 8.58, the C_L must increase with increasing altitude, hence the aircraft speed must decrease as the aircraft climbs.

If the thrust cannot be expressed by Eq. 5.22, then some numerics are required. The algorithm shown is not a practical way to the solution. In the new algorithm, we

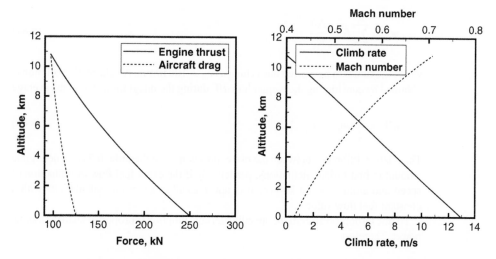

Figure 8.4 *A-300 climb profile at best C_L. Take-off mass $m = 140,000$ kg, static thrust at sea level $T_0 = 250$ kN.*

need to perform the derivative of the thrust with respect to the speed or Mach number. The optimal condition is

$$\frac{\partial}{\partial U}\left(\frac{TU - DU}{W}\right) = 0. \tag{8.60}$$

The optimality condition is valid only for a subsonic jet, whose drag equation is parabolic. A more general expression is

$$\frac{1}{W}\left[\left(\frac{\partial T}{\partial U}\right)U - \left(\frac{\partial D}{\partial U}\right)U + T - D\right] = 0, \tag{8.61}$$

which does not contain approximations. The derivative of the thrust presents little difficulty if the data are given in terms of Mach number, because

$$\frac{\partial T}{\partial U} = \left(\frac{\partial T}{\partial M}\right)\left(\frac{\partial M}{\partial U}\right) = \left(\frac{\partial T}{\partial M}\right)\frac{1}{a}. \tag{8.62}$$

8.4.2 Practical Calculation of Climb Fuel

Consider first a jet-powered aircraft. Set $f_j = TSFC$ to simplify the nomenclature in the equations. The fuel flow from take-off to service altitude is the integral

$$m_f = \int_0^1 \dot{m}_f \, dt = \int_0^1 \dot{m}_f \frac{dh}{v_c} = \int_0^1 f_j \, T \frac{dh}{v_c}. \tag{8.63}$$

The problem is solved numerically, as usual. If we advance the solution by equal steps Δh, then

$$m_f = \sum_i^n f_{j_i} T_i \frac{\Delta h}{v_{c_i}}, \tag{8.64}$$

where the sum is to be carried out as long as $v_{c_i} > 100$ ft/min. An alternative solution is obtained by considering the weight loss dW during the integration step as the average

$$dW = \frac{1}{2}\left(m_{f_i} + m_{f_{i-1}}\right) dt. \tag{8.65}$$

The relative difference between the two expressions is of the order 0.2% – a negligible amount in first-order calculations, particularly if the exact fuel flow as a function of speed and altitude is not known. It is not unusual to carry out calculations with a constant fuel flow value.

For the turboprop aircraft the specific fuel consumption is proportional to the engine power. Therefore,

$$m_f = \int_0^1 \dot{m}_f \, dt = \int_0^1 \dot{m}_f \frac{dh}{v_c} = \int_0^1 SFC \, P \frac{dh}{v_c}. \tag{8.66}$$

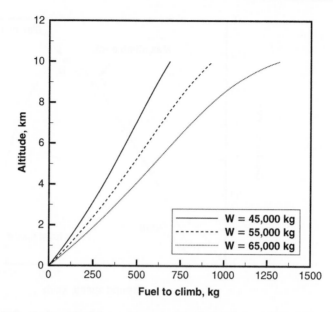

Figure 8.5 *Fuel to climb for turboprop aircraft C-130 (model B), flight profile as described in §8.3.2. Weights as indicated.*

The numerical solution is similar to Eq. 8.64. The fuel to climb for the turboprop of reference, the C-130, as discussed in §8.3.2 is shown in Figure 8.5. The increase of aircraft gross weight from 45,000 kg to 65,000 kg has the effect of nearly doubling the fuel to climb.

8.5 POLAR DIAGRAM FOR RATE OF CLIMB

The polar diagram is a summary of the aircraft steady state performance in climb, gliding and level flight (as a limiting case). Diagrams of this type are calculated at a constant mass, flight altitude and throttle setting. In general, only a few polars in this family are of interest: the polar at maximum thrust, the power-off polar, the polar at MTOW, and the polar at absolute ceiling.

Calculation of the polar is relatively straightforward if the engine thrust is not dependent on the speed; it is more elaborate for a supersonic jet aircraft. The polar is obtained by joining two segments: the climb and descent segment. An algorithm for the *positive segment* is as follows.

Computational Procedure

1. Read aircraft data (wing area, thrust angle, etc.).
2. Read aircraft charts (engine performance, aerodynamics, TSFC, etc.).
3. Set aircraft state parameters (weight, flight altitude).
4. Set Mach number equal to estimated maximum Mach.

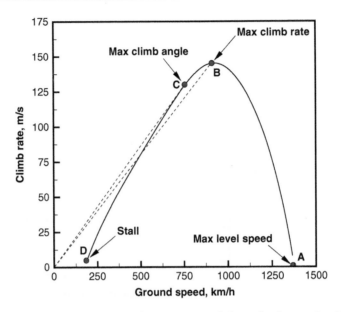

Figure 8.6   *Positive climb polar of supersonic jet fighter of reference (model C), W = 11,000 kg, starting altitude h = 2,000 m (6,562 ft), steady flight, military thrust.*

5. Calculate engine thrust, aircraft drag (see algorithm in §C.1, in Appendix C, for assembling the aircraft forces).
6. Calculate angle of attack compatible with $L = W$ (negligible normal load factor), Eq. 6.65.
7. Calculate aircraft drag.
8. Calculate specific excess power and climb rate.
9. Store values of M, v_c.
10. Reduce Mach number by dM.
11. Proceed as above from point 5, till stalling speed/Mach number is reached.

The calculation can be repeated for different aircraft weight and flight altitudes, with and without after-burning thrust (Problem 16). An example of calculation is shown in Figure 8.6. As the aircraft starts climbing, the horizontal speed must decrease. The maximum v_c is achieved at an intermediate forward speed, as indicated. At forward speeds below this point, the climb rate decreases. The climb at maximum climb angle is at yet lower speed, v_c/V. The speed of maximum climb angle is lower than the speed of maximum climb rate; this corresponds to the tangent point of the polar with the line through the center of the reference system.

The effects of starting altitude for the same problem as described in Figure 8.6 are shown in Figure 8.7. When the aircraft starts from a higher altitude, the climb profile becomes more complex, due to a combination of transonic drag rise and engine thrust. In particular, at $h = 6,000$ m (19,685 ft), there are two climb segments, including a fully supersonic envelope. The aircraft is unable to climb at speeds between $M = 1.2$ and $M = 1.4$, because of negative excess power in this range.

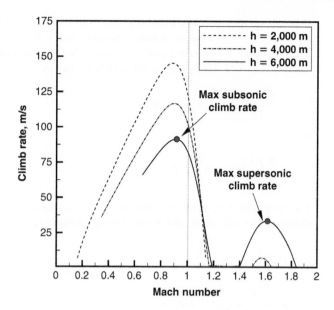

Figure 8.7   *Altitude effect on positive segment of climb polar, for the same case as Figure 8.6.*

8.6   ENERGY METHODS

The general approach to the climb problems is done by using steady state models. However, this cannot be correct, because as the climb rate and the optimal climb rate change with increasing altitude, the aircraft *must accelerate*. The difference between steady state and accelerated flight is particularly important for high-performance aircraft. Bryson and Denham[216] proved in a landmark paper that an optimal accelerated climb to ceiling requires about half the time of an optimum *quasi-steady* climb. However, the calculation of the accelerated flight is far more complicated, and is generally left to specialists in flight dynamics.

For the case of propeller-driven aircraft the quasi-steady approximation is acceptable. In fact, we have seen that the optimal climb profile of a turboprop leads to slowly changing aircraft speeds with altitudes and small climb angles. The turboprop climb shown in Figure 8.2 indicates that the aircraft accelerates by about 40 km/h in about 22.5 minutes. The average linear acceleration is clearly small.

Instead of considering forces on the center of gravity of the aircraft it is sometimes useful to write balance equations for the total energy of the aircraft in its climbing flight. Methods of this nature are called *energy methods*. The first methods based on the concept of total aircraft energy go back to the 1940s and 1950s. A widely acclaimed paper in this field is that of Rutowski[217], although the original idea seems to belong to F. Kaiser, who developed the concept of "resultant height" in Germany for the optimal climb schedule of the Messerschmidt Me 262, the first jet fighter to enter operational service. Kaiser's concept is reviewed by Merritt *et al.*[218]

Kelley[219] was among the first to understand that trading kinetic energy for potential energy can be part of the climb of a high-performance jet aircraft. These methods, although approximate, are extremely powerful, as evidenced by later research by Kelley *et al.* [220] These approximations are discussed below.

Consider the momentum equation in the flight direction multiplied by the flight speed U, Eq. 8.45. This equation can also be written as

$$m\frac{\partial}{\partial t}\frac{1}{2}U^2 = TU - DU - Wv_c,$$

(8.67)

or

$$m\frac{\partial}{\partial t}\left(\frac{1}{2}U^2 + gh\right) = TU - DU.$$

(8.68)

The term within parentheses, divided by the acceleration of gravity g, has the dimensions of a distance, and is called *energy height* h_E,

$$E = \frac{1}{2}U^2 + gh,$$

(8.69)

$$h_E = \frac{E}{g} = \frac{1}{2g}U^2 + h.$$

(8.70)

The energy in Eq. 8.69 represents the sum of the kinetic and potential energy of the aircraft per unit of mass. The energy height Eq. 8.70 represents the altitude at which the aircraft would climb if it were to convert all its kinetic energy to potential energy. The time derivative of the total energy is the work done by the power plant

$$\dot{E} = \frac{\partial E}{\partial t} = \frac{TU - DU}{m}.$$

(8.71)

The time derivative of the energy height is equal to the specific excess power

$$\frac{\partial h_E}{\partial t} = \frac{TU - DU}{W} = P_e.$$

(8.72)

Some methods used for flight path optimization use Eq. 8.69, with the additional assumption that the aircraft can *instantaneously* exchange kinetic energy with potential energy, and vice versa. In practice this approximation is a fairly good one if the short period of motion of the aircraft is neglected. However, it leads to sharp changes in direction in the flight path, which are unreasonable.

Figure 8.8 shows lines of constant energy height in the M–h plane for a unit mass ($m = 1$ kg). These lines show a knee at a point located at altitude $h = 11,000$ m (36,089 ft). This is due to the change in atmospheric conditions at the troposphere. The troposphere, in fact, is a point of discontinuity for the temperature, and hence the speed of sound. The lines of constant energy are found from

$$h = \frac{E - U^2/2}{g} = \frac{E - a^2M^2/2}{g},$$

(8.73)

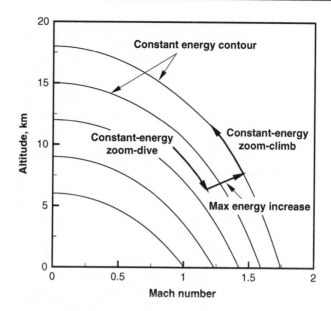

Figure 8.8   *Constant energy levels on the plane M–h.*

with E the assigned energy level. When $M = 0$, then $h = E/g$. This means that all the energy is potential energy. The physical height decreases with the increasing speed.

8.7   SPECIFIC EXCESS POWER DIAGRAMS

Diagrams of the specific excess power (Eq. 8.72) are a summary of the total performance of an aircraft in the altitude–speed plane, and complement the flight envelope that was discussed in Chapter 6.

Lines of constant specific excess power, P_e, are only valid for a fixed configuration, weight, load factor, angle of attack, engine throttle, and atmospheric conditions. In practice, only a limited number of excess power curves are drawn.

An overview of the specific excess thrust in a three-dimensional space is indicative of the acceleration capabilities of the aircraft. This is shown in Figure 8.9 for the reference fighter jet. The main complicating factor is the transonic drag rise. There are two hills, at subsonic and supersonic speeds. At subsonic speeds maximum values of P_e are found at sea level; P_e decreases rapidly with the altitude. At supersonic speeds maximum values of P_e are at intermediate altitudes (around 10 km). This interim conclusion indicates that a combat aircraft is most effective at maneuvering in the vertical plane only within a limited altitude range. Outside this range, its acceleration capabilities are severely impaired by lack of sufficient thrust, or high drag. Maneuvering outside these altitudes is obviously not recommended.

The line corresponding to zero excess power is a limiting case, because it divides the flight envelope in two regions. If $P_e < 0$, the aircraft can only decelerate, because the thrust available is less than the thrust required to overcome the drag. Therefore, $P_e = 0$ is a stationary line for the aircraft speed, and $P_e < 0$ is outside the normal flight envelope of the aircraft.

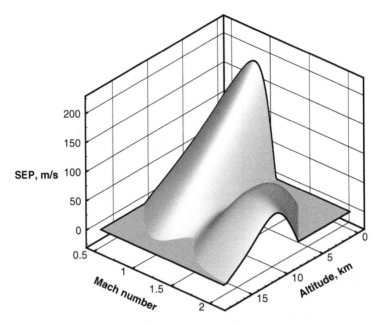

Figure 8.9 *3D view of P_e, aircraft mass $m = 12,000\,kg$; zones of $P_e < 0$ have been set to zero to emphasize the effects.*

If we use the climb rate equation and the definition of excess power, the line of zero specific excess power is found from the condition

$$U \sin \gamma + \frac{U}{g} \frac{\partial U}{\partial t} = \frac{\partial h}{\partial t} + \frac{U}{g} \frac{\partial U}{\partial t} = 0. \tag{8.74}$$

Figure 8.10 is a two-dimensional plot relative to the case of Figure 8.9. Lines of constant P_e are shown, including the line of $P_e = 0$, in addition to lines of constant energy height. As the flight altitude increases the supersonic acceleration is limited by the sharp decrease in excess thrust; for a range of altitudes, supersonic acceleration cannot be achieved at all, unless the aircraft climbs at subsonic speeds, and then zoom-dives past the speed of sound. Therefore, to achieve that speed the aircraft must climb to an altitude above the local maximum of the $P_e = 0$ contour, and then perform a zoom-dive to the starting altitude.

The zoom-dive is a peculiar maneuver: the aircraft can reach phenomenal speeds, which are limited by wing buffet, thermal and structural loads, and longitudinal stability problems.

8.8 DIFFERENTIAL EXCESS POWER PLOTS

A comparison between excess power plots of different high-performance aircraft can provide valuable information regarding their maneuver capability. One such example is shown in Figure 8.11, where two different configurations of the same aircraft have been selected. These correspond to configurations (a) and (b) in Figure 4.16. From

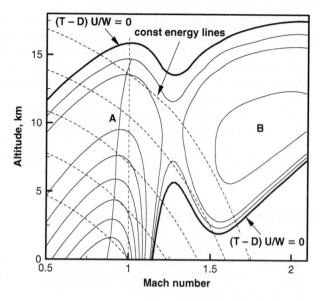

Figure 8.10   *Lines of constant specific excess power in the M–h plane with lines of constant energy level.*

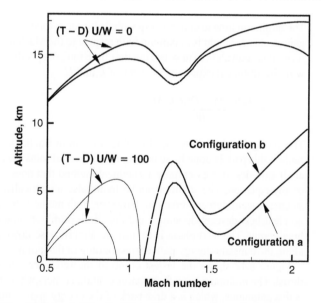

Figure 8.11   *Differential P_e plot for a given aircraft, configurations (a) and (b).*

the performance point of view, these configurations differ only in the transonic and supersonic drag characteristics; they have the same engine and the same weight. At any point in the plane *M–h* we can define the difference in excess power as

$$\Delta P_e = P_e(B) - P_e(A). \tag{8.75}$$

Charts like those in Figure 8.11 tend to get complicated, because one aircraft can have a maneuvering advantage in a region of the envelope, and a handicap in another region. In the present case, we have plotted only two specific excess power lines. One line is obviously $P_e = 0$, which expresses the limits of the maneuver envelope. Another value is $P_e = 100$ m/s. This is close to the maximum P_e for both aircraft at the given gross weight.

These comparisons become confusing if the two aircraft are very different. The choice of weights is essential, and only comparisons at similar weights make sense. By accurately reading this difference, the performance and design engineer can improve some operations; and the pilot learns to avoid the flight conditions in which he is likely to have a handicap.

8.9 MINIMUM PROBLEMS WITH ENERGY METHOD

The energy method is most suitable for the analysis of the climb performance of supersonic aircraft that accelerate past the speed of sound. A number of flight programs are of interest: fastest climb (or minimum time to climb), steepest climb, and minimum-fuel climb.

8.9.1 Minimum Time to Climb and Steepest Climb

The minimum time to climb to a specified altitude is found from the condition that the gain in energy height is maximum with respect to the Mach number. In mathematical form, it is equivalent to maximizing the climb rate, or specific excess power, Eq. 8.47, with the altitude h expressed in terms of E and U.

$$v_c = \frac{T(h, M) - D(\alpha, h, M)}{W} U. \tag{8.76}$$

The energy height is introduced in place of h from Eq. 8.69. A classical method for finding this path is done by searching the curve that joins the points of maximum P_e at all altitudes. It is essentially a graphical method that does not require the solution of any equation. The solution can be found also numerically, by advancing along a steepest ascent/descent direction. However, both methods break down when there is no clear best-descent direction, and the climb profile A shown in Figure 8.10 has a break point. The next phase is to turn the aircraft nose-down and start a zoom-dive along the constant-energy path corresponding to this point.

Figure 8.12 shows the fastest climb of the reference aircraft with after-burning thrust. The minimum-time flight paths contain corners where the control variable M is discontinuous, which is a drawback of the energy approximation.

The steepest climb condition is the maximum in the Mach–altitude plane of the specific excess power (climb rate) with respect to the Mach number,

$$max_M \left(\frac{v_c}{U}\right) = \frac{T - D}{W}. \tag{8.77}$$

The solution procedure follows the method of the fastest climb.

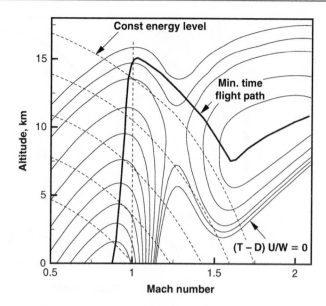

Figure 8.12 *Minimum-time climb for supersonic fighter jet calculated with energy method. The section of constant-energy dive is approximate.*

8.9.2 Minimum Fuel to Climb

A fuel climb problem is found from Eq. 8.5, which is divided by the rate of change of total energy, Eq. 8.71,

$$\frac{\dot{m}}{\dot{E}} = \left(\frac{\partial m}{\partial t}\right)\left(\frac{\partial t}{\partial E}\right) = \frac{\partial m}{\partial E} = -\dot{m}_f \frac{m}{(T-D)U}. \tag{8.78}$$

When we separate the differentials dm and dE we find

$$\frac{dm}{m} = -\dot{m}_f \frac{dE}{(T-D)U}. \tag{8.79}$$

The ratio $dm/m = dm_1$ is the *specific* change in aircraft mass, e.g. the change in mass due to fuel flow divided by the aircraft mass. A *minimum fuel climb* problem will be formulated mathematically by the condition that minimizes dm/m for a given energy level, e.g.

$$\frac{dm_1}{dE} = -\frac{\dot{m}_f}{(T-D)U}. \tag{8.80}$$

Therefore, a minimum fuel climb flight path requires the minimization of the right-hand side of Eq. 8.80, or a maximization of its inverse. We construct a function

$$f(h, M) = -\frac{(T-D)U}{\dot{m}_f}, \tag{8.81}$$

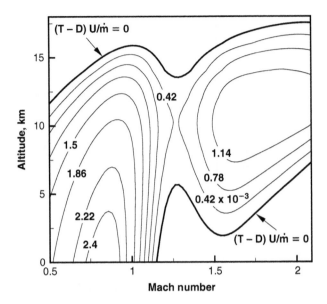

Figure 8.13 *Lines of constant function f(h, M), given by Eq. 8.81, for supersonic fighter jet aircraft.*

which is proportional to the excess thrust, and plot it as in the previous cases. Note that $f(h, M)$ is a negative function in the flight envelope of the aircraft, because $T \geq D$. The minimum fuel to climb is the locus of the maximum energy increase per unit of fuel burned at a fixed energy E. Lines of constant value of the function defined by Eq. 8.81 are shown in Figure 8.13.

The minimum-time and minimum-fuel flight paths (Figures 8.12 and 8.14) are similar. In both cases the aircraft accelerates at sea level from take-off speed to reach $M \simeq 0.85$. It then pitches down and starts a steep climb with almost constant Mach number. It reaches a maximum altitude in minimum time or minimum fuel, then it zoom-dives. The flight path up to this point can be calculated exactly. There are large approximations in the zoom-dive and zoom-climb sections.

8.9.3 Other Climb Profiles

The climb problems presented, although they may seem complicated, are no match to some real-life aircraft performance. In fact, some climb programs routinely include flight in three dimensions and acrobatic maneuvers.

Figure 8.15 shows some estimated optimal climb profiles for the F-15 Strike Eagle fighter jet aircraft. Each profile is obtained with a different gross weight. The aircraft reaches a defined altitude, at a given distance from the airport, and with a final Mach number. In all cases the aircraft accelerates at sea level to $M = 0.65$. In particular, flight programs (b), (c) and (d) involve an Immelmann at medium load factors, a pull-up at $4\,g$, and a climb at a constant climb angle $\gamma = 55$ degrees. There are two cases with supersonic acceleration at constant altitude or with a small climb, with the aircraft passing through sonic speed within half a minute from lift-off. This supersonic acceleration performance is discussed in §6.13. The time to climb varies

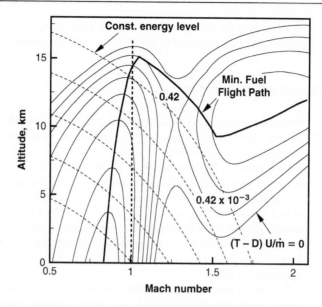

Figure 8.14 *Minimum fuel to climb for supersonic fighter jet calculated with energy method.*

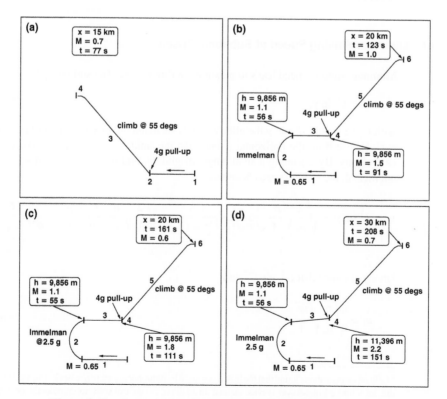

Figure 8.15 *Some climb programs of F-15 Strike Eagle.*

greatly among the climb profiles. The F-15, powered by two Pratt and Whitney F-100 engines, has a climb record to 20,000 m in 2 min, 3 s (January 19, 1975).

8.10 STEADY STATE GLIDING

A fixed-wing aircraft is said to be gliding if it maintains a stable and controlled flight with minimum or zero engine power. Under these conditions the aircraft generally loses altitude, or drifts down.

On August 24, 2001, an Airbus A-330 in flight from Toronto to Lisboa (Portugal) ran out of fuel in the middle of the Atlantic Ocean, due to a fuel leak caused by a damaged fuel feed pipe. When the crew realized the problem, they only had 5 minutes of fuel left. At that time the aircraft was about 180 nautical miles from the nearest airport. The aircraft, without engine power, glided for almost 18 minutes, descending for more than 30,000 feet (9,144 m) before landing safely at a military airport in the Azores[*]. The pilot and co-pilot received awards for the longest glide of a civilian passenger aircraft.

This was an unusual accident, nevertheless it is an eventuality that must be considered. Gliding performance is also an essential aspect of unpowered flight vehicles. In the following sections we will discuss some gliding flights and their optimal conditions.

8.10.1 Minimum Sinking Speed at Subsonic Speed

Minimum sinking speed leads to maximum airborne time. The sinking speed is

$$v_s = U \sin \gamma, \tag{8.82}$$

with U the air speed, and γ the glide angle. If we maintain the convention of positive γ clockwise from the horizontal plane, then the sinking speed is a negative vertical velocity. By assuming a small angle of gliding, and using the definition of lift coefficient, the sinking speed becomes

$$v_s = U \sin \gamma = \sqrt{\frac{2}{\rho_o}} \sqrt{\frac{W}{A}} \frac{1}{\sqrt{\sigma}} \frac{1}{\sqrt{C_L}} \frac{C_D^2}{C_L^2 + C_D^2} \simeq \sqrt{\frac{2}{\rho_o}} \sqrt{\frac{W}{A}} \frac{1}{\sqrt{\sigma}} \frac{C_D}{C_L^{3/2}} = \frac{c_1}{\sigma^{1/2}} \frac{C_D}{C_L^{3/2}}, \tag{8.83}$$

having assumed that $C_D^2 \ll C_L^2$ and

$$c_1 = \sqrt{\frac{2}{\rho_o}} \sqrt{\frac{W}{A}}. \tag{8.84}$$

---

[*] Official investigations found that the leak was caused by improper maintenance work. The mechanics did not follow a Rolls-Royce service manual. In addition, the control software mistakenly identified the fuel leak as a fuel imbalance, prompting the crew to pump more fuel to the leak.

In practice, the best glide ratio is in the range of 14 to 20, depending on the aircraft, hence C_D^2 is less than 0.5% of C_L^2. From Eq. 8.83 it appears that the condition of minimum sinking speed is equal to the condition of minimum power for a powered aircraft (Eq. 6.57), which we report here for convenience

$$U_{mp}^2 = \frac{2W}{\rho A}\sqrt{\frac{k}{3C_{D_o}}}. \tag{8.85}$$

The corresponding value of $C_D/C_L^{3/2}$ is

$$\frac{C_D}{C_L^{3/2}} = \frac{4C_{D_o}}{(3C_{D_o}/k)^{3/4}}. \tag{8.86}$$

With this value of the aerodynamic factor, the minimum v_s is

$$v_s = \left(\frac{2}{\rho_o}\frac{W}{A}\right)^{1/2}\frac{1}{\sigma^{1/2}}\frac{4C_{D_o}}{(C_{D_o}/k)^{3/4}}. \tag{8.87}$$

Equation 8.87 shows that the minimum v_s decreases as the aircraft descends – which is a good thing. Also, the air speed decreases as the aircraft descends, which is another good thing, as long as the speed can be safely maintained above stall and lateral control of the aircraft can be assured. The angle of descent is

$$\gamma \simeq \frac{v_s}{U}, \tag{8.88}$$

with U given by Eq. 8.85. A typical solution is plotted in Figure 8.16 for three values of the aircraft weight. This speed decreases with the decreasing wing loading, and with the decreasing altitude.

8.10.2 Minimum Glide Angle Versus Minimum Sinking Speed

We compare two different descent conditions: minimum sinking speed, as previously discussed, and minimum glide angle. The glide angle in absence of engine thrust is given by

$$\sin\gamma = \frac{T-D}{W} = -\frac{D}{W} \simeq -\frac{D}{L} = -\frac{1}{2}\rho\frac{A}{W}C_D U^2. \tag{8.89}$$

A minimum for $\sin\gamma$ is also a minimum for γ; this occurs for an aircraft operating at maximum glide ratio. Therefore, a suitable optimal condition at a fixed altitude is

$$\frac{\partial(\sin\gamma)}{\partial U} = 0, \tag{8.90}$$

or

$$\frac{\partial}{\partial U}\left(\frac{D}{L}\right) = 0. \tag{8.91}$$

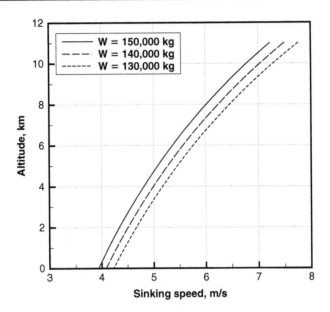

Figure 8.16 *Effect of aircraft weight on sinking speed for a minimum sinking speed glide.*

This condition was found earlier. We proved that this was a condition of minimum drag

$$U_{md}^2 = \frac{2W}{\rho A}\sqrt{\frac{k}{C_{D_o}}}. \tag{8.92}$$

The minimum sinking speed is given by Eq. 8.85, therefore the ratio between the speed of minimum glide angle and the speed for minimum descent speed is

$$\frac{U_{vs}}{U_\gamma} = \frac{1}{\sqrt[4]{3}} = 0.7598 \simeq 0.76. \tag{8.93}$$

In conclusion, *the speed of minimum descent rate is equal to 76% the speed of minimum glide angle*. This ratio is the same as the ratio between the speed of minimum power and minimum drag,

$$\frac{U_{vs}}{U_\gamma} = \frac{U_{mp}}{U_{md}}. \tag{8.94}$$

The flight path can be found from the steady state equations

$$\frac{\partial x}{\partial t} \simeq U, \quad \frac{\partial h}{\partial t} = v_s, \tag{8.95}$$

with the additional condition

$$\frac{v_s}{U} = \tan\gamma \simeq \gamma. \tag{8.96}$$

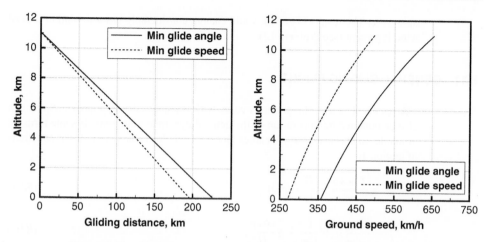

Figure 8.17   *Gliding range at two different flight conditions (left graph), and corresponding ground speed (right graph). Aircraft weight W = 145,000 kg.*

Combination of the above equations leads to the differential form

$$dx = \frac{dh}{\gamma}.$$

(8.97)

The solution to this problem is found numerically, as in most other cases. There is a simple procedure, based on the Euler integration scheme, which is the following:

Computational Procedure

1. Read the aircraft data.
2. Set the aircraft's operational data: altitude and speed.
3. The aircraft speeds from minimum sinking speed and minimum glide angle are given by Eq. 8.85 and Eq. 8.92, respectively. They are calculated at the current altitude by using σ from the ISA equations.
4. Set the amount of descent dh and calculate the glide angle from Eq. 8.96, then calculate the differential advancement $dx = dh/\gamma$.
5. At the new altitude $h-dh$ recalculate σ, and the corresponding aircraft speeds.
6. Repeat from point 3 till the aircraft will have reached sea level.

A typical result is shown in Figure 8.17. An aircraft of the class of the Airbus A-330 with a gross weight $W = 145,000$ kg was simulated without any meteorological effects. The aircraft weight was estimated at $OEW + PAY$, without usable fuel. The initial altitude was $h = 11,000$ m, with an initial Mach number $M = 0.79$.

The maximum gliding range is about 225 km (120 nm), and is obtained with a minimum glide angle program. This is enough to establish that the aircraft without fuel should be able to glide the whole distance of 100 nm without any engine power. However, the gliding time was estimated at about 32 minutes, a value perhaps too optimistic. The results obtained would change considerably if a tail wind or a head wind is present (see Problem 12).

8.11 GENERAL GLIDING FLIGHT

The results shown in the previous cases with the quasi-steady approximations can also be obtained by integration of the general dynamics equations. If some thrust is available, in absence of any weather factor (wind, down-bursts, etc.), the equations of motion are

$$\dot{U} = \frac{T\cos\alpha}{m} - \frac{D}{m} - g\sin\gamma, \tag{8.98}$$

$$\dot{\gamma} = \frac{T\sin\alpha}{Um} + \frac{L}{Um} - g\frac{\cos\gamma}{U}, \tag{8.99}$$

having assumed that the thrust is aligned with the drag force. If the angle of glide is small (around 3 or 4 degrees) we make an approximation:

$$\sin\gamma \simeq \gamma \simeq \frac{v_s}{U}, \quad \cos\gamma \simeq 1, \quad U \simeq V. \tag{8.100}$$

Therefore, the drift-down dynamics equations become

$$\dot{U} = \frac{T\cos\alpha}{m} - \frac{D}{m} - g\gamma, \tag{8.101}$$

$$\dot{\gamma} = \frac{T\sin\alpha}{Um} + \frac{L}{Um} - \frac{g}{U}. \tag{8.102}$$

These equations can be integrated as an initial-value problem, for any reasonable value of the initial conditions. The use of the full flight path equations is required to study more general gliding problems, such as gliding at constant C_L, and gliding with and without pitch damper. A glide at constant C_L may give rise to a large-amplitude oscillation (*phugoid*), if a pitch damping is not used. The phugoid oscillation occurs when a powered or an unpowered vehicle is operating at a speed and altitude away from its equilibrium, and when no adjustments are made to its flight control systems. If the vehicle is trimmed, then the oscillations are removed.

An example of the effect of constant angle of attack glide is shown in Figure 8.18. In this problem we have considered a constant C_L (or constant α) and have integrated the equations of motion, Eq. 8.98 and Eq. 8.99 by using a Runge-Kutta method, fourth-order accurate. If the C_L is relatively high, then the glider drifts down according to an oscillatory motion with a long wave length. This wave length decreases as the C_L decreases. When the glider drifts down, it increases its air speed; in doing so, it increases its lift, and therefore it pitches up. As it pitches up it loses speed, and it drifts down again. The case of $C_L = 0.30$ corresponds to a glide ratio $L/D \simeq 33$, which is at the top end of the glider performance. With a $C_L = 0.20$ we have $L/D \simeq 25$. At lower values of C_L we have poor aerodynamic performance. An application of this analysis can be done for animal flight, as well (see Problem 20).

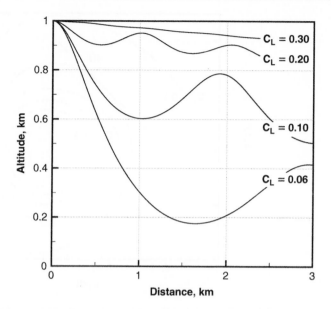

Figure 8.18 *Glide at constant angle of attack (constant C_L). Initial conditions: $U = 40\,m/s$, $\gamma = -2$ degrees, $h = 1,000\,m$. Other data: $W = 456\,kg$; $A = 17\,m^2$, $C_{D_o} = 0.0070$. No thrust.*

Generally, one can look at the differential equations in the phase plane with coordinates U, γ. This plane contains parametric plots of $\gamma(t)$ versus $U(t)$. While it may not be possible to solve the ODEs in closed form, it is always straightforward to plot the tangent vectors to the solution trajectories in the phase plane. At a point U, γ we plot the vector given by the right-hand side of the ODE system. We are able to tell immediately from such a plot whether or not the U-coordinate of a solution should be monotonic with respect to t. The trajectories in the phase plane themselves satisfy a first-order ODE. In other words, by eliminating t between U and γ, we get the ODE

$$\frac{\partial \gamma}{\partial U} = \frac{\dot{\gamma}}{\dot{U}} = \frac{T \sin \alpha / Um + L/Um - g \cos \gamma / U}{T \cos \alpha / m - D/m - g \sin \gamma}, \tag{8.103}$$

which can be studied as well. It can be verified that for the case of Figure 8.18 this derivative is oscillating, and so is the solution, by an amount that depends on the value of C_L, Figure 8.19. A non-oscillating solution requires that $\dot{\gamma}/\dot{U}$ be uniform.

The stability problems related to the phugoid have been the subject of analytical investigation since Lanchester[221]; they have been treated in virtually all books on stability and control, with approximations, linearizations and closed form solutions (for example, Perkins and Hage[33], Etkin[19], McCormick[105], Etkin and Reid[222], Nelson[223]). These books address further references on the subject. In addition, Campos *et al.*[224] discuss the speed stability of an aircraft in a dive.

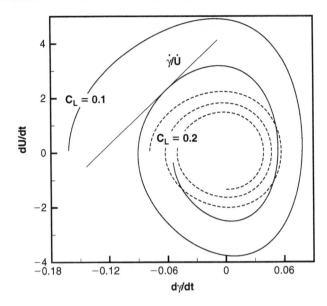

Figure 8.19 *Phase plot of the momentum equations for gliding flight at constant attitude.*

8.12 MAXIMUM GLIDE RANGE WITH ENERGY METHOD

Finally, we solve the maximum glide range for a supersonic jet fighter with the energy method, and show how the solution to this problem can be substantially different from that of subsonic flight. Take the definition of horizontal velocity component, $U = \dot{x}$, and divide by Eq. 8.71

$$\frac{\dot{x}}{\dot{E}} = \frac{U}{(TU - DU)/m}. \tag{8.104}$$

If there is no thrust, the equation can be simplified to

$$\frac{\dot{x}}{\dot{E}} = -\frac{m}{D}, \tag{8.105}$$

or

$$\frac{\partial x}{\partial E} = -\frac{m}{D}, \quad \frac{\partial E}{\partial x} = -\frac{D}{m} = -g\frac{D}{L}. \tag{8.106}$$

To maximize the gliding range, we must minimize dE/dx, subject to the constraint $L = W$. In other terms, the aircraft must descend by moving along a flight path of minimum drag or maximum glide ratio – something we already knew.

If we eliminate h in the drag force, by introducing the energy height, Eq. 8.69, then the problem is a minimum D with respect to the flight speed,

$$min_U = -D(E, U), \tag{8.107}$$

for a fixed energy level. It is possible that the initial conditions of the aircraft (speed U_o and altitude h_o for given energy level E_o) are not on the maximum range path. This

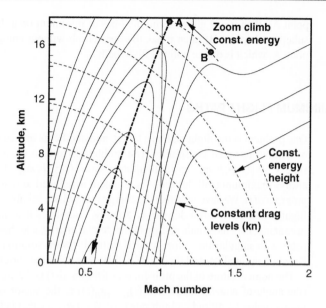

Figure 8.20   *Flight path of maximum glide range of supersonic aircraft, weight W = 12,000 kg.*

means that the drag is not minimal, and the aircraft must zoom-dive or zoom-climb to the optimal level. Once the point on the $E = E_o$ level is reached, the aircraft glides along the optimal flight path. The glide angle of this path is found from the dynamic equation along the flight path, with $T = 0$

$$\sin \gamma = -\frac{D}{W} + \frac{1}{g}\frac{\partial U}{\partial t}. \tag{8.108}$$

Since the derivative of the speed is negative, it appears that this term decreases the drag, and thus extends the glide path. The glide range is found from integrating Eq. 8.106. It consists of two terms

$$R = \int_0^2 dx = \int_{E_o}^{E_1} \frac{m}{D}\,dE + \int_{E_1}^{E_2} \frac{m}{D}\,dE, \tag{8.109}$$

where E_o is the initial energy; E_1 is the energy at ground level; and E_2 is the minimum level speed at ground level. The first term in Eq. 8.109 is due to the loss of energy from the starting point to ground level; the second term is due to deceleration at ground level, to a speed above stalling speed. Therefore, the aircraft can glide to nearly ground level. Then it does another stretch by increasing the angle of attack and extending the high-lift control surfaces. If the initial speed is supersonic, the aircraft will first dissipate its kinetic energy, while keeping nearly constant altitude; then it will glide, losing also its altitude (or potential energy).

Figure 8.20 shows the flight path of maximum glide range starting from a super-sonic speed. The test case is that of the reference jet aircraft (model C), in nominal configuration. If the aircraft finds itself already on the optimal gliding path (starting point A), then it just glides along the lines of minimum drag. If the starting point is suboptimal (point B), then it has to zoom-climb without engine thrust to an initial point A, which is on the optimal flight path.

8.13 MINIMUM FLIGHT PATHS

One problem is the flight path to a target, and another is trajectory optimization, with or without terminal constraints. The problem is to take the aircraft from its initial state $x_o = \{U, \gamma, h, m\}_o$ to its final state $x_f = \{U, \gamma, h, m\}_f$. If the engine thrust is always maximum (full throttle), the problem is reduced to finding a control variable program $\alpha(t)$. We have seen that the energy method is a suitable approach to minimum flight paths, but it has a drawback. Namely, it assumes that the aircraft can convert instantaneously its kinetic energy into potential energy, and therefore it leads to flight paths with singularities. More general methods for the solution of these problems are based on optimal control theory, of which we give a short briefing.

The nomenclature of the problem is as follows: $\alpha(t)$ is the control variable program (the angle of attack); $x(t) = x(x_1(t), \cdots, x_n(t))$ is the vector of state variables programs (velocity, altitude, climb rate $\cdots$); $\psi = (\psi_1, \cdots, \psi_p)$ is the vector of terminal constraints functions (ψ is a known function of the terminal point and the vector of state variables); finally, ϕ is the cost function, which depends on the final point and the state variables $x(t_e)$ at the final point.

Therefore, the problem is to take the aircraft to the terminal state subject to the performance criterion J

$$J = \phi[x(t_e), t_e] + \int_o^{t_e} L[x(t), u(t), t]dt \tag{8.110}$$

being a minimum or maximum. In Eq. 8.110, L is a state function. Equation 8.110 expresses the fact that the objective function depends on the final state of the aircraft and on *the course of events* between the start and the terminal point (the integral of Eq. 8.110). The aircraft can be subject to the terminal constraints

$$\psi = \psi[x(t_e), t_e] = 0. \tag{8.111}$$

The expression of the functional J looks mathematically complicated, but in many cases it can be greatly simplified. Usually, not all the final conditions are specified. For example, one may want to have a final Mach number, a final weight, a final climb angle, etc. The most important cases are:

1. *Minimum-time climb.*

$$J = \phi = t_e, \quad L = 0, \tag{8.112}$$

where L is a function of the course of events. The differential of the time can be written as

$$dt = \frac{dh}{dh}dt = \frac{dh}{v_c}, \tag{8.113}$$

where the $v_c(h)$ is the instantaneous rate of climb. Therefore, the cost function becomes

$$J = -\int_o^{te} dt = -\int_o^{he} \frac{dh}{v_c}. \tag{8.114}$$

2. *Minimum-fuel climb.* All the final values are specified, except the time

$$\phi = W_e, \quad L = 0. \tag{8.115}$$

3. *Minimum-energy climb.* The flight path angle γ_e may be specified or unspecified. The final time is unspecified. The constraint at the terminal point is

$$\psi_e = \left(h + \frac{1}{2g}U^2\right)_e. \tag{8.116}$$

The cases listed greatly simplify the expression of the objective function. To solve the problem, we resort to an extension of the Lagrange multipliers Λ, in order to adjoin the constraints Eq. 8.111. Therefore, we define the Hamiltonian function

$$H = L + \Lambda f = \Lambda f, \tag{8.117}$$

where the Lagrange multipliers are:

$$\Lambda = \{\lambda_U, \lambda_\gamma, \lambda_h, \lambda_m\}. \tag{8.118}$$

The derivation of the conditions for the Lagrange multipliers is quite elaborate. Suitable sources of information are Bryson and Ho[21] and Ashley[22], who show the entire derivation procedure. The multipliers are found from

$$\dot{\Lambda} = \frac{\partial \Lambda}{\partial t} = -\frac{\partial H}{\partial x} = -\Lambda\left(\frac{\partial f}{\partial x}\right). \tag{8.119}$$

This condition leads to

$$\dot{\lambda}_u = -\frac{\partial H}{\partial U}, \quad \dot{\lambda}_\gamma = -\frac{\partial H}{\partial \gamma}, \quad \dot{\lambda}_h = -\frac{\partial H}{\partial h}, \quad \dot{\lambda}_m = \frac{\partial H}{\partial m}. \tag{8.120}$$

These are four *differential equations*. If the change in weight due to fuel flow can be neglected, then the last condition is dropped.

8.13.1   Minimum Time to Climb

Consider a flight path in the vertical plane. When the thrust line coincides with zero-lift axis, the momentum equations for the center of gravity of the aircraft are reduced to

$$\frac{\partial U}{\partial t} = \frac{T(h, M)}{m}\cos\alpha - \frac{D(\alpha, h, M)}{m} - g\sin\gamma, \tag{8.121}$$

$$\frac{\partial \gamma}{\partial t} = \frac{L(\alpha, h, M)}{mU} - \frac{T(h, M)}{mU} \sin \alpha - \frac{g}{U} \cos \gamma. \qquad (8.122)$$

The coordinates of the aircraft will be given by

$$\dot{x} = U \cos \gamma, \qquad (8.123)$$

$$\dot{h} = U \sin \gamma, \qquad (8.124)$$

and the mass flow rate is given by

$$\dot{m}_e = -f_j T(h, M). \qquad (8.125)$$

The system has state 4, because the state variables are U, γ, h, m (or the fuel flow $\dot{m}_f$). The distance flown $\dot{x}$ can be calculated a posteriori. The control variable (or free parameter) is the angle of attack α. The adjoint equations are:

$$\dot{\lambda}_u = -\lambda_u \left[\frac{\cos \alpha}{ma} \frac{\partial T}{\partial M} - \frac{1}{ma} \frac{\partial D}{\partial M} \right] - \lambda_h \sin \gamma - \frac{\lambda_m}{a} \frac{\partial \dot{m}}{\partial M}$$

$$\qquad (8.126)$$

$$- \lambda_\gamma \left[-\frac{T}{mU^2} \sin \alpha + \frac{\sin \alpha}{maU} \frac{\partial T}{\partial M} - \frac{L}{mU^2} + \frac{1}{maU} \frac{\partial L}{\partial M} + \frac{g \cos \gamma}{U^2} \right],$$

$$\dot{\lambda}_\gamma = \lambda_u g \cos \gamma + \lambda_\gamma \frac{g \sin \gamma}{U} - \lambda_h U \cos \gamma, \qquad (8.127)$$

$$\dot{\lambda}_h = -\lambda_u \left[\frac{\cos \alpha}{m} \frac{\partial T}{\partial h} - \frac{1}{m} \frac{\partial D}{\partial h} \right] - \lambda_\gamma \left[\frac{\sin \alpha}{mU} \frac{\partial T}{\partial h} - \frac{1}{mU} \right] \frac{\partial L}{\partial h} - \lambda_m \frac{\partial \dot{m}}{\partial h},$$

$$\qquad (8.128)$$

$$\dot{\lambda}_m = -\lambda_u \left[-\frac{T \cos \alpha}{m^2} + \frac{D}{m^2} \right] + \lambda_\gamma \left[\frac{T \sin \alpha}{m^2 U} + \frac{L}{m^2 U} \right]. \qquad (8.129)$$

The initial conditions on the state parameters are all assigned. The terminal conditions can be assigned or left free. The terminal conditions on the Lagrange multipliers are

$$\Lambda(t_e) = 0. \qquad (8.130)$$

The problem formulated above is a boundary-value problem, with eight differential equations. The stopping condition is found from the altitude at the terminal point. The aircraft will reach that altitude with a time t_e, which is assured to be a minimum.

8.13.2 Solution of the Problem

The adjoint equations contain derivatives of the lift, drag, thrust and fuel flow. These quantities are given in tabulated form, as discussed earlier, hence the derivatives have

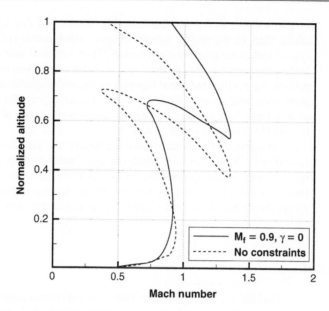

Figure 8.21   *Minimum time to climb programs, from Bryson and Denham[216].*

to be calculated numerically. It is not useful even to attempt to give an expression of the derivatives. Generally, it takes some time to program this part of the algorithm, and it is frustrating that it is neglected in most discussions of the subject.

Second, we have a system of differential equations. There is a variety of numerical methods available today, including gradient methods (steepest descent and similar), multiple shooting algorithm, and dynamic programming. Bryson and Denham[216], in their classic solution of the problem for the F-4 interceptor, used the steepest descent method. The method can be found, already programmed, in Press *et al.*[183]. It requires a first guess of the solution; it then proceeds, by local linearization around the current point, in the direction of the steepest ascent or descent.

The *multiple shooting method* of Bulirsch and Stoer[225,226] has been used by Brüning and Hahn[227] for a variety of optimal climb problems. The method of *dynamic programming* does not involve function derivatives, and belongs to a class of non-gradient methods.

One of Bryson and Denham's solutions is shown in Figure 8.21. In one case, there are no terminal constraints; in the other case, the aircraft was constrained to a minimum-time climb with a final Mach number $M = 0.9$ and horizontal flight $\gamma = 0$.

It is instructive to compare these flight programs with the ones obtained with the energy method. The main difference is that the energy methods lead to some singular points in the flight path.

8.14   ADDITIONAL RESEARCH ON AIRCRAFT CLIMB

The first relevant unsteady analysis of the aircraft climb known to the author is the one published by Miele[228,229] in the 1950s. Optimal problems have been published by Rutowski[217], Kelley and Edelbaum[230], Schultz and Zagalsky[231], Calise[232] and others. Work on the subject includes optimum climb profiles of supersonic transport aircraft with noise minimization (Berton[233]), and optimal near-guidance trajectories leading to minimum fuel, time, or cost for fixed-range (Ardema *et al.*[234]). By comparison, there is little published literature on optimal propeller aircraft climb. Ojha[235] published a brief model of the fastest climb of a piston-propeller aircraft. A discussion of the effects of the vectored thrust on the climb performance of conventional aircraft is available in Gilyard and Bolonkin[236].

An example of three-dimensional climb/turn analysis is that of Neuman and Kreindler[237,238], who derived climb-out and descent flight paths from and to runway headings, including optimization for minimum fuel consumption. These results show that the velocity profiles for straight and turning flight are almost identical, except for the final horizontal accelerating or decelerating turn.

PROBLEMS

1. Find the relationship between glide ratio and aircraft weight for aircraft model A (Appendix A) at cruise speed and cruise altitude. Calculate the maximum L/D and the corresponding weight. Draw conclusions regarding the optimal aircraft's weight.

2. A propeller-driven airplane has a mass $m = 2,000$ kg and wing area $A = 21.5$ m^2. Its drag equation is $C_D = 0.019 + 0.032C_L^2$. The maximum engine power at sea level is 1,200 kW, and the propeller efficiency is a constant value $\eta = 0.81$. Calculate the maximum rate of climb at sea level.

3. Calculate the rate of climb at constant speed on an airplane flying at an altitude where the relative air density is $\sigma = \rho/\rho_o = 0.55$. The airplane has the following characteristics: specific excess thrust $P_e = 0.250$, lift coefficient $C_L = 0.57$, drag equation $C_D = C_{D_o} + kC_L^2$, with $C_{D_o} = 0.022$, $k = 0.033$. The aircraft is initially flying with a speed $U = 400$ km/h.

4. Find the maximum rate of climb of a jet-powered aircraft in the stratosphere having the following (approximate) performance data

$$T = T_o\sigma, \quad \frac{\rho}{\rho_o} = e^{-h/H}, \quad C_D = C_{D_o} + kC_L^2,$$

where H is the limit of the stratosphere.

5. Does the aircraft speed U increase with the wing loading W/A – all other parameters being constant?

6. What is the physical meaning of *energy height*? How can this concept be applied to an aircraft to zoom past its absolute ceiling? How would you control an aircraft that has zoomed past the absolute ceiling?

7. Calculate the excess thrust of aircraft C (supersonic jet fighter) at $M = 0.8$, flight altitude $h = 11,000$ m, with an aircraft gross weight $W = 12,000$ kg.

8. Consider two high-performance aircraft, A and B, having limiting flight envelopes shown in Figure 8.22.

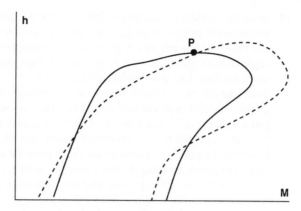

Figure 8.22 *Flight envelopes of two aircraft.*

The essential data of the aircraft are:

Data	Aircraft A	Aircraft B
Operating empty weight	9,000 kg	8,000 kg
Maximum internal fuel	4,500 kg	3,500 kg
Fuel flow at max power	14,000 kg/h	10,000 kg/h

Both aircraft reach point P with 60% fuel remaining. The questions are:

- Which aircraft has the highest excess thrust at point P?
- Which aircraft has the highest specific excess power at point P?

9. Explain the effects of aircraft mass on the descent performance at constant speed of a jet aircraft. Use the relevant equations derived in this chapter.
10. Consider a subsonic jet aircraft. Calculate the speed for maximum climb and the speed for maximum climb rate and compare the two values. Finally, assess which velocity is highest, and find an explanation for it.
11. Find the relationship between speed of minimum sinking rate v_s and the speed of minimum glide angle γ for a generic aircraft. Consider all the data known: the drag coefficients C_{D_o}, k, the weight W, the wing area A. Are these speeds equal? Discuss the results obtained. (*Use the information provided in §8.10.*)
12. Reconsider the problem of the Airbus A-330 having to glide to the nearest airport, having run out of fuel in the middle of the ocean. Repeat the simulation of §8.10.2 with a case of tail wind of 15 m/s. (A tail wind is the most recurring wind situation on east-bound North Atlantic flights.)
13. Calculate the climb polar of the supersonic jet fighter of reference. Consider the effects of aircraft weight and the effects of flight altitude. The power plant is operated without after-burning. The weights to be considered are MTOW and 0.8MTOW. Altitudes of reference are sea level and 10,000 m. The engine data are given in tabulated form in Table A.13 and the aerodynamic characteristics are given in Table A.12.

14. Solve the problem of maximum climb angle for the turboprop aircraft model B, following the method of §8.3.2. The optimal equation to be solved is Eq. 8.40 and the reference program is given in §soft:turboprop-climb. Plot the climb angle and the climb rate with respect to the flight altitude, from sea level to service ceiling. Use $m = 50,000$ kg, $M_{tip} = 0.85$.

15. Study the effects of aircraft weight on the optimal climb rate. Consider the aircraft model B, with take-off weight variable from 45,000 kg to MTOW (as given in Table A.7, Appendix A). Use the methods discussed in §8.3.2. In particular, calculate how the maximum climb rate is affected by take-off weight.

16. Calculate the positive segment of the climb polar for aircraft model C (supersonic jet fighter), by using the flight program in §8.5, at altitudes $h = 7,000$ m, and $h = 14,000$ m, with typical combat weight as indicated in Table A.9. Repeat the analysis for a flight altitude $h = 7,000$ m and a gross weight equal to MTOW. Compare the results obtained at this altitude and draw some conclusions.

17. Calculate the maximum climb rate with one engine inoperative for the turboprop aircraft of reference, assuming that the aircraft is to take off from a sea level airfield on a hot day (temperature on the ground $+30°C$) at MTOW. Use the flight program in §8.3.2.

18. Starting from a flight altitude $h = 10,000$ m, and a Mach number $M = 0.8$, study the effect of the initial climb angle on the acceleration capabilities of the supersonic jet fighter (consider very small angles). Compare the performance obtainable with and without after-burning, with particular reference to the maximum Mach number and the altitude at which it is reached.

19. Find the speed of minimum descent rate and the speed of minimum glide angle for an unpowered glider. Consider a glide ratio $L/D = 25$, a wing loading $W/A = 50$ kg/m^2, and a starting altitude $h = 1,000$ m. Compare the two speeds and draw critical conclusions.

20. The northern flying squirrel (*Glaucomis sabrinus*) is known to jump off trees 25 m high, for distances up to 45 m, and even further. Simulate the gliding flight path of the squirrel, assuming a constant lift coefficient (or altitude). Consider two weights: 0.1 and 0.2 kg. Also, consider the case of a glide with pitch damping, and compare the gliding path to the undamped path at the same gross weight. Follow the method of §8.11.

Chapter 9
Cruise Performance

The Wright Brothers' first flight at Kitty Hawk could have been performed within the 150-foot economy section of a Boeing B-747.

The Boeing Corporation

In this chapter we introduce the concepts of distance flown by a fixed-wing aircraft without stop for refueling or in-flight refueling. We will discuss a number of cruise programs at subsonic and supersonic speeds, the fuel required for a specified mission (fuel planning) and some optimal problems in long range cruise, with or without constraints. We attempt to solve such problems as: (1) the best cruise conditions from the point of view of minimum fuel consumption per unit time or distance; (2) the estimation of the fuel required for a specified mission and (3) the calculation of the best flight profile resulting in minimum fuel consumption, with or without constraints on range and time traveled.

All the models presented are based on ISA conditions. Therefore, we will use the term speed for both *true air speed* and *ground speed*. The effects of atmospheric winds on cruise performance have been published by Hale[239,54]; they include best range conditions, flight times and fuel consumption.

9.1 IMPORTANCE OF THE CRUISE FLIGHT

For most commercial aircraft the fuel consumed during the cruising phase of the flight makes up the bulk of the fuel carried, and is a key factor in the productivity and direct operating costs of an aircraft. Since the early 1970s, with the price of fuel soaring, both the airlines and the military have been concerned with energy efficient operations. For this reason, several cruise conditions have been studied. Kershner[80] reviewed the fuel costs of major international airlines from a historical point of view, and pointed out that while the DOC have been reduced considerably, the cost of the fuel is one of the major cost items, ranging from 23% in 1970 to 58% in 1980, and down to 43% in 1986. Houghton[240] showed that there are benefits in intercepting and using major jet stream winds and the global weather system to adjust the route and flight program. It was proved that fuel may be saved regardless of the wind being a tail-wind or head-wind; savings of about 1% of fuel have been calculated. The limitations of this technique are due to the high volume of aircraft on all major airways, and real-time weather forecast. The idea is not completely new, since the use of atmospheric conditions is recognized as critical to long-range bird flight.

The aircraft range, as the climb, is one of the most common flight conditions. Due to the number of free parameters involved (Mach number, altitude, lift coefficient, angle of attack, gross weight, block fuel), and number of external constraints (Air

Traffic Control, international regulations, flight corridors, atmospheric conditions), there is a variety of optimal and suboptimal conditions. ATC issues and terminal area constraints are reviewed by Visser[241]. Various ESDU's data items deal with cruise performance[17,242,243].

9.2 GENERAL DEFINITIONS

The aircraft range is the distance that can be covered in straight flight at a suitable flight altitude. The cruise cannot use all the fuel, and allowance must be made to account for the terminal phases (take-off and landing), maneuvers (loiter, holding at altitude), fuel reserves for contingency. In practice, the different flight sections are calculated separately, hence the *range equations* are limited to steady flight at altitude.

The *block fuel* is the fuel weight required to fly a specified mission, and includes the fuel to taxi at the airport. The *fuel reserve* is a contingency amount of fuel (as established by the aviation authorities), which takes into account the risk of not being able to land at the destination airport. The *mission fuel* includes (1) the fuel required to take off, accelerate and climb to the initial cruise altitude; (2) the cruise fuel; (3) the descent, terminal area approach and landing fuel and (4) maneuvering and reserve fuel. For the determination of the gross Take-off Weight (*TOW*) the taxi fuel at the departure airport is not included. The taxi fuel after landing is extracted from the reserve fuel.

The *endurance* is the time on station, e.g. the time the aircraft can be flown without landing or in-flight refueling. *Maximum endurance* performance is a basic mission requirement for search and surveillance operations. When in-flight refueling is possible, there is a limit to the number of times this operation can be done. Refueling has traditionally been a problem for the military, but also for some commercial operations (Bennington and Visser[244]), on the grounds that the amount of fuel that has to be carried by a wide-body aircraft on a long-haul flight can be four times the useful payload. Smith[245] has published a historical account of military refueling technology from 1923 onwards. This booklet highlights historical events and technological advances.

There are many performance capabilities relative to range and endurance, such as the *mission radius* (take-off, cruise out, land, deliver payload, and return to airport of origin); the *combat radius* (take-off, cruise to theater of operation, perform mission, fly back to airbase, land), and more (see the section on mission requirements). However, the calculation procedure is the same, as it just requires adding range and endurance to a basic flight segment. In cruise performance analysis all the fuel burned before reaching cruise altitude and speed is referred to as *lost fuel*.

9.3 POINT PERFORMANCE

The instantaneous conditions in aircraft cruise are called *point performance*. The basic point parameters are the glide ratio, the specific range, the figure of merit (*FoM*), the instantaneous endurance, and the glide ratio. The glide ratio was discussed in the aerodynamics chapter (§4.6). Other instantaneous parameters of interest include the product $C_L M^2$. These parameters are important because the optimal flight path or flight program for a long range cruise can be derived from integration of the point performance parameters. Generalized point performance optimization is discussed in detail by Torenbeek and Wittenberg[246].

9.3.1 Specific Range at Subsonic Speed

The *specific range* is the range flown by burning one unit of weight (or mass or volume) of fuel. For operational reasons, other units may be used. For example, to evaluate the productivity of an aircraft the airlines may quote a fuel consumption in unit of volume or weight per unit of distance per passenger. The Airbus A-330-200, operating at its speed for long range cruise uses about 2.7 liters/100 km/passenger (or 0.087 lb/nm/passenger).

In the following discussion the symbol f_j will denote the Thrust-Specific Fuel Consumption (TSFC). The elementary range dR obtainable by burning a small amount of fuel dm can be written as

$$dR = U dt = U \frac{dm}{dm} dt = \frac{U}{\dot{m}_f} dm. \tag{9.1}$$

The Specific Air Range (*SAR*) is the derivative of the range with respect to the aircraft mass (or weight)

$$SAR = \frac{\partial R}{\partial m} = \frac{U}{\dot{m}_f} = \frac{U}{f_j T}. \tag{9.2}$$

The latter equivalence is only valid for jet-powered aircraft. The physical dimensions of *SAR* are distance × unit mass of fuel (m/kgm in metric units). The specific range can be further evaluated by inserting the proper value of the engine thrust in Eq. 9.2. To have an idea of the order of magnitude of *SAR*, use the general cruise conditions of the subsonic jet aircraft of reference: $M = 0.8$, $U = 236$ m/s, $f_j \simeq 1.61 \cdot 10^{-4}$ N/s/N (0.578 lb/h/lb). These data give $SAR \simeq 17$ m/kgm. Therefore, the *SAR* is of the order of 10 m per kgm of fuel burned. As in the case of the fuel flow and the specific fuel consumption, there are about 10 different ways of expressing the specific range.

If $\dot{m}_f$ as a function of the speed and altitude is known, the *SAR* can be plotted directly from the first equivalence of Eq. 9.2. In other cases we have to calculate the drag (thrust) at cruise conditions. For a subsonic jet aircraft with a parabolic drag equation the specific range is

$$SAR = \frac{U}{f_j D} = \frac{U}{f_j} \frac{1}{c_1 \sigma (C_{D_o} + k C_L^2) U^2} = \frac{1}{c_1 \sigma f_j (C_{D_o} + k C_L^2) U}, \tag{9.3}$$

where

$$c_1 = \rho_o \frac{A}{2}. \tag{9.4}$$

Now insert the definition of C_L and find, after some algebra, the following equation

$$SAR = \frac{1}{f_j} \frac{1}{c_1 \sigma C_{D_o} U + c_2 k \, m^2 / \sigma U^3} = f(h, U, m), \tag{9.5}$$

with

$$c_2 = c_1 \left(\frac{2g}{\rho_o A} \right)^2 = \frac{2g^2}{\rho_o A}. \tag{9.6}$$

Equation 9.5 is a function of the speed, the flight altitude and the aircraft's mass. It satisfies the conditions $D = T$, $L = W$. There are several ways to rearrange Eq. 9.5. For example, the speed can be replaced by $U = aM$, and the mass can be normalized with the aircraft mass at the start of the cruise ($\xi = m/m_i$). After all this manipulation is done, the normalized specific range is

$$SAR = \frac{1}{f_j m_i^2} \frac{1}{\hat{c}_1(\sigma)M + \hat{c}_2(\sigma)\xi^2/M^3}, \tag{9.7}$$

with coefficients

$$\hat{c}_1(\sigma) = \frac{c_1 C_{D_o}\sigma a}{m_i^2}, \qquad \hat{c}_2(\sigma) = \frac{c_2 k}{\sigma a^3}, \tag{9.8}$$

and c_1, c_2 given by Eq. 9.4 and Eq. 9.6, respectively. The coefficients c_1 and c_2 depend on the flight altitude.

We plot the specific range at selected altitudes as a function of the relative mass ξ, from the start to the end of the cruise. The Mach number generally does not vary much during the cruise, therefore a family of curves at selected Mach numbers can be added to the chart. The specific range for a subsonic commercial jet with a parabolic drag is shown in Figure 9.1. It is demonstrated that: (1) at a constant flight altitude and

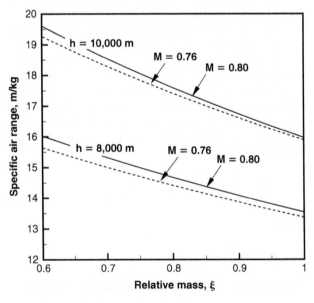

Figure 9.1 *Estimated specific range for subsonic commercial jet with parabolic drag, AUW = 160,000 kg.*

Mach number the specific range increases with the decreasing aircraft's mass and (2) at a constant aircraft weight and Mach number the specific range increases with the cruise altitude. Clearly, there is a limit to the second conclusion, because the present model requires the aircraft to operate at subcritical Mach numbers, $M < M_{dd}$.

For a flight at constant speed, one can study the effects of an aircraft's initial mass m_i on the specific range (Problem 8).

The effect of the cruise speed on the specific range can be found by studying the derivative $\partial SAR/\partial U$. In particular, the condition that gives the speed of maximum SAR is

$$\frac{\partial SAR}{\partial U} = \frac{\partial}{\partial U}\left[\frac{1}{c_1 \sigma f_j (C_{D_o} + kC_L^2)U}\right] = 0. \tag{9.9}$$

Assuming that the change in TSFC is negligible, solution of this equation leads to the speed for best range

$$U_{SAR} = \left(\frac{2W}{\rho_o \sigma A}\right)^{1/2}\left(\frac{3k}{C_{D_o}}\right)^{1/4}. \tag{9.10}$$

This speed depends on the wing loading and on the flight altitude. For a constant flight altitude, U_{SAR} decreases as the aircraft burns fuel. Figure 9.2 shows the behavior of the SAR as a function of the Mach number at selected altitudes. The AUW has been kept constant, and equal to the initial weight.

As the flight altitude increases, the Mach number for best range also increases, and eventually it reaches a value beyond the transonic drag rise. Therefore, as far as

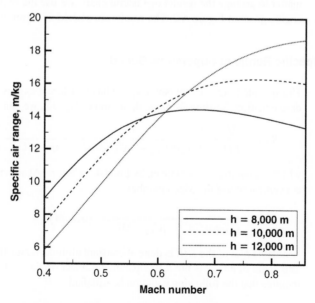

Figure 9.2 *Estimated specific range versus Mach number for subsonic jet aircraft, with AUW = 160,000 kg, at flight altitudes indicated.*

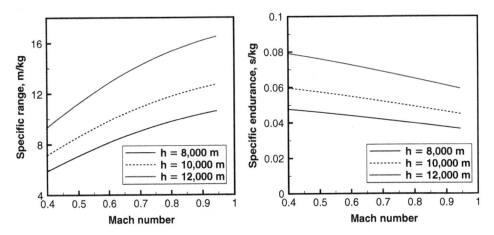

Figure 9.3 *Estimated SAR and E_s for aircraft model A, at the flight altitudes indicated. Calculations with flight data.*

optimal range is concerned, it is not convenient to fly above a certain altitude, lest the range be penalized by a drag rise.

If we compare the speed for maximum *SAR* and the speed of minimum drag, Eq. 6.50, we find that there is a factor $3^{1/4} = 1.316$ between the two. More precisely: $U_{SAR} = 1.316\, U_{md}$.

If the drag cannot be expressed by a parabolic equation, then we need to refer to Eq. 9.2 in order to calculate the *SAR*. The data required are the fuel flow as a function of the flight speed and altitude. Once again, if the data are tabulated, it is a simple matter to arrange the results in a useful chart. We use the fuel flow data for the CF6 engine given in Figure A.2. The results are plotted in Figure 9.3.

9.3.2 Specific Range at Supersonic Speed

The specific range at supersonic speed has a different expression, due to a different drag equation. Starting from the definition of Eq. 9.2, we can write

$$SAR = \frac{U}{f_j D} = \frac{U}{f_j}\frac{1}{c_1\sigma(C_{D_o} + \eta C_{L_\alpha}\alpha^2)U^2} = \frac{1}{c_1 f_j \sigma(C_{D_o} + \eta C_{L_\alpha}\alpha^2)U}, \quad (9.11)$$

where the coefficient c_1 is given by Eq. 9.4, as before. At supersonic speeds it is more convenient to use the Mach number

$$SAR = \frac{1}{c_1 f_j a \sigma(C_{D_o} + \eta C_{L_\alpha}\alpha^2)M} = f(\alpha, h, M). \quad (9.12)$$

A suitable plot of the *SAR* is done at constant altitude versus the flight Mach number. The angle of attack is not a free parameter, because the vertical equilibrium $L = W$ requires that the following equation be satisfied

$$\alpha = \alpha_o + \frac{2W}{\rho A a^2}\frac{1}{C_{L_\alpha}M^2}. \quad (9.13)$$

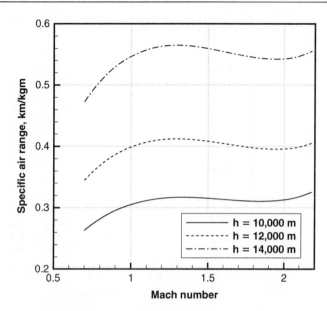

Figure 9.4 *Estimated SAR of supersonic fighter jet at the altitudes indicated; aircraft mass m = 12,000 kgm.*

The computational procedure is the following.

Computational Procedure

1. Set the flight altitude.
2. Set the Mach number (loop).
3. Calculate the angle of attack from Eq. 9.13.
4. Calculate the specific range from Eq. 9.12.
5. Increase the Mach number and iterate from point 3.

The stopping criterion is the limit of the flight data. Such a computational procedure, repeated for a number of representative altitudes, is shown in Figure 9.4.

Calculation of the speed (Mach number) corresponding to maximum specific range at transonic and supersonic speeds is more elaborate, because of the lack of an analytical expression for the glide ratio. We calculate *SAR* from Eq. 9.12, sweep the Mach number range at given flight altitude and AUW, and store its maximum value.

9.3.3   Specific Endurance, E_s

The *specific endurance* is defined as the flight time per unit of fuel mass (weight) burned, and is the inverse of the fuel flow. For a jet aircraft the change of aircraft mass with respect to time is

$$\frac{\partial m}{\partial t} = \dot{m} = -\dot{m}_f = -f_j T.$$

(9.14)

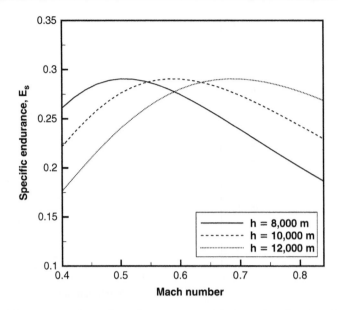

Figure 9.5   *Estimated specific endurance versus Mach number for aircraft model A, AUW = 160,000 kg, at flight altitudes indicated. (Same flight case as Figure 9.2).*

The instantaneous endurance is the inverse of the instantaneous fuel flow

$$E_s = \frac{\partial t}{\partial m_f} = \frac{1}{\dot{m}_f} = \frac{1}{f_j T} = \frac{1}{f_j D}. \tag{9.15}$$

If the fuel flow as a function of the speed (Mach number) is known, then plotting E_s is quite simple. The relationship between the specific endurance and the specific range is

$$E_s = \frac{SAR}{U}. \tag{9.16}$$

Thus, there is proportionality between E_s and SAR; maximum SAR leads to maximum E_s, and vice versa.

The specific endurance has been calculated for the reference subsonic jet aircraft (model A) and plotted in Figure 9.5 for three reference altitudes. The chart shows that E_s reaches a maximum at Mach numbers increasing with the altitude. In order words, as the aircraft flies at higher altitudes it must maintain a higher Mach number for propulsion efficiency.

For a propeller aircraft, the specific endurance is

$$E_s = \frac{1}{\dot{m}_f} = \frac{1}{SFC\,P} = \frac{TU}{SFC\eta}. \tag{9.17}$$

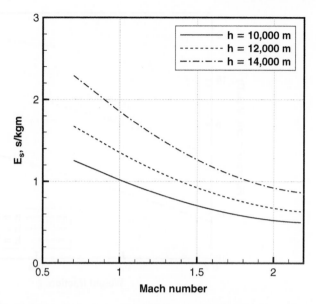

Figure 9.6 *Estimated E_s of supersonic fighter jet at the altitudes indicated, aircraft mass m = 12,000 kgm. (Same case as in Figure 9.4).*

9.3.4 Figure of Merit, *M (L/D)*

We seek conditions that lead to a constant figure of merit $FoM = M(L/D)$. Replace the parabolic drag expression, valid below the divergence Mach number M_{dd}, and find, after simplification,

$$M\left(\frac{L}{D}\right) = M\left(\frac{C_L}{C_D}\right) = M\frac{C_L}{C_{D_o} + kC_L^2} = M\frac{c_1 m/U^2\sigma}{C_{D_o} + kc_1^2 m^2/\sigma^2 U^4}, \tag{9.18}$$

$$M\left(\frac{L}{D}\right) = \frac{c_1 m/a^2 M\sigma}{C_{D_o} + kc_1^2 m^2/\sigma^2 a^4 M^4}, \tag{9.19}$$

where

$$c_1 = \frac{2g}{\rho_o A}. \tag{9.20}$$

Equation 9.19 allows us to study the effects of flight altitude, aircraft mass and flight Mach number. However, since the compressibility effects are not incorporated in the parabolic drag, the analysis is limited to a fixed Mach number, say $M = 0.8$.

A parametric study of Eq. 9.19 shows that: (1) at constant Mach number the *FoM* increases with the flight altitude and (2) at a constant Mach number and constant altitude the FoM decreases with decreasing aircraft's mass (e.g. it decreases during the cruise). Figure 9.7 shows the FoM for an arbitrary value of the aircraft weight and

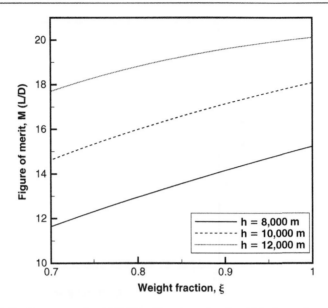

Figure 9.7   *Representative M (L/D) of jet aircraft at fixed Mach number,*
M = 0.8, initial mass m_i = 160,000 kg.

a fixed Mach number. The results show once again that it is more efficient to fly at
high altitudes.

As fuel is burned during the cruise, the weight decreases. If the Mach number is
to be maintained constant, the aircraft must climb in order to maintain a constant
FoM. In practice, continuous changes in altitude and Mach number are not allowed
by the Air Traffic Control (*ATC*); only step changes from one flight level to another
are allowed. By convention, two flight levels are 100 ft apart, up to 29,000 ft; above
this altitude the flight levels are separated by 1,000 ft.

The Mach number considered in the analysis of Figure 9.7 is realistic for modern
subsonic commercial jets. However, if a detailed analysis is required to investigate
the optimal Mach number for cruise, then the data $C_D(M, C_L)$ are required. For the
reference aircraft, some aerodynamic data are given in Appendix A, see Figure A.3,
along with Tables A.5 and A.6. The *FoM* of this aircraft is shown in Figure 9.8.
The best Mach number is slightly below $M = 0.8$ in the whole range of C_L. The best
FoM is estimated at $C_L \simeq 0.5$. At higher values of C_L, the *FoM* decreases, due to an
increase in cruise drag. This analysis demonstrates that it is necessary to cruise at low
C_L, at speeds below the transonic drag rise.

For the transonic and supersonic cruise of the fighter jet of reference, we use the
alternative drag Eq. 4.12, which represents correctly the effects of Mach number on
the aerodynamic drag of the aircraft. For such a case, we use the aerodynamic data
in Table A.12. The *FoM* becomes

$$M\left(\frac{L}{D}\right) = M\frac{C_{L_\alpha}\alpha}{C_{D_o} + \eta C_{L_\alpha}\alpha^2}. \tag{9.21}$$

As for the case of the specific range, there is no local maximum with respect to the
angle of attack. The dominant parameter in the figure of merit is the flight Mach

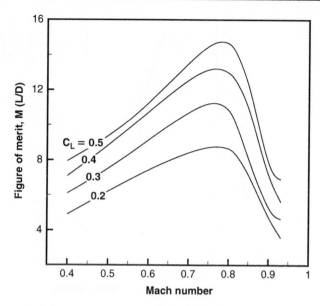

Figure 9.8   *Estimated M (L/D) of reference jet aircraft as a function of the cruise Mach number.*

number. The angle of attack in Eq. 9.21 is to be replaced with the equilibrium in the vertical direction, Eq. 9.13, and yields an involved expression of the FoM, in which the free parameters are the flight altitude and the Mach number. If

$$c_1(\sigma) = \frac{2g}{\rho_o \sigma a^2 A},$$ (9.22)

then

$$M\left(\frac{L}{D}\right) = \frac{c_1(\sigma)m/M}{C_{D_o} + \eta C_{L_\alpha} c_1^2(\sigma) m^2/M^4}.$$ (9.23)

Again, it is convenient to normalize this expression with the initial mass m_i, and plot the FoM versus Mach number at selected flight altitudes and for a given aircraft weight. The normalized equation is

$$M\left(\frac{L}{D}\right) = \frac{c_1(\sigma)\xi/(m_i M)}{C_{D_o}/m_i^2 + \eta C_{L_\alpha} c_1^2(\sigma)\xi^2/M^4} = f(\xi, h, M).$$ (9.24)

The FoM of this aircraft has two local maxima: at subsonic and supersonic Mach numbers. These maxima can be found by bookkeeping in a numerical procedure, rather than by derivation. It is enough to store in memory the local maxima as the Mach number is increased. Figure 9.9 shows that the optimal subsonic Mach number is about $M = 0.9$ to 1.0, and the optimal supersonic Mach number is $M = 1.6$ to 1.8. The exact values of these maxima depend on the flight altitude and on the relative weight of the aircraft.

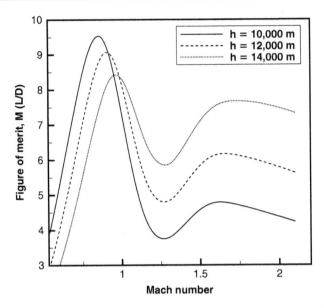

Figure 9.9 *M(L/D) of supersonic fighter jet aircraft at altitudes indicated.*

Optimal flight conditions over a long range cruise are found from the operational conditions that always give maximum specific range or maximum FoM.

9.4 THE BREGUET RANGE EQUATION

In this section we shall be concerned with the integrated performance in cruise condition, which is the basic range equation for a fixed-wing aircraft and a fixed fuel load at the start of the cruise. The Breguet range equation* is the main tool for the calculation of the cruise range of a generic aircraft, although it does not directly provide any optimal cruise conditions, and does not discriminate between flight programs. A number of approximations are used in practice to get first-order estimates useful in aircraft design.

The aircraft range is defined by the integral

$$R = \int_o^t U \, dt. \tag{9.25}$$

We multiply and divide Eq. 9.25 by dm_f, and replace $dm_f/dt = \dot{m}_f$. In order to have the effects of the fuel flow and the instantaneous aircraft mass, we find

$$R = \int_o^t U \frac{dm_f}{dm_f} \, dt = \int_i^e U \frac{dm_f}{\dot{m}_f} = -\int_i^e U \frac{dm}{\dot{m}} = \int_e^i U \frac{dm}{\dot{m}}, \tag{9.26}$$

---

* Louis Breguet (1880–1955) derived the original equation for a propeller aircraft.

because the loss of weight is due only to fuel burn. The indices i and e denote initial and end conditions, respectively. For a jet aircraft the fuel flow is replaced by the equation $\dot{m}_f = f_j T$. Now multiply and divide Eq. 9.26 by the aircraft weight and assume that at cruise conditions $L = W$. Thus, we have

$$R = \int_e^i \frac{U}{f_j T} \, dm = \int_e^i \frac{U}{f_j D} \, dm = \frac{1}{g} \int_e^i \frac{U}{f_j} \left(\frac{L}{D} \right) \frac{dm}{m}. \tag{9.27}$$

In Eq. 9.27 the fuel flow $\dot{m}_f$ is to be given in units of mass per unit of time. If other units are on input, the result would be wrong by one order of magnitude, or more. The product $g f_j$ is a fuel flow expressed in units of force per unit of time.

If a first-order approximation is sufficient, all the factors within the integral are considered constant, and the Breguet equation is reduced to

$$R \simeq \frac{U}{g f_j} \left(\frac{L}{D} \right) \ln \left(\frac{m_i}{m_e} \right) = \frac{U}{g f_j} \left(\frac{L}{D} \right) \ln \left(\frac{1}{1 - \zeta} \right), \tag{9.28}$$

where m_i is the mass at the start of the cruise, m_e is the mass at the end, and ζ is the block fuel ratio, $\zeta = m_f / m_i$

$$\frac{m_i}{m_e} = \frac{m_i}{m_i - m_f} = \frac{1}{1 - \zeta}. \tag{9.29}$$

In Eq. 9.28 L/D is an average value. In conclusion, at a fixed altitude the range increases with the flight speed, with the glide ratio and with the block fuel ratio.

More rigorously, the calculation of the cruise range requires the solution of an integral that contains several parameters of the aircraft: the initial weight, the block fuel, the flight altitude, the air speed, the specific fuel consumption, and the drag characteristics. This is indicated by the function

$$R = f \left(h, U, f_j, \frac{L}{D}, m_i, m_f \right). \tag{9.30}$$

The speed can be replaced by the Mach number. The glide ratio is a function of the aircraft's angle of attack and the Mach number, as explained in §4.7, $L/D = f(\alpha, M)$. Finally, the specific fuel consumption depends on the flight altitude and the Mach number, $f_j = f(h, M)$. The initial weight and the fuel mass can be combined into the non-dimensional block fuel ratio, Eq. 9.29. Therefore, Eq. 9.30 becomes

$$R = f(h, M, \alpha, \zeta). \tag{9.31}$$

The angle of attack is not the most useful parameter, and it is usually replaced with the C_L, because $C_L = C_{L_\alpha} \alpha$. Also, instead of the flight altitude, it is useful to have the relative air density (density altitude), which appears in the aerodynamic terms. Thus, we arrive at the expression

$$R = f(\sigma, M, C_L, \zeta), \tag{9.32}$$

which is a function of non-dimensional parameters only. Optimal solutions of the cruise conditions are of great importance, both financial and environmental. For example, a 1% fuel saving for cruising would save an estimated 6 million liters of jet fuel per day, or 4.75 million tons, or 37,700 barrels per day of refined oil[*].

We call each solution of Eq. 9.32 a *flight program*. Global optima lie in a four-dimensional space. Flight programs of interest are found with one or two parameters being constant.

9.5   SUBSONIC CRUISE OF JET AIRCRAFT

We consider three flight programs at subsonic speed: (1) cruise at constant altitude and constant Mach number; (2) cruise at constant Mach number and constant lift coefficient; and (3) cruise at constant altitude and constant lift coefficient. All these cases require a starting value of h, M and C_L, for which an optimal solution can be found.

Another case is constant altitude and constant throttle setting. The decreasing aircraft's weight will result in an increasing Mach number. The ATC rules do not favor variable flight speed, therefore this flight condition is less interesting than the previous ones. The mass of the aircraft is variable from an initial value m_i to a final value m_e.

9.5.1   Cruise at Constant Altitude and Constant Mach Number

This type of cruise condition is quite practical. During the cruise, the aircraft's drag decreases, because of the decreasing weight. As the drag decreases, also the engine thrust must decrease, and the throttle has to be stepped down. This is done automatically with the modern flight control systems. From Eq. 9.27 the range becomes

$$R = \frac{aM}{gf_j} \int_e^i \left(\frac{L}{D}\right) \frac{dm}{m}, \tag{9.33}$$

because the speed of sound depends only on the altitude (ISA conditions), and the thrust-specific fuel consumption depends on both altitude and Mach number. This problem requires the solution of an integral with a variable glide ratio. The maximum cruise range is found from always operating the aircraft at maximum glide ratio. If we use the parabolic drag equation, the glide ratio becomes a function of the aircraft mass, as shown in §4.6,

$$\frac{L}{D} = \frac{C_L}{C_D} = \frac{c_1 m}{C_{D_o} + kc_1^2 m^2}, \tag{9.34}$$

where

$$c_1 = c_1(\sigma, M) = \frac{2g}{\rho A U^2} = \frac{2g}{\rho_o \sigma A a^2 M^2}, \tag{9.35}$$

---

[*] Data for fuel consumption are about 700 million liters per day (in 2000), of which we assume 85% is used for cruise. One barrel is equal to 159 liters.

is a coefficient depending on altitude and Mach number. The angle of attack of the aircraft does not appear explicitly in Eq. 9.34. We know from the previous discussion (Figure 4.13) that the L/D reaches a maximum at one aircraft weight, then it decreases. From Eq. 9.33

$$R = \frac{aM}{gf_j} \int_e^i \frac{c_1 m}{C_{D_o} + kc_1^2 m^2} \frac{dm}{m} = \frac{aM}{gf_j} \int_e^i \frac{c_1}{C_{D_o} + kc_1^2 m^2} \, dm. \tag{9.36}$$

The solution of the integral is from the tables of indefinite integrals[247]

$$\int \frac{c_1}{C_{D_o} + kc_1^2 m^2} dm = \frac{c_1}{\sqrt{kC_{D_o} c_1^2}} \tan^{-1}\left(m\sqrt{\frac{kc_1^2}{C_{D_o}}}\right)$$

$$= \frac{1}{\sqrt{kC_{D_o}}} \tan^{-1}\left(c_1 m \sqrt{\frac{k}{C_{D_o}}}\right). \tag{9.37}$$

The definite integral between the start and end of the cruise is

$$\int_e^i \frac{c_1}{C_{D_o} + kc_1^2 m^2} \, dm = \frac{1}{\sqrt{kC_{D_o}}} \left[\tan^{-1}\left(c_1 m \sqrt{\frac{k}{C_{D_o}}}\right)\right]_e^i. \tag{9.38}$$

The final expression for the cruise range is

$$R = \frac{aM}{gf_j} \frac{1}{\sqrt{kC_{D_o}}} \left[\tan^{-1}\left(c_1 m \sqrt{\frac{k}{C_{D_o}}}\right)\right]_{m_e}^{m_i}, \tag{9.39}$$

$$R = \frac{aM}{gf_j} \frac{1}{\sqrt{kC_{D_o}}} \left[\tan^{-1}\left(c_1 m_i \sqrt{\frac{k}{C_{D_o}}}\right) - \tan^{-1}\left(c_1 m_e \sqrt{\frac{k}{C_{D_o}}}\right)\right]. \tag{9.40}$$

This equation can be reduced further by using some goniometric equivalences (see §9.5.5). From the known values of the mass at the start and end of the cruise, it is possible to find the range from Eq. 9.40. In particular, we find that at a given altitude the range increases with the flight Mach number. However, the aircraft cannot exceed the divergence Mach number, M_{dd}. The flight altitude is implicit in the coefficient c_1.

9.5.2 Cruise at Constant Altitude and Lift Coefficient

A way to look at this flight program is to consider the balance of forces in the vertical direction, $L = W$. With the expression of the C_L we find that the air speed and the Mach number are proportional to $W^{1/2}$

$$M = \left(\frac{2W}{a\rho A C_L}\right)^{1/2}. \tag{9.41}$$

Therefore, it is required that the Mach number be constantly reduced as the engines burn fuel. With the substitution of Eq. 9.41 the range equation is written as

$$R = \frac{a}{gf_j} \left(\frac{C_L}{C_D} \right) \left(\frac{2g}{a\rho A C_L} \right)^{1/2} \int_e^i m^{-1/2} \, dm, \tag{9.42}$$

where we have assumed that the changes in fuel consumption due to changes in Mach number can be neglected. Constant C_L implies constant C_D and constant C_L/C_D. By further algebraic manipulation, the cruise range becomes

$$R = \frac{1}{f_j} \left(\frac{C_L^{1/2}}{C_D} \right) \left(\frac{2a}{g\rho A} \right)^{1/2} \left[2\sqrt{m} \right]_e^i, \tag{9.43}$$

$$R = \frac{2}{f_j} \left(\frac{C_L^{1/2}}{C_D} \right) \left(\frac{2a}{g\rho A} \right)^{1/2} \left[\sqrt{m_i} - \sqrt{m_e} \right]. \tag{9.44}$$

This flight program has a number of drawbacks, namely: (1) there is a loss in jet engine efficiency with a reduced Mach number; (2) the cruise time is increased; (3) a continuous variation of throttle setting is required; and (4) the cruise range is shorter than other flight programs. Finally, a variable Mach number is not contemplated by the international regulations.

9.5.3 Cruise at Constant Mach Number and Constant C_L

The cruise of a jet aircraft with a constraint on Mach number and lift coefficient is a *cruise/climb flight*. With these constraints the C_D, the glide ratio C_L/C_D and the *FoM* are constant. Another consequence is that the factor

$$C_L M^2 = \frac{2W}{\rho A a^2} = f(m, \sigma) = const. \tag{9.45}$$

Equation 9.45 is also the condition of vertical equilibrium of the aircraft. The range equation becomes

$$R = \frac{1}{g} M \left(\frac{L}{D} \right) \int_e^i \frac{a}{f_j} \frac{dm}{m}. \tag{9.46}$$

The conditions for a constant FoM were found earlier in §9.3.4. We proved that a climb is necessary to compensate for the changes in mass. The effect of altitude is implicit in the TSFC and in the speed of sound. For flight in the stratosphere ($h > 11,000$ m), the speed of sound is also a constant. Then the cruise range can be written as

$$R \simeq \frac{a}{gf_j} M \left(\frac{L}{D} \right) \int_e^i \frac{dm}{m} = \frac{a}{gf_j} M \left(\frac{L}{D} \right) \ln \left(\frac{1}{1-\zeta} \right). \tag{9.47}$$

If the TSFC is constant, then the maximum range is found from the condition of maximum FoM throughout the cruise. The TSFC is not a constant, but depends on both temperature and Mach number. A suitable approximating function for a high-by-pass ratio turbofan is

$$TSFC = f_j = c\sqrt{\theta M}, \tag{9.48}$$

where c is a constant. Above 11,000 m the temperature is constant, therefore cruise at a constant Mach number in the lower stratosphere is an optimal condition. The global optimum range is found from the maximum of

$$f(h, M, C_L) = \frac{1}{f_j} M \left(\frac{L}{D} \right). \tag{9.49}$$

As the weight decreases, the aircraft has to adjust its altitude so as to maintain the optimal flight condition. From the definition of aerodynamic coefficients, the glide ratio can be written as

$$\frac{L}{D} = \frac{c_1 m/\sigma}{C_{D_o} + k c_1^2 (m/\sigma)^2}, \tag{9.50}$$

with $c_1 = 2g/\rho_o A$. In order to keep this ratio constant

$$\frac{m}{\sigma} = \left(\frac{m}{\sigma} \right)_i = const. \tag{9.51}$$

Therefore, as fuel is burned, the mass decreases; the relative density has to decrease, hence the aircraft has to climb. The value of Eq. 9.51 is fixed by the initial cruise conditions. Figure 9.10 shows how the flight altitude has to be increased as the aircraft flies.

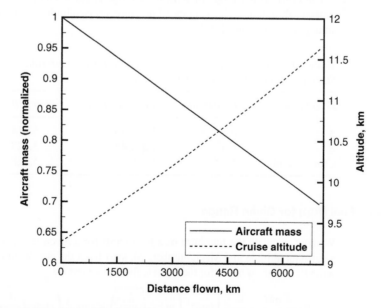

Figure 9.10 *Cruise/climb profile.*

As mentioned, a smooth cruise/climb flight path is not allowed for transport category aircraft. When a climb is required, the aircraft moves from one flight level to another in relatively short time. These levels are separated by 100 ft (30.5 m). A flight level is indicated as the flight altitude in feet divided by 100. Therefore, a flight level *FL290* corresponds to 29,000 ft. The increase in altitude shown in Figure 9.10 corresponds to several *step-ups*. The fuel required for these short segments can be calculated from energy considerations.

A further consideration is that for maximum range L/D has to be optimal at the start of the cruise. This maximum is a function of the aircraft mass; for a fixed mass (as calculated from the mission analysis), it depends on the speed and altitude. For a fixed Mach number (say $M \simeq 0.80$), this global maximum is found at altitudes exceeding the operational ceiling of the aircraft, see Figure 4.14. In conclusion, at the start of the cruise the aircraft has to be in the lower stratosphere with an optimum weight.

9.5.4 Comparison Between Cruise Programs

We are now in a position to compare the various flight programs, assuming that the aircraft has a fixed AUW, a fixed fuel mass, and the same initial conditions, namely the flight altitude, the Mach number and the lift coefficient. We consider the best range at the conditions summarized in Table 9.1.

Consider aircraft model A, with an AUW $= 145,000$ kg and a fuel fraction $\zeta = 0.138$. The starting point is $h = 11,000$ m, $M = 0.80$ for all the cruise profiles. The initial conditions require that the initial cruise lift by $C_L = 0.539$, and the glide ratio $L/D \simeq 17.98$, and $FoM \simeq 14.38$. Assume an average TSFC of the order $f_j = 1.162 \cdot 10^{-5}$ kg m/s/N. The results are summarized in the right-hand column of Table 9.1.

The cruise/climb program achieves a range about 4% higher than the cruise at constant altitude and Mach number. Therefore, cruise/climb is the best flight program among the cases considered.

Table 9.1 *Summary of subsonic cruise conditions, jet aircraft. Starting conditions: $h = 11,000$ m, $M = 0.80$.*

Flight program	Constraints	Range equation	Range (km)	E (hours)
A	h, M	Eq. 9.40	5,314	6.25
B	h, C_L	Eq. 9.44	5,328	6.27
Cruise/Climb	M, C_L	Eq. 9.47	5,528	6.51

9.5.5 Fuel Burn for Given Range

We now calculate the fuel burn of a jet aircraft for a given range (or segment) and a constant altitude and Mach number. We start from the conclusions of §9.5.1, and write the range equation for the generic flight segment *i–j*.

$$R_{ij} = \frac{aM}{gf_j} \frac{1}{\sqrt{kC_{D_o}}} \left[\tan^{-1}\left(m_j c_1 \sqrt{\frac{k}{C_{D_o}}} \right) - \tan^{-1}\left(m_i c_1 \sqrt{\frac{k}{C_{D_o}}} \right) \right]. \quad (9.52)$$

If the altitude and the Mach number are specified, to simplify Eq. 9.52, introduce the factors

$$c_3 = \frac{aM}{gf_j} \frac{1}{\sqrt{kC_{D_o}}}, \qquad c_4 = c_1 \sqrt{\frac{k}{C_{D_o}}}, \tag{9.53}$$

$$R_{ij} = c_3 [\tan^{-1}(c_4 m_j) - \tan^{-1}(c_4 m_i)], \tag{9.54}$$

which (after looking at the tables of trigonometric equivalences in the reference books) can be further reduced to

$$R_{ij} = c_3 \left[\tan^{-1} \left(\frac{c_4 m_i - c_4 m_j}{1 + c_4^2 m_i m_j} \right) + \pi \right]. \tag{9.55}$$

The term π only appears if $c_4^2 m_i m_j > 1$. Now solve Eq. 9.55 to find the value of the fuel mass during the flight segment,

$$\left(\frac{R_{ij}}{c_3} - \pi \right) = \tan^{-1} \left(\frac{c_4 m_i - c_4 m_j}{1 + c_4^2 m_i m_j} \right), \tag{9.56}$$

$$\tan \left(\frac{R_{ij}}{c_3} - \pi \right) = \left(\frac{c_4 m_i - c_4 m_j}{1 + c_4^2 m_i m_j} \right). \tag{9.57}$$

The unknowns are the aircraft mass at the end of the flight segment and the fuel mass. We assume that

$$m_i = m_j - m_f, \tag{9.58}$$

and we solve Eq. 9.57 for the unknown fuel burn. Use the known coefficient

$$c_5 = \tan \left(\frac{R_{ij}}{c_3} - \pi \right), \tag{9.59}$$

to simplify the equation. Therefore:

$$c_5 = \frac{c_4 m_i - c_4 m_j}{1 + c_4^2 m_i m_j}, \tag{9.60}$$

$$c_4 m_i - c_4 m_j = c_5 (1 + c_4^2 m_i m_j). \tag{9.61}$$

Finally, the mass of fuel burned during the flight segment from i to j is

$$m_f = \frac{c_5 + c_4^2 c_5 m_i^2}{c_4 + c_4^2 c_5 m_i}. \tag{9.62}$$

The problem is not closed, because the initial mass m_i may not be known exactly (it depends on the amount of fuel loaded, on the fuel burned so far, etc.). Thus, Eq. 9.62 represents the fuel burn during the segment i–j for given altitude, Mach number and aircraft mass at the start of the segment.

9.6   MISSION FUEL

We have discussed in preliminary analysis the aircraft's weight, the mission fuel and a number of cruise programs. We now attempt to calculate the mission fuel for a specified range. The useful expressions for the cruise of a subsonic jet aircraft are Eq. 9.40 (constant altitude and constant Mach number) and Eq. 9.47 (cruise/climb technique). Calculation of the mission fuel requires adding up the fuel required by each segment of the flight. For a passenger operation, the mission requirement is quite straightforward. We need the fuel for taxiing out, take-off, climb to cruise altitude, cruise, descent, terminal area maneuver, landing and taxiing in. In addition, we must take into account a number of contingencies, such as those required by the aviation regulations. A typical mission profile for a commercial jet is shown in Figure 9.11. It shows a climb to cruise altitude, a cruise, and a descent. It also includes the case of an aborted landing and an extension of the flight at a lower altitude. Diagrams of this nature can be made more precise by considering a quantitative scale for the distance and altitude and for each flight segment.

9.6.1   Fuel for Taxi and Take-off

The fuel for taxiing is negligible for most aircraft. Very large aircraft allow for a maximum taxi fuel of the order of 1,500 liters (1,200 kg). This is extrapolated from the difference between maximum ramp weight and maximum take-off weight, as discussed in §3.2. For an aircraft such as the Boeing B-747-400, this corresponds to about 0.8% of the maximum fuel. One case when it must be taken into account is when the aircraft is on hold at the airfield, due to local traffic and bad weather. Fuel consumption on the taxiway during holding time can be so much that the flight has to be aborted.

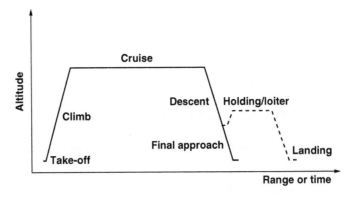

Figure 9.11   *Mission profile of passenger aircraft.*

A first-order estimate of the take-off fuel is found from energy considerations:

$$m_{f_{to}} \bar{\eta} c_p = \frac{1}{2} m_{to} U_{to}^2,$$

(9.63)

or

$$m_{f_{to}} = \frac{1}{2} \frac{m_{to} U_{to}^2}{\bar{\eta} c_p},$$

(9.64)

where m_{to} is the gross take-off mass, $\bar{\eta}$ is an average propulsion efficiency during the take-off segment, and c_p is the specific heat of combustion of the fuel. This equation contains m_{to}, which is unknown. The correct take-off speed is found according to the methods discussed in Chapter 7.

9.6.2   Fuel to Climb

The fuel to climb was calculated in Chapter 8. However, here we take a first-order approximation, following the method of Torenbeek[248], to find the climb fuel from energy methods. The fuel burned from take-off to the initial cruise altitude and Mach number (*lost fuel*) is written as

$$m_{f_{climb}} \bar{\eta} c_p = m_{to} g h + \frac{1}{2} m_{to} U^2 = m_{to} g h + \frac{1}{2} m_{to} a^2 M^2 = m_{to} g h_E,$$

(9.65)

where h_E is the energy height

$$h_E = h + \frac{1}{2g} a^2 M^2.$$

(9.66)

Therefore, the climb fuel is

$$m_{f_{climb}} = m_{to} \frac{g h_E}{\bar{\eta} c_p}.$$

(9.67)

The energy height is completely defined by the initial cruise conditions: altitude and Mach number (recall Eq. 8.70). The factor $\bar{\eta}$ is the average propulsive efficiency of the engines during the climb segment.

9.6.3   Additional Fuel

The fuel for descent, approach, terminal area maneuver and landing is generally calculated by assuming that the fuel is the same as flying at cruise altitude over the same distance. This is a conservative assumption. Additional fuel must be allowed for taxi out and other maneuvers. According to Shevell[249], the fuel for maneuvering is

$$\Delta m_f \simeq \frac{0.0025}{\eta} W_{to},$$

(9.68)

where η is the power plant efficiency at cruise conditions. Otherwise, the taxi fuel can be evaluated from the equation

$$\dot{m}_f = f_j D = f_j \left(\frac{1}{2} \rho A C_D U_{taxi}^2 + \mu W \right), \qquad (9.69)$$

where μ is the rolling resistance of the aircraft on the runway. It can be verified that the aerodynamic drag is negligible compared to the rolling resistance of the aircraft. Solution of Eq. 9.69 requires the average taxi speed, which can be inferred from the taxi distance, $U_{taxi} \simeq x/t$ and the taxi time (usually 5 minutes) at the given airport. From Eq. 9.69 the fuel required for maneuvering at the runway is estimated by

$$m_{f_{taxi}} = \dot{m}_f t. \qquad (9.70)$$

The fuel for taxi in is taken from the reserve fuel.

9.6.4 Reserve Fuel

On September 3, 1989, a Boeing B-737 aircraft ran out of fuel in Brazil, due to a navigation error. The crew, preoccupied with listening to a World Cup championship match, flew in the wrong direction. Before they realized this, the aircraft plunged into the Amazon jungle, near Sao Jose do Xingu, Brazil. There were 13 fatalities*.

Accidents of this kind are unlikely to happen, and there are strict regulations to prevent them. The Association of European Airlines (AEA) specifies a 200 nautical mile (370 km) diversion flight for short and medium range aircraft, and 250 nautical miles (463 km) for long range aircraft. In addition, it requires a 30-minute holding at 1,500 ft (457 m) altitude and a 5% mission fuel reserve for contingency. For domestic flights in the USA a 130 nautical mile (241 km) diversion and 30 minutes holding at 1,500 feet are specified.

From an operational point of view, the range may be limited by the absence of diversion airports over a predefined route. The ICAO has a particular rule (ETOPS) permitting twin-engine aircraft to fly longer routes (previously off-limits), which have no diversion airports within a 60- or 120-minute flight. This rule allows several modern twin-engine aircraft to fly over the oceans and in remote parts of the world. A discussion of ETOPS performance with OEI is available in Martinez-Val and Perez[250].

We carry out the analysis in two steps. First, we find the contingency range. This includes the specified range and the additional range as prescribed by the aviation policies. Then we calculate the fuel corresponding to this range.

- The *diversion distance* R_{div} flown with the Maximum Landing Weight (MLW) yields an increase in mission range estimated by the equation

$$\Delta R_{div} = (cR)_{div} \frac{MLW}{TOW}, \qquad (9.71)$$

---

* The pilot led the survivors to the rescue after a two-day trek. It is alleged that his first words were "Who won?".

where the coefficient c_{div} accounts for all the factors in the diversion flight that are suboptimal (lower speed, altitude, engine efficiency).

- The *holding time* (or loiter), flown at the Maximum Landing Weight (MLW) yields an increase in mission range estimated by the equation

$$\Delta R_{hold} = (cUt)_{hold} \frac{MLW}{TOW}, \tag{9.72}$$

where the coefficient c_{hold} accounts for loss of efficiency as in the diversion flight. The holding speed is generally half the cruising speed.

- The *extended duration* of the flight at the cruise speed for a time t_{exd} yields an increase in mission range estimated by the equation

$$\Delta R = Ut_{exd}, \tag{9.73}$$

- The *contingency fuel* is a percentage of the mission fuel $m_{f_{res}}/m_f = 0.05\text{--}0.10$, depending on actual policies.

The *equivalent all-out range* then is the sum of all contributions

$$R_{out} = R_{req}\left(1 + \frac{m_{f_{res}}}{m_f}\right) + [(cR)_{div} + (cUt)_{hold}]\frac{MLW}{TOW} + Ut_{exd}, \tag{9.74}$$

where R_{req} is the mission range required. The exact calculation of this parameter is essential, because it can be considerably larger that the required range.

Torenbeek[248] proposed some practical values for the coefficients in Eq. 9.74:

$$c_{div} = c_{hold} \simeq 1.1 + 0.5\eta_M, \tag{9.75}$$

where η_M is the log-derivative of the propulsive efficiency with respect to the Mach number. For a modern high-by-pass turbofan engine this quantity is estimated at $\eta_M \simeq 0.225$ for $M = 0.8$; it increases with the decreasing Mach number, so that $\eta_M \simeq 0.325$ at $M = 0.4$. Equation 9.74 can be further simplified if the reserve fuel is used for extension of the range at cruise conditions.

An alternative method consists in adding separately the contributions for hold and diversion. Clearly, a number of parameters are required for the solution of this problem: the hold time, altitude and speed; the diversion distance, altitude and speed. The fuel required from these two contingency segments can be added to the mission fuel.

9.6.5 Mission Fuel of Subsonic Jet Transport

We apply the methods discussed to the calculation of the mission fuel and gross take-off weight for the subsonic jet aircraft, model A in Appendix A, which is to fly a 1,000 km (540 nm) mission. Assume a cruise Mach number $M = 0.8$ at constant altitude $h = 11,000$ m (36,089 ft) with 20,000 kg payload.

In order to evaluate the ratio between the fuel mass required by each segment and the mission fuel, we perform a first-order analysis. We use $m_{to} \simeq 150,000$ kg, $U_{to} \simeq 300$ km/h (162 kt). The maneuver fuel is neglected.

1. Take-off fuel, with $\bar{\eta} \simeq 0.4$, $c_p \simeq 43.5$ MJ/kg. From Eq. 9.64: $m_{f_{to}} \simeq 120$ kg.

2. Climb to cruise altitude, with $\bar{\eta} \simeq 0.5$. From Eq. 9.65, we find $m_{f_{climb}} \simeq 936$ kg.
3. The aircraft reaches its cruise altitude with its cruise speed with a mass

$$m \simeq (150{,}000 - 120 - 936)\,\text{kg} = 148{,}944\,\text{kg}.$$

4. The equivalent all-out range is estimated for a flight within continental USA. Therefore, we include a diversion range of 241 km and holding for 30 minutes at $M = 0.4$. The coefficient η_M is estimated at 1.2125. The fuel reserve is 5% of the mission fuel

$$R = R_{req}(1 + 0.05) + 1.2125(241 \cdot 10^3 + 118 \cdot 30 \cdot 60)\frac{138}{150},$$

with no extension of the flight. The MLW for this aircraft is given in Table A.1. After simplifications, the equivalent all-out range is

$$R \simeq R_{req}(1 + 0.05) + 453 = 1{,}503\,\text{km} = 812\,\text{nm}.$$

5. The cruise fuel is estimated from Eq. 9.28, with an average $L/D = 17.5$, and $TSFC \simeq 8.5 \cdot 10^{-5}$ N/Ns. This equation has to be solved for the unknown block fuel ratio, ζ. The solution leads to $\zeta \simeq 0.267$. Hence the fuel mass is 40,000 kg.
6. The fuel used to cruise the distance required is about 29,000 kg (from the solution of Eq. 9.28 for specified R_{req}).

In summary: the take-off fuel is about 0.4% of the cruise fuel; the climb fuel is 3% of the cruise fuel. About 25% of the fuel is unused. This seems a large wastage, but it is due to the relatively large all-out range (50% higher than the required range).

The analysis shown emphasizes the importance of optimizing the cruise performance of the aircraft. For the case considered, we have calculated an extension of the range by about 45%, to include the various scenarios specified by the international regulations. The analysis started from an estimate of the aircraft gross weight at take-off. More rigorously, this weight is not known, and is part of the solution. We could reiterate the above procedure with the new estimate of the take-off mass. This leads to a more accurate calculation procedure, based on the computer program described below.

Computational Procedure

1. Calculate the taxi fuel, Eq 9.70.
2. Calculate the take-off fuel, Eq. 9.64 (contains unknown m_{to}).
3. Calculate climb to cruise altitude fuel, Eq. 9.65 (contains unknown m_{to}).
4. Calculate the equivalent all-out range, Eq. 9.74 (contains unknown m_{to}).
5. Calculate the fuel required to fly the equivalent all-out range, Eq. 9.62 (contains unknown m_{to}).
6. Sum all contributions to find the all-out range fuel. The mission fuel is then found from the solution of

$$m_{to} = m_e + m_p + m_f. \tag{9.76}$$

7. The take-off mass just calculated is generally different from the first guess. Then we repeat the analysis from point 1, with the new estimate of the aircraft mass, and iterate till convergence (which occurs in three to four iterations). By convergence we mean that the difference in take-off weight between two iterations is smaller than a tolerance, which we set at 0.5%.

The report of the program for a 3,000 km flight is given below.

```
Aircraft Data: airbus-A300-600.dat

Estimated TSFC

    sea level  0.8521E-04 kg_m/s/N ,    0.3068 lb/h/lb per engine

    at cruise  0.9113E-05 kg_m/s/N ,    0.0328 lb/h/lb per engine

Aerodynamics :

              CL  =  0.53493

              CD  =  0.02988

             L/D  = 17.90093

            CLto  =  2.12000

Segment analysis:

  Required range              3000.0 km    1619.9 n-miles

  Equivalent all-out range    4086.3 km    2206.4 n-miles

Fuel breakdown:

  Taxi-out fuel                      207.  kg_m

  Take-off fuel                       64.  kg_m

  Climb to cruise altitude fuel      952.  kg_m

  Maneuver fuel                      944.  kg_m

  Cruise fuel                      28656.  kg_m

     for required segment          21617.  kg_m
```

```
    Mission fuel                         30824.  kg_m

        reserve fuel                      7040.  kg_m

        actually used                    23784.  kg_m

    Block fuel ratio            xi =      0.204

    New mass estimate                   150724.  kg_m

        weight                          150724.  kg_f

Final weight breakdown (kg_f) :

    OEW   =    90900.00      0.603

    PAY   =    29000.00      0.192

    FUEL  =    30824.12      0.204

    GTOW  =   150724.12

Cruise mode: cruise-climb

Converged at iteration # 2
```

The program can provide any other output required, such as the breakdown of the aircraft weight (OEW, payload, fuel), cruise altitude, Mach number, etc. The report can be further improved by detailed calculation of the take-off (Chapter 7), climb (Chapter 8), and by more refined assumptions regarding the terminal area approach and descent.

9.7 CRUISE WITH INTERMEDIATE STOP

Is it more economical to fly a long distance non-stop or with an intermediate stop? This is the question that the wary passengers ask themselves when buying a ticket to far-away destinations. Some airlines occasionally offer cheaper seats to a destination via an off-route stop. This stop may even be in the opposite direction (example: flight from London to Washington, DC, via Frankfurt, Germany).

The aerospace industry has developed aircraft for very long range, such as the Boeing B-777-200LG (*long range*), capable of cruising a distance of about 18,000 km (9,814 nm) – a non-stop flight between London and Sydney, Australia*. This is close to the ultimate *global range*, 20,000 km, advocated by Küchemann[115]. The global

---

* In fact, the Boeing B-777-200IWG holds a no-stop cruise record of 20,045 km, 10,817 nm (1997), from Seattle to Kuala Lumpur, Malaysia.

range is half the Earth's circumference at the equator, and would allow an aircraft to fly from any point to anywhere around the world – at least in principle.

The problem is one of long range cruise in which a subsonic commercial jet has to travel from airport A to a destination airport B. The distance along the flight corridor is within the certified maximum range of the aircraft. However, here we consider the problem of an *en-route* stop at airport C. The aircraft carries enough fuel to reach airport C, refuels at this base, takes off and flies to its final destination B. We do not consider the costs of landing at B, or the direct operating costs incurred by the operator for increasing the time of the return journey. We also assume that the flight altitude and cruise speed are the same for all the cruise segments, and that the aircraft carries the same payload. For the single flight segment the take-off mass is calculated by Eq. 9.76.

The case considered for this analysis is relative to aircraft model A (Appendix A) flying at a cruise Mach number $M = 0.8$ with the same amount of payload $PAY = 29,000$ kg. The thrust-specific fuel consumption is an average $TSFC = 8 \cdot 10^{-5}$ N/Ns. If x is the distance to be flown in a direct flight, and x_1, x_2 and the flight segments, we assume that

$$x_1 + x_2 = x. \tag{9.77}$$

In other words, the en-route stop C is on the route to the final destination B. A stop with a diversion clearly increases the total range. The computational procedure uses the method in §9.6.5 as a base routine. The method consists of calculating the mission fuel for three flight segments: x, x_1, x_2. Therefore, we need a statement such as

```
call MissionFuel(x,mf)
```

which returns the mission fuel mf corresponding to a stage length x. We then calculate the relative fuel cost. This is the ratio between the fuel required by the flight with en-route stop and the direct flight,

$$m_{f_r} = \frac{m_f(x_1) + m_f(x_2)}{m_f(x)}. \tag{9.78}$$

This quantity can be plotted as a function of x_1/x (the ratio between the first flight segment and the total range). However, here we consider the case $x_1 = x_2 = x/2$. The result for a 6,000 km (3,239 nm) trip, with the same payload (29 metric tons), is summarized in Table 9.2. This shows that stopping en route does not pay off. Actually, it requires about 3.5% more fuel. There is a combination of factors that leads to these

Table 9.2 *Comparison of flight profiles, for a 6,000 km (3,238 nm) flight profile, payload weight 29,000 kg. Cruise/climb conditions. $\Delta R = R_{out} - R = $ diversion distance.*

Flight profile	Stage, km	ΔR, km	ζ	Reserve, kg	Used, kg
En-route stop	$3,000 \times 2$	1,086.3	0.204	7,030	$23,784 \times 2$
Direct flight	$6,000 \times 1$	1,121.1	0.306	6,813	$45,967 \times 1$

unexpected conclusions: (1) the flight with en-route stop requires additional fuel for take-off, landing and maneuvering and (2) the reserve fuel to be carried is nearly the same as for the longer stage length.

The present analysis does not take into account the cost of stopping at the intermediate airport (including ground services and time to get the aircraft ready for departure), therefore it is not totally indicative of the direct operating costs of long-haul flights. However, since the DOC depend greatly on fuel consumption, this analysis is realistic. Most budget airlines offer only flights from point to point to cut airport and handling costs. However, it is clear that it is not economical for any airline to stop en route, and even less so to fly to an off-route destination.

9.8 AIRCRAFT SELECTION

The next problem is one of selection of the best aircraft and mission profile for a long-haul flight. We assume that we have to connect two airports separated by a distance of 14,000 km (7,560 nm), and that we have to carry a payload of 35,000 kg.

If we use aircraft B, we can make a no-stop flight to destination. For this aircraft, we estimate a trip fuel of 77.62 metric tons (from the program `MissionFuel(..)` described earlier). We can also make the trip with aircraft A of reference by taking an intermediate stop at 7,000 km (3,780 nm). The total fuel required (including the two legs) in this case is 55.69 tons, a saving of about 25%. Therefore, it appears that the second option is more economical. If, instead, we have to carry a larger payload, say 50 tons, the aircraft of reference is not large enough. The large aircraft B, instead, can easily accommodate that payload, and the increase in fuel consumption over the previous case is only 4.5%.

The summary of our calculations (fuel and GTOW) is shown in Table 9.3. The other data are from the manufacturers.

The problem presented is a prototype of those problems that commercial operators face every day. The formulation can be further refined to include other factors, and possibly an estimate of the DOC. Green[251] showed an analysis for cargo aircraft, and recommended that from the point of view of fuel efficiency a flight segment shall not exceed 7,500 km (about 4,000 nm) – at the current level of technology.

In-flight refueling is an option that can be considered for either extending the range or increasing the payload for a fixed range. However, the calculation must include the cost of operating a tanker aircraft, and of maintaining a loiter trajectory for some time. Calculations of this type are shown by Bennington and Visser[244].

Table 9.3 *Calculated payload fuel efficiency for long-haul commercial flight. Cruise/climb condition with $M = 0.80$. All weights are in metric tons.*

Aircraft	OEW	MTOW	PAY	Range (km)	Fuel	GTOW
B	129.9	275.0	35.0	14,000	74.62	244.1
A	90.1	165.0	35.0	7,000	55.69	156.1
B	129.9	275.0	50.0	14,000	77.94	263.1

9.9 SUPERSONIC CRUISE

We now deal with a more exotic flight performance. We study the optimal conditions for cruise range at supersonic speeds for a given AUW and a given initial block fuel. A number of cases are possible, because the cruise conditions depend on the altitude, Mach number and angle of attack. The general range equation is given by Eq. 9.26.

The general problem of supersonic cruise can be quite complicated. Windhorst *et al.*[82] developed guidance techniques for minimum flight time, minimum fuel consumption and minimum DOC for a fixed range cruise. For example, a minimum fuel trajectory consists of an initial minimum fuel climb, a cruise/climb and a maximum L/D descent. The minimum DOC trajectory is nearly the same as a minimum fuel trajectory.

9.9.1 Cruise at Constant Altitude and Mach Number

For cruise in the stratosphere, $h > 11,000$ m, as is usually the case for supersonic aircraft, the speed of sound is constant. In addition, if the Mach number is also constant, the range equation becomes

$$R = \frac{aM}{gf_j} \int_e^i \frac{C_L}{C_D} \frac{dm}{m} = \frac{aM}{gf_j} \int_e^i \frac{C_{L_\alpha}(\alpha - \alpha_o)}{C_{D_o} + \eta C_{L_\alpha}(\alpha - \alpha_o)^2} \frac{dm}{m}. \tag{9.79}$$

The angle of attack at cruise condition is specified by $L = W$. The condition of the angle of attack was derived previously (Eq. 6.65). We rewrite it here for convenience

$$\alpha = \alpha_o + \frac{2W}{\rho A a^2} \frac{1}{C_{L_\alpha} M^2}. \tag{9.80}$$

When we replace this condition in Eq. 9.79 and simplify the algebra, we find

$$\frac{L}{D} = \frac{c_o/M^2}{C_{D_o} + \eta c_o^2/M^4} = \frac{c_1(\sigma, M)m}{C_{D_o} + c_2(\sigma, M)m^2}, \tag{9.81}$$

with

$$c_o(\sigma, m) = \frac{2mg}{\rho A a^2}, \quad c_1(\sigma, M) = \frac{2g}{\rho A a^2 M^2}, \quad c_2(\sigma, M) = \frac{\eta}{C_{L_\alpha}} \left(\frac{2g}{\rho A a^2 M^2} \right)^2. \tag{9.82}$$

Note that $c_2 = \eta c_1^2/C_{L_\alpha}$. The glide ratio can be optimized with respect to the Mach number or the mass. Equation 9.81 shows the dependence of the glide ratio from the Mach number and the aircraft's mass. When we use the coefficients defined by Eq. 9.82, the range equation becomes

$$R = \frac{aM}{gf_j} \int_e^i \frac{c_1(\sigma, M)}{C_{D_o} + c_2(\sigma, M)m^2} \, dm. \tag{9.83}$$

The integral is rather involved, but it is similar to an expression seen before (Eq. 9.39)

$$\int_e^i \frac{dm}{C_{D_o} + c_2 m^2} = \frac{1}{\sqrt{C_{D_o} c_2}} \left[\tan^{-1} \left(m \sqrt{\frac{c_2}{C_{D_o}}} \right) \right]_e^i, \tag{9.84}$$

$$R = \frac{aM}{gf_j} \frac{c_1}{\sqrt{C_{D_o} c_2}} \left[\tan^{-1} \left(m \sqrt{\frac{c_2}{C_{D_o}}} \right) \right]_e^i, \tag{9.85}$$

$$R = \frac{aM}{gf_j} \frac{c_1}{\sqrt{C_{D_o} c_2}} \left[\tan^{-1} \left(m_i \sqrt{\frac{c_2}{C_{D_o}}} \right) - \tan^{-1} \left(m_e \sqrt{\frac{c_2}{C_{D_o}}} \right) \right], \tag{9.86}$$

or

$$R = \frac{aM}{gf_j} \sqrt{\frac{C_{L_\alpha}}{C_{D_o} \eta}} \left[\tan^{-1} \left(m_i \sqrt{\frac{c_2}{C_{D_o}}} \right) - \tan^{-1} \left(m_e \sqrt{\frac{c_2}{C_{D_o}}} \right) \right]. \tag{9.87}$$

Equation 9.86 is similar to Eq. 9.40, valid for the same flight program at subsonic speeds.

Figure 9.12 shows the cruise performance of the supersonic fighter jet of reference, calculated according to Eq. 9.86, for a fixed weight. A given range can be obtained by subsonic cruise at relatively low altitude, or supersonic cruise in the lower stratosphere. At a constant Mach number, the range increases with the altitude.

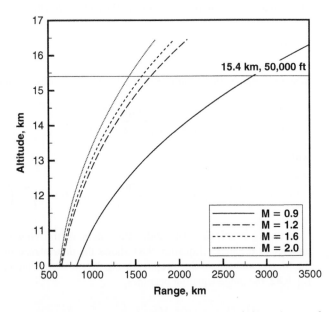

Figure 9.12 *Estimated cruise range for supersonic jet fighter showing lines of constant Mach number. Take-off weight W = 10,000 kg.*

9.9.2 Cruise at Constant Mach Number and C_L

Consider a cruise in the lower stratosphere at fixed Mach number and C_L. The corresponding angle of attack α is constant, and is set at the start of the cruise, according to Eq. 9.13, and is calculated with a mass $m = m_i$. As a consequence, $M(C_L/C_D)$ is a constant. To maintain a constant C_L and α with the mass decreasing, the aircraft must climb, according to the same principle illustrated in §9.5.3. In fact, from Eq. 9.13 α is constant if $m/\sigma = const$. Therefore, this flight program is a *supersonic cruise/climb*. With these considerations, the range equation becomes

$$R = \frac{a}{g}M\left(\frac{C_L}{C_D}\right)\int_e^i \frac{1}{f_j}\frac{dm}{m} \simeq \frac{a}{gf_j}M\left(\frac{C_L}{C_D}\right)\ln\left(\frac{1}{1-\zeta}\right). \tag{9.88}$$

The changes in f_j must be considered. If an expression like Eq. 9.48 can be used, then the calculation of the optimal range is simplified, otherwise we can use the TSFC from tabulated data.

For a given block fuel ratio, the range in the lower stratosphere calculated from Eq. 9.88 is maximized if

$$f(\sigma, M) = \frac{M}{f_j}\frac{C_L}{C_D} = max. \tag{9.89}$$

If the TSFC were a constant, then the supersonic cruise would be optimal with a maximum glide ratio. The condition for maximum C_L/C_D has been discussed in §4.7, while the *FoM* was discussed in §9.3.4. A useful method to study the behavior of Eq. 9.89 is to construct a three-dimensional plot with the following procedure.

Computational Procedure

1. Set the flight altitude (outer loop).
2. Set the Mach number (inner loop).
3. Calculate the initial angle of attack from Eq. 9.13.
4. Calculate the lift and drag coefficients.
5. Interpolate the flight data to find the correct specific fuel consumption f_j.
6. Calculate the function $f(\sigma, M)$, Eq. 9.89.
7. Increase the Mach number, and go over this loop as many times as required to reach the limit of the flight data.
8. Increase the altitude and proceed in the same way.

The function $f(\sigma, M)$ is shown in Figure 9.13. It shows that there are two local maxima. One maximum can be at subsonic or supersonic speed, depending on the altitude. The other maximum is at about $M \simeq 1.6$, and does not vary much with the altitude. The result depends on the aerodynamic performance of the aircraft and on the TSFC. If the aircraft starts cruising at an altitude of 10,000 m (32,808 ft), at some point it will climb to 12,000 m (39,370 ft). At the new altitude the range factor can be suboptimal, but it is in general a higher value than at the starting altitude.

Figure 9.14 shows a comparison between two flight programs: a constant M–h and a cruise/climb (constant M–C_L). The cruise/climb leads to a longer range, a result

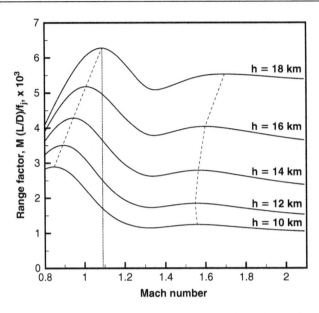

Figure 9.13 *Range factor for flight at constant Mach number and constant C_L.*

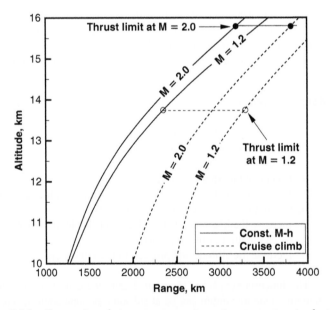

Figure 9.14 *Comparison between supersonic cruise programs at selected Mach numbers. The thrust limit denotes altitude at which $T = D$, hence the maximum speed.*

already known from subsonic cruise. Also, the cruise range increases uniformly with the altitude. However, there are additional constraints to take into account. One of those is the thrust limit, e.g. the altitude at which there is no excess power, and the maximum engine thrust is equal to the drag. Calculation of this limit is done at constant Mach number. This means that the aircraft cannot be operated to cruise at $M = 1.2$ at altitudes above $h = 13{,}750$ m (45,112 ft), while a supersonic cruise at $M = 2.0$ is possible at an altitude just below $h = 15{,}800$ m (51,837 ft). To fly faster, the aircraft must fly higher.

9.10   CRUISE RANGE OF PROPELLER AIRCRAFT

We now get down to more mundane speeds and altitudes, to deal with propeller-driven aircraft. These can be powered by gas turbine or reciprocating engines.

The range equation, Eq. 9.26, is valid in the general case. For a propeller-driven aircraft we need to replace the fuel flow equation: $\dot{m} = SFC\,P$. The range becomes

$$R = \int \frac{1}{SFC} \frac{U}{P}\, dm. \tag{9.90}$$

We replace the power equation $\eta P = TU$ in Eq. 9.90 to obtain

$$R = \int \frac{\eta}{SFC} \frac{1}{T}\, dm = \frac{1}{g} \int \frac{\eta}{SFC} \frac{L}{D} \frac{dm}{m}. \tag{9.91}$$

The latter equation contains the equilibrium conditions $L = W$, $D = T$. It appears that the range of a propeller aircraft is not dependent explicitly on its speed. However, there is a new variable: the propeller's efficiency. This, as we have seen, depends on the advance ratio J, which is a combination of rotational speed Ω and forward speed U. If the engine power depends on the AUW (all other parameters being the same), then the propeller's efficiency is also dependent on the AUW by way of the shaft power.

The types of cruise condition for the propeller-driven aircraft are similar to the jet engine aircraft. In particular, we note that the range is a function of this type:

$$R = f\left(\sigma, \eta, U, SFC, \zeta\right). \tag{9.92}$$

We discuss the solution of the cruise flight at constant speed and altitude. The remaining cruise programs can be found with an analysis similar to the subsonic jet aircraft (Problem 10).

9.10.1   Cruise at Constant Altitude and Speed

The range equation for this flight condition becomes

$$R = \frac{\eta(J,\theta)}{g\,SFC} \int_e^i \frac{L}{D} \frac{dm}{m}, \tag{9.93}$$

where we have assumed that the propulsive efficiency η is a function of the propeller's advance ratio J and pitch setting θ, as described in §5.9; the SFC is only a function of h and U, therefore it is taken out of the integral. If we write the glide ratio as a function of the aircraft mass, as done in §9.5.1, then

$$R = \frac{\eta(J,\theta)}{g\,SFC} \int_e^i \frac{c_1 dm}{C_{D_o} + kc_1^2 m^2}. \tag{9.94}$$

The solution of the integral is similar to Eq. 9.39, therefore the range can be written as

$$R = \frac{\eta(J,\theta)}{g\,SFC} \tan\left(c_1 m \sqrt{\frac{k}{C_{D_o}}}\right)_e^i. \tag{9.95}$$

If the block fuel and the aerodynamics are fixed, then the range is maximized by the function $\eta(J,\theta)/SFC$. This factor is a combination of engine and propeller performance. The solution of Eq. 9.95 inevitably needs propeller charts. An analytical solution for a constant-speed propeller was given by Cavcar and Cavcar[252] for a piston-prop aircraft at constant altitude and cruise speed.

9.11 ENDURANCE

Endurance is the flight time of an aircraft. If the speed and flight altitude are constant, then the endurance is the ratio between the speed and the range

$$E = \int_e^i \frac{1}{f_j} \frac{1}{D}\,dm = \frac{1}{g}\int_{m_1}^{m_2} \frac{1}{f_j}\left(\frac{L}{D}\right)\frac{dm}{m}. \tag{9.96}$$

For a propeller-driven aircraft, instead, the corresponding equation is

$$E = \int_e^i \frac{1}{SFC}\frac{\eta}{DU}\,dm = \frac{1}{g}\int_e^i \frac{1}{SFC}\frac{\eta}{U}\left(\frac{L}{D}\right)\frac{dm}{m}. \tag{9.97}$$

It is useful to find a relationship between the range and the corresponding endurance, because the calculation of the former performance will lead to the latter in a straightforward way. The general expression has an integral form

$$U = \frac{dR}{dE}, \qquad dE = \frac{1}{U}dR, \qquad E = \int_e^i \frac{1}{U}\,dR. \tag{9.98}$$

If the air speed is constant, then the range is proportional to the endurance. Engineering problems of interest are those of maximum endurance at cruise conditions*. For general conditions other than cruise (including loiter, surveillance, flight within prescribed area, etc.) see Sachs[253].

---

* The current endurance record for non-refueled flight is the round-the-world flight of the Global Flyer in 2005 (67 hours, 1 minute): 19,894 nautical miles. The aircraft was a single seater turbofan-powered airplane, with an estimated L/D = 37.

9.12   EFFECT OF WEIGHT ON CRUISE RANGE

Consider a subsonic jet airplane. Its range equation is given by Eq. 9.28. Assume that the take-off mass increases from a reference value of m_i to a value m_1. Also assume that the aircraft is to fly at the same speed and altitude, and that the engines' specific fuel consumption is not affected by the increased mass. With these assumptions, the ratio between ranges is

$$\frac{R_1}{R_o} = \frac{\ln (m_{i1}/m_{e1})}{\ln (m_{io}/m_{eo})}, \tag{9.99}$$

where the index "e" denotes the conditions at the end of the cruise, and "i" is the initial point of the cruise. The initial mass can be written as

$$m_i = m_e + m_p + m_f. \tag{9.100}$$

OEW and PAY are the fixed portion of the mass (or weight). If we call f the ratio between the fuel mass and the constant portion of the mass $m_c = m_e + m_p$, then

$$f = \frac{m_f}{m_e + m_p} = \frac{m_f}{m_c}, \tag{9.101}$$

$$m_i = (1 + f)m_c, \qquad m_e = m_c, \qquad \frac{m_i}{m_e} = 1 + f. \tag{9.102}$$

With the above, Eq. 9.99 becomes

$$\frac{R_1}{R_o} = \frac{\ln (1 + f_1)}{\ln (1 + f_o)}. \tag{9.103}$$

The change in f due to change in aircraft fixed mass (due to either changes in OEW or PAY) can be written as

$$f_1 = \frac{m_f}{m_c + \Delta m_c} = \frac{m_f/m_c}{1 + \Delta m_c/m_c} = \frac{f_o}{1 + \Delta m_c/m_c}. \tag{9.104}$$

From here we can find the change in air range due to a relative change in aircraft fixed $\xi = \Delta m_c/m_c$

$$\frac{R_1}{R_o} = \frac{\ln [1 + f_o/(1 + \xi)]}{\ln [1 + f_o]}. \tag{9.105}$$

We have assumed that the fuel mass is the same for the two configurations. A plot of the range ratio is shown in Figure 9.15 as function of the fixed mass ratio, for a parametric value of f.

9.13   EFFECT OF THE WIND ON CRUISE RANGE

The atmospheric winds and global jet streams play a key role in long range cruise. They are particularly noticeable on the busy North Atlantic route. They can cut the

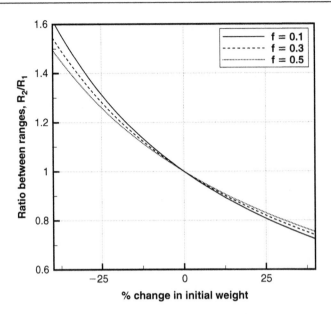

Figure 9.15 *Effect of changes in the aircraft weight on the cruise range, at specified fuel ratios, as indicated.*

eastbound journey time by 1 hour, and add up to 1 hour to the westbound flight. This is a considerable time. We now look briefly at these winds from the point of view of range and fuel consumption. A detailed analysis of this problem is available in Hale[54]. The range equation is written in terms of true air speed $U_{TAS} = U \pm U_w$, where U denotes the ground speed; U_w is the wind velocity.

$$R = \frac{1}{g} \int_e^i \frac{U_{TAS}}{f_j} \left(\frac{L}{D} \right) \frac{dm}{m}. \tag{9.106}$$

Again, a variety of cruise conditions can be prescribed. If we want to maintain constant altitude and air speed, then the problem is solved as previously described in ISA conditions, §9.5.1; we just replace U with U_{TAS}. All other parameters being the same, the ratio between the cruise range with a head-/tail-wind and the cruise range in ISA conditions is

$$\frac{R}{R_{ISA}} \simeq \frac{U_{TAS}}{U} = 1 \pm \frac{U_w}{U}. \tag{9.107}$$

This equivalence is only approximate, because there may be changes in drag characteristics between the two flight conditions. Clearly, for a case of head-wind, the ground speed is reduced by U_w, and the aircraft needs a longer time to reach its destination (vice versa with a tail-wind). For example, an $M = 0.80$ flight at the troposphere means a true air speed $U_{TAS} = 850$ km/h (458.7 kt). A 20 kt (37 km/h) head-wind reduces the range by 1%. The ground speed is $850 - 10 = 840$ km/h.

9.14 ADDITIONAL RESEARCH ON AIRCRAFT CRUISE

Early studies of range problems of jet aircraft appeared in the 1940s. These include the work of Page[254], Ashkenas[255], Perkins and Hage[33], and Edwards[256]. It has been established that steady state cruise is not fuel optimal. Speyer[257] demonstrated that an oscillating control leads to a decrease in fuel consumption. Gilbert and Parsons[258] applied periodic cruise control on the McDonnell–Douglas F-4 interceptor, and Menon[259] showed mathematically the conditions under which oscillating solutions occur in a long range cruise. Sachs and Christodoulou[260] analyzed the benefits of cyclic flight, with dolphin-type climbs and glides, and showed that (at least theoretically) the range can be increased with such a flight program. Nowadays, there is a general agreement that a cruise/climb technique yields maximum cruise range. Schultz and Zagalsky[231] used an energy method and control theory to solve problems such as a minimum-fuel cruise/climb from a given altitude and speed. The range and the final altitude and speed were part of the solution. Disagreement still exists on other issues, such as the best starting altitude and the best ratio of cruise speed over minimum drag speed. Torenbeek[248] proposed a unified treatment for the subsonic cruise performance for gas turbine power plants, and optimum cruise performance. Advances in aircraft design have made it possible to cruise over very long distances. AGARD CP-547[261] discusses some fundamental issues.

9.15 FORMATION FLIGHT

Formation flight is the cruise of two or more fixed-wing aircraft in close proximity. Interest in this aspect of flight mechanics stems from the flight of birds. Indeed, a number of publications has addressed this aspect of flight, and the issue continues to draw attention, see Lissaman and Shollenberger[262]. The problem is a complicated one, and has been solved for a general case, including ground effect, by King and Gopalarathnam[263]. The latter method uses a vortex lattice formulation to find the optimal distribution of lift and downwash. Lissaman[264] indicated that a solution of the problem can be derived in the Treffz plane. Methods used for the evaluation of formation flight endurance make use of some basic aerodynamic theorems, originally due to Munk (stagger theorem; reciprocity theorem; cosine theorem). An elementary discussion of this theory is given by Milne-Thompson[265].

The idea is to find a method capable of providing the lift and drag of each wing in a formation flight. An optimum solution (minimum total drag; best distribution of drag among wings, etc.) can then be found from the aerodynamic performance.

A general theory of interference for three-dimensional lifting surfaces can be established from the concept of acceleration potential doublets (Ashley and Landhal[121]), or the use of vortex lattice methods, King and Gopalarathnam[263]. However, these methods are relatively complicated, and can be replaced by substantial (and powerful) simplifications, such as a method that makes use of Munk's interference theorems*. This method has some limitations, namely on the separation between wings, which cannot be overlapping.

---

* These theorems originally were part of M.M. Munk's dissertation at the University of Göttingen, and were presented with an abundance of integrals[266]. Milne-Thompson's account[265] is easier to read.

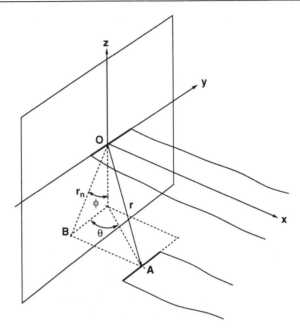

Figure 9.16   *Reference system and arrangement of two generic wings (or lifting lines).*

The essential assumption is that the wing is a lifting line. Consider the reference system and the nomenclature of Figure 9.16. Two lifting wings are assumed to be flying steady state in level flight. The first lifting wing is centered in O; the second lifting wing is centered in A. The distance vector between the centers of the two lifting wings is called **r**. The projection of **r** on the plane normal to the flight (plane *y–z*) is called r_n. The stagger angle ϑ is defined on the plane *x–y*, as the angle between the vector through the center of the lifting lines, OA, and its projection onto the plane *y–z*. If $\vartheta = 0$ the system is said to be *unstaggered*. This means that the lifting lines are co-planar on the reference plane *y–z*. In other words we say that the wings fly *line-abreast*.

Finally, the angle ϕ is the angle between the plane *x–z* and the vector r_n. We call this quantity the *offset angle*, because when the two wings are above one another (biplane case), the offset angle is zero. If the wings are on a horizontal plane, then $\phi = \pi/2$. In practice, for any given two lifting wings not lying on the same plane two offset angles are defined, ϕ_i, ϕ_j.

A generic lifting wing is reduced to a single vortex line having a known distribution of circulation. This vortex is closed with the tip vortices and a starting vortex at infinity (horseshoe model). This model replaces the wakes with tip vortices, and the circulation with a bound vortex running spanwise.

Under the above linearizing assumptions, the aerodynamics arising from the existence of an arbitrary number of lifting surfaces can be solved rapidly. Starting from the results of Milne-Thompson[265], we find that the *total drag mutually induced by a pair of wings i and j* is given by

$$D_{ij} = \frac{1}{2\pi} \rho \int_{-b/2}^{b/2} \int_{-b/2}^{b/2} \Gamma_i \Gamma_j \frac{\cos(2\phi_{ij})}{r_n^2} \, dy_i \, dy_j, \tag{9.108}$$

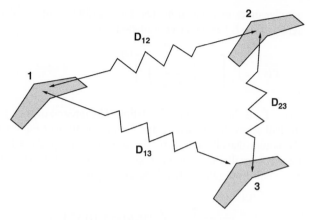

Figure 9.17   *Mutual interference drag and interference matrix for three lifting wings.*

where Γ denotes the circulation, b is the wing semi-span, and ρ is the air density. The interference drag is a perturbation quantity that decreases as the mutual distance of the elements of a given pair increases.

Equation 9.108 does not contain the stagger angle. One conclusion is that the *total induced drag* of a multi-element lifting system does not change if the elements are translated in a direction parallel to the flight, provided that the circulation is kept constant (*Munk's stagger theorem*). However, if the system is unstaggered, the drag induced by one element i onto an element j is equal to the vice versa

$$\frac{1}{2}D_{ij} = \frac{1}{2}D_{ji}, \tag{9.109}$$

which expresses *Munk's reciprocity theorem*. Another conclusion is that if the number of lifting surfaces is greater than 2, then an interference matrix must be computed, as indicated in Figure 9.17.

A number of other considerations are in place. First, for the computation of the total induced drag, the system can be unstaggered, e.g. the wings can be translated so as to lie on a plane normal to the free stream. However, unstaggering produces a different distribution of loads among the wings. Second, the expression giving the induced drag is singular for some configurations, and cannot be used to calculate the self-induced drag. For example, if two wings are co-planar, one behind the other, then $r_n = 0$, and Eq. 9.108 is indeterminate. However, for positive separations (e.g. when one wing is *outside* the span of the other wing), then it is appropriate to use the present theory.

Equation 9.108 can be solved in closed form if the lift distribution is elliptic*, which results in a great advantage from a computational point of view. From the

---

* This is not a restriction, just an assumption to find a result in a closed form. Moreover, elliptical lift distribution is not optimal, in the sense that there exist other distributions that yield a lower induced drag.

aerodynamics of isolated bodies, the distribution of circulation for a wing whose symmetry axis lies on the coordinate y_o is

$$\Gamma(y) = \Gamma_o \left[1 - \left(\frac{y - y_o}{b/2} \right)^2 \right]^{1/2}, \tag{9.110}$$

where Γ_o is the maximum circulation. The integral of such a function is the area of the half ellipse having axes b and $2\Gamma_o$. Integration of Eq. 9.108 is done numerically. We assume that all wings carry the same weight, e.g. $\Gamma_{oi} = \Gamma_{oj} = \Gamma_o$. Also, the lifting lines are parallel to each other, $\phi_i = \phi_j = \phi = \pi/2$, hence $\cos(\phi_i + \phi_j) = -1$. For a given pair i and j the mutual induced drag is

$$D_{ij} = c_1 \sum_n \sum_m \left[1 - \left(2 \frac{y(i, n) - y_o(i)}{b} \right)^2 \right]^{1/2}$$

$$\times \left[1 - \left(2 \frac{y(j, m) - y_o(j)}{b} \right)^2 \right]^{1/2} dy_i \, dy_j, \tag{9.111}$$

with

$$c_1 = \frac{1}{2\pi} \rho \Gamma_o^2. \tag{9.112}$$

The induced drag from Eq. 9.111 is a negative value, as it can be verified. The indices n and m denote the number of elements in which each wing is subdivided; $y(i, n)$ is the coordinate of the mean point of the nth panel on wing i; likewise, $y(j, m)$ is the mth panel on wing j. The calculation of the drag matrix $\mathbf{D} = \{D_{ij}\}$ is done for all $j > i$, which greatly simplifies the matter. We define the ratio of the drag of the formation flight to that of the in-solo flight (denoted by a nought)

$$D_{interf} = 1 + \frac{D}{D_o}, \tag{9.113}$$

The drag of the wing with elliptic loading is found from low-speed aerodynamics (see Katz and Plotkin[137] for full derivation), and the result is

$$D_o = \frac{1}{8} \pi \rho \Gamma_o^2. \tag{9.114}$$

In the first instance, assume that the wings are flying line abreast.

Figure 9.18 shows the total system drag for 2, 5, and 25 wings. For more than 50 wings, the results do not change significantly, due to the fact that the sum over the terms of Eq. 9.108 yields a series rapidly convergent. Therefore, there is an asymptotic value to the drag saving in a formation flight. By looking at the results of Figure 9.18, for a single pair flying abreast with a separation $s/b = 0.001$, the elliptic loading yields a drag fraction 0.730; 5 wings flying abreast with the same separation give a drag fraction 0.511, and 25 wings give a result of 0.350 – a massive drag reduction.

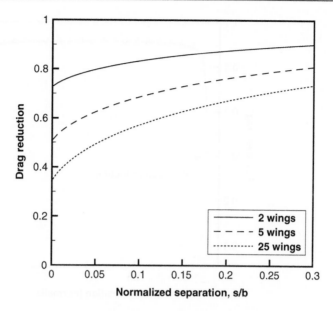

Figure 9.18 *Drag savings in formation flight, elliptic loading, no overlapping.*

Hummel[267] reported some results of formation flight testing, done with Dornier Do-28 airplanes, and implied that a 15% power reduction can be achieved by an aircraft flying astern a leading aircraft. Only the rear airplane had equipment installed to measure the power consumption. Therefore, no data are available for the front airplane.

Ray *et al.*[268] and Vachon *et al.*[269] measured the drag and fuel consumption performance of two F/A-18s in formation flight, as a function of the lateral, longitudinal and vertical position of the trail aircraft. Both aircraft were instrumented for accurate positioning, in and out of the trailing vortices. The maximum drag reduction was about 20% for a longitudinal separation $x = 3b$, and a lateral separation between 0.1 and 0.2 wing spans, with the trail aircraft slightly below the lead aircraft. The maximum reductions in fuel consumption were measured at 18%.

If the system is unstaggered, line abreast flight produces maximum induced lift at the central position. The wings at the tip gain just half of these benefits. If the system is staggered, and the wings are flying in a *vee* formation, the wings at the tips gain most, due to the fully developed upwash field created by the leading wings.

With the total drag at a minimum, we may want to seek the distributions that lead to the most uniform distribution of drag among the wings. If there are n wings, then the optimization problem involves $n-1$ parameters. The solution is a *vee* formation flight.

If the wings are staggered by an angle ϑ, Munk's theory yields a drag from i to j

$$D_{ij} = \frac{1}{4\pi}\rho \int_{-b/2}^{b/2} \int_{-b/2}^{b/2} \Gamma_i \Gamma_j \left[\frac{(1 + \sin\vartheta)\cos(\phi_i + \phi_j)}{r_n^2} \right.$$
$$\left. + \frac{\sin\vartheta\cos(\phi_i + \phi_j)}{2r^2} \right] dy_i \, dy_j. \qquad (9.115)$$

The drag from j to i requires the change of the sign of ϑ.

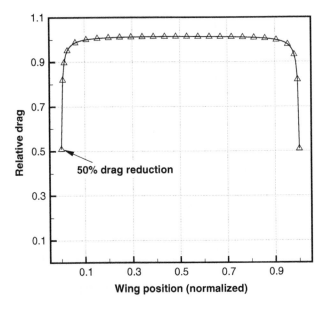

Figure 9.19 *Drag distribution for line abreast flight.*

9.15.1 Range and Endurance in Formation Flight

We are now in a position to calculate the effects on range and endurance of the formation flight. The drag experienced by a given wing consists of its profile drag and its induced drag. The power required to drive the wing is

$$P = (D_v + D_i)U = (c_1 U^2 + D_i)U. \tag{9.116}$$

We have seen that the power is at a minimum when the profile drag is equal to the induced drag. Due to the interference effect of the formation flight, the induced drag decreases, hence also the cruise speed for minimum power shall decrease.

Consider a formation flight made of five wings, such as one that can be envisioned in a mission of unmanned air vehicles. The induced drag saving is estimated at 49%, for closely spaced (but not overlapping) wings. Therefore

$$\frac{P}{P_o} = 0.51 \frac{U}{U_o}. \tag{9.117}$$

If the fuel flow is proportional to cruise power (such as in a propeller-driven aircraft), and the cruise speed is maintained constant, then the endurance increases by about 96% ($E \sim 1/\dot{m}$), and the range increases by about the same amount. If the cruise speed in formation flight is reduced by 20%, then the relative power would be $P/P_o \simeq 0.41$, and the endurance would be increased by a factor 2.45 (245%).

These considerations indicate that even with a simple model it is possible to estimate the increment in endurance and range of a set of closely spaced fixed wings, and that the benefits of flying in formation are formidable in terms of cruise performance.

Figure 9.20   *Vee formation flight of swans.*

Figure 9.21   *Vee formation flight of swans with branching out.*

One problem that remains in the present discussion is the optimal arrangements of wings that leads to a uniform distribution of drag. This is an optimization problem that requires staggering the wings in a *vee* formation, as shown in Figure 9.20. For a formation flight with n wings, the optimal problem contains $n - 1$ free parameters (the position of the wings in the flight direction).

Figure 9.21 shows a *vee* formation of swans. In this photo the birds join at different points. Branching out is beneficial in the same way as a side of a single *vee* formation. Therefore, flight optimization can be carried out by using one parameter at a time. In other words, one can set the position of a leading wing and optimize the position of the second wing; then the third wing is optimized with respect to a frozen combination of the previous wings.

It is believed that formation flight and atmospheric conditions combine to provide enhanced benefits to long distance cruise of most birds. Another important effect is due to the presence of updrafts, or thermals. One consequence is that a vertical gust increases the lift and rotates the lift vector forward, thus providing an additional thrust component. Drag will is also affected, but for efficient wings the effect of the lift is predominant[270].

As an example, the bristle-thighed curlew (*Numenius tahitiensis*) is known to fly from Western Alaska to its winter home in the Polynesian islands, in a 10,000 km flight. The longest no-stop stretch is 3,000 km. The lesser golden plover (*Pluvalis dominica*) may travel 4,000 km without en-route stop from the Aleutian islands to Hawaii and mid Pacific islands, with a 3,300 km flight in about 35 hours. The hummingbird (*Archilochus colubris*) migrates across the Gulf of Mexico to Central America breeding grounds in a 1,000 km stretch. This is an amazing feat for a bird weighing about 3 grams, with high metabolic rate, high wing beat frequency, and without food supply during the flight.

PROBLEMS

1. Calculate and plot the specific range for the supersonic aircraft model C as a function of Mach number, for a fixed angle of attack $\alpha = 3$ degrees, flight altitude $h = 12,000$ m and aircraft mass $m_i = 10,000$ kg and 12,000 kg. Follow the procedure in §9.3.2.
2. Calculate the thrust-specific fuel consumption (in N/Ns) for the Boeing 747-400 during a long range cruise at altitude $h = 10,500$ m, knowing that the engines burn 3% of the aircraft gross weight at take-off (390,000 kg) every hour.
3. The Boeing B-747-400 uses up to 5.5 tons of food and drink supplies during a long range international flight. Calculate and discuss in general terms how this additional load affects the operating costs of the aircraft.
4. The subsonic commercial jet aircraft A, whose data are given in Appendix A, flies at a cruise altitude with its economic cruise speed. Assuming that the aircraft takes off at MTOW, and that initial fuel load is 30% of its gross weight, calculate the aircraft's range (using the Breguet equation), and the fuel remaining when the aircraft has traveled two thirds of this distance.
5. Calculate the maximum range of airplane model A (subsonic transport aircraft) at maximum PAY, PAY/2, PAY/4, using the approximate range equation, Eq. 9.28. Consider a cruise at $h = 10,500$ m, with Mach number $M = 0.78$.

6. Some budget airlines have started reducing the baggage allowance that can be carried in the cargo hold (checked-in baggage), from 20 kg to 15 kg per passenger. Calculate how this change in policy can affect the range of aircraft model A (Appendix A). Or, for a given range of 2,000 km, how much fuel can be saved by flying lighter? Perform calculations in the case of (a) the aircraft is full and (b) the aircraft is 50% full. Additional data required.

7. Calculate the effect on the range and endurance of the commercial jet aircraft model A, Appendix A, due to an increase in operating empty weight by 10%. Assume that the block fuel ratio is 25%, and that the aircraft is to fly at the most economic cruise speed and altitude.

8. Calculate the specific range of aircraft model A flying at cruise conditions and cruise altitude. The flight distance is 2,000 km, and the initial fuel weight is 80% of the maximum fuel. The gross take-off weight is 90% of MTOW. Start the solution procedure from Eq. 9.5 and plot the specific range as a function of the distance flown.

9. Calculate the increase in endurance due to formation flight of two closely spaced airplanes. Assume that the drag saving is 25% compared to the case when the airplanes are flying with a large lateral separation.

10. Calculate the range at subcritical speed ($M < M_{dd}$) for aircraft model A (Appendix A), considering three cruise conditions, as shown in Table 9.1. Assume an initial mass $m_i = 145,000$ kg, and cruise altitude $h = 11,500$ m. Find the maximum cruise range.

11. Investigate the impact of fuel consumption on the direct operating costs of a typical subsonic transport aircraft. Do the necessary research to find data published from the airlines, the aviation authorities, the ministry of transportation, and the professional organizations. Produce a chart that shows the incidence of fuel relative to other cost items. (*Problem-based learning: additional research is required.*)

12. Calculate the specific range of the supersonic aircraft model C at a fixed altitude $h = 13,000$ m, as a function of the aircraft mass and the Mach number, for $M > 1.0$. Assume that the initial mass is $m_i = 10,000$ kg. Produce a suitable plot of the data showing the impact of both flight Mach number and aircraft weight. Also, calculate and plot the cruise angle of attack and drag coefficient. Discuss the results. Use the method presented in §9.3.2.

13. Calculate the specific range for the subsonic jet aircraft of reference (Appendix A), and plot it as a function of its speed. Consider a cruise altitude $h = 10,500$ m, a block fuel ratio $\zeta = 0.30$, an initial mass $m_i = 155,000$ kg, and three representative aircraft masses: at the start of the cruise, at the end, at half the cruise, and at the end of the cruise. Calculate the speed for maximum range using Eq. 9.10, and verify that the latter result conforms with the plots obtained.

14. Estimate the specific range of the Boeing B-747 at cruise conditions, $h = 11,500$ m, $M = 0.82$, with the $AUW = 370,000$ kgf. Assume that the fuel flow at these conditions is 8,950 kg/h.

15. Consider the two aircraft listed in Table 9.3. Calculate the fuel saving from carrying a payload of 40 metric tons (40,000 kgf) to a distance of 12,000 km, by direct flight with aircraft A, and by en-route stop with aircraft B (the refueling stop is at 6,000 km). Calculate also the GTOW for both aircraft. Assume the same Mach number $M = 0.81$ at an initial altitude $h = 10,700$ m. Consider

a cruise/climb flight program. (For the calculation of the fuel expended the reserve fuel is not counted, although it must be included in the estimation of the GTOW.)

16. Write the range equation for a propeller-driven aircraft that is to fly at constant altitude and constant lift coefficient. Discuss the parameters involved, with reference to optimal cruise range.

17. Consider the event of a vertical gust on a fixed wing. The wing is flying at a speed U; the vertical gust has a speed w_g. Make a sketch that shows the velocities, the inflow angle, the lift and drag force. Discuss how thrust can be generated.

Chapter 10
Maneuver Performance

The fighter pilots have to rove in the areas allotted to them in any way they like, and when they spot an enemy they attack and shoot him down. Anything else is rubbish.

Attributed to Baron M. von Richthofen

The term *maneuver* refers to any change of the flight path. There are at least three basic types of maneuvers: (1) a turn in the horizontal plane, when the aircraft changes heading (lateral maneuver); (2) a turn in a vertical plane (*pull-up*), when the aircraft increases its altitude; and (3) a roll around the aircraft's longitudinal axis, when the aircraft rotates around itself, whilst following a straight flight path (longitudinal maneuver). During a turn on the horizontal plane the aircraft gradually changes its banking angle from zero to a maximum, and then back to zero. A pull-up occurs with a variable radius of curvature and speed. Therefore, a turn maneuver is always unsteady. Banking is always associated with large centrifugal accelerations. Banking at large angles and high speed leads to an increased risk of stalling the aircraft.

In addition to the cases listed, there are a large number of maneuvers, performed only by military aircraft, aerobatic aircraft and demonstrators. These include the Immelmann* (half loop and half roll), the Herbst maneuver (a post-stall 180 degree turn at minimum turn radius), the split-S (a 180-degree turn in the vertical plane with the aircraft rolling on its side, or Herbst in reverse), the cobra, and others. These maneuvers are essentially due to pilots' skills, and involve complicated three-dimensional flight paths.

A concept of interest in this context is *aircraft agility*. The definition given by AGARD[271] describes the agility as the ability to maneuver rapidly along an arbitrary flight path. *Systems agility* is the capability to change mission requirements. However, there are several other definitions, including the capability to move from one maneuver to another, the capability to move the aircraft to a shoot position, etc.

10.1 BANKED LEVEL TURNS

A *banked level turn* (or coordinated turn) is a turn during which all the forces on the aircraft are balanced. Therefore, the aircraft flies with a constant air speed and constant altitude. However, due to the change of heading, there is a centrifugal acceleration. The nomenclature and the reference systems for a banked turn are shown in Figure 10.1. This is a view of the forces in the vertical plane $\{y, z\}$; ϕ is the bank angle, e.g. the angle between the vertical plane and the aircraft's symmetry plane;

---

* Max Immelmann was a German pilot of WWI who invented the maneuver, flying a Fokker airplane. In one battle Fokker's interrupter gear malfunctioned and he shot his own propeller.

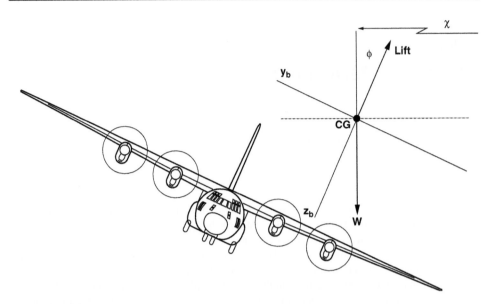

Figure 10.1 *Nomenclature and reference quantities for a banked turn.*

χ is the radius of curvature of the flight trajectory. In the following discussion, the subscript "t" will denote quantities during turn.

In a correctly banked turn the balance of forces along the horizontal and vertical direction, are, respectively

$$L \cos \phi - W = 0, \tag{10.1}$$

$$L \sin \phi = m \frac{U^2}{\chi}. \tag{10.2}$$

The right-hand term of Eq. 10.2 denotes the force due to the centrifugal acceleration of the CG. Combination of the two equations provides the bank angle ϕ,

$$\tan \phi = \frac{1}{g} \frac{U^2}{\chi}. \tag{10.3}$$

The bank angle increases with speed; it also increases as the radius of turn decreases. When the aircraft operates at the smallest values of χ, it is said to perform a *tight* turn. During maneuvers of this type, the lift is higher than the aircraft's weight. We define the normal load factor n as

$$n = \frac{L}{W}. \tag{10.4}$$

In a straight level flight $n = 1$. This value, also called "1g flight", is the neutral value. Now the bank angle, Eq. 10.3, can be written in terms of n, by using Eq. 10.1.

$$n = \frac{1}{\cos \phi} = \sec \phi. \tag{10.5}$$

From Eq. 10.3, the radius of turn expressed in terms of the load factor becomes

$$\chi = \frac{U^2}{g}\frac{1}{\tan\phi} = \frac{1}{g}\frac{U^2}{\sqrt{n^2-1}},$$ (10.6)

where we have used the trigonometric equivalence $\sec\phi = 1 + \tan^2\phi$. In civil aviation, the bank angle is kept within 25 degrees, which leads to a normal load factor of about 1.1 – largely tolerable by most passengers.

From the definition of turn radius χ, Eq. 10.6, the centrifugal acceleration is

$$a = \frac{U^2}{\chi} = g\sqrt{n^2-1}.$$ (10.7)

This acceleration can be normalized with g, to provide the g-factor, or the relative centrifugal acceleration during a banked turn

$$\frac{a}{g} = \sqrt{n^2-1}.$$ (10.8)

A normal load factor $n = 1$ corresponds to zero centrifugal acceleration, which proves that $n = 1$ is a flight with no acceleration.

10.2 BANKED TURN AT CONSTANT THRUST

We consider a case in which the turn is performed at constant thrust. The drag coefficient in the turn is

$$C_{D_t} = C_{D_o} + kC_L^2 = C_{D_o} + k\left(\frac{2nW}{\rho AU^2}\right)^2.$$ (10.9)

There is an increase in lift-induced drag proportional to n^2, and hence an increase in overall drag, $\Delta D = D_t - D > 0$. If we write the conservation of total energy of the aircraft, then

$$\frac{\partial h_E}{\partial t} = \frac{T-D}{W}U.$$ (10.10)

where h_E is the energy height, defined by Eq. 8.70; the right-hand side is the specific excess power. This equation expresses the fact that an excess power can be used to increase the speed or the altitude, or both. A change in aircraft drag dD can be introduced in Eq. 10.10, so that

$$d\left(\frac{\partial h_E}{\partial t}\right) = \frac{T}{W}dU - \frac{dD}{W}U - \frac{D}{W}dU,$$ (10.11)

$$\frac{dD}{W}U = -d\left(\frac{\partial h}{\partial t} + \frac{U^2}{2g}\right) + \frac{T}{W}dU - \frac{D}{W}dU = -d\left(\frac{\partial h}{\partial t} + \frac{U^2}{2g}\right),$$ (10.12)

because $T = D$. If the flight altitude is maintained constant ($dh/dt = 0$), an increase in drag corresponds to a decrease in speed, and vice versa,

$$\frac{dD}{W} U = -\frac{U}{2g} dU. \tag{10.13}$$

If, instead, the speed is maintained constant ($dU = 0$), then there must be a decrease in altitude

$$\frac{dD}{W} U = -d\left(\frac{\partial h}{\partial t}\right) = v_s < 0, \tag{10.14}$$

where v_s denotes the sinking speed. Its absolute value can be found from

$$v_s = \frac{D_t - D}{W} U = \frac{\rho A U^2}{2W}(C_{D_t} - C_D) = \frac{2W}{\rho A U^2} k(n^2 - 1). \tag{10.15}$$

For a given speed, the rate of descent increases with n^2. The load factor can be expressed in terms of the bank angle, turn radius, or the turn rate.

The change in drag produces a change in the speed of minimum drag, U_{md}; this can be found by using the same arguments as in §6.8. The C_L corresponding to minimum drag in a banked turn is found from the condition

$$\frac{\partial}{\partial C_L}\left(\frac{C_{L_t}}{C_{D_o} + kC_{L_t}^2}\right) = 0. \tag{10.16}$$

This condition leads to

$$C_L = \sqrt{\frac{1}{n}\frac{C_{D_o}}{k}}. \tag{10.17}$$

Therefore, from Eq. 6.50 the ratio of minimum drag speeds between banked turn and level flight is

$$\frac{U_{md_t}}{U_{md}} = n^{1/2}. \tag{10.18}$$

If the aircraft's speed before the turn is $U > U_{md}$, the drag increase due to the turn will slow down the aircraft; this will contribute to a decrease in drag to the point when the forces on the aircraft are balanced, and the aircraft can sustain a turn at constant altitude, albeit at a lower speed. If, instead, the aircraft's speed before the turn is $U < U_{md}$, a decrease in speed will produce a further increase in drag. If the drag rise cannot be matched by the engine thrust, the aircraft will descend.

10.3   POWER REQUIREMENTS

Assume that the C_L during a turn is maintained constant, and equal to the value in straight level flight,

$$C_L = C_{L_t}, \qquad \text{or} \qquad \frac{2W}{\rho A U^2} = n\frac{2W}{\rho A U_t^2}. \tag{10.19}$$

This leads to the equation

$$\frac{1}{U^2} = n\frac{1}{U_t^2}, \qquad \text{or} \qquad U_t = U\sqrt{n}. \tag{10.20}$$

which expresses the equivalence between aircraft speeds, at *constant lift coefficient*. If the turn is done at *constant speed*, then the relationship between lift coefficients becomes

$$C_{L_t} = nC_L. \tag{10.21}$$

Next, consider a turn at constant power, $P = P_t$. This requires

$$TU = T_t U_t. \tag{10.22}$$

This condition can be achieved by increasing the thrust when the speed decreases. The power in steady state flight is found from the equilibrium of forces in the flight direction multiplied by the flight speed,

$$P = \frac{1}{2}\rho A C_D U^3. \tag{10.23}$$

Eliminate the velocity in Eq. 10.23, by using the definition of C_L

$$U^2 = \frac{2L}{\rho A C_L}, \qquad U = \left(\frac{2L}{\rho A C_L}\right)^{1/2}, \qquad U^3 = \left(\frac{2L}{\rho A C_L}\right)^{3/2}. \tag{10.24}$$

Insert Eq. 10.24 into Eq. 10.23, and find

$$P = \frac{1}{2}\rho A C_D \left(\frac{2L}{\rho A C_L}\right)^{3/2} = \sqrt{\frac{2}{\rho A}}\frac{C_D}{C_L^{3/2}}L^{3/2} = \sqrt{\frac{2}{\rho A}}\frac{C_D}{C_L^{3/2}}(nW)^{3/2}. \tag{10.25}$$

Therefore, the power required for a steady banked turn grows with $(nW)^{3/2}$. If $C_D/C_L^{3/2}$ is held constant, the ratio between power required in a turn and the steady state power is

$$\frac{P_t}{P} = n^{3/2}. \tag{10.26}$$

There is another useful expression for the turning power. For a parabolic drag at subsonic speeds, Eq. 10.23 becomes

$$P = \frac{1}{2}\rho_o \sigma C_D U^3 = \frac{1}{2}\rho_o \sigma C_{D_o} U^3 + k\frac{2}{\rho_o A}\frac{(nW)^2}{U}. \tag{10.27}$$

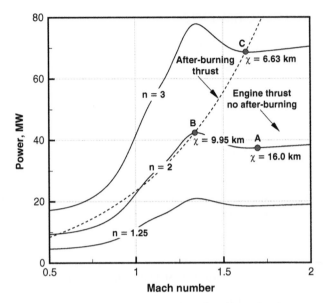

Figure 10.2 *Power for supersonic turn, with and without after-burning, aircraft weight W = 12,000 kg, constant angle of attack α = 2 degrees, altitude h = 8,000 m (26,246 ft).*

Equation 10.27 can be plotted versus the flight speed at different values of n. Matching this power with the available engine power will give the operation point. The minimum speed that the aircraft can sustain is the stall speed (within a safety margin), hence the lower limit of the curves will be given by the stall curve.

We consider in more detail the more complicated problem of high-speed turn performed by the reference supersonic aircraft. The lift and drag characteristics of this aircraft are given in Figures A.8 and A.9. We consider a turn at the flight altitude $h = 8,000$ m (26,246 ft), with a gross weight $W = 12,000$ kg.

The algorithm consists in solving Eq. 10.25 over a range of Mach numbers with the C_D from Eq. 4.12 and the C_L from Eq. 4.6. The results of the analysis are shown in Figure 10.2, for normal load factors up to $n = 3$. Also shown is the available engine thrust versus Mach number, with and without after-burning.

The power required to make a supersonic turn is phenomenal. Even with after-burning engines, capable of developing over 100 kN of thrust, a supersonic turn cannot be performed at high load factors, at $h = 8,000$ m. However, the performance is strongly dependent on the altitude.

10.4 EFFECT OF WEIGHT ON TURN RADIUS

The radius of turn is given by Eq. 10.6. Therefore, the ratio between radii of turn corresponding to load factors n_1 and n, respectively, becomes

$$\frac{\chi}{\chi_1} = \left(\frac{n_1^2 - 1}{n - 1}\right)^{1/2}. \tag{10.28}$$

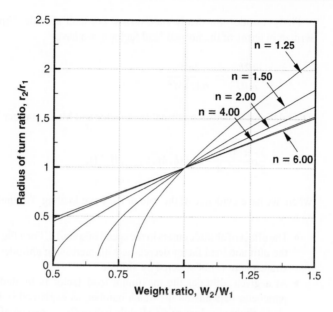

Figure 10.3   *Turn radius versus aircraft weight, normalized data.*

If we write the relationship between load factor and aircraft weight

$$n_1 W_1 = nW,$$

(10.29)

then Eq. 10.28 becomes

$$\frac{\chi}{\chi_1} = \left(\frac{n^2 (W/W_1)^2 - 1}{n - 1} \right)^{1/2}.$$

(10.30)

The result of this equation is shown in Figure 10.3. In the horizontal axis we have plotted the weight ratio W/W_1; in the vertical axis we have plotted the corresponding ratio between the radii of turn, χ/χ_1. Each curve represents a constant value of the load factor, as indicated. As the load factor increases, Eq. 10.30 becomes linear; at a load factor $n \sim 6$, the weight effect can hardly be distinguished from the curve at $n = 4$. In conclusion, for a banked turn at constant altitude and constant engine thrust, a weight increase will increase the radius of turn – all other parameters being the same.

10.5   MANEUVER ENVELOPE: n–V DIAGRAM

The maneuverability of an aircraft depends on its ability to perform turns. This capability has some limits, imposed by the aerodynamic performance, by the engines, the structural characteristics, and – increasingly – on the resistance of the human body. The essential parameters are the thrust ratio T/W, the wing loading W/A, and the maximum C_L. There are two types of envelope: one is the load factor versus Mach number; the other is the turn rate versus Mach number (or speed).

First, consider the steady state turn at constant altitude. From the lift equation, written in terms of the normal load factor n, we have

$$C_L = \frac{2L}{\rho A U^2} = \frac{2nW}{\rho A a^2 M^2}.$$ (10.31)

The relationship between normal load factor and Mach number becomes

$$n = \frac{1}{2}\rho \frac{a^2 M^2 C_L}{W/A} = f(h, M, C_{L_{max}}, W/A),$$ (10.32)

where we have evidence of the role of the wing loading. We find that:

- The effect of altitude enters in the speed of sound a. For a flight in the troposphere, the ultimate load factor decreases with increasing altitudes. A turn in the lower stratosphere is not affected by altitude.
- At a given altitude, the maximum load factor is limited by the $C_{L_{max}}$. This parameter depends on the Mach number, as explained in Chapter 4. However, also a flight at values of C_L slightly below $C_{L_{max}}$ may be of concern, because of buffeting associated with the unsteady separated flow, or because of control and stability problems. Therefore, the usable C_L for turning must be a fraction of $C_{L_{max}}$. In order to plot Eq. 10.32, we need the function $C_{L_{max}}(M)$. This function is given in Figure A.12 for the supersonic jet aircraft of reference.

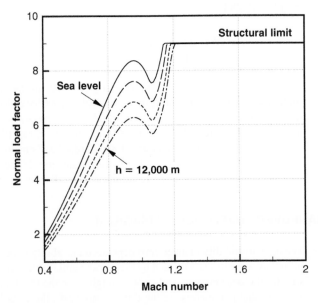

Figure 10.4 *Maneuver limits imposed by $C_{L_{max}}$, $W = 14,000$ kg, altitudes as indicated, ISA conditions.*

- The third maneuver limit is due to the maximum engine thrust. One can think of moving along the horizontal solid line of Figure 10.4 (indicated as *structural limit*) and increasing the Mach number as much as the engine thrust allows. The calculation of this limit is more elaborate, because the maximum Mach number during a turn also depends on the angle of attack of the aircraft. The governing equation is

$$nT = \frac{1}{2}\rho A(C_{D_o} + \eta C_{L_\alpha}\alpha^2)a^2 M^2, \tag{10.33}$$

or

$$M^2 = \frac{n}{c_1}\frac{T(M)}{C_{D_o} + \eta C_{L_\alpha}\alpha^2}, \tag{10.34}$$

with $c_1 = \rho A a^2 / 2$ a constant coefficient at a constant turn-altitude. Equation 10.34 is an implicit relationship between M and α.

This turn must be made within the limits of maximum speed and flight altitude, and also within the limits of energy levels. Therefore, the rate of change of specific excess power is one of the most important factors in maneuverability.

- Finally, the load factor can exceed the structural limit at certain speed. Therefore, if other factors do not intervene, the maneuver envelope has a flat top.

The maneuver limits due to aerodynamics are shown in Figure 10.4 at a number of selected altitudes. This result was found by solving Eq. 10.32, with the appropriate value of $C_{L_{max}}(M)$ for a given weight. The result shows that the load factor increases sharply up to $M \sim 0.9$, then it decreases, due to transonic drag rise. Finally, it regains magnitude, because the Mach number increases, while $C_{L_{max}}(M)$ stagnates. The horizontal line in the graph illustrates the maneuver limit imposed by the structural integrity of the aircraft. This is a flat top, as anticipated.

A qualitative example of maneuver envelope for a high-performance aircraft is shown in Figure 10.5. For a given gross weight, the envelope depends on the flight altitude. The lift is limited by the subsonic buffet and supersonic horizontal tail deflection limit. The maximum g-factor depends on the structural limits. The maximum level speed depends on the altitude, therefore the speed limit in the flight envelope can move backward or forward, depending on the aircraft. The maneuver envelope at sea level is indicated by the shaded area.

10.6 TURN RATES

The *rate of turn* is the angular velocity of the aircraft. Consider again the case of a coordinated turn. The turn rate is

$$\dot{\psi} = \frac{U}{\chi} = \frac{g}{U}\sqrt{n^2 - 1}, \tag{10.35}$$

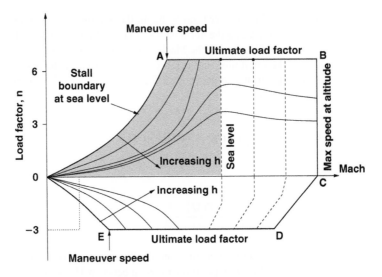

Figure 10.5 *n–V diagram for a supersonic jet aircraft.*

where we have used the definition of turn radius χ from Eq. 10.6. If we express the velocity of turn in terms of the load factor,

$$U = \sqrt{n \frac{2W}{\rho_0 \sigma A C_L}}, \tag{10.36}$$

the turn rate becomes

$$\dot{\psi} = g\sqrt{\frac{\rho A C_L}{W}}\sqrt{\frac{n^2 - 1}{n}}. \tag{10.37}$$

Collect all the constant terms in Eq. 10.37 into a coefficient c_1

$$c_1 = g\sqrt{\frac{\rho_0}{2}}. \tag{10.38}$$

Therefore,

$$\dot{\psi} = c_1\sqrt{\frac{\sigma C_L}{W/A}}\sqrt{\frac{n^2 - 1}{n}}. \tag{10.39}$$

Equation 10.39 shows how the turn rate is dependent on altitude, C_L, and W/A. When maximum turn rates are quoted, it is essential to specify at least the altitude and weight.

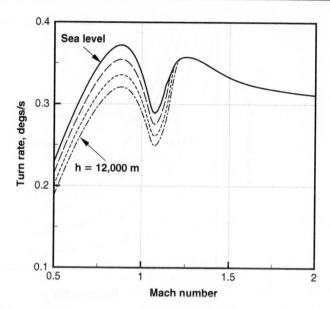

Figure 10.6 *Calculated turn rates for a supersonic jet aircraft of reference, at selected flight altitudes.*

Figure 10.6 shows the turn rate for the same supersonic aircraft at different altitudes, using the function $C_{L_{max}}(M)$. After an initial increase with the Mach number, the turn rate is affected by transonic effects. The transonic dip shown in this simulation might not occur in practice, and a progressive decrease in turn radius may be experienced at supersonic speeds[111]. The other detail to be noted in Figure 10.6 is that at full supersonic speeds, the altitude has virtually no effect on the turn radius, as all the curves collapse into a single curve.

10.6.1 Corner Velocity

The curves shown in the chart of Figure 10.6 denote performance at constant load factor and variable C_L. It is also possible to calculate curves of constant C_L and variable load factor. These curves will have to intersect the former ones at some points that define the corner speeds at that particular C_L.

The *corner velocity* is the minimum velocity at which the aircraft can reach its maximum load factor. It is found at the intersection between the curve of maximum constant n-factor and the curve corresponding to a maneuver at maximum C_L. In analytical terms, the relationship between the g-factor and the Mach number is found from the definition of C_L,

$$U^2 = a^2 M^2 = \frac{2nW}{\rho A C_{L_{max}}}.$$

(10.40)

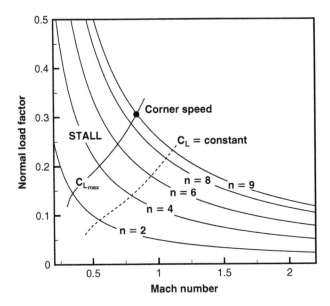

Figure 10.7 *Turn rates and corner speed for supersonic jet aircraft of reference, sea level, ISA conditions.*

In the case of the reference supersonic jet aircraft,

$$n = \frac{a^2}{2} \frac{\rho}{W/A} C_{L_{max}}(M) M^2.$$

(10.41)

The turn performance in terms of Mach number is shown in Figure 10.7. We calculated a corner Mach number $M = 0.84$ at sea level, corresponding to a turn rate $\dot{\psi} = 0.31$ rad/s (17.5 deg/s) for a g-factor limit equal to 9.

Performance charts of actual airplanes look like that of Figure 10.7 with additional information, such as lines of constant turn radius, sustained turn radius, etc. Since this chart is calculated for a given weight and flight altitude, to analyze the aircraft performance during a high-speed turn we need charts in a range of gross weights and flight altitudes. Each of these charts defines a flight envelope in the n–M plane, bounded by the structural limits of the aircraft, by the lift coefficient (aerodynamics) and engine thrust (propulsion).

General three-dimensional turn problems for minimum time and minimum fuel have been solved by Hedrick and Bryson[272,273], using the energy method. These authors considered powered accelerated turns, coasting decelerating turns, climb turn from take-off, and subsonic and supersonic speeds. Kelley[274,275] has solved the problem of differential turning for the out-climbing and out-turning strategies, which are the basic evasive tactics of a combat aircraft.

10.7 SUSTAINABLE *g*-LOADS

In the preceding sections we have discussed the maneuver limits from the point of view of the aircraft. There is one more factor that limits the centrifugal accelerations

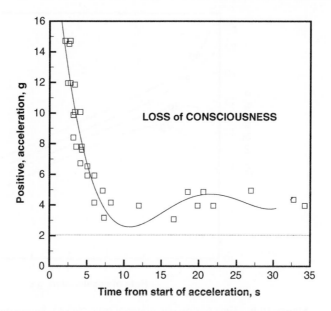

Figure 10.8 *Sustainable g-loads map, adapted from NASA's Bioastronautics Data Book[278].*

of the aircraft: the pilot. The large accelerations usually encountered by an aircraft in a tight turn can reach values beyond the sustainability of a pilot. LeBlaye[276] discussed aspects of aircraft agility and sustainable loads. Jaslow[277] studied the problem of loss of spatial orientation in a coordinated turn and its effects on accident rates on various classes of aircraft. An important reference is NASA's *Bioastronautics Data Book*[278]. All the studies confirm that posture is critically important. Pilots of high-performance aircraft can tolerate forces of $+9g$ with the use of special suits, as discussed in reference 279. People generally lose consciousness when the acceleration is in the $+g$ direction, when blood is drawn away from the brain toward the feet*.

Figure 10.8, adapted from Webb[278], shows the boundaries of sustainable g-loads as a function of the exposure time in seconds. It shows how the human body has not evolved for such an abuse. The solid line is a curve fit of the experimental points, which are a selection of symptoms, ranging from confusion to grayout, blackout and total loss of conscience.

A sustainable g-load is an acceleration that does not affect the pilot in any serious way (orientation, vision, heart beat, blood pressure). At high g-loads even at a relatively low Mach number at the highest load factors the pilot's breathing must be aided by oxygen reserve. A load factor of 4 sustained for more than 5 seconds is potentially dangerous. A load factor of 2 can seemingly be endured for a long time.

---

* Research into long-term sustainable g-forces is a subject of imaginative research. Hyper-g studies were done in the 1970s on chickens that were placed in 2.5g accelerations for several months at a time, see Smith[280]. After 6 months in the *chronic* centrifuge, the chickens developed stronger bones and adapted to the new environment.

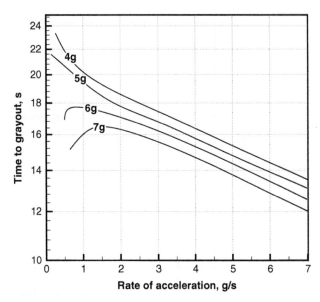

Figure 10.9 *Time to grayout, adapted from NASA's Bioastronautics Data book[278].*

The time to grayout as a function of the acceleration rates (in g per second) is shown in Figure 10.9. Unconsciousness sets in at variable times, depending on the g-level, and it varies from 1 second to 0.1 seconds. By comparison, the red-headed woodpecker is estimated to "suffer" up to 12,000 accelerations per day at $10g$, while pecking. There is no evidence that this leads to brain damage[179].

10.8 UNPOWERED TURN

Now we go back to more mundane speeds, and discuss the turning performance of unpowered gliders. A steady state turn of a glider causes a loss of altitude. The calculation of the sinking speed v_s, the radius of turn χ and the turn rate $\dot{\psi}$ can be done if the bank angle ϕ is assigned. The dynamics equations in the direction of the flight path and in the vertical direction are, respectively,

$$D - W \sin \gamma = 0, \tag{10.42}$$

$$L \cos \phi - W \cos \gamma = 0. \tag{10.43}$$

The sinking rate is

$$v_s = \frac{\partial h}{\partial t} = -\frac{DU}{W}. \tag{10.44}$$

If we replace the parabolic drag equation in Eq. 10.44, we have

$$v_s = -\frac{\rho A}{2W}(C_{D_o} + kC_L^2)U^3 = -\frac{\rho A C_{D_o}}{2W}U^3 - \frac{\rho A k}{2W}\left(\frac{2L}{\rho A U^2}\right)^2 U^3. \qquad (10.45)$$

The lift is found from Eq. 10.43. With this expression Eq. 10.45 becomes

$$v_s = -\frac{\rho A}{2W}C_{D_o}U^3 - \frac{2W}{\rho A}\frac{k}{U}(n\cos\gamma)^2. \qquad (10.46)$$

The solution of Eq. 10.46 requires the descent angle γ. This can be calculated from the glide ratio of the unpowered airplane, L/D, that is found by dividing Eq. 10.42 by Eq. 10.43,

$$\tan\gamma = \frac{D}{L\cos\phi}, \qquad \gamma = \tan^{-1}\left(\frac{n}{L/D}\right). \qquad (10.47)$$

Equation 10.47 requires the L/D, which is dependent on the speed. The turn rate is found from

$$\dot{\psi} = \frac{U}{\chi} = \frac{g}{U}\frac{L}{W}\sin\phi = \frac{g}{U}\cos\gamma\tan\phi, \qquad (10.48)$$

where we have used the combination of Eq. 10.42 and Eq. 10.43. The radius of turn is found from the first of the equivalences in Eq. 10.48.

Computational Procedure

1. Read the aircraft parameters, including aerodynamic data.
2. Set the bank angle ϕ.
3. Calculate the relative air density.
4. Set the air speed U equal to the stall speed.
5. Calculate the glide ratio.
6. Calculate the descent angle γ from Eq. 10.47.
7. Calculate the turn rate from Eq. 10.48.
8. Increase the speed and repeat calculations from point 4.

The solution to this algorithm is shown in Figure 10.10 for a sailplane starting a turn at an altitude of 1,000 m (3,048 ft), for selected bank angles. The problem's data are: $C_{D_o} = 0.007$, $k = 0.022$, $W = 450$ kg, $A = 17.0$ m^2.

For the same case and for a load factor $n = 1.25$, we have calculated the best gliding speeds. This is shown in Figure 10.11, along with the turn radius. Further analyses can be done (see Problem 11). Some gliding trajectories, obtained from the integration of the flight path equations, are shown in Figure 10.12.

10.9 SOARING FLIGHT

Soaring is a flight maneuver in which a vehicle gains altitude while performing a turn. This term is generally associated with unpowered vehicles such as gliders, but

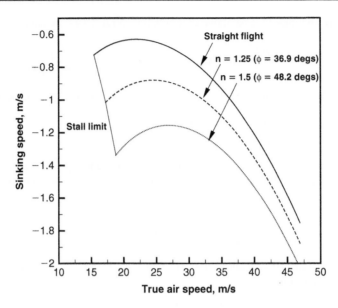

Figure 10.10 *Sinking speed for glider turning at constant load factors (as indicated).*

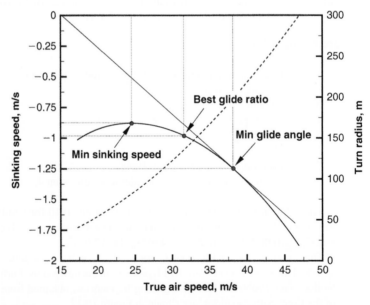

Figure 10.11 *Sinking speed for the glider of Figure 10.10, with $n = 1.25$. Best gliding speeds and corresponding turn radius.*

its theories stem from extensive analysis of the flight of birds. We have shown in §10.8 that an unpowered vehicle drifts down in a coordinated turn. In order to gain height, the vehicle must take advantage of particular atmospheric conditions, namely a vertical wind or a gusty horizontal wind. The vertical wind is often created by

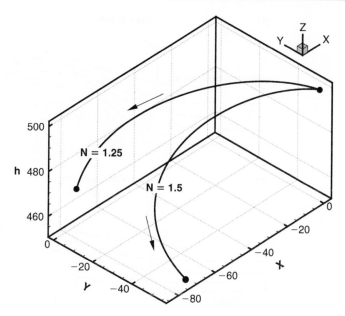

Figure 10.12   *Gliding trajectories relative to the case of Figure 10.10.*

warm air (*thermal*) that rises from the Earth's surface. A horizontal wind is not necessary, although its presence can be used to fly forward while soaring. The latter case leads to what is called *dynamic soaring*. Vertical winds are due to a combination of factors, including rising of warm air (*thermal soaring*), presence of hills (*slope soaring*), water/land boundaries and other features (trees, ridges, weather patterns) as sketched in Figure 10.13. The vertical velocity gives rise to *declivity currents*. The relative importance of thermals and horizontal winds depends on the presence of water and land. Flight over land uses mainly thermal soaring; flight over water is almost exclusively the domain of bird flight, and is dominated by dynamic soaring, e.g. the presence of gusts and turbulence.

In addition to the full turn soaring, there are *essing maneuvers* (with alternate left- and right-hand turns, not complete turns) and *dolphining* (a cross-country flying technique that uses gliding and soaring, with zero altitude loss or gain over a straight course). The current record for altitude gain for open class gliders is nearly 13,000 m (achieved in 1961), and the longest distance flown is over 3,000 km (2003)*. These flights have been obtained using essing and dolphining, with relatively little time spent on thermal soaring.

The importance of soaring flight research in the development of the powered aircraft is highlighted by Short[281], who published an account from the American point of view. A detailed bibliography on the subject is available in this paper. A good introductory discussion is given by Alexander[282] (for bird flight) and Irving[283] (for sailplane design and operation). These books provide additional references. The observation that the air must have a vertical motion goes as far back as Rayleigh (1883), but had many

---

* These aviation records are kept by the FAI. There is no official endurance record to this date.

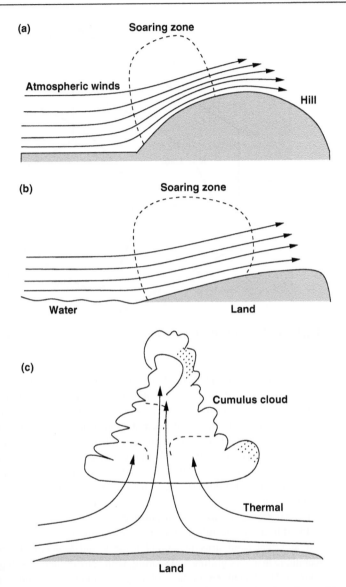

Figure 10.13 *Declivity currents due to hills (a), land boundary layer (b), and thermals (c).*

people puzzled for years. Early theories of soaring flight are attributed to Prandtl[284], Klemperer[285] and others. A flight path problem has been considered by Metzger and Hedrick[286], who optimized the performance of a glider for maximum cross-country distance by using optimal control theory. The variational problem includes all the modes of soaring discussed above. These authors discuss alternative flight modes and the knowledge of the atmospheric conditions. The application of innovative concepts (winglets, super-circulation) derived from soaring flight is discussed by Bannasch[287].

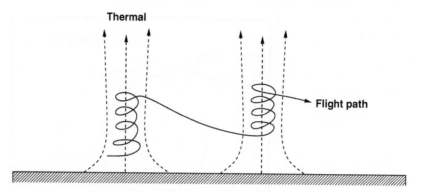

Figure 10.14 *Soaring from one thermal to the next.*

The research on birds is equally extensive. Cone[288] discussed the thermal soaring of birds from extensive observations of vultures, hawks, herons and other birds. Cone highlighted the importance of the slotted wing tip for reducing the drag. Hedenstroem[289] discussed the role of soaring in migrating birds.

The general issues of soaring flight are exceedingly complex, therefore we will limit ourselves to some basic cases. Consider first the case of static thermal soaring, in which there is a column of air rising vertically from the Earth's surface. In still air the vehicle sinks at a rate $v_s = U \sin \gamma$. Therefore, the vehicle loses energy at a rate $W v_s$. In soaring flight the vehicle will be sinking with the same speed with respect to the air. Hence, if the upwash is equal to the sinking speed the vehicle would be flying parallel to the ground. If the air were rising with a speed greater than v_s, the vehicle would be gaining altitude.

Next, consider the case of dynamic thermal soaring. The vertical current is super-imposed to a horizontal wind U_w. We consider the motion of the vehicle with respect to a *thermal bubble* and then the motion of the bubble with respect to the Earth. The basic stages in the formation of a warm shell are quite complex, and require a lengthy discussion of meteorological fluid mechanics. In brief, a protuberance of warm air forms on the ground. This slowly evolves into a mushroom-shaped vortex in which cool air is drawn from the surroundings. The warm vortex core grows until it bursts; then it is carried away by the prevailing winds or the jet stream. Obviously, there can be more than one thermal bubble, so the vehicle can descend from one thermal to the other, as indicated in Figure 10.14.

Assume that the vehicle is already inside the thermal bubble. The vehicle is turning in a level plane with respect to the bubble. However, the bubble is rising from the ground and drifting according to the prevailing winds, just like a balloon. Therefore the combined motion will consist of a roughly circular motion within the bubble, a vertical motion due to the rising of the thermal, and a horizontal motion due to the prevailing winds. If the vehicle enters the thermal bubble at a low level, it will gain altitude at relatively high speed, by means of full circles. However, as it climbs its soaring rate decreases. When the upwash is equal to the sinking speed, the vehicle will have reached its vertical equilibrium. It can be shown that the flight trajectory projected on the plane parallel to the ground is a *trochoid*. Mathematically, the trochoid is the locus of a point at a fixed distance from the center of a circle rolling on a fixed

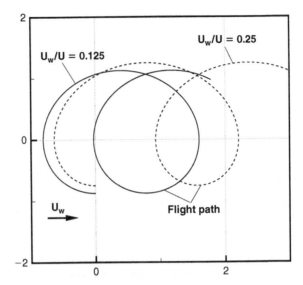

Figure 10.15 *Trochoidal flight path in the horizontal plane.*

line. The parametric equations of this curve are

$$\frac{x}{\chi} = \frac{U_w}{U}\psi - \sin\psi,$$ (10.49)

$$\frac{y}{\chi} = \frac{U_w}{U} - \cos\psi,$$ (10.50)

where χ is the radius of curvature of the flight path in the thermal, U_w is the horizontal component of the wind, U is the air speed of the vehicle, and ψ is angular displacement of the trochoid from the initial point. The curve is shown in Figure 10.15 for two values of the velocity ratio $U_w/U < 1$. With the parametric values used, the flight path is more exactly a prolate cycloid. The maximum width of the trochoidal flight path is equal to 2χ. The radius of the trochoid circle is $(U_w/U)\chi$ (see Wolfram[290] for further details).

The vertical component of the velocity is variable. In particular, the vehicle gains more altitude on the windward side of the circle. On the leeward side the altitude gain is considerably smaller. Actually, it is possible to drift down in the leeward side, if the wind speed is large, or the upwash is small. This is explained by considering the direction of the vector velocity. For a fixed attitude, the velocity is highest when the vehicle flies into the wind and smallest when it flies away from it. Therefore, for a fixed attitude, the C_L corresponding to a head-wind flight is higher, and the vehicle can climb. A bird can obviously adjust its wings in an infinite variety of ways, and can take full advantage of the changing air speed.

In this discussion we have assumed that the size of the bubble is much larger than the radius of the circle, and that the upwash is basically uniform. This is a restrictive condition. The ability to exploit the whole range of thermal sizes and upwash velocities is dependent on the ability to perform tight turns. The smaller the radius of turn, the greater the duration of soaring. Equation 10.46 gives the sinking speed as a function of the main flight parameters (aerodynamics, structural characteristics, speed, altitude). The sinking speed increases as the radius of curvature decreases. This is easy to prove if the flight path is a circle due to a coordinated turn. We can assume it is valid also for a trochoidal flight path in which the altitude gain or loss is small compared to the radius of turn, and also when the wind speed is small. If we replace the expression of the radius of curvature of the flight path (Eq. 10.6) into Eq. 10.46, we find

$$v_s \simeq -\frac{1}{2}\frac{\rho A C_{D_o}}{W}U(n^2 - 1)^{1/2}g\chi - \frac{2W}{\rho A}\frac{k}{U}(n\cos\gamma)^2. \tag{10.51}$$

Therefore, a minimum sinking speed requires a low profile drag, a large radius of curvature, a low wing loading, and a low induced drag. In soaring flight the induced drag can be considerably larger than the profile drag. The most important geometrical factors affecting the induced drag are the aspect-ratio of the wing and the wing tip design. The radius of turn and the corresponding banking angle have to be adjusted, so that at any point the sinking speed is at least equal to the upwash of the thermal. Then the circling flight becomes automatic.

We have observed turkey vultures (*Cathartes aura*) soaring in absence of horizontal winds. At each new circle the birds gained height with a progressively smaller radius of turn. The birds entered the thermal bubble from its periphery and slowly circled with tighter turns. The upwash of the thermal is minimum at the periphery and maximum at the center. However, the opposite is also possible. In fact, as the birds gain height, their sinking speed increases due to density effects (see Eq. 8.88 and Figure 8.16). The upwash of the thermal decreases with the altitude, therefore as height is gained, the sinking speed has to decrease. This can be done by increasing the radius of turn – all other parameters being constant.

Dynamic soaring is mastered by birds such as the albatross (a water bird) and the condor (a land bird), although mechanisms by which drag is reduced are different. The albatross takes off against the wind, and gains altitude without flapping its slender wings, until it loses momentum in the faster-moving layers of air further from the water. The bird then turns and dives back to the water, gaining so much speed that, after turning again, it can once more climb against the wind.

The condor has a wing aspect-ratio of about 6, which makes it relatively inefficient. The comparatively high-induced drag is offset by a combination of optimal spanwise loading and its fan-like wing tips.

10.10   ROLL PERFORMANCE

In this section we consider the one-degree of freedom roll of a high-performance aircraft. During a roll the aircraft rotates around its longitudinal axis while maintaining a straight course. The free parameter is the roll angle ϕ. Roll rates and roll accelerations for these aircraft can be relatively high, and are a key factor to their overall agility. Roll performance and requirements, as from Military Standards*, generally refer to: (1) maximum roll rate; (2) time to roll to 90 degrees; and (3) time to execute a full roll

---

* Information provided by the United States Department of Defense Single Stock Point, DODSSP.

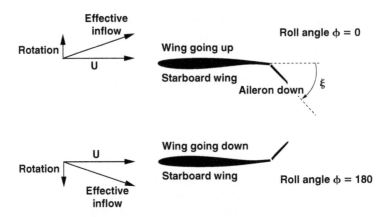

Figure 10.16 *Sketch of inflow conditions on starboard wing, assuming an anti-clockwise roll in the flight direction.*

(360 degrees). For other types of airplanes, roll capability is not essential. Nevertheless, it is important to understand how the airplane responds to rolling moments created by thrust asymmetry (§10.11) and atmospheric turbulence.

The roll discussed below is difficult to achieve at high speeds, due to a number of unwanted effects. One such effect is the yaw response. If the aircraft is turning left in roll, it will cause a yaw (side-slip) in the opposite direction. This effect is called *adverse yaw*. Another important effect is the *inertial coupling* (or roll coupling). When an aircraft rotates around an axis that is not aligned with its longitudinal axis, the inertial forces tend to swing the aircraft out of the rotating axis, with potentially lethal consequences. The problem is compounded by low aspect-ratio wings, high speeds, high altitudes, and consequently high inertial forces compared to the aerodynamic forces. This problem has been treated theoretically by Phillips[291], before it was actually encountered, and is a complex subject for stability and control. Phillips pointed out that it is a precise relationship among the moments of inertia that defines the limits of inertial stability and instability in roll. Pinsker[292] established the critical flight conditions for yaw divergence, pitch divergence and autorotational rolling.

In the present case, we consider an aircraft that does not have problems with inertial coupling. Seckel[151] noted that a steady state roll is a maneuver that cannot be performed, but must involve cyclic variations of angles of attack α and side-slip β. To begin with, when the aircraft starts a roll ($\phi = 0$), its longitudinal axis is not aligned with the rotation axis (though by a small angle). This generates a side-slip β. When the aircraft is on its side ($\phi = 180$ degrees), some side-slip force has to be generated to balance the weight. The combination of yaw and roll can lead the aircraft into inertial coupling instability. Methods of prevention of inertial coupling are discussed by Seckel[151] (Chapter XVII). See also McCormick[104] for directional control analysis.

Consider now the situation illustrated in Figure 10.16. After 180 degrees, the aircraft will find itself upside-down. The inflow on the wing is made of two components: one is due to the forward speed U; the other is due to the roll itself. If we assume that the former component does not change during the aircraft rotation, then the starboard

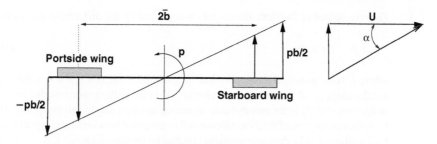

Figure 10.17   *Sketch of inflow conditions on wing, seen in the y–z plane. Anti-clockwise roll in the flight direction.*

wing will find itself with the same angle of attack α_o of the wing-level flight. This angle is generally small.

The rotational component of the speed changes the inflow according to the concept illustrated in Figure 10.16. The maximum angle of attack is at the tip, and is estimated from

$$\tan \alpha = \frac{U}{pb/2} = \frac{2U}{pb},$$
(10.52)

where $p = \dot{\phi}$ is the rotational speed around the longitudinal axis. It can be verified that in many cases the forward speed is much higher than the rotational speed, therefore: $\tan \alpha \simeq \alpha$. Figure 10.17 shows the arrangement with exaggerated angles. In general, the higher the roll rate, and the lower the forward speed, the higher the risk of stall. The problem is complicated by the presence of a spanwise component of the velocity. Delta wings with various aerodynamic arrangements can improve the stall behavior at high angles of attack. Wings for high roll rates must be able to operate at high angles of attack without stalling.

A rolling moment can be created by operating on the control surfaces of the wing. For reference, the ailerons are set at an angle $\pm \xi$ on either side of the symmetry plane ($y_b = 0$), although in practice the deflection angle can be different. In the latter case, the angle ξ to consider is a mean value. Linear response cannot be guaranteed in the general case, because the suction and pressure on the wing tips distort the flow. The vertical tail creates additional flow separation and damping.

The effect of the aileron deflection is almost impulsive. In practice, a deflection speed of about 60 degrees/second is achieved. With a maximum aileron authority of 30 degrees (up and down), about half a second is required for the aileron to reach its limit position. In most roll analyses, a step change in aileron angle is considered. However, even in the event of a finite deflection speed, the flexibility of the wing may reduce the control effectiveness.

To calculate the variation of angle of attack during the rotation around the x axis we have to add the effect of the rotational speed. Therefore, the inflow direction at spanwise position y will have the following expression

$$\alpha(y) \simeq \tan^{-1}\left(\frac{py}{U}\right) \simeq \left(\frac{py}{U}\right).$$
(10.53)

The one-degree of freedom aileron roll is described by the differential equation

$$I_x \dot{p} = \mathcal{L}_\xi + \mathcal{L}_p, \tag{10.54}$$

where I_x is the principal moment of inertia around the roll axis x, $\dot{p} = \ddot{\phi}$ is the angular acceleration, $p = \dot{\phi}$ is the angular velocity, $\mathcal{L}_\xi$ is the rolling moment due to aileron deflection, and $\mathcal{L}_p$ is the aerodynamic damping moment due to roll rate $p = \dot{\phi}$. Therefore, the one-degree of freedom aileron roll is governed by a forcing moment (created by the aileron) and a damping moment (created by the aerodynamics). It is convenient to rewrite the moments in the following form

$$\mathcal{L}_\xi = \left(\frac{\partial \mathcal{L}}{\partial \xi} \right) \xi, \tag{10.55}$$

$$\mathcal{L}_p = \left(\frac{\partial \mathcal{L}}{\partial p} \right) p. \tag{10.56}$$

These derivatives express the moment curve slopes, e.g. the moment response due to a small change in aileron angle and roll speed, respectively. With these definitions of the moments, the roll Eq. 10.54 becomes

$$\dot{p} = \frac{1}{I_x} \left(\frac{\partial \mathcal{L}}{\partial \xi} \right) \xi + \frac{1}{I_x} \left(\frac{\partial \mathcal{L}}{\partial p} \right) p. \tag{10.57}$$

The parameter defined by

$$L_\xi = \frac{1}{I_x} \left(\frac{\partial \mathcal{L}}{\partial \xi} \right). \tag{10.58}$$

is the *aileron effectiveness*. The effectiveness L_ξ denotes the scale of lateral response to a unit input in aileron deflection. Likewise, we can define the *damping in roll* parameter L_p

$$L_p = \frac{1}{I_x} \left(\frac{\partial \mathcal{L}}{\partial p} \right), \tag{10.59}$$

which is the derivative of the rolling moment with respect to the roll rate, normalized with the moment of inertia. By using the definitions of effectiveness and damping in roll, the roll equation finally becomes

$$\dot{p} = L_\xi \xi + L_p p. \tag{10.60}$$

The parameters L_ξ and L_p are integral properties of the aircraft (or the wing system) and depend on the Mach number, density altitude, and several geometrical details.

If C_l is the rolling moment coefficient, then we have:

$$\frac{\partial \mathcal{L}}{\partial \xi} = \frac{\partial}{\partial \xi} \left(\frac{1}{2} \rho A b U^2 C_l \right) = \frac{\rho A U^2 b}{2} \left(\frac{\partial C_l}{\partial \xi} \right) = \frac{\rho A U^2 b}{2} C_{l_\xi}, \tag{10.61}$$

where C_{l_ξ} denotes the derivative of the rolling moment coefficient with respect to the aileron deflection. In other words, C_{l_ξ} is a non-dimensional aileron effectiveness. The response to the aileron deflection ξ is

$$L_\xi \xi = \frac{\rho A U^2 b}{2 I_x} C_{l_\xi} \xi, \tag{10.62}$$

from which we find the relationship between C_{l_ξ} and L_ξ,

$$C_{l_\xi} = \frac{2 I_x}{\rho A U^2 b} L_\xi. \tag{10.63}$$

The damping in roll can be written as

$$
\begin{aligned}
L_p &= \frac{1}{I_x} \left(\frac{\partial \mathcal{L}}{\partial p} \right) = \frac{1}{I_x} \frac{\partial}{\partial p} \left(\frac{1}{2} \rho A b U^2 C_l \right) \\
&= \frac{\rho A U^2 b}{2 I_x} \left(\frac{\partial C_l}{\partial p} \right) = \frac{\rho A U b^2}{4 I_x} \frac{\partial C_l}{\partial (pb/2U)}.
\end{aligned}
\tag{10.64}
$$

The damping moment is associated with the damping-in-roll coefficient

$$C_{l_p} = \frac{\partial C_l}{\partial (pb/2U)} = \frac{\partial C_l}{\partial \alpha} = \frac{2U}{b} \left(\frac{\partial C_l}{\partial p} \right). \tag{10.65}$$

The values of C_{l_p} are dependent on the speed regime (subsonic or supersonic) and on the wing geometry (taper ratio, aspect-ratio, wing sweep), therefore a generalization cannot be made. Roll damping is also affected by aeroelasticity. The order of magnitude for the damping coefficient is $C_{l_p} = -0.5$ to -0.1. However, this value generally decreases when approaching the speed of sound, and increases again at low supersonic speeds. From Eq. 10.65 and Eq. 10.64 we find

$$C_{l_p} = \frac{4 I_x}{\rho A U b^2} L_p. \tag{10.66}$$

With substitution of Eq. 10.63 and Eq. 10.65 in the roll equation, we have

$$\dot{p} = \frac{\rho U^2 A b}{2 I_x} C_{l_\xi} \xi + \frac{\rho U A b^2}{4 I_x} C_{l_p} p, \tag{10.67}$$

or

$$\dot{p} = \frac{\rho A U b^2}{2 I_x} \left(C_{l_\xi} \frac{U}{b} \xi + \frac{1}{2} C_{l_p} p \right). \tag{10.68}$$

Equation 10.68 is a differential equation that must be solved with the initial condition

$$p(t = 0) = \dot{\phi}(t = 0) = 0. \tag{10.69}$$

Note that the factor

$$\tau = \frac{2I_x}{\rho A U b^2},$$ (10.70)

has the dimension of a time; it is a time constant characteristic of the aircraft. The solution of Eq. 10.68 with the boundary conditions Eq. 10.69 is

$$p = \dot{\phi} = \frac{2U}{b} \frac{C_{l_\xi}}{C_{l_p}} \xi (1 - e^{-t/\tau}).$$ (10.71)

The asymptotic value of the roll rate is

$$p_{max} = \frac{2U}{b} \frac{C_{l_\xi}}{C_{l_p}} \xi.$$ (10.72)

These expressions hide the complexity of the problem. First, the moment response due to aileron deflection is not always linear; second, the damping and aileron derivatives depend on a fairly large number of parameters, including the geometry of the aircraft (wing, tail, aileron) and operational conditions (altitude, Mach number and aileron deflection). The essential relationships are

$$U = a(h)M, \quad C_{l_\xi} = f(M, h, \text{geometry}), \quad C_{l_p} = f(M, h, \text{geometry}).$$ (10.73)

These laws have to be studied separately for each aircraft. The ESDU report some useful charts for roll analysis[293,294]; in particular, they have relevant charts to interpolate the values of L_p and L_ξ for a given wing geometry, aileron deflection and Mach number. Some experimental data are available in Sandhal[295] (delta wings of rocket-propelled vehicles at supersonic speeds up to $M = 2.0$), Myers and Kuhn[296,297] (swept-back wings at Mach numbers below 0.8), Anderson et al.[298] (straight wings at transonic speeds). The effect of Mach number is marginal for most wings up to the point of transonic drag rise, when there is a drop in the aileron effectiveness.

Figure 10.18 shows the roll coefficients for a swept-back wing from wind tunnel testing, Myers and Kuhn[297]. There is a marked increase in magnitude in the roll damping C_{l_p} with the Mach number, and a more modest drop in aileron effectiveness.

A numerical solution of the aileron roll is shown in Figure 10.19. The parameters used are: mass $m = 9,300$ kg, at a flight altitude $h = 9,000$ m (29,528 ft), and Mach number $M = 0.8$. The roll rate reaches a maximum of 235 degrees/s after about 0.2 seconds. A full turn is completed in about 1.6 seconds; the time to turn 90 degrees is about 0.45 seconds. This is done without overshooting the steady state value, and the aileron is said to *command* the roll rate. However, the roll rate is maintained for quite some time, and unless additional damping is applied, the aircraft will make several rounds, with possible loss of control.

A study on the roll performance is the weight effect on the roll rate – all other parameters being the same. The results are shown in Figure 10.20. Increasing the aircraft mass from 9,300 kg to 15,300 kg has the effect of almost doubling the response time. The maximum roll rate is not affected by the weight. From the results obtained with the present theory, we conclude that our reference aircraft is capable of performing a roll in steady flight with a rate of the order of 2/3 rounds per second at $M = 0.8$.

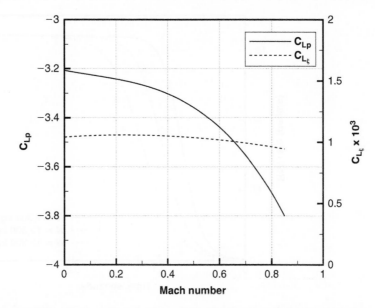

Figure 10.18 *Roll coefficients C_{l_p} and C_{l_ξ} as a function of the Mach number; wing of aspect-ratio 4, airfoil section NACA 65A006, quarter-chord sweep of 46.7 degrees, angle of attack $\alpha = 3.45$ degrees. Elaborated from Myers and Kuhn[297].*

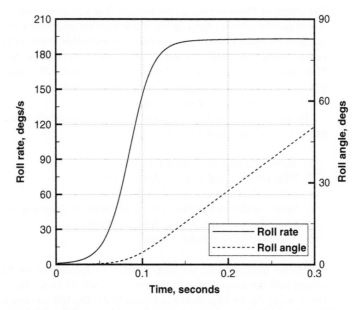

Figure 10.19 *Solution of the roll equation, with $m = 9,300$ kg, $h = 9,000$ m (29,528 ft), $M = 0.8$.*

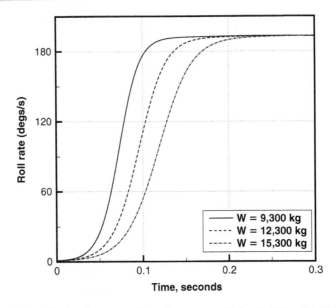

Figure 10.20 *Weight effects on roll performance, with h = 9,000 m (29,527 ft), M = 0.8, weights as indicated.*

10.10.1 Effects of Mach number

The effects of Mach number on the roll rate depend on the behavior of the damping-in-roll and the aileron effectiveness. These have specific values for each aircraft, but can also be extrapolated from existing general tables, such as ESDU[293,294]. Roll performance, with particular reference to damping characteristics as a function of Mach number has been investigated by Stone[299], Sanders[300] and others. Theoretical derivations of the damping-in-roll coefficients have been done by Malvestuto *et al.*[301] and Jones and Alskne[302] at supersonic speeds.

The extrapolation of data was done for our reference fighter aircraft (Tables A.10 and A.11), with further assumptions, such as:

1. Aileron chord at semi-span, 0.58 m;
2. Wing chord at mid-span of aileron, 3.85 m;
3. Section thickness at aileron's mid-span, 0.1541;
4. Spanwise distance at inboard limit of aileron at hinge line, 0.20;
5. Sweep-back of aileron hinge line, 19.23 degrees;
6. Aileron deflection angle, 5 degrees.

With the aircraft's data, we have calculated a function of the Mach number describing both the damping-in-roll and the aileron effectiveness. The result of the calculation, using Eq. 10.72 is shown in Figure 10.21. The roll rate reaches a maximum at about $M = 0.82$, then it suffers a sharp drop, due to various transonic effects on the wing (lift stall, loss of aileron effectiveness, drag rise). Flight data on this aircraft are not available.

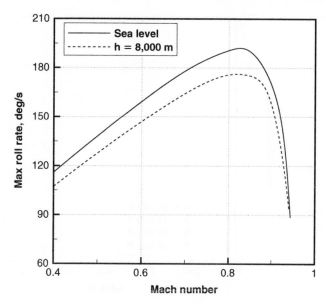

Figure 10.21   *Mach number dependence of the maximum roll rate for the reference aircraft.*

Figure 10.22 shows roll rates from flight testing of the F-16XL (data from Stachowiak and Bosworth[303]). This aircraft is a research derivative of the US Air Force. It has a wing radically different from the F-16. This is, in fact, a cranked-arrow double delta wing, with an aspect-ratio $\mathcal{AR} = 1.7$. The ESDU data do not cover the parametric space of this wing, therefore a simulation cannot be done. However, some flight data are available, as shown in Fig. 10.22. The maximum roll rates are almost independent of flight altitude; they reach a maximum of 180 degrees/s at a speed of about 400 kt (700 km/h), then decrease to 80 degrees/s at high subsonic speeds.

Some roll data for early fighter airplanes were collected by Toll[304]. These data proved that the roll rate reaches a maximum value before decreasing more or less sharply – depending on the airplane. There are exceptions, such as the North American XP-51 Mustang (1940), for which the roll rate increases in the whole speed range. Some of these data are shown in Figure 10.23, and are relative to standard weights, altitude 10,000 ft (about 3,000 m), and a 50 pound (22.7 kg) stick force.

The best roll rates up to the end of World War II were the Spitfire with clipped wings*(150 degrees/s) and the Focke Wulf Fw-190 (160 degrees/s). A Spitfire with clipped wings had a lower radius of gyration, hence a lower roll moment of inertia, hence a higher roll rate. It is reported (see, for example, Skow[305]) that the original version of the Spitfire-I was unable to compete with the agility of the Focke Fw-190, because this had a higher roll rate. As a consequence, the Spitfire pilots were often unable to assume an advantage to establish a gun-tracking position.

The highest roll rates of acrobatic aircraft are in the range of 300 degrees per second. The McDonnell–Douglas A-4 Skyhawk (1954), a delta wing with $\Lambda = 40$ degrees,

---

* The Spitfire with clipped wings had a wing span of 9.93 m; the normal wing span was 11.23 m.

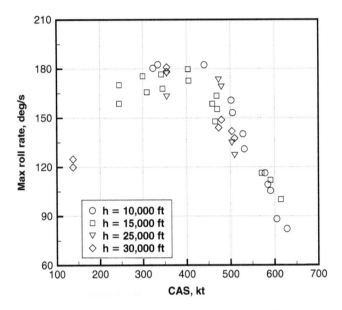

Figure 10.22 *Maximum roll rates versus calibrated air speed (in knots) for experimental aircraft F-16XL, from Stachowiak and Bosworth[303]. Flight altitudes as indicated. Data from flight testing. Inertias and weight not known.*

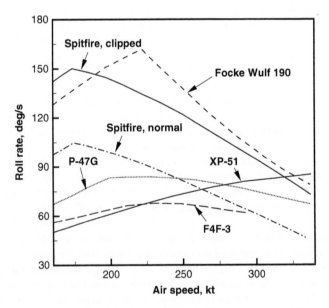

Figure 10.23 *Roll rates of WWII airplanes, adapted from Toll[304].*

claimed an outstanding roll performance of two rounds per second. So, how fast should a high-performance aircraft roll? This is a question that Azbug and Larrabee[306] have tried to answer.

By comparison, the roll rate of the barn swallow (*Hirundo rustica*) is estimated at 5,000 degrees/s. The swallow is able to make sharp turns while flying at high speed, enabling it to catch flies and other insects for feeding. The all-movable wings of the swallow (as well as many other birds) are a way to control wing stall.

10.10.2 Dihedral Effect

An explanation of why a yaw/side-slip generates a rolling moment is given below. Consider a wing with dihedral φ operating with a small amount of side-slip β.

The wing will have a higher angle of attack on the forward side than the rearward side. If the local chord at spanwise position is c, the fluid particles will travel inward or outward by an amount $c\varphi$. These fluid particles will travel toward a lower point on the forward side, and toward a higher point on the rearward side. The amount by which these particles are displaced depends on the geometry of the wing, as shown in the sketch of Figure 10.24. If the trailing edge is a straight line, then the spanwise offset does not depend on the side of the airplane, and is equal to $c\varphi\beta$. In any other case, there is a difference in spanwise offset. Since the fluid particles have fluctuated by an amount $c\varphi\beta$ over the chord c, the change in inflow conditions will be

$$\Delta\alpha = \pm\frac{c\varphi\beta}{c} = \pm\varphi\beta. \tag{10.74}$$

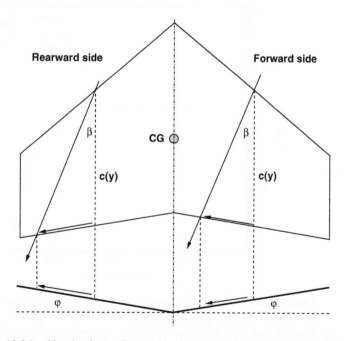

Figure 10.24 *Sketch of wing flow with side-slip (angles are exaggerated).*

Clearly, this condition tends to generate different values of the lift on the forward and rearward half-wing, and hence a rolling moment.

The Aileron may be necessary at high speeds for roll control. In fact as the speed increases, there is another unwanted response: an increase in effective dihedral angle (*wing dropping*, or uncommanded roll). The rolling moment starts appearing at transonic speeds in steady level flight. This originates from an original asymmetry in the airplane or in the transonic flow. A right side-slip (in the flight direction) causes a left roll-off, and vice versa. In general, wing drop refers to abrupt and irregular lateral motion of the aircraft, due ultimately to an asymmetric wing stall. If uncontrolled, it can result in a large unstable roll. This is shown for a case of a straight wing in Figure 10.25, elaborated from Anderson *et al.*[298] C_{l_β} is the rolling moment derivative with respect to the side-slip angle β,

$$C_{l_\beta} = \frac{\partial C_l}{\partial \beta}.$$
(10.75)

This problem appeared in the first generation of transonic aircraft in the 1950s, but it continues to be poorly understood. Wing drop has affected most high-performance military aircraft to this day, including the McDonnell–Douglas F-18 E/F, the F-111 and the British Sea Harrier. See, for example, Rathert *et al.*[307] for research relative to the North American F-86 Sabre, Chambers and Hall[308] for an updated historical review of aircraft affected, and Hall *et al.*[309] and Owens *et al.*[310] for advanced research programs.

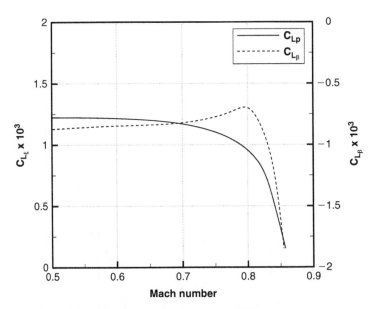

Figure 10.25 *Aileron effectiveness and rolling moment derivative with respect to side-slip at transonic speeds for airplane with a straight wing. Airfoil section: Republic R-4, 10% thick, $\alpha = 0$, twist $= -2$ degrees. Elaborated from Anderson et al. (NACA RM-A51B28).*

10.11 AIRCRAFT CONTROL UNDER THRUST ASYMMETRY

Assume a jet aircraft that, due to an engine failure, has a thrust asymmetry. For reference, consider a twin-engine aircraft, such as model A in Appendix A. The control of the aircraft is achieved by a combined rudder and aileron deflection, Figure 10.26.

In the first analysis, we assume that there is no thrust vectoring, and no external forces, in order to simplify the equations. Also, assume that the response of the aircraft is linear, e.g. it can be calculated as a linear combination of the rudder and aileron deflections, for any reasonable value of the side-slip and bank angle. With these assumptions, the static side force coefficient is

$$C_Y = C_{Y_\beta}\beta + C_{Y_\xi}\xi + C_{Y_\theta}\theta + \frac{W\sin\phi}{\rho A U^2/2} = 0, \tag{10.76}$$

where

$$C_Y = \frac{Y}{\rho A U^2/2}, \tag{10.77}$$

is the side force coefficient, β is the side-slip, ϕ is the roll angle, ξ is the aileron deflection (assumed symmetrical), θ is the rudder deflection; C_{Y_β}, C_{Y_ξ} and C_{Y_θ} are the derivatives of the side-force coefficient with respect to the yaw angle β, aileron angle ξ and rudder angle θ, respectively. The last term in Eq. 10.76 is the side-force coefficient created by the weight component along the spanwise axis. It can also be written as $C_L\sin\phi$. The static pitching moment coefficient is

$$C_l = C_{l_\beta}\beta + C_{l_\xi}\xi + C_{l_\theta}\theta = 0. \tag{10.78}$$

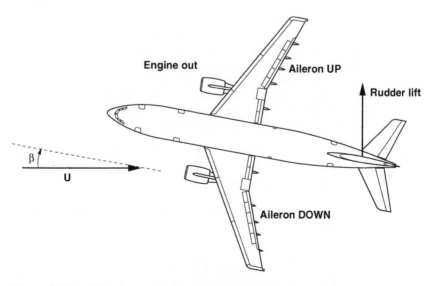

Figure 10.26 *Side-slip nomenclature, view from above.*

The static yawing moment coefficient is

$$C_N = C_{N_\beta}\beta + C_{N_\xi}\xi + C_{N_\theta}\theta + \frac{\Theta T b_t}{\rho A U^2 b/2} + C_{D_e}\frac{b_t}{b} = 0. \tag{10.79}$$

where ΔT is the thrust asymmetry, b_t is the moment arm of the thrust asymmetry, and C_{D_e} is the idle engine's drag coefficient.

Since Eqs 10.76–10.79 reflect static conditions for the aircraft, they represent a limit case. From Eq. 10.76, the side-slip angle in terms of the remaining parameters is

$$\beta = -\frac{1}{C_{Y_\beta}}\left[C_{Y_\xi}\xi + C_{Y_\theta}\theta + \frac{2W\sin\phi}{\rho A U^2}\right]. \tag{10.80}$$

The side-slip angle increases with the weight and decreases with the increasing speed. However, the value of the aerodynamic coefficients is important in limiting the yaw effects. If the engine out is at the starboard side, the aircraft will veer to the starboard. The rudder will have to deflect in the opposite direction to create a restoring moment. The convention is that $\beta < 0$ and $\theta > 0$ (the opposite happens for a port-side engine failure). The bank angle ϕ depends on the position of the thrust with respect to the center of gravity of the aircraft.

Now the problem is this: given the aircraft data (weight, wing area, flight altitude), and the stability coefficients (calculated by other means), we need to calculate the minimum speed that guarantees full control (longitudinal, lateral, and directional). This is the so-called *minimum control air speed*, VMC.

Calculation of the control derivatives is quite elaborate, but there are computer programs designed for this purpose, for example the US Air Force Stability and Control Digital Datcom[311], and a number of more recent vortex lattice aerodynamic programs. These coefficients are sensitive to the aircraft configuration and may change considerably from one aircraft to another. Some stability derivatives for the Boeing B-747-100, Lockheed C-5A, Grumman F-104A and other representative aircraft are available in Heffley and Jewell[154].

For a commercial jet liner, FAR regulations 23.149 require that control is assured in the worst possible scenario, e.g. with one engine at the take-off thrust, the aircraft at MTOW, the CG in the aft position, the flaps at take-off position, and the landing gear retracted. A maximum bank angle of 5 degrees is allowed. The unknown parameters in Eqs 10.76–10.79 are: the minimum control speed $U = VMC$, the side-slip angle β, the aileron deflection ξ, and the rudder deflection θ. For a fixed rudder deflection $\theta = \theta_{max}$, the solution system is written as

$$\begin{bmatrix} C_{l_\beta} & C_{l_\xi} & 0 \\ C_{N_\beta} & C_{N_\xi} & 2\Delta T b_t/\rho A b \\ C_{Y_\beta} & C_{Y_\xi} & 2W\sin\phi/\rho A \end{bmatrix} \begin{bmatrix} \beta \\ \xi \\ 1/U^2 \end{bmatrix} = -\begin{bmatrix} C_{l_\theta}\theta \\ C_{N_\theta}\theta + C_{D_e}b_t/b \\ C_{Y_\theta}\theta \end{bmatrix}, \tag{10.81}$$

where the unknowns are β, ξ, $1/U^2$. For a fixed aileron deflection $\xi = \xi_{max}$, the solution system is

$$\begin{bmatrix} C_{l_\beta} & C_{l_\theta} & 0 \\ C_{N_\beta} & C_{N_\theta} & 2\Delta T b_t/\rho A b \\ C_{Y_\beta} & C_{Y_\theta} & 2W\sin\phi/\rho A \end{bmatrix} \begin{bmatrix} \beta \\ \theta \\ 1/U^2 \end{bmatrix} = -\begin{bmatrix} C_{l_\xi}\xi \\ C_{N_\xi}\xi + C_{D_e}b_t/b \\ C_{Y_\xi}\xi \end{bmatrix}. \tag{10.82}$$

Both systems assume that the drag coefficient of the idle engine is not dependent on the speed. The proper value of C_{D_e} can be found with the method described by Torenbeek[312],

$$C_{D_e} = 0.0785d_i^2 + \frac{2}{1 + 0.16M^2}A_n\frac{U}{U_j}\left(1 - \frac{U}{U_j}\right), \tag{10.83}$$

where U_j is the velocity of the jet at the nozzle, d_i is the fan diameter, and A_j is the area of the nozzle. Typical values of U/U_j are 0.92 (high-by-pass turbofan), 0.42 (low-by-pass jet engine), 0.25 (straight turbojet and turboprop). Equation 10.83 shows that the C_{D_e} depends on the Mach number, and hence on the aircraft's speed. A suitable computational model is the following.

Computational Procedure

1. Read the aircraft data $(W, A, b, b_t, \Delta T, \ldots)$.
2. Read the aerodynamic coefficients $(C_{l_\xi}, C_{l_\theta}, \ldots)$.
3. Set the flight altitude.
4. Solve the problem defined by the system Eq. 10.81, using as a parameter the weight.
5. Solve the problem defined by the system Eq. 10.82, using as a parameter the weight.

To maintain the linearity of the system, we can calculate the C_{D_e} a posteriori, or iteratively. For example, guess the VMC, calculate the Mach number, calculate the C_{D_e} from Eq. 10.83 and solve the system again.

The solution of the system requires the inversion of the matrix on the left-hand side. This is done by one of the standard methods available in most software platforms. A calculation was done for the Boeing B-747-100, whose aerodynamic coefficients are known from Nelson[223] and summarized in Table 10.1.

Figure 10.27 shows the side-slip angle of the aircraft for varying aileron deflection at a fixed rudder deflection, at a fixed weight and air speed, as indicated in the graph. The side-slip is linear and weakly affected by the rudder deflection.

Figure 10.28 shows the calculated VMC as a function of the aircraft's GTOW. It shows the VMC limited by the maximum rudder and aileron deflection. Both speeds are above the stall speed, calculated with a $C_L = 1.8$. With a GTOW of 180 tons or less, it is not possible to ensure a full control of the aircraft. In some cases the rudder deflection has an opposite trend to the one shown, and a branched solution can be found.

Table 10.1 *Stability derivatives for calculation of aircraft response to asymmetric thrust, model Boeing B-747-100.*

C_{l_β}	−0.2210	C_{N_β}	0.1500	C_{Y_β}	−0.9600
C_{l_ξ}	0.0460	C_{N_ξ}	0.0064	C_{Y_ξ}	0.0000
C_{l_θ}	−0.0070	C_{N_θ}	0.1090	C_{Y_θ}	−0.1750

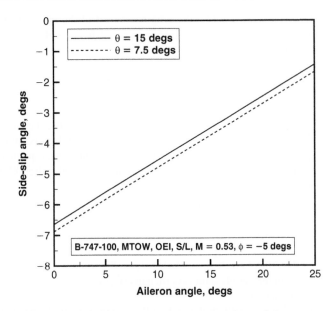

Figure 10.27 *Calculated side-slip angle versus aileron deflection at max rudder for the Boeing B-747-100, weight and speed as indicated in the graph.*

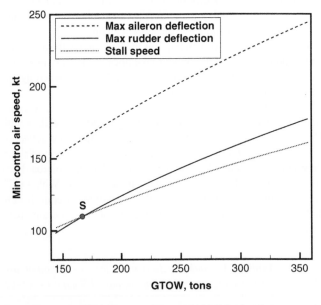

Figure 10.28 *Calculated VMC versus gross weight for the Boeing B-747-100. Bank angle $\phi = -5$ degrees.*

10.12 PULL-UP MANEUVER AND THE LOOP

The pull-up maneuver is a turning flight in the vertical plane, with a zero bank angle, and flight path inclined by a variable angle γ. In the complete loop the flight path is vertical at one point, horizontal, vertical and finally horizontal. The dynamic equations of the aircraft are the same as in the case of climb flight (Chapter 8), which we rewrite here for convenience:

$$\frac{\partial h}{\partial h} = U \sin \gamma, \tag{10.84}$$

$$m\frac{\partial U}{\partial t} = T - D - W \sin \gamma, \tag{10.85}$$

$$mU\frac{\partial \gamma}{\partial t} = L - W \cos \gamma. \tag{10.86}$$

We introduce the normal load factor $n = L/W$, so that Eq. 10.86 becomes

$$\frac{\partial \gamma}{\partial t} = (n - \cos \gamma)\frac{g}{U}. \tag{10.87}$$

The rate of change of γ with altitude is introduced in Eq. 10.87 by recalling that

$$\frac{\partial \gamma}{\partial t} = \frac{\partial \gamma}{\partial h}\frac{\partial h}{\partial t} = \frac{\partial \gamma}{\partial h}v_c = \frac{\partial \gamma}{\partial h}U \sin \gamma. \tag{10.88}$$

Therefore,

$$\frac{\partial \gamma}{\partial t} = \frac{\partial \gamma}{\partial h}U \sin \gamma = (n - \cos \gamma)\frac{g}{U}. \tag{10.89}$$

By eliminating the speed U from Eq. 10.89 with the definition of C_L,

$$U^2 = \frac{2nW}{\rho A C_L}, \tag{10.90}$$

we find

$$\frac{\partial \gamma}{\partial h} = \frac{g\rho A C_L}{2nW}\left(\frac{n}{\sin \gamma} - \frac{1}{\tan \gamma}\right). \tag{10.91}$$

Equation 10.91 expresses the change of heading with respect to time. This rate of change is dependent on the wing loading, on the flight altitude, and on the lift coefficient. The aircraft speed is implicit. A maximum of the turn rate is found by deriving Eq. 10.91 with respect to n and C_L. The turn rate is variable from 1 to a maximum within the loop, and cannot be calculated directly from the discussion above. Additional conditions are required.

A relatively simple solution can be found if the flight path is assigned, and if the total energy of the aircraft remains constant. A suitable flight path is found from

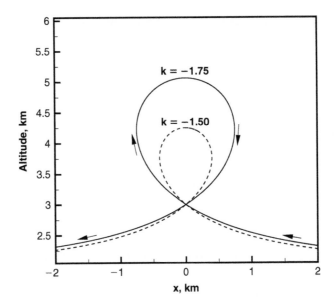

Figure 10.29 *Conchoid of de Sluze, with a fixed value a = 1, and k = −1.5 and k = −1.75.*

the *conchoid of de Sluze**, which is a curve defined in polar coordinates by the equation

$$a(r \cos \vartheta - a) = k^2 \cos^2 \vartheta. \tag{10.92}$$

In this equation a and k are two constant parameters, and r and ϑ are the polar coordinates. With appropriate values of the parameters, the conchoid makes a loop defined by the coordinates

$$h = r \cos \vartheta + h_o, \qquad x = r \sin \vartheta. \tag{10.93}$$

In Eq. 10.93 h_o is an arbitrary starting altitude (or the horizontal asymptote of the conchoid). This curve is shown in Figure 10.29.

If we use the energy methods, and assign the speed and the altitude at the start of the pull-up, Eq. 10.92 and Eq. 8.71 can be used to solve for the acceleration along the loop. The result is shown in Figure 10.30. The maximum acceleration of the tightest loop is about 2.5g. In both cases the Mach number is practically constant.

Minimum-time loop maneuvers have been solved by Uehara *et al.*[313] by using the calculus of variations. The control variables are the thrust and the lift coefficient. These authors considered the effects of density altitude, the effects of Mach number on the engine thrust and constraints on the normal accelerations, due to pilot's physiological limits. Other simulation models for turning in the vertical plane are discussed by Shinar *et al.*[314]

---

* From the Belgian mathematician René F.W. de Sluze (1622–1685).

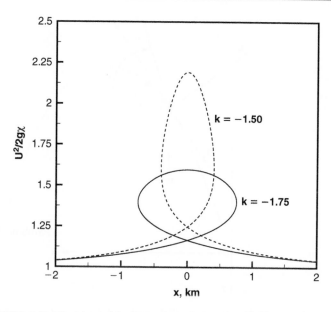

Figure 10.30 *Acceleration of the aircraft performing a loop along a prescribed conchoid (Figure 10.29).*

10.13 ZERO-GRAVITY ATMOSPHERIC FLIGHT

A well-known method to create a zero-gravity environment in the Earth's atmosphere is a flight maneuver known as *parabolic flight*. More specifically, this term refers to a wing-level maneuver in the vertical plane with pull-ups and pull-outs. Aircraft used for zero-gravity research include the Boeing B-727-200, the Douglas KC-135 (a derivative of the DC-10), the Airbus A-300, the Ilyushin Il-76, and the Cessna Citation II. Generations of NASA astronauts have been trained with parabolic flights. There are now perspectives into scientific research under zero- and micro-gravity conditions, also using unmanned aircraft.

A full cycle consists of three phases. First, the aircraft is at a minimum altitude in horizontal flight. Next, the aircraft is pulled up to a steep climb angle of 45 to 50 degrees with a 1.8g normal acceleration. At a fixed altitude, the parabolic climb starts. Past the apex, the aircraft accelerates downward with the acceleration of gravity, and is recovered in descent with a 1.8g pull-out maneuver. The micro-gravity effect lasts between 20 and 25 seconds. The cycle can be repeated a number of times. The use of sounding rockets, capable of reaching altitudes of 120 km, is an interesting alternative for scientific research, because the zero-gravity effect can last several minutes.

The following considerations apply to the parabolic flight, and are aimed at showing the principles of generating micro-gravity effects. Consider the Earth reference system, which has the vertical axis z pointing downward, see Figure 10.31. We consider a flat, non-rotating Earth. This is a good approximation for the brevity of the parabolic flight.

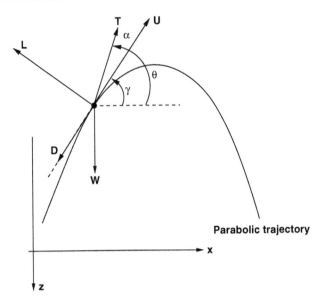

Figure 10.31 *Parabolic flight path with forces and nomenclature.*

The properties of the parabola have been known for centuries, and have been applied to a variety of ballistic problems. A point mass moving along a parabola has a constant acceleration along its axis, and zero acceleration (constant speed) along the normal axis. The acceleration changes sign when the point travels through the apex. Likewise, a zero-g parabolic flight has a constant speed in the horizontal direction and a $+g$ acceleration in the vertical direction, when the aircraft travels *downwards*.

We assume that a jet aircraft climbs along a parabola from an initial condition x_o, h_o, U_o, γ_o to a maximum altitude h_a at the apex of the parabola. A zero-gravity flight requires that the acceleration along z be equal to $+g$ in the descent phase. If the aircraft climbs along a parabola with acceleration $-g$, then its equation along z is

$$m\ddot{h} = -mg. \tag{10.94}$$

The acceleration in the horizontal direction is zero,

$$m\ddot{x} = 0. \tag{10.95}$$

The velocity components of the center of gravity of the aircraft along the trajectory are found from integration of Eq. 10.94 and Eq. 10.95,

$$w(t) = \dot{h} = -gt + U_o \sin \gamma_o, \tag{10.96}$$

$$u(t) = U_o \cos \gamma_o. \tag{10.97}$$

The total velocity U is

$$U(t) = (u^2 + w^2)^{1/2} = [(gt)^2 + U_o^2 - 2U_o g \sin^2 \gamma_o]^{1/2}. \tag{10.98}$$

The acceleration along the trajectory is

$$\dot{U} = \frac{d}{dt}(u^2 + w^2)^{1/2} = \frac{u\dot{u} + w\dot{w}}{(u^2 + w^2)^{1/2}} = \frac{u\dot{u} + w\dot{w}}{U} = \frac{w\dot{w}}{U} = -\frac{gw}{U}. \tag{10.99}$$

The position of the center of gravity is found from integration of Eq. 10.96 and Eq. 10.97,

$$h(t) = -\frac{1}{2}gt^2 + U_o \sin \gamma_o t + h_o, \tag{10.100}$$

$$x(t) = U_o \cos \gamma_o t + x_o. \tag{10.101}$$

The time to the apogee, t_a, is calculated analytically from Eq. 10.96, assuming $w(t) = 0$,

$$t_a = \frac{U_o \sin \gamma_o}{g}. \tag{10.102}$$

The apogee is

$$h_a = h_o + \frac{1}{2g}\left(\frac{U_o \sin \gamma_o}{g}\right)^2. \tag{10.103}$$

The flight path angle γ is calculated from

$$\tan \gamma = \left(\frac{\dot{h}}{\dot{x}}\right) = \tan\left(\frac{w}{u}\right), \tag{10.104}$$

or

$$\gamma(t) = \tan^{-1}\left(\frac{-gt + U_o \sin \gamma_o}{U_o \cos \gamma_o}\right). \tag{10.105}$$

The rate of change of the climb angle is

$$\dot{\gamma}(t) = -\frac{gU_o \cos \gamma_o}{(gt)^2 + U_o^2 - 2gU_o \sin \gamma_o t}. \tag{10.106}$$

Finally, the radius of curvature of the parabola can be found from the equation

$$\chi = \frac{(\dot{x}^2 + \dot{h}^2)^{3/2}}{\dot{x}\ddot{h} - \ddot{x}\dot{h}} = \frac{1}{g}\frac{(u^2 + w^2)^{3/2}}{u}. \tag{10.107}$$

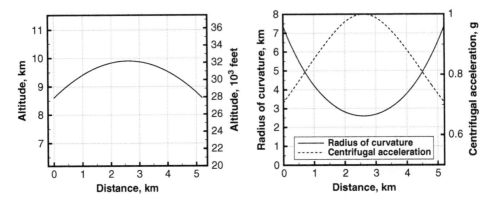

Figure 10.32 *Theoretical parabolic trajectory from starting altitude 8,600 m (27,922 feet), and corresponding radius of curvature and centrifugal acceleration (right).*

All the quantities in Eq. 10.107 have already been calculated.

Alternatively, we can set four boundary conditions for the trajectory: speed and angle of climb at the start, and speed and altitude at the apex. This requires a flight trajectory described by a polynomial of order 3.

The centrifugal acceleration associated to a point mass traveling along this trajectory is U^2/g. At the apex it is at a minimum, exactly $1g$. The trajectory is shown on the left of Figure 10.32. A number of conclusions are in place:

- When the aircraft climbs, its vertical acceleration is about $-1g$ (opposite to the acceleration of gravity). This corresponds to an effective $2g$ flight upward (a steady level flight is $1g$, from Eq. 10.8). The passengers feel heavier.
- When the aircraft is at the apex, the centrifugal acceleration of the CG is $-1g$ (upward). Since the aircraft is climbing at about $-1g$, the total feeling is $-1g$, to compensate for the acceleration of gravity. In this case most people feel upside-down.
- When the aircraft starts its zoom-dive, the vertical acceleration is $+1g$. In this phase the passengers feel weightless.
- There is no zero-gravity effect during the climb section of the parabola. This point is often misunderstood. The idea that there is a zero-gravity feeling during the climb contravenes the laws of physics.

For the zero-gravity feeling to last a fixed amount of time Δt, the altitude loss must be

$$\Delta h = \frac{1}{2}g(\Delta t)^2. \tag{10.108}$$

For example, if the zero-gravity effect is to last 25 seconds, the loss of altitude must be about 3,000 m (10,000 ft). Therefore, the aircraft must zoom along the parabola, past its starting altitude, down to about 6,700 m (21,981 ft).

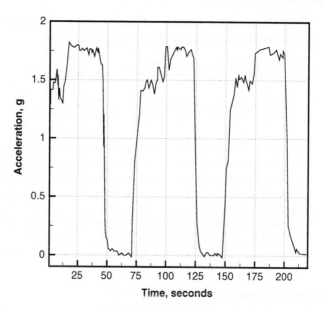

Figure 10.33 *Average accelerometer response for the KC-135 in micro-gravity parabolic flight, from Smith and Workman*[315].

The parabolic flight trajectory requires that the effective acceleration of the CG of the aircraft varies sharply between 2g (zoom-climb) and 0 (zoom-dive). This is the case for the aircraft KC-135, as shown from accelerometer measurements in Figure 10.33. The large-frequency oscillations are due to external factors, such as atmospheric turbulence and vibrations.

The next problem is to find the angle of attack and thrust programs that are compatible with the trajectory defined by Eq. 10.100 and Eq. 10.101. The acceleration of the CG of an aircraft of mass m is

$$a = \frac{T+A}{m} + g. \tag{10.109}$$

The acceleration felt by an accelerometer in the aircraft will be

$$a^* = a - g = \frac{T+A}{m}. \tag{10.110}$$

If we want to maintain this acceleration constant and equal to a fraction ξ of the acceleration of gravity, then $a^* = \xi g$. If $\xi = 0$, then we have a zero-g flight.

In order to calculate the net effects on the passenger flying on such a trajectory, we must consider some apparent forces. One is the centrifugal force, due to the prescribed trajectory. Then there is the acceleration along the trajectory. An additional acceleration occurs when the passengers are off the center of gravity of the aircraft, due to the pitch rate (nose up and nose down). This rotational acceleration is of the order of 0.03g for a point 10 m (30 feet) off the CG, as discussed by Bahr and Schulz[316]. We neglect this effect.

The resulting accelerations felt by a passenger on the aircraft along the tangential and normal direction to the parabola are, respectively,

$$\frac{f_t(t)}{m} = \frac{T \cos \alpha - D - W \sin \gamma}{m} = \boldsymbol{a} \cdot \boldsymbol{t} = a_t = -\dot{U}, \qquad (10.111)$$

$$\frac{f_n(t)}{m} = \frac{T \sin \alpha + L - W \cos \gamma}{m} = \boldsymbol{a} \cdot \boldsymbol{n} = a_n = -\frac{U^2}{\chi}. \qquad (10.112)$$

where $\boldsymbol{t}$ and $\boldsymbol{n}$ denote unit vectors parallel and normal to the trajectory. Note that there is no acceleration of the point mass m normal to the flight path. However, the passengers in the cabin will feel the effects of the centrifugal acceleration U^2/χ. Essentially, the acceleration components a_t and a_n are assigned by the constraint of parabolic flight. Equation 10.111 and Eq. 10.112 are a system of two equations in the unknowns α and T. Contrary to Menon et al.[317], there is no need to assume that the drag remains constant and equal to the level flight of the trimmed aircraft. The aircraft's drag is given by Eq. 4.12. The algorithm is described below.

Computational Procedure

1. Set the initial conditions at the start of the parabolic flight: U_o, h_o, x_o, γ_o.
2. Advance the time by Δt, so that $t = t + \Delta t$.
3. Calculate the new position vector $x(t)$, $h(t)$ from Eq. 10.101 and Eq. 10.100, respectively.
4. Calculate the air density ρ at the current altitude from the standard atmosphere.
5. Calculate the new velocity components $u(t)$, $w(t)$ and air speed $U(t)$ from Eq. 10.97, Eq. 10.96, and Eq. 10.98.
6. Calculate the acceleration along the parabola, $\dot{U}(t)$, from Eq. 10.99.
7. Calculate the radius of curvature, $\chi(t)$, from Eq. 10.107.
8. Calculate the angle of climb $\gamma(t)$ from Eq. 10.105.
9. Calculate the angle of attack $\alpha(t)$ from the system Eq. 10.111 and Eq. 10.112, by using a Newton-Raphson method from an initial guess α.
10. Calculate the engine thrust $T(t)$ from Eq. 10.111.
11. End loop.

Additional calculations include the glide ratio $L/D = C_L(t)/C_D(t)$. As an afterthought, we can calculate the pitch rate required, and the pitch acceleration. From the definition of pitch attitude,

$$\theta = \gamma + \alpha, \qquad (10.113)$$

we find

$$q = \dot{\theta} = \dot{\gamma} + \dot{\alpha}, \qquad \dot{q} = \ddot{\theta} = \ddot{\gamma} + \ddot{\alpha}. \qquad (10.114)$$

The analysis presented serves to prove that a parabolic flight provides a zero-g feeling, at least in principle. In practice, the analysis requires a consideration of the aircraft trim, and the flight controls to achieve the parabolic maneuver.

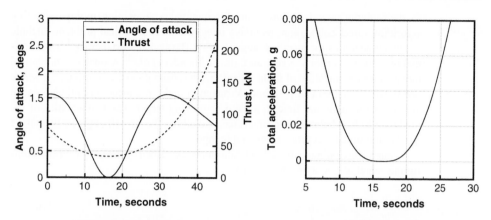

Figure 10.34 *Parabolic trajectory from starting altitude 8,600 m (27,922 feet): angle of attack and thrust programs (left), and corresponding acceleration (right).*

Lunar and Martian parabolas can be flown as well. This requires *stretching* the flight paths, in order to provide the gravity effects required (on the Moon $g = 1/6$ of Earth; on Mars $g = 3/8$ of Earth).

Example

Consider the aircraft Douglas KC-135, whose aerodynamic data are shown in Figure 4.9. The aircraft has a gross weight $W = 130,000$ kg. The aircraft has four engines, each delivering a maximum 98 kN (22,000 lb) at sea level. At the start of the parabolic flight it will be at 27,922 ft (8,600 m), with an $M = 0.74$, and a climb angle $\gamma_o = 45$ degrees.

The results of the present simulation are shown in Figure 10.34. The aircraft climbs to its apogee, where it has a minimum Mach number $M = 0.55$, a minimum engine thrust of about 40 kN, and nearly zero angle of attack. From this point it starts its zoom-dive along the parabola, and the passengers feel a near-zero gravity. Both the angle of attack and the thrust increase. The zero-gravity effect is equivalent to $0.05g$ or less for about 19 seconds. If we consider that an object positioned off the CG of the aircraft is subject to an increase of up to $0.03g$, because of pitch-down accelerations, then we can consider the $0.05g$ a practical limit. We will refer to such a flight as a zero-g, although this is not strictly correct.

10.14 FLIGHT PATH TO A MOVING TARGET

Flight paths to a moving target are of interest for interceptor aircraft and a number of weapon systems (air-to-air, surface-to-air). A similar class of problems include tracking, e.g. minimum deviation from a specified flight path. These problems cover three-dimensional flight, with a combination of climb and turn. The methods required for their solution include optimal control (sequential programming, multiple shooting algorithm, direct collocation, recursive integration, etc.). Typical problems include optimal trajectories for specified range (Lee[318], Barman and Ertzberger[319]) and optimal time-maneuver for V/STOL aircraft, Lichtsinder *et al.*[320] A survey of numerical methods for optimization of flight trajectories has been published by Betts[321], where additional references can be found. Virtanen *et al.*[322] produced an optimal control algorithm for minimum time to a fixed or moving target. The mathematics of the method is hidden behind a user-interface, and allows the aircraft engineer to analyze quickly multiple trajectories from simple inputs.

A general three-dimensional flight path can be described by the following equations for the center of gravity of the aircraft

$$\frac{\partial x}{\partial t} = U \cos\gamma \cos\psi, \tag{10.115}$$

$$\frac{\partial y}{\partial t} = U \cos\gamma \sin\psi, \tag{10.116}$$

$$\frac{\partial h}{\partial t} = U \sin\gamma, \tag{10.117}$$

$$\frac{\partial U}{\partial t} = \frac{1}{m}[T_{max}(h, M) - D] - g\sin\gamma, \tag{10.118}$$

$$\frac{\partial \gamma}{\partial t} = \frac{g}{U}(n\cos\phi - \cos\gamma), \tag{10.119}$$

$$\frac{\partial \psi}{\partial t} = \frac{g}{U}\frac{n\sin\phi}{\cos\gamma}, \tag{10.120}$$

$$\dot{m}_f = -f_j T_{max}(M), \tag{10.121}$$

where, in addition to the known symbols, we have ψ as the heading angle. The formulation above can be written in compact vector form as

$$x = f[x(t), u(t), t]. \tag{10.122}$$

In a three-dimensional flight, the minimum time path to a moving target is defined by the function

$$\begin{aligned}\Psi &= \Psi(t, x, y, h) = \Psi[x(t), y(t), h(t)] \\ &= \alpha_1(x - x^*)^2 + \alpha_2(y - y^*)^2 + \alpha_3(h - h^*)^2,\end{aligned} \tag{10.123}$$

where the asterisk denotes the position of the moving target, and the right-hand side of the equation is the square of the distance between the aircraft and the target at time t; α_i ($i = 1, 2, 3$) are three weight coefficients. Usually, there will be a constraint on minimum velocity, on maximum altitude, on maximum normal load factor, in addition to initial and final conditions. A moving target is defined by its position, velocity, flight path angle and heading.

PROBLEMS

1. An aircraft executes a correctly banked turn at an altitude where the relative density of the air is $\rho/\rho_o = 0.35$. The aircraft's wing loading is $W/A = 250\,\text{kg/m}^2$. The limit load factor is 2.5.

   (a) Prove that the banking angle ϕ is related to the radius of turn χ by the relationship

   $$\chi = \frac{2}{\rho}\frac{W}{A}\frac{1}{C_L}\frac{n}{\sqrt{n^2 - 1}}.$$

   (b) Calculate the minimum radius of turn in horizontal flight.

2. Consider the aircraft model C (Appendix A), flying at an altitude where $\sigma = \rho/\rho_o = 0.55$ with weight $W = 11{,}000\,\text{kg}$. For this aircraft find the minimum radius of turn.

3. Consider the subsonic transport aircraft model in Appendix A, and calculate the turn power as a function of the speed in a banked turn. Assume the following load factors: $n = 1.25, 1.50, 2, 2.5$. The aircraft weight is 80% of its MTOW, and the flight altitude is $h = 1{,}000\,\text{m}$. Assume that the engine power is not dependent on the speed.

4. For the propeller-driven aircraft model in Appendix A study the change in radius of turn, turn rate and power required for turn as a function of the flight speed, from sea level to absolute ceiling. Find the altitude at which the turn radius is minimum. The limiting load factor on the aircraft is 2.5.

5. For the same aircraft as in Problem 4 find an expression for the stall speed in terms of the normal load factor, and describe how the stall speed is affected by a normal load factor n. The weight is $50{,}000\,\text{kg}$.

6. Calculate the maneuver limit due to aerodynamics for the supersonic jet aircraft of reference (data in Appendix A), for the aircraft in combat configuration, $W = 12{,}000\,\text{kg}$, at altitudes $h = 3{,}000, 9{,}000$ and $15{,}000\,\text{m}$. Study the effect of aircraft weight on the maneuver curves, by considering the following weights: $W = 10{,}000\,\text{kg}$, $W = 14{,}000\,\text{kg}$.

7. A high-performance, high-speed aircraft with low aspect-ratio wings is to perform a rapid roll around its longitudinal axis. The aircraft is likely to experience a phenomenon called *inertial coupling*. Describe the problem of inertial coupling when it occurs, why it occurs, and how it can be prevented, by either design or operation. Make use of sketches, find or derive the relevant equations of motion. (*Problem-based learning: additional research is required.*)

8. Consider a high-performance aircraft, with low-aspect-ratio wings, and large leading-edge sweep angle. This aircraft is capable of maneuvering at high angles of attack. A form of lateral instability may occur, called *wing rock*. Investigate the nature of wing rock, under which circumstances it is likely to occur, and the flight mechanics problems involved, such as loss control. (*Problem-based learning: additional research is required.*)

9. In general, roll due to aileron deflection is done with an unequal deflection of the ailerons: the aileron going up is deflected by a larger angle than the aileron moving down, contrary to the hypothesis of §10.10. Investigate the reasons why

this is done and what the advantages are. (*Problem-based learning: additional research is required.*)

10. Consider an unpowered glider with the following characteristics: Wing span $b = 12.2$ m, wing area $A = 15$ m^2, empty weight $W = 180$ kg. Calculate the sinking speed as a function of the true air speed at sea level, for a straight flight, and for banked turns at 30 and 45 degrees. For all cases, calculate the glide ratio L/D, the radius of turn χ, the turn rate $\dot{\psi}$ and plot the quantities versus the air speed. Finally, write a program that calculates the trajectory of the glider for a given bank angle, for a generic starting point U, h.

11. Write a computer program to calculate the flight trajectory of an aircraft performing a loop in the vertical plane. Select the data of the reference supersonic jet in Appendix A, at a nominal weight $W = 10,000$ kg. The loop is to start from a suitable altitude, for example $h = 8,000$ m.

12. Assume a generic twin-engine jet aircraft, whose drag equation is parabolic, and whose thrust is approximately $T = T_o \sigma^{0.8}$. For this aircraft find the condition (angle of attack, speed and altitude) of minimum fuel to turn. Assume a balanced turn of $\Delta \psi = 180$ degrees at constant altitude.

13. Investigate the possible causes of declivity currents, why these are important for soaring flight, and how they can be used. Investigate how thermal shells are created. Discuss separately the case of static and dynamic soaring. (*Problem-based learning: additional research is required.*)

14. Consider a twin-engined aircraft, such as that shown in Fig. 10.26, with engines below the wings. The aircraft suffers the loss of port-side engine thrust while cruising at altitude. Produce a sketch to highlight the direction of the bank angle, and the movements required to the rudder and ailerons to re-establish full control of the aircraft. In particular, answer whether the aircraft banks to the port or starboard side.

15. Calculate the time to perform a loop along a prescribed conchoid (Eq. 10.92), assuming that the maneuver is done at constant total energy. For reference, the start and the end of the maneuver are defined by a normal load factor $n = 1.1$. Assume that the starting altitude is $h = 3,000$ m and the starting Mach number is $M = 0.70$. For the conchoid assume $a = 1$, $k = -2$. Also, calculate n as a function of time and the apogee of the trajectory. Verify whether the true air speed is maintained constant along the trajectory.

16. Is it possible to perform a zero-gravity atmospheric flight along the ascent path of a parabola? If a parabolic flight is assigned to a sounding rocket that reaches an apogee of 300 km from sea level, is it possible to perform a zero-gravity ascent? (Hint: you need to consider changes in gravitational acceleration with flight altitude.)

Part II
Rotary-Wing Aircraft Performance

Chapter 11

Rotorcraft Performance

Whereas the airplane was developed by young men not previously known as inventors ... the fortunes of the helicopter have been more in the hands of those with reputation previously acquired elsewhere.

E.P. Warner[323], 1922

This chapter introduces the direct-lift vehicles, e.g. vehicles that generate lift by the action of one or more rotors. This category includes helicopters, autogiros and convertiplanes. The chapters that follow are a survey of performance calculation methods that make use of simplified models (one-dimensional axial flow). However, most of the discussion is limited to the conventional helicopter.

The flight conditions of a conventional helicopter include hover, axial climb and descent, forward (inclined) climb and descent, level forward flight, sideways, backward and yawed flight, turning and maneuvering, autorotation, and flight conversion. The flight condition between hover and level flight at moderate speeds is called flight *transition*. Besides these conditions, we must consider off-design flight operations, such as the vortex ring state, autorotation and cornering flight. Therefore, the number of possible flight conditions of a rotary-wing aircraft exceeds that of a fixed-wing aircraft.

A full performance analysis, aimed at selecting or designing a helicopter, planning a mission, or upgrading a vehicle, must include at least range, endurance, hover in- and out-of-ground, climb rates at full power and OEI, fuel consumption, and performance in non-standard atmosphere (cold or hot day).

11.1 FUNDAMENTALS

It often happens that the rotor technology dominates any discussion on the helicopter. Indeed, some essential geometrical parameters, such as the airfoil section, the shape of the blade tips, the blade twist, and the articulation of the hub, are very important in terms of overall rotorcraft performance.

Rotary-wing vehicles are far more complex than fixed-wing aircraft, because of the presence of rotating parts, their aerodynamics and the interference with the airframe. There are additional stability, control, vibration and noise problems, and a variety of flight conditions peculiar to the rotary-wing aircraft. Furthermore, the blades are highly flexible, and are required to perform complicated movements as they travel around the hub (flapping, pitching/feathering, lagging).

The original idea of the helicopter pioneers was to develop a direct-lift aircraft that would be able to hover and to convert its flight from vertical to horizontal, and vice versa. This concept, simple as it looked, required over 30 years of experiments by engineers and scientists around the world. The history of the development of the practical helicopter is one of the most fascinating and controversial subjects in aviation. One of the contentious issues is whether stable hovering is enough to grant anyone the right to the invention. Individuals who have claimed to have invented the helicopter include *at least* Etienne Oehmichen, Louis Breguet, Corradino D'Ascanio, Nicolas Florine, Anton Flettner, Heinrich Focke and Igor Sikorsky. However, the helicopter was at a dead end until Juan de la Cierva invented the cyclic pitch control. His system provided the means for handling the structural loads, and was a major step forward in the development of both the helicopter and the autogiro. For a detailed history on this subject, we recommend reading the work of Leishman[324] (with extensive bibliography), then Pirie and Lambermont[325] (illustrated with amazing vintage photos), Liberatore[326] (with plenty of discussion of technical ideas), Everett-Heath[327] (about Soviet helicopters) and Boulet[328] (about recollections of the pioneers).

Maneuverability takes forms not seen in any other flight vehicle. The helicopter can turn around itself on the spot, pull up, do a split-S maneuver, and fly backward and sideways at great speeds. For example, the Kamov Ka-52 Alligator claims a maximum backward speed of 90 km/h, and the Boeing-Sikorsky RAH-66 Comanche would be capable of a backward speed of 130 km/h. Therefore, the helicopter is a highly versatile flight vehicle that has reached phenomenal levels of sophistication.

For reference, Jane's[329,330] publishes detailed listings on all helicopters in service, under development and upgrading, with plenty of technical data. A wealth of information is made available by the FAA/JAA to support the helicopter certification standards and procedures (FAR 29 of the Airworthiness Standards: transport category rotorcraft). These standards are becoming increasingly important, because they are used also for the certification of military helicopters. The proof of compliance is specified by FAR §29.21.

11.2   HELICOPTER CONFIGURATIONS

Before introducing a general description of the helicopter types we need to address a fundamental flight problem: torque reaction. As the rotor spins in one direction, it does work on the air, and needs a torque to drive it. This torque is provided by the engine, and it must be reacted by the airframe. Torque reaction causes the airframe to spin in the opposite direction, unless the torque is balanced by other means: a direct moment or a force through an offset with respect to the rotor shaft. In the first case a counter-rotating rotor is required; in the latter case full control is achieved with a tail rotor, as shown in Figure 11.1. Calculations of torque balance will be done in §13.2.6.

Helicopters are primarily classified on the basis of their rotor system, on their gross weight and the type of operation they perform (civil, commercial, military). There is also a convention regarding the weight of the aircraft. A light helicopter has an MTOW below 12,000 lb (5,450 kg); a medium helicopter has an weight up to 45,000 lb (10,000 kg), although this category includes several heavy helicopters; heavy-lift helicopters have an MTOW above this value. It is estimated that 95% of the helicopters operating worldwide are light helicopters; most of them have the conventional tail rotor configuration, as in Figure 11.1.

Figure 11.2 shows the most common configurations. Most helicopters have a single main rotor, slightly inclined forward, and a tail rotor, as illustrated in Figure 11.1. This is nearly vertical and nearly aligned with the longitudinal axis of the airframe for torque balance, stability, control and maneuver. Under this class there are all

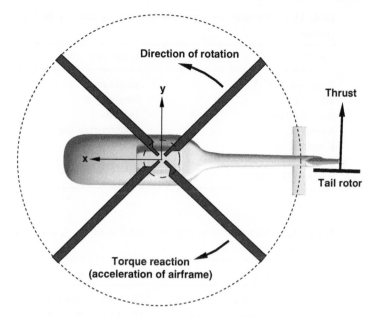

Figure 11.1 *Torque reaction phenomenon; torque reaction is in the opposite direction.*

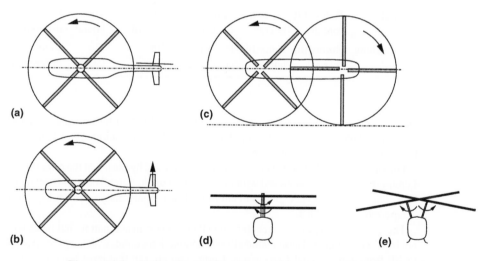

Figure 11.2 *Helicopter configurations, with various anti-torque systems: (a) conventional helicopter; (b) tandem; (c) NOTAR; (d) coaxial; (e) intermeshing.*

the general utility vehicles, heavy-lift vehicles, light helicopters and most of the high-performance military helicopters.

Another class of vehicle is characterized by two main rotors: these can be coaxial and counter-rotating, tandem (two separate axes), and intermeshing (with the rotor disks partially overlapping and the shafts inclined outward). Other helicopters may

Table 11.1 *Summary of main helicopter configurations, based on yaw control systems.*

Vehicle	Rotor type	Tail rotor
Eurocopter EC-365	Single rotor	Fenestron
Sikorsky S-76	Single rotor	Yes
Boeing CH-47	Tandem	No
Kamov Ka-52	Counter-rotating	No
Kaman K-max	Intermeshing	No
Boeing-Sikorsky V-22	Tiltrotor	No
Boeing MD-500	NOTAR	No

Table 11.2 *Summary of helicopter control methods; p, q, r are pitch/roll/yaw rates, respectively.*

Configuration	x	y	z	p	q	r
Single rotor	MR tilt fore/aft	MR tilt lateral	MR thrust collective	MR tilt fore/aft	MR tilt lateral	TR thrust
Tandem	Rotors tilt fore/aft	Rotor tilt lateral	Rotor thrust collective	Rotors differential	Tilt lateral	Rotors differential
Coaxial	Rotors tilt fore/aft	Rotor tilt lateral	Rotor thrust collective	Tilt fore/aft	Tilt lateral	Rotors differential
Intermeshing	Rotors tilt fore/aft	Rotor tilt lateral	Rotor thrust collective	Tilt fore/aft	Tilt lateral	Rotors differential

lack the tail rotor, and owe their control and stability to other technologies, such as NOTAR, which is based on jet propulsion. Table 11.1 is a summary of helicopter types based on the torque reaction system.

For all types of helicopters vertical control is achieved by means of the rotor thrust. Longitudinal control is achieved with the collective pitch. Lateral control is achieved by a combination of cyclic pitch and tail rotor thrust. A summary of control systems is reported in Table 11.2.

There are other types of helicopters, mostly of experimental nature, but these will not be discussed here. These include the compound helicopter, which derives part of its lift from conventional fixed wings. Compound aircraft that have been designed, tested and built include the McDonnell XV-1 (1954), the Fairey Rotodyne (1957), the Hiller X-18 (1959), the Lockheed AH-56 Cheyenne, the Kamov Ka-22 and more recently the Sikorsky-Piasecki X-49. See Deal and Jenkins[331] for some flight research on this type of aircraft, and Stepniewski and Keys[110] for a preliminary design analysis.

Figure 11.3 shows NASA's research aircraft RSRA (1980s). The aircraft was designed to study the effects arising from stopping the main rotor in flight. The lift would then be provided with the conventional wings mounted low on the fuselage. This concept gave the aircraft the vertical stability of a helicopter, and the horizontal

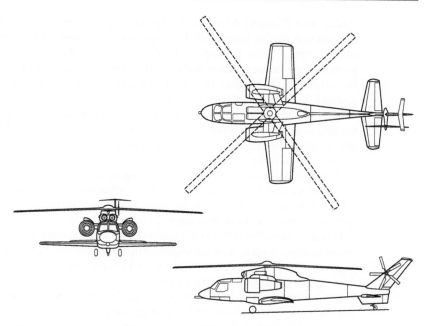

Figure 11.3 *NASA's RSRA compound helicopter configuration.*

cruise capability of a fixed-wing aircraft, as documented by Erickson *et al.*[332] A critical problem in flight stability and vibration is the off-loading of the rotor and the reduced tip speed (see also Bergquist[333]).

The autogiro is another hybrid vehicle that derives its lift from a rotating blade and its forward thrust from a conventional propeller. This type of aircraft (the original idea of Juan de la Cierva[334]) has some flaws, and it has never been successful beyond the realm of aviation enthusiasts. Nevertheless, the autogiro continues to stir attention in the aviation community. For a given GTOW, modern autogiros require the same power levels as conventional helicopters at intermediate flight speeds. For reference to the reader, Leishman[335] discussed the technical development of the autogiro, and Houston[336] its flight dynamics. The vast literature on the subject is cited in these references.

11.3 MISSION PROFILES

The mission profiles of the helicopter include many of the operations of the fixed-wing airplane. Due to its peculiar flight characteristics, the helicopter can carry a wide range and type of payloads. This is true both in the civil and military arena. Civilian operations include: scheduled flight services between airports and heliports, search and rescue, disaster relief, traffic monitoring, policing, and executive services. A review of the roles of the civil helicopter is available in Fay[337].

Many civil helicopters are approved for operation from/to challenging heliports, such as off-shore platforms (even in severe weather conditions), aircraft carriers, and rooftops of buildings. They are allowed to fly over areas where there are no emergency landing sites (Category A). For such operations the helicopter must be able to take off with One Engine Inoperative (OEI), and achieve at least a 100 feet-per-minute (0.5 m/s) rate of climb. Helicopters that do not fall under this category are certified to operate in safer areas, where landing is always possible (Category B). In any case, the certification authorities (see, for example, FAR Part 29) require a height/velocity diagram, both with single and dual engine operation, to be included in the flight manual. The height/velocity diagram is a flight envelope in which the operation of the aircraft is possible and safe. It will be discussed in some detail in §14.5. Take-off operations are affected by the atmospheric conditions, helipad size and position, and community constraints.

Many helicopters currently operate within large metropolitan areas, and have access to a good network of helipads. Back in 1951, when the first helicopter operations started at the heart of Manhattan, the press was stunned by helicopters "taking off and landing from 16-storey skyscrapers".

Manufacturers offer different helicopters in the same nominal series, depending on applications. Each of them has slightly different performance charts and different flight manuals. An example of this is the Sikorsky S-76C, available in utility/off-shore configuration, or executive configuration.

Helicopters for sea and naval operations include Anti-Submarine Warfare (ASW), anti-surface ship warfare, medical evacuation (*medevac*), logistics, search and rescue, and mine-sweeping*. Other military operations include long-range missions in hostile territory, to be conducted day and night, in adverse weather, and extreme environments.

Mission planning for commercial and passenger traffic can be slightly more complicated than a flight mission of a fixed-wing aircraft, essentially because helicopters tend to fly at lower altitudes around congested and built-up corridors. Flight planning in these cases needs permission to fly over these areas. In addition, it may require studying different flight corridors to minimize community noise, which depends heavily on local weather conditions. The flight altitude for each segment should be the maximum allowed by the ATC, in order to minimize noise disturbances and maximize safety. In London this altitude is 2,500 ft (770 m).

11.4 FLIGHT ENVELOPES

The flight envelopes circumscribe the entire issue of rotary-wing performance. All the rotary-wing aircraft in Figure 11.2 are capable of vertical take-off and landing; they are also capable of transitioning to low-speed forward flight and to turn around themselves. Therefore, the helicopter represents a unique flight vehicle that extends the spectrum of powered flight possibilities, although it is not in direct competition with the fixed-wing airplane, as far as speed is concerned.

In 1987, Drees[339] wrote that *speed will certainly be increased to the 450 kt level.* That has not happened yet, nor does there seem to be serious attempts at reaching these flight performances with a rotary-wing aircraft. Nevertheless, research has continued

---

*The US Army has different designations for the same class of vehicles: OH is an observation helicopter; UH is a general utility, CH is a cargo/heavy-lift helicopter; and AH is an attack/combat helicopter. See also the *International Countermeasures Handbook*[338].

at various institutions. Scott[340] in an exhaustive report at Sikorsky concluded that for commercial transports, tilt-wings and variable diameter tilt-rotor concepts have a better performance. For a military/attack role, the variable-diameter helicopter was best. A design speed of 375 to 425 kt was found to be the maximum desirable for transport missions. Research in high-speed rotorcraft at McDonnell–Douglas is documented by Rutherford *et al.*[341] These authors discuss options for rotorcraft speeds in the range of 450 kt with rotor-wing and tilt-wing options. Alternative configuration options are reviewed by Talbot *et al.*[342] Lynn[343] concluded that, with the exception of autorotative performance and the large download at low speeds, the compound helicopter has several advantages at high speed.

The level flight speed record is owned by the Westland Lynx helicopter, with 400.87 km/h (216.3 kt), thanks to a combination of installed power and the helicopter's stability at high speeds[344]. The actual speeds of the common helicopter are considerably lower, in the range of 250 to 280 km/h (e.g. up to 150 kt).

The maximum speed is limited by a large number of factors, not directly related to the installed engine power. These factors include the tip speed, the dynamic stall, vibration, stability, maximum transmission torque, noise emission, etc. These issues will be discussed in more detail in §14.1.

At present, high-altitude helicopter flight has a limited commercial market, although it is of some interest for mountain rescue operations, for military operations and atmospheric research. In fact, most helicopter flights take place at low altitudes, up to 1,000 m (3,000 ft) from ground level.

11.5 DEFINITIONS AND REFERENCE SYSTEMS

To properly understand the operation and performance of the helicopter, it is necessary to explain the nomenclature and the reference systems commonly used. Figure 11.4 shows the nomenclature in the conventional symmetry plane of the aircraft, x–z. We note that the aircraft is generally not symmetric on this plane for several reasons (presence of tail rotor, asymmetric tail planes and control surfaces, blade orientations, windows, probes, etc.). To start with, we will have a reference system that moves with the helicopter. This system is centered at the center of gravity (CG), which is a variable point. The x axis is on the symmetry plane, and runs from the tail to the nose of the aircraft. The y axis is at 90 degrees with the y axis and pointing to starboard (from the point of view of the pilot); the z axis forms a right-hand orthogonal reference with x and y.

Separate reference systems are required for the rotors. The main rotor is mounted on a shaft, slightly tilted forward. The tilt angle (or mast angle), calculated with respect to the vertical axis on the ground, is called α_T. As they rotate, the blade tips travel along a flight path that lies on a plane, generally called Tip Path Plane (TPP). There can be considerable differences between the TPP and the rotor disk. However, in the performance calculations discussed in the next chapters the TPP is assumed to coincide with the rotor disk plane. Figure 11.5 shows the plane traced by the blade tips in their rotation around the shaft.

Another important plane from the control point of view is the swashplate plane. This is the reference plane for the cyclic pitch commanded by the pilot. The swashplate is a mechanical system consisting of two disks placed as a sandwich between some

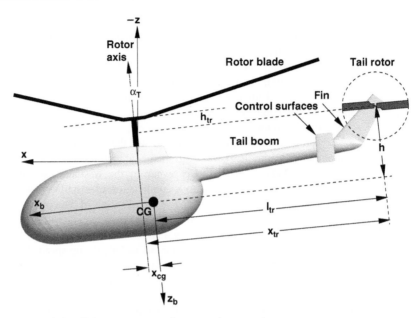

Figure 11.4 *Helicopter nomenclature and conventions.*

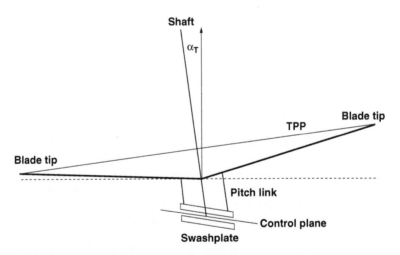

Figure 11.5 *Tip Path Plane (TPP), or rotor disk.*

roll bearings. The upper disk rotates around the lower disk, and both can assume an arbitrary orientation, commanded by some actuators; the vertical movement of the swashplate commands the cyclic pitch of the blades via a series of pitch links.

The rotor blades are flexible and are subject to periodic loads that create lead/lag, flapping and torsion. For reference, in normal operations these movements are of the order of a few degrees, and are important for the blade's dynamics. In addition, there can be a complex hinge system (in articulated rotors), as sketched in Figure 11.6, that allows for a number of degrees of freedom.

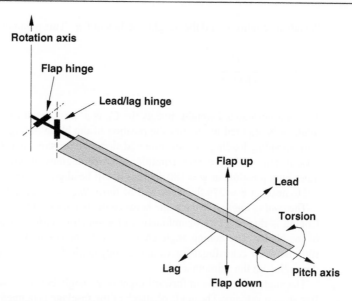

Figure 11.6 *Articulated rotor with flap and lead hinges. The actual sequence of hinges depends on the particular rotorcraft.*

The tail rotor is nearly vertical, and nearly aligned with the conventional symmetry plane. In fact, the rotor plane is inclined forward and upward by a small angle. A separate reference system centered at the hub of the tail rotor is required for detailed tail rotor calculations (aerodynamics, aeroelastic response, noise emission).

The balance of forces and moments is a more complex matter, since the main rotor thrust does not act at the CG. Furthermore, there is a set of moments acting on the aircraft, namely the rolling moment and the hub moment; additional global forces are created by the horizontal tail plane. The full balance equations and the practical means to achieve the balance in flight is called *trim*. Figure 11.4 shows the relevant distances required for preliminary performance and stability calculations.

11.5.1 Rotor Parameters

The nomenclature used for the blades and blade sections is similar to the propellers. In particular, the blade pitch θ, blade chord c, rotor solidity σ, tip speed U_{tip} and tip Mach number M_{tip} have the same definition. The definitions of thrust, power, and torque coefficients are, respectively,

$$C_T = \frac{T}{\rho A(\Omega R)^2}, \quad C_P = \frac{P}{\rho A(\Omega R)^3}, \quad C_Q = \frac{Q}{\rho A(\Omega R)^2 R}. \tag{11.1}$$

Since the power is $P = Q\Omega$, from the definition it follows that $C_Q = C_P$. Parameters that are specific to the helicopter include: the blade azimuth ψ, the coning angle β, the advance ratio μ, and the weight coefficient C_W. This parameter is defined as

$$C_W = \frac{W}{\rho A (\Omega R)^2}.$$ (11.2)

This is a normalized weight, just as the C_L is a normalized lift force. The *azimuth angle* ψ is required to identify the position of the blade during its rotation, and the corresponding loading. A blade oriented along the principal axis of the aircraft, from nose to tail, will have zero azimuth. After 90 degrees, it will be in the so-called advancing position; at $\psi = 180$ degrees it will be aligned with the longitudinal axis, tail to nose; at $\psi = 270$ the blade will be *retreating*, see Figure 13.1.

The *coning angle* β is the angle between the blade's mean line and the rotor plane. Coning is due to a cross-combination of geometric settings, flapping and pitching of the blades. The coning angle changes as the blades rotate, because the equilibrium between centrifugal forces and aerodynamic forces is a differential equation depending on the azimuth angle.

The angle of attack α in forward flight is the angle between the rotor disk and the free stream velocity. The angle of attack of the fuselage is defined in the same way as an airplane; this can be different from the rotor's angle of attack, depending on flight conditions. The angle $-\alpha_T$ is the angle of inclination of the thrust in the direction of the force of gravity.

The *advance ratio* is the ratio between the air speed parallel to the TPP and the tip speed, i.e.

$$\mu = \frac{U \cos \alpha}{U_{tip}} = \frac{U \cos \alpha}{\Omega R} = \frac{U_T}{\Omega R},$$ (11.3)

where $U_T = U \cos \alpha$ is the in-plane velocity component, and α is the angle of attack of the rotor. The effective rotor loading is defined in terms of the blade loading coefficient C_T / σ,

$$\frac{C_T}{\sigma} = \frac{1}{\sigma} \frac{2T}{\rho A U_{tip}^2}.$$ (11.4)

This is an average measure of blade loading. A parameter that influences the stall limit is the propulsive force coefficient,

$$\frac{C_{D_e}}{\sigma} = \frac{1}{\sigma} \frac{2D_e}{\rho A U_{tip}^2},$$ (11.5)

where D_e is the effective drag in the flight direction, and C_{D_e} is the effective drag coefficient. Estimated values of the rotor solidity are shown in Figure 11.7. These have been elaborated for sets of rotor data. These data can also be plotted as a function of the MTOW. We conclude that most values of σ fall within the range 0.04 and 0.14.

Helicopter rotor rpm are constant. The main reason why the rpm is kept constant is that even a small change over the design rotational speed can lead some subsystems

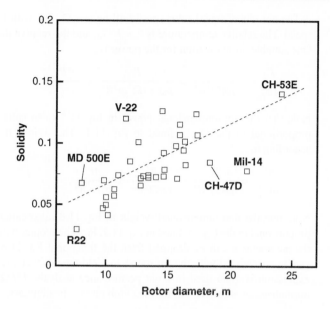

Figure 11.7   *Estimated rotor solidity for selected helicopters. Note the scatter among various helicopters.*

into resonance problems. In fact, it can be proved (see Bramwell[345] and Prouty[346]) that there is only a narrow range over which the rpm can be varied without operating within the natural bending and flapping frequencies of the blades. Until effective methods for avoiding resonance are found, constant rotor speeds will be commonplace in helicopter engineering[†].

11.6   NON-DIMENSIONAL PARAMETERS

An effective method for first-order analysis of helicopter performance is the use of a few non-dimensional parameters, as first proposed by Knowles[347]. Cooke and Fitzpatrick[4] use extensively this method for flight test analysis. The power required by the helicopter in any flight condition is a function of a number of parameters, namely

$$P = f(W, R, v, v_c, h, \rho, T, \Omega), \tag{11.6}$$

with the usual meaning of the symbols. Helicopter performance must be calculated and tested over a range of weights, rotor speeds, forward speeds, climb rates, altitudes and atmospheric conditions. Therefore, the parametric space is large.

[†] As a matter of fact, the rpm does change within a limited range, due to engine power transients. These changes, however, are usually contained within 10 rpm.

The non-dimensional rotor speed will be $\omega = \Omega/\Omega_o$, with Ω_o the nominal rotor speed. The relative temperature is $\theta = T/T_o$, and the relative density is $\sigma_1 = \rho/\rho_o$.[‡] One suitable normalization for the power is

$$\frac{P}{\rho A U_{tip}^3} = \frac{P}{\rho A \Omega^3 R^3} = \frac{P}{\rho_o \sigma_1 \pi \Omega^3 \omega^3 R^3} \propto \frac{P}{\sigma_1 \omega^3 R^5}. \tag{11.7}$$

Note that the non-dimensional power in Eq. 11.7 (also called *referred power*) is proportional to the C_P, defined in Eq. 11.1. The weight W can be normalized according to

$$\frac{W}{\rho A U_{tip}^2} = \frac{W}{\rho A \Omega^2 R^2} = \frac{W}{\rho \pi \Omega^2 R^4} \propto \frac{W}{\sigma_1 \omega^2 R^4}. \tag{11.8}$$

Note that the non-dimensional weight in Eq. 11.8 (also called *referred weight*) is proportional to the C_W, defined in Eq. 11.2. For a helicopter with a fixed rotor radius, the parameter R can be dropped from the functional Eq. 11.6. The forward speed can be normalized with the tip speed, to yield the advance ratio. However, another expression is often used in flight performance analysis: $U/\Omega R \propto U/\omega$. The same normalization is carried out for the climb rate v_c. In summary, we have

$$\frac{P}{\sigma_1 \omega^3} = f\left(\frac{W}{\sigma_1 \omega^2}, \frac{U}{\omega}, \frac{v_c}{\omega}, h, \theta, \omega\right). \tag{11.9}$$

The parameters U/ω and v_c/ω are called *referred true air speed* and *referred climb rate*, respectively. Since the direct measurement of density is not possible, σ_1 is generally replaced by δ/θ. An alternative expression of the referred weight is

$$\frac{P}{\delta/\sqrt{\theta}} = f\left(\frac{W}{\delta}, \frac{U}{\omega}, \frac{v_c}{\omega}, h, \frac{\omega}{\sqrt{\theta}}\right). \tag{11.10}$$

The parameter $\omega/\sqrt{\theta}$ gives the effects of the air temperature on the rotor speed. In conclusion, the helicopter performance can be referred to a limited number of non-dimensional parameters. Not all of them will be present at all times. For example, the hovering analysis requires that $U/\omega = v_c/\omega = 0$. For a forward flight analysis $v_c/\omega = 0$. A forward climb flight will depend on all the parameters. A flight procedure requiring a constant referred weight W/δ requires aircraft to climb as fuel is burned. In fact, as the weight decreases, a constant referred weight is achieved by climbing to a lower pressure altitude.

11.7 METHODS FOR PERFORMANCE CALCULATIONS

The fundamentals of propulsion and aerodynamics of the rotor are discussed in a number of textbooks. Glauert[167] was one of the first to treat the aerodynamics of the air-screw from a general point of view, followed by Theodorsen[168], although the original ideas go back to Rankine (1865) and Froude (1889). Glauert's book is

---

[‡] We use σ_1 in place of the standard σ to avoid confusion with the symbol used earlier in the fixed-wing aircraft analysis; θ is not to be confused with the blade pitch.

recommended for its clarity. Since the publication of Glauert's theory, a number of other developments have taken place. For example, a comprehensive treatment of the momentum theory for helicopter performance (hover, climb, descent, forward flight, flight restrictions) was published by Heyson[348].

The first relevant studies on helicopter performance were due to Bailey[349] and Gessow and Tapscott[350]. These authors, along with Tanner[351] and others, show the rotorcraft performance in terms of the rotor's non-dimensional coefficients. Gessow and Myers[352] have included chapters on performance in their classic book on helicopter aerodynamics, and are the authors of several technical papers on the subject, dating from the 1940s. More modern books on the subject include Stepniewski and Keys[110], Prouty[353] and Leishman[324]. All of these textbooks are geared in part toward rotorcraft performance. Cooke and Fitzpatrick[4] have a good chapter on helicopter performance in their book devoted to helicopter test and evaluation. Stepniewski and Keys devote several chapters on a case study of helicopter preliminary design and performance analysis. The book by Johnson[354] is also recommended, although it does not show many calculation methods. However, Johnson himself[355] is the author of a comprehensive rotor code, CAMRAD II, that is widely used in the industry for rotorcraft performance calculations. Another comprehensive code is GENHEL, of NASA/US Army[356,357]. This code uses a six-degree of freedom model for the aircraft, with rigid-body airframe and rotor dynamics. The rotor aerodynamics are modeled with the blade element theory. A similar approach is followed by Eurocopter's STAN flight mechanics code. Some standard methods for the calculation of helicopter performance are available in the ESDU publications, particularly on hover and forward flight[358,359].

Since all the calculation methods are necessarily complicated by the unsteady effects of the rotating blades, and by the interference between various components, we need to address a key point regarding suitable methods for the evaluation of the rotorcraft performance. We will show how simple methods based on the combined momentum and blade element theories yield steady state solutions suitable for most performance calculations. Obviously, we can improve on these methods, but the step forward requires taking into account the unsteady aerodynamics of the rotor, the full rotor dynamics, the aeroelastic response, and the interaction between rotors and airframe. Therefore, we find it appropriate to start with the basic momentum theory to calculate the vertical flight performance of the helicopter.

PROBLEMS

1. A helicopter rotor in hover encounters aerodynamic forces that oppose its rotation. When a torque is applied from the engines to maintain a rotation at a constant speed, the airframe starts rotating in the opposite direction. Describe methods to counteract this torque and sketch the forces and moments involved in the control of the helicopter.

2. In a conventional helicopter stability control can be achieved with both a pusher or a tractor tail rotor. A pusher tail rotor produces thrust away from the vertical fin; a tractor rotor produces thrust against the fin. Discuss, by further investigations, the merits of each configuration. *(Problem-based learning: additional research is required.)*

Chapter 12

Rotorcraft in Vertical Flight

If a man is in need of rescue, an airplane can come in and throw flowers on him [...], but a direct-lift aircraft could come in and save his life.

Igor Sikorsky

This chapter discusses methods for the calculation of performance of the rotorcraft in vertical flight, including hover, axial climb and descent, and ground effects. Hover is a flight condition in which the aircraft maintains its altitude at a fixed position. For both Category A and B helicopters, the hover performance must be determined over the ranges of weight, altitude, and temperature for which take-off data are scheduled (FAR Part 29.49).

The hover condition and the vertical climb are the easiest to model theoretically. Nevertheless, the rotor encounters difficult operating regimes at all descent speeds. At these conditions all the simple theories fail, and we shall see the reason.

12.1 HOVER PERFORMANCE

The simplest method for deriving a first-order estimate of the power required to hover is the one-dimensional axial momentum theory. We use the expressions found in Chapter 5, with some changes in the reference systems, as shown in Figure 12.1. The stream tube has a vertical axis through the rotor center. The inflow is from the top of the figure. The contraction of the stream tube is explained with the same concepts used for the propeller.

When we take into account these changes for application to the helicopter in hover, the one-dimensional axial momentum theory provides the induced velocity at the rotor in hover,

$$v_h = \sqrt{\frac{T}{2\rho A}} = \sqrt{\frac{W}{2\rho A}}. \tag{12.1}$$

The ideal power required to hover is given by the product between the thrust and the induced velocity, i.e.

$$P_h = Tv_h = W\sqrt{\frac{W}{2\rho A}} = \frac{W^{3/2}}{\sqrt{2\rho A}} = \sqrt{\frac{W}{2\rho}}\sqrt{\frac{W}{A}}, \tag{12.2}$$

where the disk area is $A = \pi R^2$. We conclude that the induced power to hover increases with the aircraft weight, and decreases with the increasing rotor radius (all other parameters being constant). This would imply that increasing the diameter at constant weight would be beneficial. This is not always the case.

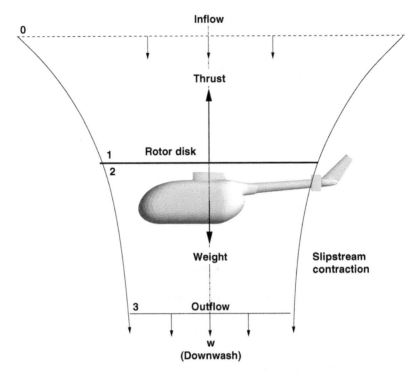

Figure 12.1 *Hover condition, with stream tube around the rotor disk, and airframe.*

A number of factors are to be considered, in order to take into account effects such as the non-uniform downwash, the tip losses (tip vortices, blade/vortex interaction, turbulence), and the rotation of the slipstream. Equation 12.2 was derived with a one-dimensional flow approximation that uses the momentum, energy and continuity equations. We collect all these exceptions in an induced-power factor k,

$$P = k \frac{W^{3/2}}{\sqrt{2\rho A}} = k \sqrt{\frac{W}{2\rho}} \sqrt{\frac{W}{A}}. \tag{12.3}$$

The disk loading $D_L = W/A$, which appears in Eq. 12.3, is an important design and performance parameter. We have elaborated the disk loading for several helicopters. Figures 12.2 and 12.3 show a summary of disk loading for military helicopters and for general utility helicopters.

The issue is more complicated than it appears. In fact, the disk loading is controlled by both the radius and the weight. For a given disk area, slender blades are lighter, and so is the hub and the transmission group. Hence, the disk loading would decrease. On the other hand, the weight of the transmission group depends on the limit torque. For a given tip speed, the induced torque does not depend on the radius, but the profile torque decreases with the radius. A reduction in blade radius would decrease the limit torque, hence the weight of the transmission group and the disk loading. Engine and transmission ratings are expected to grow over time; thus, disk loadings tend to grow as well.

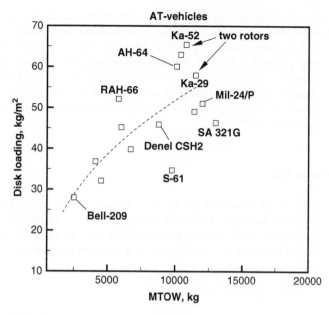

Figure 12.2 *Disk loading of some military helicopters.*

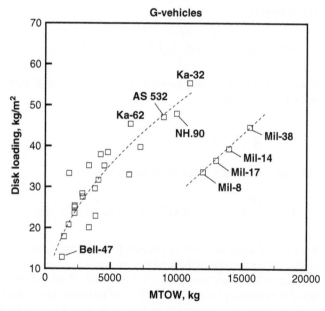

Figure 12.3 *Disk loading of some general utility helicopters.*

For reference, the CH-46 *Sea King* started at 7,940 kg (17,500 lb) and reached 9,980 kg (22,000 lb). This occurred without much change in the weight of the rotating parts.

In the case of military helicopters, generally the vehicles of the Kamov series show a larger disk loading; this is partially due to the fact that these helicopters use coaxial counter-rotating blades. Coaxial rotors have higher disk loading, because of interference effects. Coleman[360] reviewed the aerodynamic theory of the coaxial helicopter rotors. Single-rotor vehicles, such as the AH-64 and the RAH-66, have a high disk loading at their gross weight. The data in Figure 12.3 are more susceptible to a single line correlation, except for the Mil helicopters. These are characterized by large rotor diameters, with a larger number of blades (up to seven) and relatively low disk loading. Data on Mil helicopters can be found in Everett-Heath[327].

12.1.1 Profile Power

The induced power of Eq. 12.3 must be augmented with the power P_o required to overcome the profile drag of the blades. This power component is found by taking the drag of a blade element and integrating over the span. The element of drag dD created by a blade element of span dy at spanwise position y is

$$dD = \frac{1}{2}\rho C_D U^2 (dA). \tag{12.4}$$

The element of blade area is $dA = cdy$, and the local blade speed can be expressed in terms of tip speed as $= rU_{tip} = \Omega y$. The contribution of the blade element to the profile power is

$$dP_o = dD\,U = \frac{1}{2}\rho\,cdy C_{D_o} U^3 = \frac{1}{2}\rho c\,C_D \Omega^3 y^3 dy. \tag{12.5}$$

If we normalize the radial position y by the blade radius R, then $r = y/R$, $dy = dr/R$, and Eq. 12.5 becomes

$$dP_o = \frac{1}{2}\rho c\,C_D \Omega^3 R^4 r^3 dr. \tag{12.6}$$

The total power requirement for N blades becomes

$$P_o = \int_0^R dP_o = \frac{1}{2}\rho c N\,\Omega^3 R^4 \int_0^1 C_D(\alpha, Re, M) r^3 dr. \tag{12.7}$$

A number of issues have to be resolved before calculating this integral. The profile drag coefficient is a function of the local Reynolds and Mach numbers, $C_D = C_D(\alpha, Re, M)$, as we know from basic aerodynamics. Additional effects include the centrifugal pumping of the boundary layer and the effects of turbulent transition. To solve the integral in closed form, it is customary to take an average value of the C_{D_o} over the speed range involved. Most effects are neglected by the performance engineer. However, it is important to take into account the compressibility effects if the blades operate above the divergence Mach number M_{dd}.

The second aspect is the limit of integration, because as we reach the blade tips the flow is affected by three-dimensional effects and various losses that are not explicitly accounted for in Eq. 12.7. Various semi-empirical corrections exist to derive an *effective* blade radius, as discussed in §5.9.2. In the present context, we resolve all these parameters into an average value of the profile drag coefficient, and allocate all the effects mentioned to the $\overline{C}_{D_o}$. With this simplification, the solution of the integral in Eq. 12.7 yields

$$P_o = \frac{1}{8}\overline{C}_D\rho\sigma A U_{tip}^3, \tag{12.8}$$

and in terms of power coefficient

$$C_{P_o} = \frac{P_o}{\rho A(\Omega R)^3} = \frac{\sigma\overline{C}_D}{8}. \tag{12.9}$$

Since the C_D depends on the angle of attack of the blade section, in hover we have the following approximation:

$$C_D(r) = C_{D_o} + C_{L_\alpha}\alpha^2 = C_{D_o} + C_{L_\alpha}\left(\theta + \frac{\lambda}{r}\right)^2 \simeq C_{D_o} + C_{L_\alpha}\theta^2, \tag{12.10}$$

so that the local C_D grows like θ^2. If there is a linear distribution of twist,

$$\theta(r) = \theta_o + \theta_1\frac{y}{R} = \theta_o + \theta_1 r, \tag{12.11}$$

then the average C_D is found from integration of Eq. 12.10 with Eq. 12.11,

$$\begin{aligned}\overline{C}_D &\simeq \overline{C}_{D_o} + \frac{1}{R}\int_0^1 C_{L_\alpha}\,(\theta_o + \theta_1 r)^2\,dr \\ &= \overline{C}_{D_o} + C_{L_\alpha}\left(\theta_o^2 + \theta_1^2\frac{R^2}{3} + \theta_o\theta_1 R\right).\end{aligned} \tag{12.12}$$

Therefore, $\overline{C}_D$ increases quadratically with the collective pitch. When the rotor is heavily loaded, the approximation is not valid, because the local induced velocity considerably changes the effective angle of attack of the blade section.

To this end, the total power required to hover is estimated from adding the profile power to the induced power and the power losses due to the transmission group, P_t,

$$P = k\frac{W^{3/2}}{\sqrt{2\rho A}} + \frac{1}{8}\overline{C}_D\rho\sigma A U_{tip}^3 + P_t. \tag{12.13}$$

The transmission losses are proportional to the torque and power transmitted, and obviously on other peculiar aspects of the machine, such as number of stages and oil churning in the group. Equation 12.13 does not contain the tail rotor power, which will be calculated separately at a later time.

12.1.2 Blade Element Analysis in Hover

The combined blade-element and momentum theory, which we have applied to the propeller analysis (Chapter 5; computational method on page 107), is frequently applied to the helicopter rotor in hover, climb and forward flight. The method requires limited numerical calculations, but it is powerful in providing engineering data for a variety of cases. An example of application to the hover performance is shown in Figures 12.4 and 12.5. These figures show the variation of the C_T and C_P versus the collective pitch for a two- and a three-bladed rotor (with solidity $\sigma = 0.0424$ and 0.0626, respectively). The calculations are compared with the experimental data of Knight and Hefner[361]. The rotor of reference had a constant-chord NACA 0015 blade section. The comparison is excellent for both blades in the whole range of available data.

The estimate of the profile power coefficient C_P allows the calculation of the Figure of Merit, FoM, of the rotor. This is a measure of efficiency of the rotor in converting engine power into useful thrust to stay airborne. This parameter was first introduced by Glauert (1927) to measure the lifting efficiency of a rotor that is not moving in axial flight:

$$FoM = \frac{P_i}{P_i + P_o} = \frac{C_{P_i}}{C_{P_i} + C_{P_o}} = f(C_T/\sigma). \tag{12.14}$$

The FoM is a useful efficiency metric at a given disk loading W/A. A comparison with experimental data obtained by Knight and Hefner[362] is shown in Figure 12.6 for a two-bladed rotor. We have inferred an average $C_D \simeq 0.0115$, over the whole range

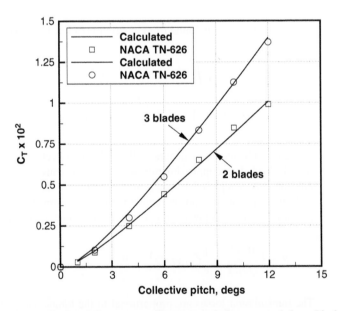

Figure 12.4 *Calculated C_T versus collective pitch for two- and three-bladed hovering rotors. Calculations compared with the experiments of Knight and Hefner[361]; $d = 1.524\,m$, $c = 0.051\,m$, $960\,rpm$; airfoil NACA 0015.*

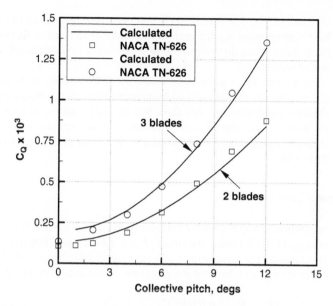

Figure 12.5 *Calculated C_Q versus collective pitch for two- and three-bladed hovering rotors. Calculations compared with the experiments of Knight and Hefner[361]; d = 1.524 m, c = 0.051 m, 960 rpm; airfoil NACA 0015.*

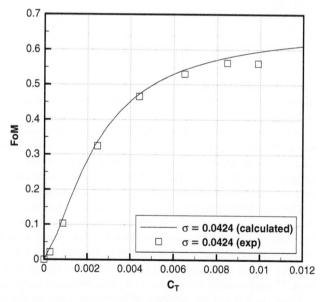

Figure 12.6 *Two-bladed helicopter rotor in hover: FoM versus C_T. Comparison with the experimental results of Knight and Hefner[362]. Rotor identical to that of Figure 12.4.*

of measured C_T. The correlation is good, except at the highest values of C_T, which have been obtained at pitch angles of the order of 12 degrees. At this incidence, is it likely that a strong tip vortex leads to separation at the outer board, with consequent loss of sectional lift.

For a given disk loading and solidity, the *FoM* depends on the profile drag coefficient. The C_D can be reduced by proper design of the blades and the tips. The simulation and reference data show that the *FoM* increases monotonically with the C_T, therefore the rotor efficiency in hover increases with the weight – all other parameters being constant. A realistic value of the C_T for a real-life helicopter is between 0.05 and 0.1.

12.1.3 Power Loading

An alternative way of plotting the rotor coefficients is based on the concept of power loading. This is defined as the ratio T/P, e.g. the amount of thrust (lift) per unit power. This parameter is a measure of efficiency on its own, because the lower the engine power for a given thrust, the more economical the aircraft is. In hover conditions

$$PL = \frac{T}{P} \simeq \frac{W}{P} = \frac{\rho A (\Omega R)^2 C_T}{\rho A (\Omega R)^3 C_P} = \frac{1}{\Omega R} \frac{C_T}{C_P}. \tag{12.15}$$

For a given engine power, the *PL* is highest at the maximum take-off weight. For a given weight, the *PL* increases with the decreasing tip speed; this, in turn, can be reduced by reducing the diameter and the rpm. Another useful expression of the power loading is

$$PL = \frac{T}{P} = \frac{T}{T^{3/2}/\sqrt{2\rho A}} = \sqrt{\frac{2\rho A}{T}} = \frac{1}{v_i} \simeq \sqrt{\frac{2\rho}{W/A}}. \tag{12.16}$$

Therefore, the *PL* is equal to the inverse of the average induced velocity at the disk; the maximum *PL* is obtained with a minimum induced velocity, or with a minimum disk loading.

In Figure 12.7 we have plotted the ratio C_T/C_P (proportional to the power loading) versus the thrust coefficient. The calculations are compared with the experimental data of Knight and Hefner[362] for two rotor configurations. The parameters of the calculation are the same as before. Again, there is a loss of accuracy with a high C_T; otherwise the results are excellent, considering all the approximations of the theory.

The importance of the geometrical details must not be underestimated. In particular, the tip geometry plays a critical role in rotorcraft performance; as a consequence it has been the subject of intensive research for many years. A review of the effects of tip geometry on the rotor characteristics is available in Philippe *et al.*[363] A number of other studies of interest for the performance of the tip geometry include Scott *et al.*[364], who evaluate five different tip geometries, including the BERP.

Figure 12.8 shows the influence of the tip geometry on the rotor power in hover conditions[365], elaborated from flight testing of the Aérospatiale SA 365N (now Eurocopter EC-365N). The swept back tip increases the MTOW of the aircraft by 70 to 80 kg.

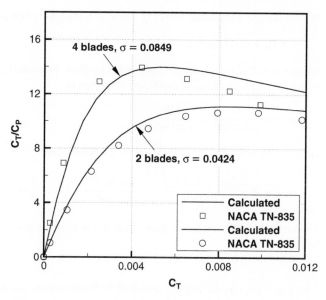

Figure 12.7 *Helicopter rotor in hover: C_T/C_P versus C_T. Comparison with the experimental results of Knight and Hefner[362] for a two- and a four-bladed rotor.*

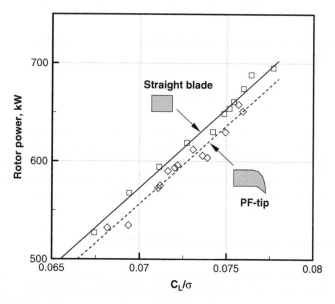

Figure 12.8 *Rotor power in hover. Influence of a parabolic swept-back tip on the power requirements. Data from flight test, adapted from Vuillet[365].*

12.2 EFFECT OF BLADE TWIST

Most helicopter blades are twisted, so as to decrease the pitch toward the blade tips, which rotate at high speeds. There are blades with both linear and non-linear twist (example: Eurocopter EC-135). Therefore, the local pitch can be described by

$$\theta = \theta_o + \theta_1 r + \theta_2 r^2 + \cdots, \tag{12.17}$$

where $r = y/R$ is the non-dimensional radial position. From the blade element theory, §5.9.2, the element of thrust coefficient is written as

$$dC_T = \frac{dT}{\rho A (\Omega R)^2} \simeq \frac{N dL}{\rho A (\Omega R)^2}. \tag{12.18}$$

Hence, if C_l is the local lift coefficient of the blade, then

$$dC_T = N \frac{\rho (\Omega y)^2 C_l (cdy)/2}{\rho A (\Omega R)^2} = N \frac{y^2 C_l (cdy)/2}{AR^2} \cdots = \frac{1}{2} \sigma C_l(r) r^2 dr. \tag{12.19}$$

The sectional lift coefficient is found from linear aerodynamics,

$$C_l(r) = C_{l_\alpha} (\alpha - \alpha_o) = C_{l_\alpha} (\theta - \alpha_o - \phi), \tag{12.20}$$

where α_o is the zero-lift angle of attack of the blade section, and ϕ is the inflow angle. Since $\phi = \lambda/r$ (Eq. 5.56), we also have

$$C_l(r) = C_{l_\alpha} \left(\theta - \alpha_o - \frac{\lambda}{r} \right). \tag{12.21}$$

In hover conditions, λ is calculated from Eq. 5.73. Assuming $\lambda_c = 0$, after simplifying the equation we have

$$\lambda = \frac{\sigma C_{L_\alpha}}{16} \left(1 + 32 \frac{\theta r}{\sigma C_{L_\alpha}} \right)^{1/2}. \tag{12.22}$$

The total thrust coefficient is found from integration of Eq. 12.19 with Eq. 12.21, Eq. 12.17 and Eq. 12.22. This integration is best done numerically. The integrated C_T grows with the pitch distribution θ_1. It also grows with θ_o^2, as it can be inferred from Figure 12.4. A fair comparison between thrust distributions is done at a constant C_T, because this is directly proportional to the helicopter's weight.

12.3 NON-DIMENSIONAL HOVER PERFORMANCE

The hover performance can be plotted in terms of the non-dimensional parameters derived in §11.6. The general power equations in hover become

$$\frac{P}{\sigma_1 \omega^3} = f \left(\frac{W}{\sigma_1 \omega^2}, h, \frac{\omega}{\sqrt{\theta}} \right), \qquad \frac{P}{\delta/\sqrt{\theta}} = f \left(\frac{W}{\delta}, h, \frac{\omega}{\sqrt{\theta}} \right). \tag{12.23}$$

Rotor speeds do not vary much, and it is safe to assume that $\omega = 0.95$ to 1.05. For changes in temperatures contained within ± 15 degrees around the sea level standard, also θ ranges from 0.95 to 1.05. Therefore, the parameter $\omega/\sqrt{\theta}$ is mostly important at high altitudes and on cold/hot days, when a 10% variation around the nominal point must be considered. If neither the referred rotor speed, nor the altitude are important in a particular analysis, then

$$\frac{P}{\sigma_1 \omega^3} = f\left(\frac{W}{\sigma_1 \omega^2}\right), \qquad \frac{P}{\delta/\sqrt{\theta}} = f\left(\frac{W}{\delta}\right). \tag{12.24}$$

The momentum theory discussed earlier, along with the profile power calculation, allow us to elaborate a bit further on the actual relationships in Eq. 12.24. If we normalize Eq. 12.13, then

$$\frac{P}{\sigma_1 \omega^3} = \frac{Q}{\sigma_1 \omega^2} + c_1 \left(\frac{W}{\sigma_1 \omega^2}\right)^{3/2} + c_2, \tag{12.25}$$

where Q is the profile torque, and c_2 is a coefficient dependent on the transmission losses. Equation 12.25 indicates that the correlation between the referred power and the referred weight to 3/2 is linear if the transmission losses are constant. This relationship is calculated from Eq. 12.2 to prove the point. The reference case is the helicopter model D (Appendix A) at the nominal rotor speed and standard atmospheric conditions. Figure 12.9 shows this relationship at three reference altitudes. If the rotor operates at its nominal speed, $\omega \simeq 1$, a non-dimensional performance chart would be W/σ_1 versus P/σ_1.

The tip effects on the referred power deserve a special mention. From Eq. 12.2, abstracting from the transmission losses, we have

$$\frac{P}{\sigma_1 \omega^3} \propto \frac{P}{\rho \Omega^3} = \left[\left(\frac{W}{2\rho A}\right)^{3/2} \frac{1}{\rho \Omega^3} + \frac{1}{8} \frac{\sigma \rho A (\Omega R)^3}{\rho \Omega^3} \overline{C}_D\right], \tag{12.26}$$

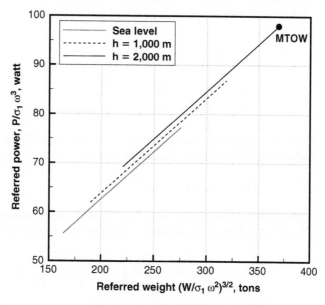

Figure 12.9 *Non-dimensional analysis of hover flight at three reference altitudes.*

$$\frac{P}{\sigma_1 \omega^3} \propto \left[\frac{1}{(2A)^{3/2}} \left(\frac{W}{\rho \Omega^2} \right)^{3/2} + \frac{\pi}{8} \sigma R^5 \overline{C}_D \right],$$
(12.27)

$$\frac{P}{\sigma_1 \omega^3} = c_1 \left(\frac{W}{\sigma_1 \omega^2} \right)^{3/2} + c_2 \overline{C}_D,$$
(12.28)

where c_1 and c_2 are two constants. Now the referred power is clearly dependent on the $\overline{C}_D$. The tip effects can be assessed from the variation of referred power at constant referred weight.

12.4 VERTICAL CLIMB

From the axial momentum theory we can calculate the basic requirements for the helicopter climb in vertical flight. If v_c is the climb rate, the mass flow rate through the disk is

$$\dot{m} = \rho A (v_c + v_i).$$
(12.29)

If we use the same considerations done for the propeller in axial flight (§5.9.1), we find that the relationship between the downwash velocity w and the induced velocity at the disk is $w = 2v_i$. The rotor thrust is equal to the mass flow rate times the change in axial velocity, e.g.

$$T = \dot{m}w = 2\dot{m}v_i = 2\rho A(v_c + v_i)v_i.$$
(12.30)

For a climb at constant v_c, we have $T \simeq W$. Solving Eq. 12.30 in terms of the induced velocity, we find

$$v_i^2 + v_c v_i - \frac{W}{2\rho A} = 0,$$
(12.31)

$$v_i = -\frac{1}{2} v_c \pm \frac{1}{2} \sqrt{v_c^2 + \frac{2W}{\rho A}}.$$
(12.32)

Only the positive solution applies to the present case. This solution is valid for all climb rates. The corresponding climb power is

$$P = T(v_c + v_i) = Tv_c + Tv_i.$$
(12.33)

The first term is the useful power; the second term is the induced power. We substitute Eq. 12.32 to find a suitable expression for performance analysis,

$$P = Wv_c + Wv_i = Wv_c + \frac{1}{2} W \left(-v_c + \sqrt{v_c^2 + \frac{2W}{\rho A}} \right),$$
(12.34)

$$P = \frac{1}{2} Wv_c + \frac{W}{2} \sqrt{v_c^2 + \frac{2W}{\rho A}}.$$
(12.35)

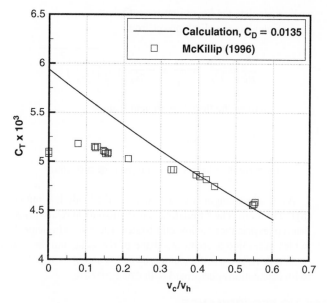

Figure 12.10   *Axial climb performance of four-bladed rotor. Reference data from McKillip[366]. Calculation parameters: $d = 4.88\,m$; rpm $= 215$, $\sigma = 0.0663$; $\theta_o = 9.3$ degrees; $\overline{C}_D = 0.0135$.*

At very small climb rates, $v_c^2 \ll 1$, we have

$$P \simeq \frac{1}{2}Wv_c + \sqrt{\frac{W^3}{2\rho A}} = \frac{1}{2}Wv_c + P_h. \tag{12.36}$$

This is an important result: at low climb rates, the climb power is the sum of the hover power and half the power required to lift the rotor at the same climb rate.

The maximum climb rate in vertical flight is found by solving Eq. 12.34 in terms of v_c with the maximum engine power in place of P. To the power calculated from Eq. 12.34 we need to add the profile power. We assume that there are no substantial changes compared to the hovering case, therefore we use Eq. 12.8.

Experimental data on axial performance of a rotor are available in McKillip[366], and include data for both climb and descent. These refer to a four-bladed rotor, with a NACA 0015 blade section. The tip speed is $U_{tip} = 55$ m/s, which corresponds to a Reynolds number 2.4×10^5, the same as Knight and Hefner[361]. Figure 12.10 shows a comparison between calculated and reference data in axial climb. The C_D was taken from Knight and Hefner[361] at an average effective angle of attack of 5 degrees. The basic calculation procedure is the following.

Computational Procedure

1. Read the rotor's parameters.
2. Set the climb rate, v_c.
3. Calculate the hover power, P_h, using Eq. 12.2 and Eq. 12.8.

4. Calculate the hover power coefficient, C_{P_h}.

5. Calculate the climb power from

$$\frac{P}{P_h} = \frac{C_P}{C_{P_h}} = \frac{v_i}{v_h} = -\frac{v_c}{2v_h} + \sqrt{\left(\frac{v_c}{2v_h}\right)^2 + 1}.$$

6. Calculate the thrust coefficient from

$$C_T = C_P \, U_{tip}.$$

7. Increase the climb rate and repeat the procedure.

As an afterthought, we can calculate the blade loading coefficient C_T/σ, and the figure of merit. The simulation is realistic only at relatively high climb rates, and falls short of expectations at low climb rates. The results indicate that a more refined model, which takes into account the shape of the wake and the non-uniform downwash, is required. This was further shown by Felker and McKillip[367], who compared the results with calculations performed with CAMRAD/JA.

12.5 CEILING PERFORMANCE

The hover ceiling *Out of Ground Effect* (OGE) is reached when the power required by hover is matched by the maximum engine power, and the helicopter is far from the ground. The power required to hover (Eq. 12.13) increases with the flight altitude. The available engine power, instead, decreases with altitude. There will be a point at which a further increase in hover power cannot be matched by the engine. In first approximation the power of a gas turbine engine decreases with the altitude according to

$$P_e \simeq P_{e_o} \delta \sqrt{\theta} \simeq P_{e_o} \sigma_1^{1.35}, \tag{12.37}$$

In Eq. 12.37 δ is the relative pressure, θ is the relative temperature and σ_1 is the relative density. This result is derived from the expression of the corrected fuel flow

$$\dot{m}_f = \frac{\dot{m}_{f_o}}{\delta \sqrt{\theta}} = SFC \, P. \tag{12.38}$$

Therefore, the ceiling condition is written as

$$P_{e_o} \delta \sqrt{\theta} = k \frac{W^{3/2}}{\sqrt{2\rho A}} + \frac{1}{8} \overline{C}_D \rho \sigma A U_{tip}^3 + P_{other}, \tag{12.39}$$

where P_{other} includes all other power components (tail rotor power, transmission losses, auxiliary systems). A closer inspection of the power components indicates that

$$P_h \sim \frac{1}{\sqrt{\sigma_1}}, \qquad P_o \sim \sigma_1, \qquad P_e \sim \sigma_1^{1.35}. \tag{12.40}$$

Therefore, the induced power increases; the profile power and the engine power decrease with the altitude. There are several methods to solve Eq. 12.39. The parameters required are the average profile drag $\overline{C}_D$ of the blade sections and the engine power. The easiest method is the one based on the analysis of the available excess power. The climb rate is found from the specific excess power

$$v_c = \frac{P_e - P_{req}}{W}. \tag{12.41}$$

The solution algorithm is straightforward:

Computational Procedure

1. Read the aircraft's data.
2. Set altitude, and calculate atmospheric parameters.
3. Calculate available engine power, Eq. 12.37.
4. Calculate hover power requirements, Eq. 12.39.
5. Calculate specific excess power, Eq. 12.41.
6. If $v_c > 0.5$ m/s (service ceiling), increase altitude and reiterate until the aircraft has reached the ceiling.

The procedure can be repeated for a selection of GTOW, disk loading, atmospheric conditions, blades aerodynamics, rotor solidity, engine power, OEI, etc. A parametric study of Eq. 12.39 allows to study the best configuration for high-altitude hover at a fixed AUW (see Problem 3). A typical solution is shown in Figure 12.11.

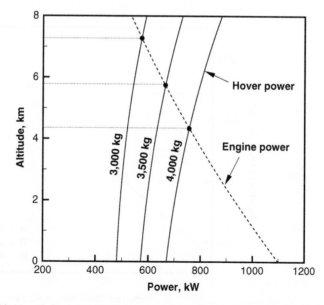

Figure 12.11   *Calculated OGE hover performance for the main rotor of reference (aircraft Model D) at the take-off weights indicated; standard conditions.*

For Category A helicopters, the FAR regulations (Part 29) specify that in the critical take-off configuration the steady OGE climb rate, 200 feet above the ground, must be at least 100 feet/min (0.5 m/s) at all weights, altitudes, and temperatures. The operating engine may use the 2 minute OEI power (generally called *emergency power*).

Practical values of the OGE hover ceiling are in the range of 5,000 m (16,400 feet) for most conventional helicopters. The OGE ceiling record is held by the Aérospatiale SA-315 *Lama*, at an altitude of 12,442 m (40,820 feet, June 1972). The helicopter was powered by a single Turbomeca engine delivering 650 kW; it had a tubular steel frame and a Plexigas cabin, to reduce its OEW. The ultimate take-off and landing record has been established by a Eurocopter AS350 B3 *Ecureil*, which landed on top of Mount Everest (8,850 m, 29,035 feet) in May 2005, although this achievement has been disputed. This helicopter is certified as a single-engine aircraft, and is powered by a Turbomeca Arriel B2 gas turbine.

The non-dimensional climb performance is specified by the relationship

$$\frac{P}{\sigma_1 \omega^3} = f\left(\frac{W}{\sigma_1 \omega^2}, \frac{v_c}{\omega}, h, \frac{\omega}{\sqrt{\theta}}\right). \qquad (12.42)$$

There are several ways of plotting the non-dimensional performance. For example, for a selected number of AUW we can plot the referred weight versus the referred rate of climb – at a fixed altitude; or we can plot the referred power versus the referred weight at constant referred climb rates. Figure 12.12 shows the referred power versus the referred climb rate at selected AUW.

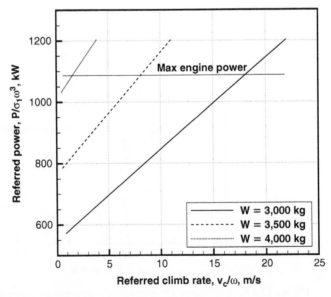

Figure 12.12 *Non-dimensional analysis of vertical climb (aircraft model D) at the AUW indicated; standard conditions at altitude h = 0.5 km.*

12.6   GROUND EFFECT

The hover ceiling *In Ground Effect* (IGE) is somewhat different than the OGE ceiling calculated earlier. The ground proximity affects the downward movement of the slip-stream, and creates conditions that prevent the system of vortices traveling away from the rotor disk. In fact, under some conditions, compounded by atmospheric winds, the wake interacts strongly with the blades.

Physically, the downwash is blocked by the ground, therefore part of the flow may bounce back onto the rotor disk and spread outwards, as a fountain. Curtiss *et al.*[368,369] described a series of complex flow states in ground effect that are dependent on both clearance and forward speed. The jet-fountain effect is experienced at low or zero advancing speeds; by contrast, a horseshoe vortex from the blade tips is experienced in fast forward flight, similar to a free flight condition. At intermediate speeds, a series of wake patterns is evidenced. Some numerical studies on ground-effect wakes include that of Brown and co-workers[370]. A comprehensive theoretical analysis based on a modified momentum theory is due to Heyson[371].

From the point of view of general performance, it is useful to have semi-empirical correlations that provide changes in power and thrust for take-off, approach and landing operations. Over the years, experimental and theoretical results have produced a set of criteria (not always in agreement) that allow for sensible corrections to the rotor power. These include the formulas given by Betz[372], Cheeseman and Bennett[373], Hayden[374], and Knight and Hefner[362]. The data point out that the ground effect becomes noticeable at about one rotor diameter from the ground. It increases to the minimum ground clearances commonly encountered by helicopter rotors, $h/d \simeq 0.25$.

Cheeseman and Bennett[373] provided an equation that is simple and useful. At a constant engine power output, the rotor thrust is found to be increasing with the rotor moving closer to the ground. However, it is more useful to have a constant thrust, because this force balances out the weight of the aircraft. At constant thrust, the power required to hover decreases with the rotor approaching the ground. The correction to the rotor power, to account for the ground proximity, is done via the coefficient k_g, defined by

$$\left(\frac{P_{IGE}}{P_{OGE}} \right)_{T=const} = k_g = \frac{1}{1 + d/8h},$$ (12.43)

which is found valid at ground clearances $h/D > 1/4$. Hayden proposed the formula

$$\left(\frac{P_{IGE}}{P_{OGE}} \right)_{T=const} = \frac{1}{0.9926 + 0.03794(d/h)^2},$$ (12.44)

to correlate some experimental data, as shown in Figure 12.13. The figure also shows some data in closer proximity to the ground, obtained by Fradenburgh[375] at Sikorsky. The relationship between rotor diameters needed to hover at the same AUW is found from

$$\frac{P_{IGE}}{P_{OGE}} = \frac{d_{OGE}}{d_{IGE}}.$$ (12.45)

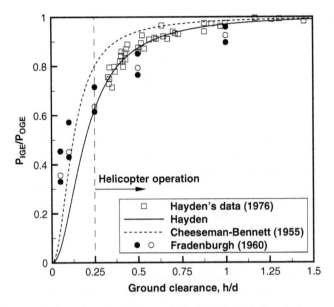

Figure 12.13 *Reference data of Hayden and Fradenburgh for a helicopter rotor in ground effect, compared with two semi-empirical correlations, as indicated. Fradenburgh's data are at three values of C_T/σ.*

The condition of hover ceiling in ground proximity is a correction of Eq. 12.39

$$P_{e_o}\delta\sqrt{\theta} = k_g\left(k\frac{W^{3/2}}{\sqrt{2\rho A}} + \frac{1}{8}\overline{C}_{D_o}\rho\sigma A U_{tip}^3\right) + P_{other}. \tag{12.46}$$

where k_g is a suitable ground correction factor. Note that this factor multiplies both the induced and the profile power. This is because the empirical power relationships refer to the total hover power. The solution procedure follows the method of page 328. Figure 12.14 shows calculated IGE and OGE hover performance of the reference helicopter given in Appendix A (aircraft D). The results were obtained with a constant tip speed $U_{tip} = 230$ m/s, $\overline{C}_{D_o} = 0.015$, and a constant additional power equal to 12% of the hover power, to account for transmission losses and tail rotor power.

12.7 VERTICAL DESCENT

Vertical descent is not the normal operation of the helicopter, because a smooth flight is difficult to maintain. The pilot will always try to descend and land in oblique flight, as far as it is possible. In fact, the rotor advances in the same direction as the slipstream and interacts with a complicated vortex system. In general, vertical descent can be powered or unpowered. In the latter case the helicopter operates in autorotation. This problem is described separately.

The use of the momentum theory is necessarily contrived, because in its basic form it only gives an average value of the induced velocity at the disk; in reality, the flow below the disk can be directed both upwards or downwards. The rotor flow is always

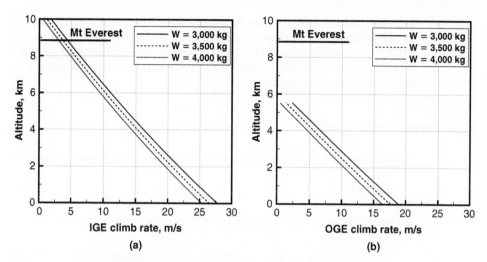

Figure 12.14 *Estimated IGE and OGE hover ceiling for the reference helicopter (model D) at three reference weights.*

unsteady. For a powered climb or descent, the power relative to the hover case is

$$\frac{P}{P_h} = \frac{v_c + v_i}{v_h} = \frac{v_c}{v_h} + \frac{v_i}{v_h}.$$ (12.47)

The first terms represents the power required to change the altitude of the aircraft; the second term is the induced power, and represents a loss of useful energy. If the descent velocity is relatively large, e.g. $|v_c/v_i| > 2$, a solution can still be found from the momentum theory. In this case it is possible to prove that the relationship between the induced velocity at the disk and the downwash w is the same as before: $w = 2v_i$. Furthermore, the thrust is

$$T = -\dot{m}w = -2\rho A(v_c + v_i)v_i,$$ (12.48)

where the sign $-$ indicates that the average flow has changed direction: when the rotor sinks at a relatively high rate, the flow through the disk is from below. Simplification of Eq. 12.48 yields

$$v_i^2 + v_c v_i + \frac{T}{2\rho A} = 0.$$ (12.49)

The solutions are

$$v_i = -\frac{v_c}{2} \pm \frac{1}{2}\sqrt{v_c^2 - \frac{2T}{\rho A}},$$ (12.50)

or in a non-dimensional form,

$$\frac{v_i}{v_h} = -\frac{v_c}{2v_h} \pm \sqrt{\left(\frac{v_c}{2v_h}\right)^2 - 1}.$$ (12.51)

Since the climb rate is negative and the induced velocity is positive, only the negative solution is legitimate. In any case, the term under square root implies that $v_c/2v_h \geq 1$, which is the result that we have anticipated. When this case is verified, the rotor is said to operate in the *windmill brake state*; energy is extracted from the air to maintain the rotation, like in a wind turbine.

12.8   HOVER ENDURANCE

We have seen in Chapter 5 that the fuel flow for a power plant delivering a power P is

$$m_f = SFC\, P. \tag{12.52}$$

The weight loss due to fuel consumption can be written as

$$\frac{dW}{dt} = -\frac{dW_f}{dt} = -g\,\dot{m}_f = -g\,SFC\,P = -g\,SFC\,\rho A(\Omega R)^3 C_P. \tag{12.53}$$

We have the factor g in order to maintain the convention used for the fixed-wing aircraft that the fuel flow is given in units of mass per unit of time. Since in hover the thrust is equal to the weight, we also have

$$\frac{dW}{dt} = -\frac{dT}{dt} = \rho A(\Omega R)^2 \frac{dC_T}{dt}. \tag{12.54}$$

If we combine Eq. 12.53 with Eq. 12.54 we find

$$\frac{dC_T}{C_P} = -g\,SFC\,(\Omega R)dt. \tag{12.55}$$

Integration of Eq. 12.55 between an initial state "i" and a final state "e" yields the hover endurance

$$E = \frac{1}{g\,SFC}\,\frac{1}{(\Omega R)}\int_e^i \frac{dC_T}{C_P}. \tag{12.56}$$

The factor g disappears if the fuel flow is given in units of weight per unit of time. Solution of this integral requires knowledge of the relationship between the power and thrust coefficients. This is generally found by a numerical solution. For example, we can use the simulated data of Figure 12.7. The performance is sometimes given in charts, using as a parameter the C_T (see, for example, Makofski[376]). For a given engine, and for a given rotational speed, the endurance is only a function of the rotor performance, and is quantified by the integral of dC_T/C_P.

To conclude this chapter, it is interesting to discuss briefly the vertical flight of the hummingbird. This is the only bird known to hover steadily, and to convert its flight to forward, backward and upside-down. High-speed photography has indicated that hover is performed by a complex wing beat, at a rate of about 80 cycles per second. Vertical lift is created with an almost rigid movement of the wings that trace a figure of 8. Lift is created during both the downstroke and the upstroke. However, the amount of energy expended for hovering is very high. It has been estimated that the energy output per unit of mass would be about ten times that produced by a fit athlete running a marathon at a speed of 15 km/h. Relevant studies in this field include Weis-Fogh[377], along with the book by Greenwalt[378]. Hummingbirds have been observed hovering at altitudes above 4,300 m (14,100 ft) in the Peruvian Andes, in spite of the reduced air density and limited oxygen availability (Altshuler and Dudley[379]). For its amazing vertical flight capabilities, the hummingbird has become the logo of the American Helicopter Society.

PROBLEMS

1. Calculate the difference in power required to take a passenger from the ground to the top of a 200-m-tall building by helicopter and by elevator. Consider that the passenger (payload) has a weight of 85 kg. Assume that the climb rate will be the same for the helicopter and the elevator. For the helicopter, take the data of the reference model in Appendix A.

2. Calculate how the power requirements for hover out of ground change with altitude. Consider a helicopter with a gross weight $W = 2,000$ kg, ISA conditions. Plot the ratio P/P_{sl}, e.g. the power required to hover relative to sea level conditions, and discuss the practical limits of operation. In first analysis consider only the induced power; then add the profile power of the rotor. Finally, discuss if there is a market for high-altitude helicopter operations.

3. Discuss how the hover ceiling out-of-ground effect can be increased. Write the equation of the power requirements in hover and match it to the engine power. Assume that the gas turbines deliver a power at altitude defined by Eq. 12.37. Discuss how the OGE ceiling changes with the engine power, the rotor diameter, the rotor solidity and the profile drag coefficient. Assess which parameter has the largest effect. Consider an aircraft gross weight $W = 2,000$ kg.

4. Find an analytical expression for the ratio between induced power and induced velocity ratio P_h/λ_h for a helicopter in hover out-of-ground effect. The parameter λ_h is defined by $\lambda_h = v_h/\Omega R$. Then consider the helicopter of reference (Appendix A), with a gross weight $W = 3,000$ kg, sea level, ISA conditions, and calculate the value of P_h/λ_h.

5. Consider a conventional helicopter, with a main rotor rotating anticlockwise, and a tail rotor rotating clockwise. The tail rotor is installed at the end of a thin tail boom. Study the effectiveness of a pusher versus tractor tail rotor, by considering the various inflow conditions created by the interaction between main and tail rotor, when the helicopter is hovering. Consider side winds at ± 45 degrees and ± 90 degrees. Study how these performances may limit the sideways speed of the helicopter.

6. For the helicopter model reported in Appendix A (aircraft D, general utility helicopter), calculate the hover ceiling *In-Ground Effect* (IGE), by using the same considerations as in §12.1 and the same values of the relevant parameters. In particular, consider that the rotor plane is placed at a distance of 3.0 m from the ground. Use Eq. 12.46 and perform a parametric study aimed at increasing the IGE hover ceiling. Also find the rotor diameter that gives maximum IGE ceiling at the design rotor solidity σ.

7. Consider a rotor with N blades, with a constant chord c and a radius R. The rotor is in hover with a rotational speed Ω. Assume also that the blade's C_L and C_D are constant along the radial direction, with $C_L \gg C_D$. Calculate the thrust and power from integration of the aerodynamic forces, and find the corresponding non-dimensional coefficients C_T/σ and C_P/σ in terms of the aerodynamic coefficients of the blade.

8. Calculate the thrust and torque coefficient C_T and C_P of the CH-47 rotor (data in Table A.17 in Appendix A) for a range of collective pitch $\theta_o = 1$ to 12 degrees. Use the axial momentum theory. Plot the data in the form shown in Figure 12.4. Assess whether the approximation $C_T \propto \theta_o^2$ is acceptable.

Chapter 13
Rotorcraft in Forward Flight

The vehicle appeared so reluctant to fly forward that we even considered turning the pilot's seat around and letting it fly backward.

Igor Sikorsky, speaking about his helicopter VS-300, 1940

Take-off and stable hover flight took the helicopter pioneers a few decades to achieve. The conversion from hover to forward flight, although not as straightforward as some media imagined, required more innovative ideas. The invention of the swashplate, the cyclic pitch and the direct moment control by shaft tilt proved to be essential for the practical development of the rotary-wing aircraft. Basically, the tilting of the main rotor provides a thrust component on the horizontal plane. This force component can be used for advancing the aircraft, but then new trim and stability issues arise. On one occasion during trial flights of the VS-300, Sikorsky rebutted the press by saying "that is a minor problem we have not solved yet".

In this chapter we derive suitable expressions for the power P required for the helicopter to fly level at a fixed altitude. Knowledge of the function $P(U)$, or $P(\mu)$, will allow us to carry out a number of other performance calculations. A detailed analysis of the tail rotor is also presented.

13.1 ASYMMETRY OF ROTOR LOADS

The torque reaction problem described earlier is general to all flight conditions. When the rotor is in forward flight, the inflow changes as the blades rotate. As they rotate, the blade sections are subject to sinusoidal changes of the in-plane velocity and the effective angle of attack.

Consider the blade at azimuth $\psi = 90$ and $\psi = 270$ degrees, as indicated in Figure 13.1. In the first case the blade is advancing; therefore, it generates a relatively large lift force. After travelling half a revolution, the blade will find itself in a *retreating* position, and it generates considerably less lift, unless some pitch corrections are applied. One can assume that the forcing deriving from this rotation has a 1/rev frequency. If the rotor disk is inclined by an angle α with respect to the forward speed U_∞, the nominal inflow at a radial position y is found from

$$U(\psi, y) = \Omega y + U_\infty \cos \alpha \sin \psi. \tag{13.1}$$

The total in-plane velocity experienced by the blade section is a sum of the forward in-plane velocity and the rotational velocity. The combination of these components creates a region of *reverse flow*. With this term we indicate that the blade sections

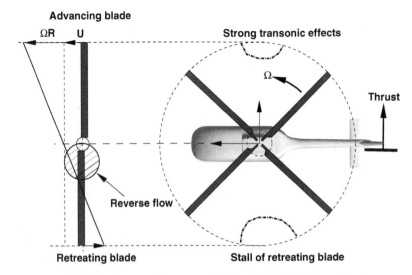

Figure 13.1 *Asymmetric blade loads for helicopter in level forward flight.*

face the free stream with their trailing edge. The amount of reverse flow is governed by the advance ratio μ, Eq. 11.3. The locus of reverse flow is found from

$$\Omega y + U_\infty \cos \alpha \sin \psi = 0, \tag{13.2}$$

and hence

$$r = \frac{y}{R} = -\mu \sin \psi. \tag{13.3}$$

This radius increases with the advance ratio.

13.2 POWER REQUIREMENTS

The total power required for forward flight is the sum of the main rotor power, the tail rotor power, the power required to run the auxiliary systems, and a number of power losses. The power share between the rotors and the auxiliary systems depends on the flight condition. The total power required to fly forward is

$$P = P_i + P_o + P_p + P_{tr} + P_t, \tag{13.4}$$

where P_i is the induced power, P_o is the profile power of the blades, P_p is the power required to overcome the drag of the airframe, P_{tr} is the tail rotor power, and P_t is the power due to transmission losses. We will not consider the power of the auxiliary systems.

We will assume that the overall drag of the helicopter at trim conditions can be calculated without taking a detailed look at the various components and interference factors, and we set the problem of finding the power required for level flight of the given helicopter.

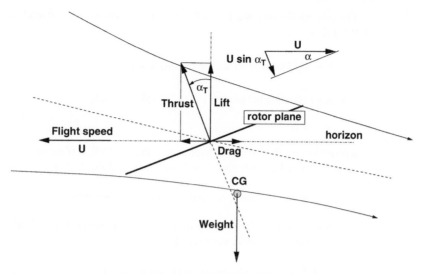

Figure 13.2 *Nomenclature and conventions for forward flight.*

13.2.1 Induced Power

The calculation of the induced power in forward flight requires full knowledge of the momentum theory, a few tricks, and empirical corrections with flight data. Since we are mostly concerned with the performance of the helicopter, we cannot derive all these expressions in the present context. For further details, we refer to the specialized literature, in particular Leishman[324], Stepniewski[110] and Johnson[354].

When the rotor disk is inclined by an angle α with respect to the true air U speed, the total velocity at the disk will be

$$U = \left[(U \cos\alpha)^2 + (U \sin\alpha + v_i)^2\right]^{1/2}. \tag{13.5}$$

In level forward flight, the angle of attack of the rotor is equal to the tilt angle, $\alpha = \alpha_T$, as shown in Figure 13.2. This equation was first derived by Glauert, who assumed that the mass flow rate through the disk has essentially the same expression as in axial flight,

$$\dot{m} = \rho A U. \tag{13.6}$$

Equation 13.5 has the correct asymptotic limits in hover and high-speed flight. Amazingly, this simple assumption shows coherence with flight measurements, first done by Brotherhood and Stewart[380] on the Sikorsky R-4B, using the smoke filament technique.

The next problem is to find an expression for the induced velocity v_i, as it was done in the case of the rotor in hover (or propeller in axial flight). If w is the downwash velocity created by the disk (e.g. the velocity component normal to the disk far downstream), the momentum equation in the direction normal to the rotor disk is

$$T = \dot{m}(U \sin\alpha + w) - \dot{m}U \sin\alpha = \dot{m}w. \tag{13.7}$$

The value of the downwash w is found from the conservation of total energy through the disk,

$$P = T(U \sin \alpha + w) = \frac{1}{2}\dot{m}(U \sin \alpha + w)^2. \tag{13.8}$$

This is an equation between v_i and w. Its solution yields

$$w = 2v_i, \tag{13.9}$$

a result known from the axial momentum theory, §5.9.1. If we replace Eq. 13.9, the mass flow rate Eq. 13.6, and the value of the speed Eq. 13.5 in Eq. 13.7, we find

$$T = \dot{m}w = (\rho A U)2v_i = 2\rho A v_i \left[(U \cos\alpha)^2 + (U \sin\alpha + v_i)^2\right]^{1/2}. \tag{13.10}$$

The thrust in Eq. 13.10 is replaced by the thrust in hover, Eq. 12.2. This is required to provide the lift force to the aircraft

$$2\rho A v_h^2 = 2\rho A \, v_i \left[(U \cos\alpha)^2 + (U \sin\alpha + v_i)^2\right]^{1/2}. \tag{13.11}$$

The latter equation is a relationship between the induced velocity in forward flight, v_i, and the induced velocity in hover, v_h. Solving in terms of v_i, it yields

$$v_i = \frac{v_h^2}{\sqrt{(U \cos\alpha)^2 + (U \sin\alpha + v_i)^2}}, \tag{13.12}$$

which is in implicit form. Before attempting to solve this equation, it is convenient to normalize the velocities. We define the induced velocity ratio in hover and level forward flight ($\alpha = \alpha_T$), respectively,

$$\lambda_h = \frac{v_h}{\Omega R}, \quad \lambda = \frac{U \sin\alpha + v_i}{\Omega R}, \tag{13.13}$$

so that Eq. 13.12 becomes

$$v_i = \frac{\lambda_h^2}{\sqrt{(U \cos\alpha)^2/(\Omega R)^2 + \lambda^2}}. \tag{13.14}$$

Finally, we introduce the advance ratio (Eq. 11.3) to eliminate the forward speed from Eq. 13.14. The result is

$$v_i = \frac{\lambda_h^2}{\sqrt{\mu^2 + \lambda^2}}. \tag{13.15}$$

Equation 13.15 is useful for the calculation of the induced power of the isolated rotor in forward flight. In fact, the forward flight induced power for the isolated rotor is

$$P = TU \sin\alpha_T + Tv_i. \tag{13.16}$$

The first term in Eq. 13.16 is the power required to advance at a speed U; the second term is a form of loss, due to the field of induced velocities around the rotor (the downwash field). It will be useful to work in terms of the ratio between the power in forward flight and the power in hover (already calculated). Therefore, we divide Eq. 13.16 by P_h, which is given by Eq. 12.2,

$$\frac{P}{P_h} = \frac{TU \sin \alpha_T + Tv_i}{Tv_h} = \frac{U \sin \alpha_T + v_i}{v_h} = \frac{(U \sin \alpha_T + v_i)/\Omega R}{v_h/\Omega R} = \frac{\lambda}{\lambda_h}.$$

(13.17)

Now introduce Eq. 13.15 in Eq. 13.17, along with the definition of advance ratio (Eq. 11.3),

$$\lambda = \mu \tan \alpha_T + \frac{\lambda_h^2}{\sqrt{\mu^2 + \lambda^2}}.$$

(13.18)

The ratio between induced powers in forward flight and hover is

$$\frac{P}{P_h} = \frac{\lambda}{\lambda_h} = \frac{\mu \tan \alpha_T}{\lambda_h} + \frac{\lambda_h}{\sqrt{\mu^2 + \lambda^2}}.$$

(13.19)

The procedure for calculating the induced power in forward flight is the following. If the angle of attack of the rotor is assigned, then Eq. 13.18 can be solved for the unknown λ. In the practical flight environment, the tilt and the rotor thrust depend on the collective and cyclic pitch, and can be controlled directly by the pilot. Therefore, α_T can change in flight as a function of the speed.

The collective pitch stick moves up and down, pivoting about the aft end, through a series of mechanical linkages. The amount of movement of the stick determines the amount of blade pitch change. The effective angle of attack α_b of the blades changes as the pitch is adjusted. Since the drag D is proportional to α_b^2, an increase or decrease in α_b leads to an increase or decrease in drag, and hence to an increase or decrease of the engine rpm. However, the rpm must be maintained constant, so there must be a way to make changes in engine power output as the blades drag changes. This coordination of power with blade pitch is controlled through a collective pitch lever/throttle linkage, that automatically increases power when the collective pitch is increased, and decreases power when the collective pitch is decreased.

There is no easy way to evaluate α_T from the power. The solution of the problem will be based on the solution of Eq. 13.18 with a given α_T, and on the iterative adjustment of α_T so that the vertical and horizontal equilibrium of the aircraft are satisfied. This is a basic trim operation.

The solution of the non-linear Eq. 13.18 is done iteratively, by using a Newton method. An analysis of the derivative $\partial \lambda / \partial \alpha$ indicates that $\partial \lambda / \partial \alpha$ is positive and nearly constant in the range of most likely angles of attack of the rotor (up to 30 degrees forward). The task is to find the root of the residual

$$f(\lambda) = \lambda - \mu \tan \alpha_T - \frac{\lambda_h^2}{(\mu^2 + \lambda^2)^{1/2}} = 0.$$

(13.20)

In this case the function derivative with respect to the independent variable λ can be calculated analytically:

$$\frac{\partial f}{\partial \lambda} = 1 + \frac{\lambda_h^2 \lambda}{(\mu^2 + \lambda^2)^{3/2}}. \tag{13.21}$$

In summary, we have the following algorithm.

Computational Procedure

1. Assign the tilt angle.
2. Solve Eq. 13.18 for the induced velocity ratio λ. For this purpose, use a Newton-Raphson procedure. Start from a first-guess value $\lambda = \lambda_h$. At iteration i update the values of λ according to

$$\lambda_i = \lambda_{i-1} - \left[\frac{f(\lambda)}{(\partial f / \partial \lambda)} \right]_{i-1}.$$

The calculation of the residual is done from Eq. 13.20 and the function derivative is taken from Eq. 13.21.
3. Calculate the induced power from Eq. 13.19.
4. Increase the air speed, and repeat the calculation.

In Figure 13.3 we show the calculated induced power at three constant tilt angle settings, with the aircraft at a gross weight $W = 3,500$ kg flying at sea level conditions.

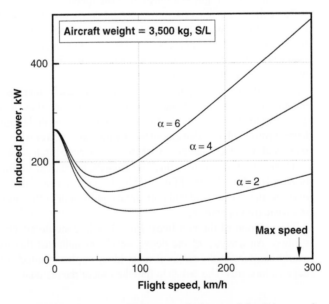

Figure 13.3   *Helicopter power (kW) versus flight speed (km/h) at sea level, for the tilt angles indicated (helicopter model D, Appendix A).*

However, this is the rotor power alone, and does not even account for the trim constraints. For the helicopter as a whole we must take into account the other sources of power, as written in Eq. 13.4.

13.2.2 Blade Profile Power

The profile power is calculated as in the case of pure hover. Consider a blade of constant chord c. The blade profile drag of a blade element at radial position y in a free stream of speed U is

$$dD = \frac{1}{2}\rho dA \, \overline{C}_D U^2(y), \tag{13.22}$$

where the element of blade area is $dA = c\,dy$, and the local blade speed can be expressed in terms of the advance ratio $\mu \simeq U/\Omega R$. The velocity component of interest is the one along the chord of the blade element, therefore

$$U(y) = \Omega y + U \sin \psi = \Omega y + \mu \Omega R \sin \psi = \Omega R \left(\frac{y}{R} + \mu \sin \psi \right). \tag{13.23}$$

Normalize the radial coordinate by introducing the definition $r = y/R$, with a differential $dr = dy/R$. With this new parameter, the contribution of the blade element to the profile power is

$$dP_o = dDU = \frac{1}{2}\rho \, cdy \overline{C}_D U^3(r) = \frac{1}{2}\rho c \, \overline{C}_D \Omega^3 R^3 \, (r + \mu \sin \psi)^3 \, Rdr. \tag{13.24}$$

To calculate the total profile power due to N blades we need to integrate Eq. 13.24 over the radius and for a full rotation 2π of the rotor.

$$P_o = \int_0^{2\pi} \int_0^R NdP_o = \frac{1}{2}N\rho c \, \overline{C}_D \, \Omega^3 R^4 \int_0^{2\pi} \int_0^1 (r + \mu \sin \psi)^3 dr \, d\psi. \tag{13.25}$$

The solution of the integral is[*]

$$\int_0^{2\pi} \int_0^1 (r + \mu \sin \psi)^3 \, dr \, d\psi = \frac{\pi}{2} \left(1 + 3\mu^2 \right). \tag{13.26}$$

The profile power is found from

$$P_o = \frac{\pi}{4} N c \rho \, \overline{C}_D \, \Omega^3 R^4 (1 + 3\mu^2). \tag{13.27}$$

---

[*] For the solution we need to expand the terms under the power. The integrals of the odd powers of $\sin \psi$ are identically zero.

Finally, with the introduction of the rotor solidity, $\sigma = Nc/\pi R$, the profile power becomes

$$P_o = \frac{1}{4}\rho\sigma\overline{C}_D\,\Omega^3 R^3 (1 + 3\mu^2),\qquad(13.28)$$

which corresponds to a power coefficient

$$C_{P_o} = \frac{P_o}{\rho A\Omega^3 R^3/2} = \frac{\sigma\overline{C}_D}{8}(1 + 3\mu^2).\qquad(13.29)$$

There are different corrections to improve on Eq. 13.29. Johnson[354] proposed to calculate the profile power from

$$C_{P_o} = \frac{1}{8}\sigma C_{L_\alpha}\left(1 + 3\mu^2 + \frac{3}{8}\mu^4\right),\qquad(13.30)$$

to take into account the effects of reverse flow on the retreating blade.

From Eq. 13.29 the power increases parabolically with μ. If we calculate the profile power of the reference helicopter, we find a first-order estimate between 180 and 230 kW, depending on the C_D. This power amounts to over 20% of the installed power at sea level. The effects of rotor radius and the rotational speed on the profile drag can be relatively high (Problem 5). Although the profile power increases with U_{tip}^3, its contribution is small compared to the drag of the airframe.

13.2.3 Compressibility Effects

Substantial compressibility effects appear on the advancing blades, when the Mach numbers of the inflow velocity exceed the point of transonic drag divergence. As a matter of fact, compressibility effects start to become noticeable at high subsonic Mach numbers, say about $M = 0.5$ to 0.6. These speeds are generally shock-free, or with weak shocks.

The shape of the airfoil section is essential in producing steady and unsteady lift, drag and pitching moment on the blade section, in addition to more local features, such as trailing-edge separation, boundary layer/shock wave interaction, and noise emission. Airfoils for rotorcraft applications have benefited from progress in advanced design and optimization methods. Therefore, they can be designed *ad hoc*, even with multiple constraints. There are still a large number of helicopters using NACA airfoils, particularly NACA 0012 (about 1/3 of all helicopters), and NACA 23012 with modifications. These airfoils have a relatively low drag divergence point, compared to modern rotorcraft airfoils. Power savings with optimized airfoil sections can be in excess of 5%.

A typical transonic drag rise is shown in Figure 13.4, elaborated from wind tunnel data of the rotorcraft airfoil Vertol 23010-1.58. The slight decrease in profile drag at intermediate Mach number is associated to beneficial Reynolds number, which concurs in reducing the boundary layer thickness.

A calculation of the rotor's profile power can be done with the combined blade element momentum theory purpose, as described in Chapter 5. The method uses the

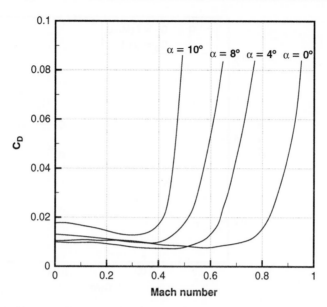

Figure 13.4 *Vertol 23010-1.58 drag divergence at selected angles of attack, and flight Reynolds numbers.*

tabulated data of the airfoil, and takes advantage of the transonic drag rise over a wide range of angles of attack. We have calculated the hover power for the isolated CH-47 rotor for a range of disk loading C_T/σ, as shown in Figure 13.5. Two sets of results are shown: one obtained with the actual aerodynamic characteristics of the Vertol blade sections, Figure 13.4; the other set of results was obtained with an estimated $\overline{C}_{D_o} = 0.01$. In both cases we assumed an $M_{tip} = 0.63$, well below the M_{dd} of this airfoil. The difference in the results is attributed to the choice of $\overline{C}_D$.

Nitzer and Crandall[381] studied the problem from the point of view of changes in the pressure rise over the suction side, and showed correlations between drag divergence and Mach numbers for several NACA airfoils. An additional summary of data is available in Abbott and von Doenhoff[122], who include the pitching moment divergence. Transonic data for more advanced rotor airfoils are available in Dadone[382], Bingham and Noonan[383] and Bousman[384].

Gessow and Crim[385] were the first to investigate these effects on the overall rotor power requirements. The empirical corrections proposed by these and other authors are perhaps a bit out of date, because simple numerical solutions, based on the combined blade element and axial momentum theory, with the aid of tabulated data, yield better solutions, at a comparatively low cost.

Experimental data showing the Mach number effects at high transonic speeds are available in Powell[386,387], Shivers and Carpenter[388] and Jewel[389]. An empirical correlation between $C_{L_{\max}}$ and M_{dd} for several rotorcraft airfoils is shown in Figure 13.6.

At low compressible speeds, the Prandtl-Glauert correction (Eq. 4.27) on the C_{l_α} can be applied. Figure 13.7 shows the *FoM* measured by Jewel at Mach numbers up to 0.98 on a rotor having a NACA 0012 section, a linear twist $\theta_1 = -10$ degrees, a solidity $\sigma = 0.099$, and a taper ratio 0.5. An estimate of the *FoM* for this rotor is also shown.

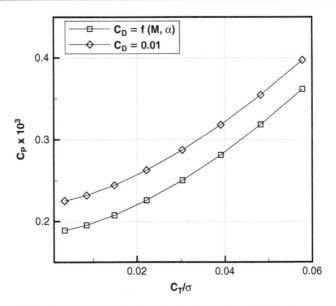

Figure 13.5 *Rotor power calculated with the combined blade element momentum theory for the CH-47 rotor in hover at sea level.*

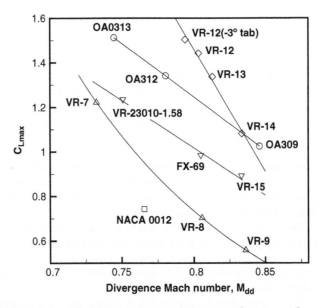

Figure 13.6 *C_{Lmax} for some common airfoil sections for rotorcraft application.*

13.2.4 Vehicle Drag

The aircraft's parasite drag includes the drag of the fuselage (with several non-streamlined subsystems) and the interference created by the rotor systems with the

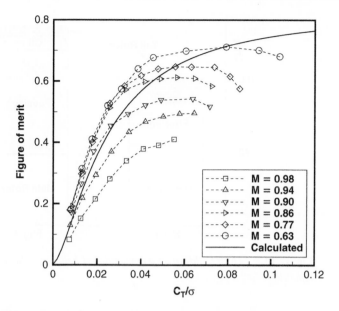

Figure 13.7 *Figure of merit at selected Mach numbers. Experimental data of Jewel[389] and comparisons with the present calculations. Drag coefficient used: $\overline{C}_D = 0.0087$, no compressibility correction.*

airframe (main rotor and fuselage, tail rotor and fuselage, main to tail rotor). However, the effective thrust must match the overall drag, and the engine power must be equal to Eq. 13.4. For a modern general utility or commercial helicopter, whose airframe is well streamlined, up to 40% of the drag derives from the airframe and its subcomponents; the remaining drag is attributed to the rotor system and the hub. Figure 13.8 shows a qualitative chart with the contribution of various factors at hover and cruise speed.

The airframe drag is one of the most important limiters to helicopter speed. A wide body of research exists on the subject, including the effects of up-sweep of the rear fuselage, angles of attack, scale effects, aerodynamic interference effects, external weapons, sponsons, undercarriage, tail surfaces, etc. A modern general utility helicopter like the Eurocopter EC-365 Dauphin has a streamlined body, and an enclosed tail rotor (*fenestron*). The experimental approach remains a tool of considerable importance, but computational methods are being developed that reduce the wind tunnel and flight costs. This is an open area of research.

Methods for optimal design of a helicopter fuselage are discussed by Wilson and Ahmed[390], who also provide a detailed bibliography. Seddon[391] discusses briefly some concepts of drag reductions, including the minimization of internal space, landing gear retraction, streamlining, etc. Stepniewski and Keys[110] provide examples of how the various components of drag (hub, landing gear, airframe) can be estimated, although the problem of interference is difficult to evaluate in general cases. Hoerner[392] in his essential book on aerodynamic drag provides data for a great variety of subsystems.

The drag of the aircraft depends on the geometry itself and on the angle of attack and side-slip. When the helicopter is flying at high speeds, its longitudinal axis

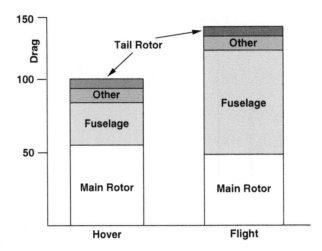

Figure 13.8 *Drag build-up on a conventional helicopter, hover and forward flight conditions (arbitrary scale).*

is inclined by a certain angle on the horizontal plane (-5 degrees is the practice), which creates additional shape/pressure drag. A particular case of airframe drag is the so-called vertical drag (*download*). This is the drag due to vertical flow from above. For a clean fuselage in vertical flow, Stepniewski and Keys[110] estimate a $C_D = 0.4$, which is essentially the C_D of a circular cylinder in cross-flow at high Reynolds numbers. The C_D increases to values above unity for complex geometries, including horizontal lifting surfaces, external weapons, etc.

Figure 13.9 shows the drag polar of three helicopter fuselages, characterized by different rear up-sweep angles, from Ahmed and Amtsberg[393]. The incidences extend up to ± 30 degrees. For one of the geometries the drag coefficient decreases with increasing angles of attack, up to the limit of the wind tunnel data (20 degrees). These and other wind tunnel results indicate that there is a relationship between the fuselage drag and its pitch attitude α. A suitable drag equation is

$$f_e = \frac{D}{q} = f_0 + f_1\,\alpha^2, \tag{13.31}$$

where f_0 is the zero-pitch equivalent flat plate area and f_1 is a coefficient that depends on the pitch angle and the tail plane; $q = \rho U^2/2$ is the dynamic pressure. This equation is conceptually important, because as the speed is increased, the aircraft must increase its pitch attitude for a proper trim. As the pitch attitude increases, the fuselage drag also increases. In general, power due to the vehicle's drag can be written as

$$P_p = \frac{1}{2}\rho A_{ref} C_{D_f} U^3, \tag{13.32}$$

where A_{ref} is a suitable reference area and C_{D_f} is the estimated drag coefficient of the aircraft. There are various methods to define the reference area, including the wetted area, the rotor disk area, the frontal area, etc. The use of the wetted area is somewhat dubious, because there are too many interference factors on the aircraft to

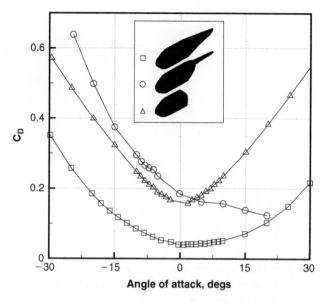

Figure 13.9   *Drag polars for three helicopter fuselages; DLR's wind tunnel data, adapted from Ahmed and Amtsberg[393]. Reference area unspecified.*

take into account. From Eq. 13.32 the airframe drag is proportional to the cube of the flight speed. We can see how this drag component increases to the point of being the dominant resistance at high speeds. The product

$$f_e = A_{ref} C_{D_f} \tag{13.33}$$

is called *equivalent flat plate area* (it has the dimensions of an area, m^2). In engineering analysis it is normalized with the rotor disk area, f/A. The only way to calculate the drag of the airframe is to find the value of the equivalent flat plate area, and this can only be done in the wind tunnel or in flight test measurements. A third way is to look at existing data from other helicopters in the same category and make a well-educated guess: f_e/A is found to be in range between $4 \cdot 10^{-3}\, m^2$ for clean designs and low gross weight, and $2.5 \cdot 10^{-2}\, m^2$ for heavy-lift helicopters. For a clean design at the weight range of our reference helicopter, the equivalent flat plate area is estimated at $f_e/A \sim 5 \cdot 10^{-3}\, m^2$.

On average, for the helicopter model reported in Appendix A, we will have about 30% of the total drag count from skin friction, 40% from the systems interference (main rotor, fuselage, tail rotor, hub), 10% from the landing gear; the remaining drag is due to all other causes. The rotor hub interference drag increases considerably at high speeds, therefore cowling systems, as in the design of the Eurocopter EC-365, are highly beneficial.

Modern methods for estimating the airframe drag are based on sophisticated computational methods (*computational fluid dynamics*), and rely on the prediction of the aerodynamics of the vehicle[394]. Even then, some allowance must be made for details of the airframe that cannot be easily modelled (gaps, probes, external stores, landing

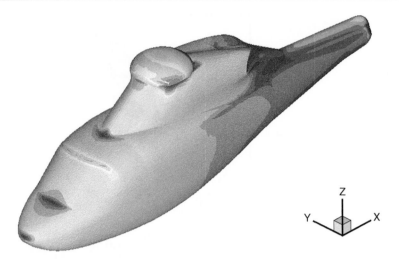

Figure 13.10   *Distribution of skin friction coefficient for the Eurocopter EC 365 Dauphin at $Re = 60 \cdot 10^6$ (based on the fuselage length). Fenestron and tail surfaces excluded. ISA conditions.*

gear, horizontal fin, etc.), and the additional information must be extracted from extensive wind tunnel testing. An example of such a computational model is shown in Figure 13.10. The integration of the skin friction coefficient and the surface pressure yield an estimate of the profile drag of the airframe. The wetted area can be readily computed from the model. The data found from the simulation are: $\overline{c}_f = 2.895 \cdot 10^{-3}$, wetted area $A_w \sim 31.9\,\mathrm{m}^2$ (this value does not include the tail surfaces). The calculated drag coefficient is $C_D \sim 0.0029$.

For a case of vertical flight, the calculations were performed under the assumption of constant true air speed, $U = 15\,\mathrm{m/s}$ for the full-scale fuselage. The result of the calculation indicates that there is a total vertical drag of the order of 450 N, or about 45 kg. For a GTOW $= 3,000\,\mathrm{kg}$, this corresponds to a download of about 1.5%, a relatively low value.

13.2.5   Interference Effect of the Airframe

The downwash velocity created by the helicopter rotor impinges on the upper part of the airframe and leads to an increased downward force, as described earlier. The short distance between the rotor disk and the airframe creates an interference effect, in which the aerodynamics of the rotor and of the airframe are mutually affected. Also, the downwash velocity is dependent on the radial position on the rotor disk, and the airframe is a three-dimensional body. For military helicopters with high disk loading, exposed surfaces (sponsons, external stores), the vertical drag can be as much as 10% of the aircraft's weight. This drag clearly limits the performance of the aircraft, particularly on hot days and at high altitude. For clear fuselages the loss does

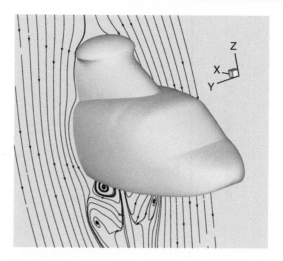

Figure 13.11   *Vertical flow on a plane normal to the longitudinal axis of the aircraft; the computed streamlines show massively separated flow with a constant downwash velocity of 15 m/s.*

not exceed 5%. Some light helicopters, namely Robinson R-22 and R-44, have a high mast that greatly reduces the rotor/airframe interference.

One simple (and very approximate) method to take into account this effect is to consider an average drag coefficient of the airframe in vertical motion, and correct the downward force by the additional drag. Therefore, the hover condition will become

$$T = W + D_v,$$ (13.34)

where D_v is the vertical drag, a function of the rotor thrust. One way to estimate the term D_v is to make an integration of the downwash below the rotor disk.

Extensive experimental research exists on the aerodynamic interference. For example, Balch *et al.*[395] have published a large study on rotor/fuselage/tail rotor interference for the Sikorsky UH-60 Black Hawk and S-76 models with different rotor blades. Berry and Bettschart[396] published experimental and numerical results on the interaction between the rotors and the fuselage of the Eurocopter EC-365N Dauphin (our reference model in Appendix A). The interaction between the T-tail and the main rotor wake on the Boeing-Sikorsky RAH-66 helicopter was studied by Gorton *et al.*[397]

13.2.6   Tail Rotor Power

In conventional helicopters, the tail rotor is needed to balance the rolling moment created by the main rotor in the different phases of flight. The tail rotor plane is set at a small angle with respect to the vertical plane, and also at a small angle with respect

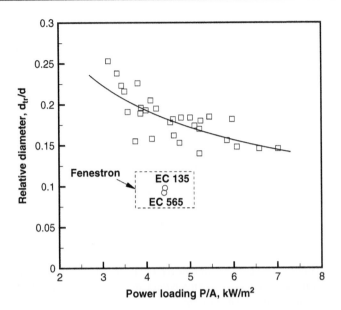

Figure 13.12 *Rotor diameter ratio versus power loading. Solid line denotes power fit, excluding the fenestron helicopters (dashed window).*

to the longitudinal plane of the helicopter. Therefore it gives a contribution to the overall thrust and the lift. The amount of thrust required by the tail rotor depends on the type of flight. It is highest during high-powered climb and high forward speed. It is relatively low at moderate-level flight speeds. A chart showing statistical data of rotor diameter ratios, d_{tr}/d, versus power loading is shown in Figure 13.12, elaborated from several production helicopters. The fenestron has a smaller diameter than the conventional tail rotor at comparable power loading.

If x_{tr} is the distance between the shafts of the main and tail rotor, the tail rotor thrust required for torque balance in a forward flight climb is

$$Q \simeq \frac{P_i + P_o + P_p + P_c}{\Omega} = x_{tr} \, T_{tr}, \tag{13.35}$$

from which we find

$$T_{tr} = \frac{P_i + P_c + P_o + P_p}{\Omega x_{tr}}. \tag{13.36}$$

From the analysis of Eq. 13.36, the tail rotor power increases with the thrust required for torque balance. Let us consider first the conventional tail rotor. The total thrust T_{tr} is derived by the action of the tail rotor blades, in the same way as the main rotor or an ordinary propeller. Part of the thrust is from the induced power; the remaining thrust is due to the action of the blades in the real viscous flow, hence to the action of lift and drag, as explained in the combined blade element theory. If we assume that T_{tr} is derived by the induced power, then

$$P_{tr} \simeq T_{tr}\, v_{tr}^i,$$ (13.37)

where v_{tr}^i is the average induced velocity on the tail rotor disk. In pure hover conditions, we find

$$v_{tr}^i = \sqrt{\frac{T_{tr}}{2\rho A_{tr}}},$$ (13.38)

where A_{tr} is the tail rotor's disk area. Thus, Eq. 13.37 becomes

$$P_{tr} \simeq \sqrt{\frac{T_{tr}^3}{2\rho A_{tr}}}.$$ (13.39)

We must take into account the power loss due to a number of factors, including the non-uniform loading, the tip losses, the presence of the vertical fin, and the influence of the main rotor. We collect all these exceptions into the factor k_{tr}, whose value is estimated at $k_{tr} = 1.20$ to 1.25. The effects of non-uniform inflow and tip losses were previously estimated at 15%. The fin itself has a blockage effect estimated at up to 20% in forward flight.

If we replace the torque balance condition in hover, Eq. 13.36, the rotor power becomes

$$P_{tr} \simeq \frac{k_{tr}}{\sqrt{2\rho A_{tr}}}\sqrt{\left(\frac{P_i + P_o}{\Omega\, x_{tr}}\right)^3}.$$ (13.40)

It is useful to verify the ratio of the tail rotor power to the main rotor power for this flight condition

$$\frac{P_{tr}}{P_{mr}} \simeq \frac{k_{tr}}{\sqrt{2\rho A_{tr}}}\sqrt{\frac{P_i + P_o}{\Omega\, x_{tr}}}.$$ (13.41)

The resulting expression of Eq. 13.41 can be plotted as a function of a number of parameters, including the aircraft's weight, the distance x_{tr} between rotors, and the tail rotor's diameter (see Problem 9). If we take the reference helicopter (model D, Appendix A), the tail rotor power in OGE hover at sea level is about 8% of the main rotor power.

Now consider the helicopter in forward level flight. The tail rotor operates in the same way as the main rotor. Its induced power in forward flight is found from the solution of

$$\lambda = \mu_{tr} \tan \alpha_{tr} + \frac{\lambda_h^2}{(\mu_{tr} + \lambda^2)^{1/2}},$$ (13.42)

where in this case

$$\lambda_h = \frac{v_{tr}^i}{(\Omega R)_{tr}}.$$ (13.43)

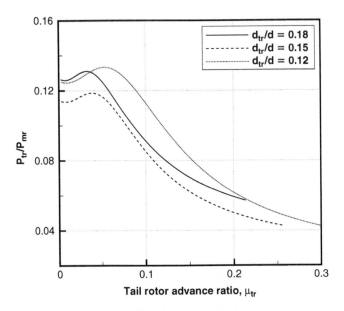

Figure 13.13 *Tail rotor power versus advance ratio, for three values of the tail rotor diameter, relative to the main rotor. Aircraft weight $W = 3,000\,kg$, sea level.*

The induced velocity v_{tr}^i is calculated from Eq. 13.38,

$$v_{tr}^i = \sqrt{\frac{T_{tr}}{2\rho A_{tr}}} = \frac{1}{\sqrt{2\rho A_{tr}}}\sqrt{\frac{P_i + P_o}{\Omega x_{tr}}}. \tag{13.44}$$

The induced power of the tail rotor becomes

$$P_{tr} = T_{tr}v_{tr}^i. \tag{13.45}$$

To calculate the total tail power requirements we need to add the profile power. We can use Eq. 13.28, with the appropriate value of the advance ratio and solidity for the tail rotor. We can also estimate the ratio between profile power of the tail and main rotors from

$$\frac{P_o}{P_{tr}^v} \simeq \left(\frac{\Omega}{\Omega_{tr}}\right)^3 \left(\frac{d}{d_{tr}}\right)^4, \tag{13.46}$$

having assumed that the solidity is the same for the two rotors*. As a result of the forward speed, the total power requirements decrease, as shown in Figure 13.13. The results of Figure 13.13 show that there is relatively large tail rotor power requirement at hover conditions; this power decreases marginally at very low advance ratios, it

---

* Tail rotor speeds are in the range of 1,000 to 3,000 rpm. A ratio $\Omega_{tr}/\Omega \simeq 5$ is not unusual. This is one of the main causes of helicopter noise.

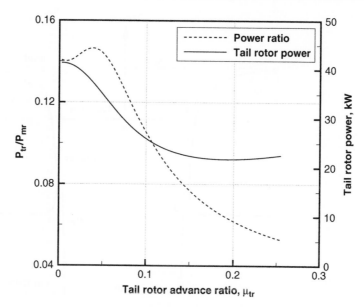

Figure 13.14   *Tail rotor power versus advance ratio, similar to the case of Figure 13.13, W = 3,000 kg, $d_{tr}/d = 0.18$.*

increases again to reach an absolute maximum in transitional flight; then it decreases progressively as the forward speed increases. The analysis of the results indicates that the dominant source of tail rotor power is the induced component; the profile power is only a few percent of the total tail rotor power. This result is coherent with the flight performance, an indication that the first-order theory is correct. The maximum value of the tail rotor thrust exceeds 15% in the cases calculated, and for heavy lift helicopter it means a lot of power. Therefore, other methods have been explored[398].

The effect of the weight on the tail rotor power is marginal in the whole speed range, as it can be verified. In absence of more specific data on the tail rotor configuration, we will assume that the tail rotor power is a percentage of the main rotor power, and the figure of 5% in forward flight and 10 to 12% in hover should suffice in preliminary calculations.

Early work on the tail rotor power includes that of Amer and Gessow[399]. Lynn *et al.*[400] and Cook[401] discussed the fundamentals of tail rotor design, and the interference problems created by the fin and the horizontal tail. The calculation for the fenestron is different, because the fan has a larger number of blades and a larger solidity*. Furthermore, this rotor has a smaller diameter (half the conventional tail rotor), and is not affected by interference and blockage effects in the same amount as the conventional tail rotor[402,403].

An estimate of induced power is found from the one-dimensional axial momentum theory, derived from assuming that the fenestron has many blades and operates in a duct with an inlet and a diffuser. Consider the stations from 1 to 3 in Figure 13.15;

---

* The fenestron powers several Eurocopter vehicles; it has 10 to 13 blades.

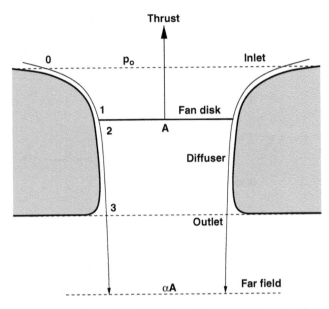

Figure 13.15 *Nomenclature for calculation of fenestron's thrust and power.*

v_i will be the induced velocity at the fan disk, and w the flow velocity at the far downstream in the stream tube, as in the earlier discussion; A is the fan disk area, and α the contraction ratio of the slipstream ($\alpha < 1$). In this configuration, thrust is generated by the fan *and* the shroud. The Bernoulli equation at the inlet will be

$$p_1 + \frac{1}{2}\rho v_i^2 = p_0. \tag{13.47}$$

The Bernoulli equation at the diffuser will be

$$p_2 + \frac{1}{2}\rho v_i^2 = \frac{1}{2}\rho w^2 + p_0. \tag{13.48}$$

The difference between these equations will give the fan's thrust

$$T_{fan} = (p_2 - p_1)A = \frac{1}{2}\rho A w^2. \tag{13.49}$$

The momentum equation, applied between sections 0 (inlet) and 3 (outlet), yields

$$T = T_{fan} + T_{shroud} = (\rho A v_i)w. \tag{13.50}$$

Therefore, the fan's thrust, relative to the total fenestron's thrust is

$$\frac{T_{fan}}{T} = \frac{\rho A w^2/2}{(\rho A v_i)w} = \frac{1}{2}\frac{w}{v_i}. \tag{13.51}$$

The velocity w will be found from the continuity equation. If αA is the cross-sectional area of the stream tube through the fan, then

$$\rho A v_i = \rho \alpha A w, \qquad w = \frac{v_i}{\alpha}. \tag{13.52}$$

If we substitute the last result in Eq. 13.51, then

$$\frac{T_{fan}}{T} = \frac{1}{2\alpha}. \tag{13.53}$$

The induced velocity at the fan's disk will be found from a combination of the continuity Eq. 13.52 and the momentum Eq. 13.50

$$v_i = \sqrt{\frac{\alpha T}{\rho A}}. \tag{13.54}$$

Finally, we find the induced power required for the fenestron to produce a total thrust $T = T_{tr}$, as specified by the torque balance Eq. 13.36

$$P_i = T_{fan} v_i = \sqrt{\frac{T^3}{4\alpha \rho A}}. \tag{13.55}$$

There are two ways to compare the power performance of the fenestron. One method is to plot the ratio

$$\frac{P_i}{P_{tr}} = \frac{1}{\sqrt{2\alpha}}. \tag{13.56}$$

Since the contraction ratio of the slipstream is less than 1, then for a given tail rotor thrust, and given disk loading, the fenestron requires less power than the conventional tail rotor, by the amount $1/\sqrt{2\alpha}$. However, the conventional tail rotor thrust must be augmented for the interference created by the fin and the main rotor.

13.3   ROTOR DISK ANGLE

Often the analysis of the induced power is done with a fixed tilt angle, as shown in Figure 13.3. This assumption is not correct, because the balance equations in the horizontal and vertical direction are not likely to be satisfied. We now calculate the correct tilt angle to fly at a speed U, assuming that the rotor thrust is applied at the CG of the aircraft (a rough approximation, but useful for performance analysis).

To create useful thrust in the flight direction, the rotor disk (or the TPP) has to be inclined by an angle α_T, see Figure 13.2. In level flight the balance of forces on the aircraft in the vertical and horizontal direction will be

$$T \cos \alpha_T - W = 0, \tag{13.57}$$

$$T \sin \alpha_T - D = 0. \tag{13.58}$$

Combination of the two momentum equations yields the value of the tilt angle compatible with level flight conditions.

$$\tan \alpha_T = \frac{D}{W} \simeq \frac{D}{T}. \tag{13.59}$$

The last equivalence is valid if the tilt angle is small. Since the helicopter's drag is a function of the flight speed, also the tilt angle will be a function of the true air speed. The next problem is to calculate the aircraft drag in forward flight. Not all the power is expended for overcoming the resistance of the aircraft. In fact, the tail rotor delivers thrust for longitudinal stability, although a small portion can be used for further thrust. The transmission losses do not contribute directly to the aerodynamic drag, and only a fraction of the induced power is used for advancing. A suitable expression for the total power required to overcome the aerodynamic resistance of the helicopter is

$$P = DU = TU \sin \alpha_T + P_p + (P_o - P_o^h).$$ (13.60)

The first term is found from the momentum theory, and gives the contribution of the induced power that serves to advance the rotor. Part of the induced power is used to create useful thrust in the vertical direction, and does not contribute to the drag. The second term in Eq. 13.60 is the parasitic drag of the helicopter; this is essentially the drag of the fuselage and all its subsystems (landing gear, probes, rotor hub, etc.). The third term is the difference in main rotor profile power between the forward flight and the hover condition. The profile power is not zero when the helicopter is in hover; it increases with μ^2. From Eq. 12.8 we find

$$P_o - P_o^h = \frac{3}{8} \rho \sigma \overline{C}_D (\Omega R)^3 \mu^2.$$ (13.61)

An alternative expression for the drag, found in other textbooks, is $D = P/U$, which becomes singular as the forward speed tends to zero. Therefore, the tilt angle is found from Eq. 13.59 with the drag calculated from Eq. 13.60. With the air speed approaching zero, $D \rightarrow 0$, and the tilt angle would have to be zero. This is unlikely to occur in practice, because most helicopters have a pre-fixed inclination of the rotor on the horizontal plane. From Eq. 13.60, the drag is

$$D = T \sin \alpha_T + \frac{P_p + (P_o - P_o^h)}{U},$$ (13.62)

and the tilt angle will be

$$\tan \alpha_T = \frac{D}{W} = \frac{T}{W} \sin \alpha_T + \frac{P_p + (P_o - P_o^h)}{WU}.$$ (13.63)

The power components in Eq. 13.63 decrease as U^3, therefore the power term tends to zero. In hover the only solution is $\alpha_T = 0$. Since the rotor shaft is generally inclined forward by a small angle, in hover the aircraft will be inclined by a small angle backward. Equation 13.63 is an implicit equation in α_T, and can be solved with a numerical method, for example Newton-Raphson. For this purpose we need to calculate the derivative of the residual

$$f(\alpha_T) = \tan \alpha_T - \frac{T}{W} \sin \alpha_T - \frac{P_p + (P_o - P_o^h)}{WU} = 0.$$ (13.64)

This derivative is

$$\frac{\partial f}{\partial \alpha_T} \cong \frac{1}{\cos^2 \alpha_T} - \frac{T}{W} \cos \alpha_T - \frac{\sin \alpha_T}{W} \frac{\partial T}{\partial \alpha_T}. \tag{13.65}$$

The only way to calculate this derivative is to use some numerics, with a small perturbation around the current value of α_T. The tilt angle is updated according to

$$\alpha_T^{n+1} = \alpha_T^n - \left[\frac{f(\alpha_T)}{\partial f / \partial \alpha_T} \right]^n. \tag{13.66}$$

The balance of the aircraft (trim) is not complete without writing the conditions of equilibrium for the moments. In fact, the thrust line does not generally go through the center of gravity of the helicopter (as we have assumed), and a pitching moment is generated. This has to be counterbalanced with additional control surfaces at the tail and proper distribution of the mass. The FAR/JAA regulations establish that the OEW and corresponding CG must be determined by weighing the aircraft without the crew and payload, but with fixed ballast, unusable fuel and oils.

Computational Procedure

1. Set the air speed U and a first guess α_T.
2. Calculate the parasitic power P_p from Eq. 13.32.
3. Calculate the increase in profile power P_o from Eq. 13.61.
4. Calculate the induced power P_i and the induced velocity ratio λ from Eq. 13.19.
5. Calculate the resulting thrust from

$$T = 2\rho A \lambda^2 (\Omega R)^2.$$

6. Calculate the derivative of the residual, Eq. 13.65.
7. Update the tilt angle with Eq. 13.66.
8. Repeat from point 4 until convergence.

13.4 CALCULATION OF FORWARD FLIGHT POWER

We are now in a position to calculate the forward flight power of the helicopter, from hover to the maximum speed. The method uses the trim solution highlighted in the previous section. The computational model is summarized below:

Computational Procedure

1. Set the flight altitude h.
2. Set the aircraft true air speed U.
3. Calculate the tilt angle using the procedure described in §13.3.
4. Assemble the helicopter power requirements, by doing the following calculations:
   - Main rotor's hover induced power, P_h, from Eq. 13.19.
   - Main rotor's forward flight induced power, P_i, from Eq. 13.19.

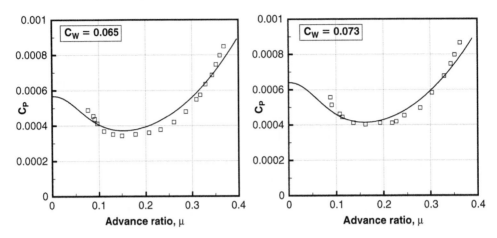

Figure 13.16 *Total power (kW) required for level flight versus flight speed at sea level, ISA, for the helicopter model UH-60A.*

- Main rotor's profile power, P_o, from Eq. 13.28.
- Airframe power, P_p, from Eq. 13.32.
- Tail rotor power
  - Induced power, from Eq. 13.45.
  - Profile power, from Eq. 13.28.
- Level flight power with transmission losses.

5. Calculate the remaining quantities of interest: L/D, fuel flow, specific range, specific endurance, etc.

The entire procedure can be implemented into a routine called `AssemblyPower`. This routine returns all the power components and the correct tilt angle for any reasonable air speed and flight altitude.

A validation of the computational procedure is shown in Figure 13.16 for the general utility helicopter Sikorsky UH-60A, in basic configuration. The calculation was done with two values of the weight coefficient, $C_W = 0.0065$ and $C_W = 0.0073$. The weight corresponding to these coefficients is found from Eq. 11.2. The other data are: rpm = 258, tail rotor rpm = 1,200, $x_{tr} = 9.90$ m, $\overline{C}_{D_o} = 0.0085$. The experimental data and the equivalent flat plate area f_e have been extrapolated from Yeo et al.[404], who have used data from flight testing at Sikorsky. Data are available for a standard rotor and for a rotor with an advanced tip geometry. For the parameter f we have used

$$ f = \frac{D}{q} = 3.324 + 0.0152 \, (0.1575 \alpha_T)^2, \tag{13.67} $$

in international units. This equation implies that the value of f is $3.324 \, \text{m}^2$ at zero pitch attitude, and it increases parabolically with α_T.

The contribution of the main power components (excluding the tail rotor power, not shown for clarity) is reported in Figure 13.17. The induced power has a local minimum at intermediate speeds, after which it starts growing. Both the parasitic

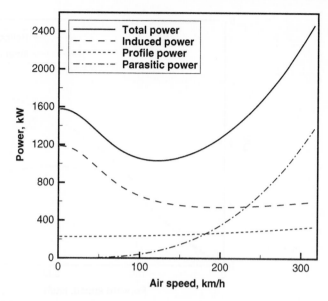

Figure 13.17 *Calculated power components for the UH-60A with $C_W = 0.0073$. Sea level, standard conditions.*

power and the profile power grow as U^3. In particular, the parasitic power is over 50% of the level power at the maximum level speed, which is relatively high.

The calculation was repeated for the reference helicopter with a fenestron. We have used the data on Table A.15 in Appendix A, in addition to an equivalent flat plate area $f/A = 3 \cdot 10^{-4}$ (very streamlined fuselage), and a constant transmission power equal to 4%.

The analysis shown is relatively crude, and it can be improved by using a number of additional assumptions, to calculate the forward flight power of a trimmed helicopter. The step is not straightforward, because there are several effects to take into account, for example: the actual geometry of the rotor (tip, airfoil section) and its aerodynamic performance, the rotor dynamics and the trim conditions. Calculation of the induced power can be improved by using more advanced methods that take into account the rotation of the slipstream and the asymmetry of the flow at the rotor disk. This is done in more sophisticated comprehensive codes, such as CAMRAD[355] and GENHEL[356].

13.5   L/D OF THE HELICOPTER

As in the case of the fixed-wing aircraft, it is possible define a ratio L/D. The definition given in other textbooks and in engineering practice is

$$\frac{L}{D} = \frac{T \cos\alpha_T}{D} = U\frac{T \cos\alpha_T}{P} = U\frac{W}{P}, \tag{13.68}$$

where D is the drag of the aircraft required for level flight. If the drag is calculated as previously indicated, then L/D tends to infinity as the speed tends to zero. If the drag

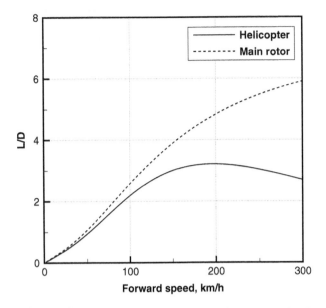

Figure 13.18 *Calculated L/D ratios for reference helicopter, sea level,*
$W = 3,500\,kg.$

is calculated from the equation $P = DU$, where P is the level flight power, then L/D
tends to zero as the speed tends to zero. This definition of drag is throughly perplexing,
but for coherence we have calculated the L/D for our reference helicopter according
to the engineering practice. This is shown in Figure 13.18.

The maximum value of L/D is well below 9, which is about half the value achieved
by the best fixed-wing aircraft. This leads to the conclusion that the direct-lift vehicle
is less efficient than the fixed-wing vehicle. Examples of L/D ratios are shown in
Figure 13.19, elaborated from Vuillet[365] and other sources. The maximum values of
L/D are obtained at large advance ratios, say $\mu = 0.3$ to 0.4.

If one wants to calculate the L/D of the rotor alone, then the airframe drag and the
tail rotor drag must be excluded in the computation of Eq. 13.68. The transmission
losses can be cut down by half, and perhaps neglected altogether. Clearly, this leads
to a larger L/D, as defined by

$$\left(\frac{L}{D}\right)_{rotor} = U\frac{T\cos\alpha_T}{P_i + P_o}. \tag{13.69}$$

13.6 FORWARD FLIGHT ANALYSIS

We now study the effects of the main parameters on the forward flight performance.
In all cases we consider the reference helicopter, model D. For a given engine installation, the maximum speed is found when the forward flight power is equal to the
maximum engine power. The engine power is dependent on flight altitude and speed,

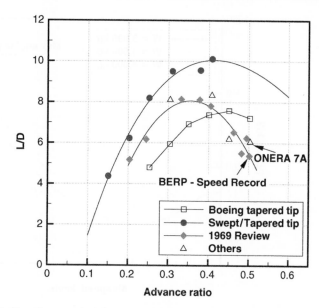

Figure 13.19 *Estimated glide ratios of helicopters, elaborated from Vuillet[365].*

as discussed in Chapter 5. The non-dimensional analysis of §11.6 has shown the relevant parameters for a comprehensive evaluation of the level flight performance:

$$\frac{P}{\sigma_1 \omega^3} = f\left(\frac{W}{\sigma_1 \omega^2}, \frac{U}{\omega}, h, \frac{\omega}{\sqrt{\theta}}\right). \tag{13.70}$$

A number of analyses are in place: (1) the referred power versus the referred air speed, at a constant $W/\sigma_1\omega^2$ in standard atmosphere; and (2) the referred power versus the referred air speed, at constant, non-standard, referred rotor speed $\omega/\sqrt{\theta}$.

13.6.1 Effect of Gross Weight

The calculated effect of the aircraft's weight on the power required to fly level is shown in Figure 13.20. The weight adds a roughly constant amount of power, e.g. an additional power that is nearly independent on the flight speed. However, the most evident effect of the AUW is in the low to medium speeds.

The maximum speed is obviously limited by the installed engine power. However, a number of other factors intervene, such as the performance of the rotor blades at relatively high advance ratios. Beyond a certain combination of thrust, forward speed, and rotor speed, the helicopter performance is limited by a rapidly increasing stall of the retreating blade, as indicated by Gustafson and Myers[405].

The non-dimensional analysis of the level flight performance is shown in Figure 13.21, at a fixed referred weight. The points of minimum power are indicated. A referred weight can be found from a combination of AUW, rotor speed and

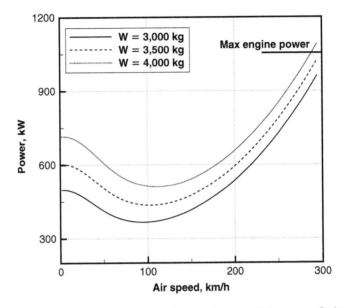

Figure 13.20 *Weight effect on required power for level flight versus flight speed (km/h) at sea level, ISA conditions (helicopter model D, Appendix A).*

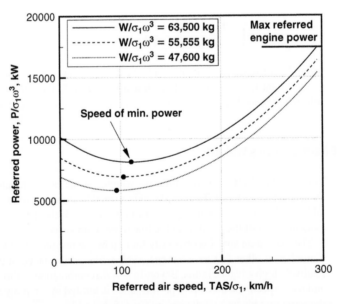

Figure 13.21 *Referred power versus the referred air speed, at selected referred weights, as indicated; sea level, ISA conditions.*

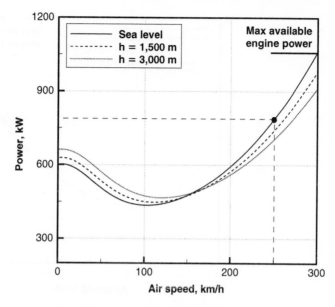

Figure 13.22 *Effect of flight altitude on required power for level flight versus flight speed; ISA conditions (helicopter model D, Appendix A).*

density altitude. The corresponding referred power required to fly at a fixed referred speed is found from a combination of shaft power, rotor speed and density altitude. In other words, two rotorcraft having the same referred weight can be compared either at a fixed referred speed or at a fixed referred power.

13.6.2 Effect of Flight Altitude

The effect of flight altitude is represented by the density altitude in the power equation. The profile power and the parasite power increase with σ_1. The induced power is proportional to $1/\sqrt{\sigma_1}$. The tail rotor power depends in a more complicated way on σ_1, but at a high speed it has limited importance. However, the density plays differently on the power components. Such an effect is shown in Figure 13.22 for three flight altitudes. At low and medium speeds there is a negligible difference, but as the speed increases the power lines tend to diverge. The results of Figure 13.22 have been obtained from the solution of the power equation at three flight altitudes.

13.6.3 Effect of Atmospheric Conditions

Most helicopter performance data are given in charts under ISA conditions and deviations thereof. These can be extreme low or high temperatures. Examples of performance charts for $\pm25°C$ from the ISA values are given in Appendix A for the Eurocopter EC-365 Dauphin. The difference in performance can be considerable, therefore the atmospheric conditions must be clearly specified. Now assume that the temperature at a given flight altitude is not the standard value. From the equation

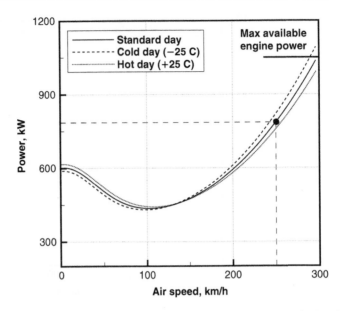

Figure 13.23 *Effect of atmospheric conditions on required power for level flight versus flight speed (helicopter model D, Appendix A).*

of density altitude (Eq. 2.34, Chapter 2) we find the density corresponding to the actual temperature. This new value of the air density must be used in the appropriate equations, namely Eq. 13.19, Eq. 13.28, Eq. 13.32, and the remaining power components. The effect of a $\pm 25°C$ on the flight performance of the reference helicopter is shown in Figure 13.23. For example, on a hot day the power required to hover is increased by 20 kW; the power for high-speed level flight is reduced by about 25 kW.

The non-dimensional forward flight analysis of the reference helicopter is shown in Figure 13.24. The extreme values of $\omega/\sqrt{\theta}$ are obtained with a temperature $\pm 25°C$ around the standard value, and a rotor speed $\pm 5\%$ of the nominal speed.

13.7 PROPULSIVE EFFICIENCY

We have seen that a suitable performance index of the rotor in hover is the figure of merit. When the aircraft is in forward flight, it can be useful to define a propulsive efficiency, to evaluate the fraction of the engine power going into useful power. The propulsive efficiency is defined as

$$\eta_p = \frac{TU \sin \alpha_T}{P_{level}} = \frac{WU \tan \alpha_T}{P_{level}}. \tag{13.71}$$

If the aircraft is climbing, we must add the effects of climb rate:

$$\eta_p = \frac{W(U \tan \alpha_T + v_c)}{P_{level}}. \tag{13.72}$$

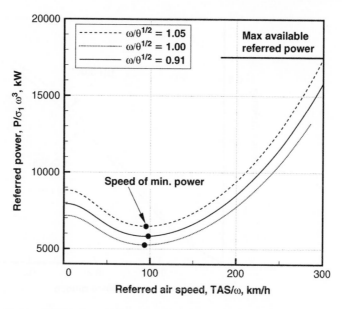

Figure 13.24 *Referred power for level flight speed at non-standard conditions (helicopter model D, Appendix A).*

According to this definition, the propulsive efficiency of a hovering helicopter is zero. In fact, the engine power is expended to create the thrust to hover. This power is used in large part to increase the energy of the air around the aircraft, and to compensate for the profile drag.

The propulsive efficiency was calculated for two helicopters in level flight, and is shown in Figure 13.25. This figure demonstrates that η_p increases with the flight speed, and reaches a value of about 0.7.

13.8 CLIMB PERFORMANCE

For the helicopter to climb, the engine power must be greater than the power required to hover, P_h. The excess power will generate the necessary conditions for climb against the helicopter's own weight and the overall drag. A number of different climb profiles can be described. Besides the vertical climb (already discussed), there will be a few special climb profiles, such as the fastest climb in forward flight, the climb profile for minimum perceived community noise, and climb profiles constrained by obstacles of all kinds. Intermediate climb envelopes between any set of altitudes, as well as more arbitrary climb programs in three dimensions, can be prescribed.

The FAR regulations (Part 29, §29.64) require that compliance be demonstrated at all weights, altitudes, and temperatures within the operational limits established, with the most unfavourable position of the CG, also with OEI.

In general, the total climb power will be written as

$$P(U, v_c, W, h) = P_i + P_o + P_p + P_c + P_{tr} + P_t. \tag{13.73}$$

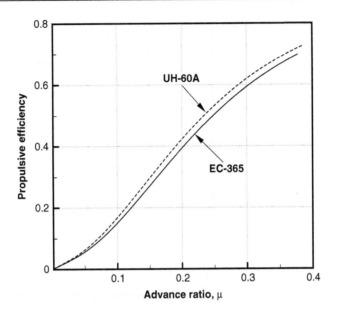

Figure 13.25 *Propulsive efficiency of two helicopters, UH60A and EC-365, at level forward flight, sea level, standard conditions.*

These power components are calculated separately. For a fixed AUW, they are a function of the true air speed; in addition, P_c is proportional to the climb rate. This is found from

$$v_c(U,h) = \frac{P_e(U,h) - P_{level}(U,h)}{W}, \tag{13.74}$$

where P_e is the available engine power at the current altitude. Note that $P_e - P_{level}$ is the excess power. The maximum rate of climb can be calculated from Eq. 13.74, by using the results previously obtained from the forward flight analysis.

Some additional considerations are in place. First, as the aircraft climbs, the airframe suffers from increased drag, due to increased download effects. The climb angle varies from 90 degrees (vertical flight) to a small amount at high forward speeds. The longitudinal axis of the aircraft, though, maintains a nearly constant and relatively small angle α_T. As a consequence, the fuselage operates at very high angles of attack if the air speed is relatively low. There is a massive flow separation on the after-body, as shown in Figure 13.11.

Second, for a climb at maximum power, the blade pitch has to be increased. This implies that the profile power increases as well. An approximation of P_o can be found from $P_o \propto \theta_o^2$ (see also Eq. 12.12).

Third, the correct angle of attack of the rotor is required to calculate the total induced power. For a climb at low speeds, this angle of attack is considerably different from the tilt angle α_T.

Fourth, the tail rotor power must increase, because at the maximum power the increased induced and profile power require an increase in torque balance. Cooke and Fitzpatrick[4] showed that the excess power Eq. 13.74 is accurate enough at medium

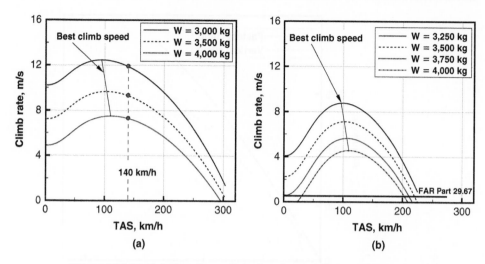

Figure 13.26 *Calculated rate of climb in forward flight, AEO (left graph) and OEI (right graph), GTOW as indicated; sea level standard conditions.*

to high air speed, in spite of all the changes in the power components. Nevertheless, the low speed climb is underpredicted.

We solve the problem for assigned tilt angle α_T and true ground speed. The relationship between the climb angle γ, the rotor's angle of attack α and the rotor tilt α_T is

$$\alpha = \gamma + \alpha_T. \tag{13.75}$$

From the triangle of velocities, the climb angle γ is

$$\gamma = \sin^{-1}\left(\frac{v_c}{U}\right) = \tan^{-1}\left(\frac{v_c}{V}\right). \tag{13.76}$$

The vertical flight program is unlikely to occur in practice, unless the rotorcraft is constrained on all sides. In fact, the power requirements are higher than other flight programs. Also, because of the poor autorotational performance in hover (Chapter 14), the vertical climb is to be avoided.

An example of forward rate of climb calculation is shown in Figure 13.26. When all the engines are operating, we have used the maximum continuous power; with one engine inoperative, we have used the maximum 30-second emergency power (680 kW). The graphs show also the speeds for best climb. The speed of maximum climb rate is around 100 km/h (54 kt), for a GTOW ranging from 3,000 to 4,250 kg. In case of OEI, the aircraft must be able to take off at its MTOW. This is potentially a serious issue, because insufficient power or a too heavy aircraft would not be certified by the authorities.

Fastest climb refers to the maximum climb rate, v_c. This is achieved at the speed of minimum power. Figure 13.27 shows the time to climb, in vertical flight and fastest inclined flight. Fuel consumption for the vertical flight is estimated as 50% higher than the fastest climb.

The climb rate of the helicopter is often given in meters per minute or feet per minute. These are odd units, but they are used in practice. As the helicopter climbs,

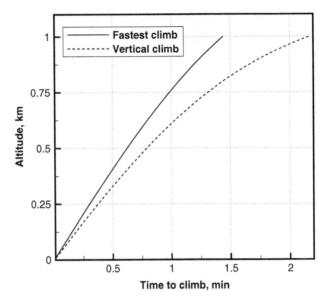

Figure 13.27 *Time to climb to specified ceiling from sea level, vertical and fastest climb; reference helicopter, GTOW = 3,500 kg.*

the rate of climb decreases. If the helicopter operates at maximum power, the engines are not capable of delivering the extra power to offset the reduction in climb rate. If the helicopter climbs at low climb rates, then it is possible to increase the power and maintain the climb rate for a longer time.

The maximum climb rates provided by the manufacturers vary greatly, the largest being a factor 4 of the lowest. However, some customers (particularly the military) prefer to have specifications that say what the maximum climb would be in 1 minute, with specified power output and aircraft weight. Some climb performance data are shown in Figure 13.28, according to the helicopter type. We have converted the data from feet/minute to m/s. It is important to mention that the performance analysis in this figure is dependent on the actual engine power. If the engine is upgraded, the same helicopter can have a point performance better than the one shown.

13.9 PERFORMANCE OF TANDEM HELICOPTERS

We now consider the tandem helicopter. This is a helicopter with two counter-rotating rotors, which can be partly overlapping. We propose to find the power required to hover, climb and fly forward, and the hover ceiling in- and out-of-ground effect, as in the case of the single-rotor helicopter. The performance of the tandem rotor will be compared with that of the single-rotor helicopter at the same gross weight.

The main aerodynamic problem of the tandem helicopter is the interference between the rotors. The aft rotor (in the forward flight configuration) operates in the wake of the fore rotor; the blockage effect of the airframe complicates the matter even

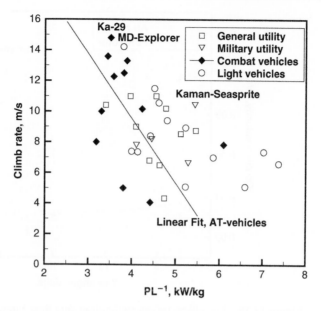

Figure 13.28 *Maximum rate of climb versus power loading for different classes of helicopters, as indicated.*

further. From the point of view of the power requirements, the problem is two-fold: to derive first-order expressions for the total power and thrust of the tandem rotor; and to quantify the amount of aerodynamic interference, depending on configuration and flight condition.

The performance data available in this field are mostly of an experimental nature. A large body of research is available from the development of the Boeing-Vertol 107/CH46 Sea King, and CH-47 Chinook, in addition to the Piasecki YF-21/H-21/CH-21/UH-21 Workhorse, and the Bristol 192 Belvedere. Although this class of helicopters is not as common as the single-rotor helicopter, the use of the tandem helicopter for military operations is widespread. Zilliac *et al.*[406] estimated that by 1998 the CH-46 had performed over 210,000 landings on carrier decks alone. However, the use of the CH-47 Chinook for civilian operations (including transport of emergency supplies) has increased. This helicopter continues to be the workhorse of the VTOL aircraft.

Other helicopters with two main rotors include several Kamov helicopters in the Soviet Union, and Kaman helicopters in the United States. Performance and aerodynamic data for these vehicles are less accessible.

Some data on the aerodynamic interference between rotors on a model helicopter were published by Halliday and Cox[407]. These authors proved that the front rotor produces more thrust at forward speeds, but a backward tilt of the rear rotor gives a good degree of compensation. Yates[408] published some data that show the vibratory effects of some critical flight conditions (vortex ring, vertical descent, yawed flight) on a tandem rotor, particularly in the case of transitional flight. This is the regime of flight between hover and forward flight at low speed.

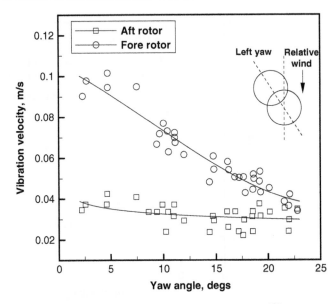

Figure 13.29 *Yaw effects on rotor vibration, data from Yates[408]. The solid lines denote the best fit of each series of data.*

Figure 13.29, elaborated from Yates, shows the vibratory effects on the rotors created by a yaw angle. Vibration is essentially created by the aerodynamic interference between rotors. The downstream rotor ingests turbulent and vertical flow from the forward rotor, and is more subject to vibrations. The conclusion is that rotor vibration created by a yaw angle can be considerably reduced by flying the tandem helicopter in yawed flight. The benefit is more substantial for the aft rotor. The yaw angle of current tandem rotor helicopters for minimum forward flight power is estimated at 12 to 16 degrees.

We assume that the two rotors have the same diameter d, and that the distance between rotor hubs is d_r. The vertical separation, as shown in Figure 13.30, is the gap d_v; the distance between rotor axes along the ground is the so-called stagger distance, d_h. The relationship between these distances is $d_r = (d_h^2 + d_v^2)^{1/2}$. The rotors are inclined by an angle α_T on the vertical. These angles may not be the same.

13.9.1 Assembling the Power Requirements

To assemble the power required to hover, climb and forward flight of the tandem helicopter, we follow the method of the single-rotor helicopter. The total power is

$$P = P_i + P_o + P_p + P_t, \tag{13.77}$$

where in this case P_i is the total induced power, P_o is the total rotor profile power, P_p is the parasite power of the airframe, and P_t is the power loss due to the transmission system. The overall interference effects (between rotors and airframe) are accounted

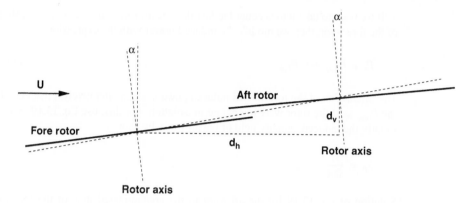

Figure 13.30   *General arrangement and nomenclature for tandem rotor helicopter.*

for in the calculation of the total induced power and in the airframe drag. If the helicopter is to climb, then we need to add the climb power P_c to Eq. 13.77. There is no tail rotor. Torque balance is achieved thanks to a combination of counter-rotation and differential cyclic pitch. This is because each rotor absorbs a different amount of power and torque. A different torque means that torque reaction cannot be completely neutralized by counter-rotation alone.

For the induced power we need to take into consideration the effect of overlapping rotors, the effect of the fuselage on the rotor downwash, and the ground effect (if any). The problem is clearly complicated by the interference between rotors, which is further complicated by the presence of the ground during take-off and landing. Another essential aspect is the distribution of loads between fore and aft rotors. The aft rotor operates in the slipstream (or downwash) of the fore rotor, therefore its power requirements would tend to be higher, as it was found experimentally by Heyson[409]. This distribution of loads, and its relation to the center of gravity of the aircraft, is essential in helicopter stability. A simplified analysis, based on the axial momentum theory exists (see Payne[410], Stepniewski and Keys[110], Harris[411], Leishman[324], among others). It is reasonable to assume that the thrust is equally split, even though the powers may be different. The difference in power requirements between rotors is an important consideration. Since the power is delivered by a shaft connected to the gas turbine engines by means of a gearbox, the split of fore/aft rotor power has to be considered in the design phase.

The induced power of an isolated rotor in hover out-of-ground effect is given by Eq. 12.3, and for a tandem rotor it can be written as

$$P_i = k_o k_i \sqrt{\frac{W^3}{2\rho A}}, \tag{13.78}$$

where W is the total aircraft weight, and k_o is the *overlap factor*; k_o is an index of the loss of efficiency due to the aerodynamic interference between rotors. This factor depends strongly on the relative stagger, e.g. the ratio between the stagger distance and the rotor diameter. Values of the overlap factors have to be found experimentally, although the momentum theory yields a good approximation[324].

If we want to take into account the fact that the aft rotor operates in the slipstream of the fore rotor, then we modify the induced power with the expression

$$P_i = P_{i_{fore}} + k_o P_{i_{aft}}. \tag{13.79}$$

The calculation of the fore and aft induced powers is done iteratively. First, calculate the $P_{i_{fore}}$ as if the rotor were to operate in isolation. For this, use Eq. 13.19. Second, modify the induced velocity field of the aft rotor, by using

$$\lambda_h = \frac{v_i}{\Omega R} + \lambda_{fore}. \tag{13.80}$$

Solution of Eq. 13.19 for the aft rotor in the contaminated flow of the fore rotor will yield a higher power. One can reiterate the procedure, by calculating the fore rotor power, after modifying the induced velocity field due to the aft rotor. We have verified that this operation seldom yields an improvement in the estimates, because the difference in fore rotor power in the presence of aft rotor downwash in forward flight is marginal. A theoretical discussion of downwash interference between rotors was published by Castles and De Leeuw[412], who concluded that the interference induced velocity on the aft rotor due to the presence of the fore rotor is of the same order of magnitude of the self-induced velocity. The interference effect of the aft rotor onto the fore rotor is of much lower magnitude at cruise speeds.

The data reported in Figure 13.31 have been adapted from a number of sources, including Stepniewski and Keys[110]. The curve fit of these data is given at the top left of the graph.

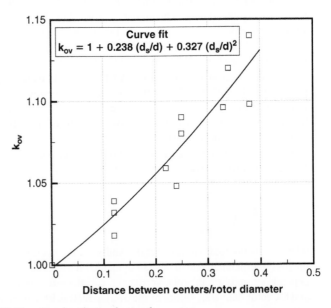

Figure 13.31 *Overlap factor for tandem rotors.*

To take into account the interference of the fuselage, we introduce another correction factor, k_f

$$P_i = k_f k_o k_i \sqrt{\frac{W^3}{2\rho A}}.$$ (13.81)

The interference of the fuselage gives rise to a change in downwash velocity field below the rotors, and this again has to be evaluated experimentally, or with sophisticated aerodynamic simulation methods. Whereas the flow around the isolated streamlined fuselage is a problem that can be solved, the effect of the rotors is far more complicated, and the best solution for preliminary performance analysis seems to be the use of an actuator disk in place of the actual rotor.

Whatever the origin of the data, we assume that k_f is a known factor. The rotor profile power is calculated in the same way as in the single rotor helicopter: we just need to remember that we have two rotors of the same diameter, rotating at the same speed, see Eq. 12.8.

Finally, the transmission losses are taken as a percentage of the total power. The transmission system consists of a shaft that delivers power to both rotors. If there is an engine failure, the other engine can still deliver an emergency power to the rotors, although only for a limited amount of time (typically, 30 minutes).

Other power components that should be taken into account include: exhaust hot gas reingestion, compressor bleed, inlet pressure losses to viscous drag, etc.

13.9.2 Tandem Helicopter: Example of Calculation

We have considered the tandem helicopter CH-47SD, whose relevant performance data, as provided by Boeing, are given in Table A.18. The tilt angles of the rotors have been estimated at 5.5 degrees (fore) and 4.5 degrees (aft).

In addition to the data provided, we have calculated the rotor solidity from the blade diameter, the blade chord ($c = 0.81$ m) and the number of blades: $\sigma = 0.08458$. The blades are made of fiberglass and have an airfoil section V-23010-1.58. The average drag coefficient of this section as a function of Mach is shown in Figure 13.4. A suitable average value at Mach numbers below the drag divergence is $\overline{C}_D \sim 0.015$ at constant angles of attack $\alpha = 0$ to 10 degrees.

Finally, the two rotors are fixed at hubs of different height, but because of the tilt arrangement, they are almost planar when rotating, therefore the axial momentum theory applies; the two rotors are assumed to carry half the weight and develop half the thrust required for flight.

The performance data published by Boeing for the standard model CH-47SD are given in Table A.18. For the Allied Signal engine an average specific fuel consumption of $8.014 \cdot 10^{-4}$ N/s/watt (0.483 lb/h/lb) has been assumed for a maximum continuous power at sea level ISA conditions.

Computational Procedure

1. Set flight altitude.
2. Set the advance ratio (or flight speed).

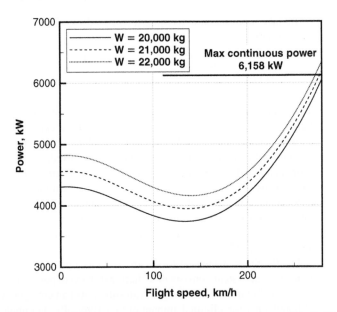

Figure 13.32 *Estimated forward flight power for the helicopter CH-47-SD, sea level, ISA conditions.*

3. Calculate the ideal induced power from Eq. 12.3.
4. Correct the ideal induced power for three-dimensional effects, non-uniform downwash and partial overlapping, using the appropriate coefficient for this case: $k_i = 1.15$.
5. Calculate the profile power from Eq. 12.8, and multiply by 2, to account for the two rotors.
6. Calculate the airframe power from Eq. 13.32.
7. Assemble the forward flight power, Eq. 13.77, and include the estimated transmission power losses.
8. Calculate the specific excess power from Eq. 13.74, and hence the maximum v_c.

Figure 13.32 shows the calculated power requirements for level flight at sea level at three aircraft gross weights, as indicated. For reference, we have added the maximum continuous power from the engine. The results show that the aircraft is capable of reaching a maximum speed of the order of 280 km/h (151 kt) at sea level with a gross weight of 22,000 kg, which is not too far off the performance data provided by Boeing.

Figure 13.33 shows the calculated fore/aft power split between rotors versus the average advance ratio of the helicopter*. The aft rotor power is higher than the fore rotor power, in the whole range of forward speeds, as expected. At hover conditions

---

* The advance ratio is defined by Eq. 11.3. If the two rotors have tilt angles that differ more than 1 degree, then we should calculate the advance ratios μ_F and μ_R, and plot the power versus one of these quantities.

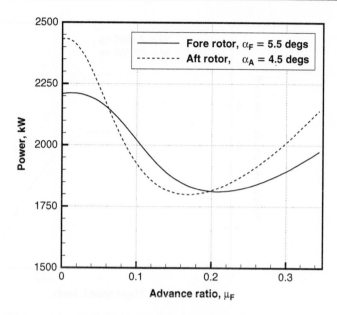

Figure 13.33   *Estimated fore/aft rotor power split for the CH-47SD, assuming half weight on each rotor; AUW = 20,000 kg, ISA, sea level.*

the power is equally split. This comes from our assumption that each rotor generates the same thrust, and that each thrust is equal to half the AUW. Furthermore, since there is no additional downwash interference, other than the one calculated with the overlap factor, the power requirements are the same.

The difference in power split between the rotors depends on the tilt angles. If we increase the fore rotor tilt, while keeping a fixed aft tilt rotor, the distribution of thrust changes, and for high forward speeds it is possible that the power requirements of the fore rotor are larger than the aft rotor.

To find the maximum climb rate we calculate the specific excess power. For two reference weights we have plotted the maximum sea level capability of the CH-47SD resulting from the calculation in Figure 13.34. The performance reported by the manufacturer cannot be reproduced by the simulation at a weight $W = 22,680$ kg, but the rate of climb is basically the same if the aircraft gross weight is decreased to 20,000 kg.

Finally, we calculated the service ceiling of the aircraft, by simply increasing the flight altitude. The results of Figure 13.35 show the distributions of climb velocity and specific excess thrust of the CH-47. The calculated service is between 3,300 and 3,400 m, which is in line with the performance data provided in Table A.18.

13.10   SINGLE OR TANDEM HELICOPTER?

After showing simple models for the tandem helicopter, one may wonder how a tandem helicopter performs compared with a single rotor helicopter, and whether

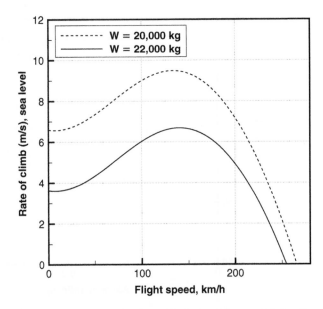

Figure 13.34 *Rate of climb of CH-47-SD for level forward flight. Simulated data, sea level, ISA conditions.*

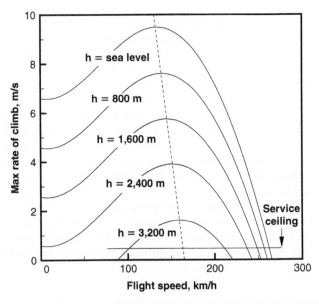

Figure 13.35 *Rate of climb and service ceiling of CH-47-SD for level forward flight. Simulated data, W = 20,000 kg, ISA conditions.*

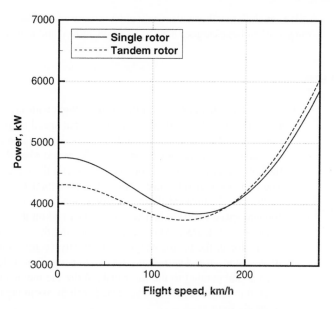

Figure 13.36   *Forward flight power for single and tandem rotor helicopters. Simulated data, sea level, ISA conditions, GTOW = 20,000 kg.*

there is a range of speeds, gross weights, and ranges that make the tandem helicopter more appropriate. The tandem helicopter is still an exotic aircraft, as only a few types have been designed and built, whereas the conventional helicopter is found in several configurations and sizes.

To make a fair comparison, we consider the same CH-47SD model that we have used in our discussion, and we "replace" the rotor system with a single rotor plus a tail rotor (this would be extravagant, because it would require the redesign of the aircraft). The main rotor of the conventional helicopter has the same diameter and the same blades – nothing has changed. A tail rotor is needed for lateral stability, therefore as part of the power is apparently reduced because of the single rotor, the power itself has to be augmented by a few percent to take into account the presence of the tail rotor. The aircraft is to take off and fly level at nominal sea level conditions, with a gross weight of 20,000 kg. The calculation methods for the forward flight power are presented for the conventional helicopter, therefore no further discussion is needed. The results of the calculations are shown in Figure 13.36.

The profile power of the conventional helicopter is half that of the tandem helicopter, because the blades rotate at the same speed, and the advance ratio is the same. The power induced by the conventional helicopter does not require the correction for overlapping rotors. The result shows that at low to medium speeds the power required for forward flight is considerably lower for the tandem helicopter. At higher speeds, the conventional helicopter is more efficient. However, the tandem helicopter power can be further reduced by flying the helicopter with a small yaw angle. At the highest speeds, this change would make the forward flight performance of the tandem helicopter nearly equivalent to that of the conventional helicopter.

We conclude that the tandem helicopter is more useful from hover to medium speed ranges, when large cargoes have to be lifted, moved and relocated.

PROBLEMS

1. A two-bladed helicopter rotor flies in level flight with an advance ratio $\mu = 0.30$. Its diameter is $d = 11.0$ m, and its rotational speed is rpm $= 340$. Neglect the rotor tilt. Calculate the radius of the reverse flow region and produce a sketch indicating how this zone of reverse flow is created when the blade moves around the hub. Describe how this zone increases with the increasing forward speed.

2. The helicopter of reference (Appendix A), with a gross weight of 3,000 kg, is climbing in forward flight. Its instantaneous climb rate is 15 m/s, and its forward speed is 60 km/h. The rotor is inclined on the horizontal plane by an angle $\alpha_T = 3$ degrees. Make a sketch showing the direction of the drag force, the lift force, the thrust, weight and velocity vector. Find the angle of attack of the rotor, after determining the component of the velocity in the direction normal and parallel to the rotor disk. Write the momentum equations for the aircraft in the horizontal and vertical direction, assuming that the climb is done in steady state.

3. Consider the main rotor of the reference helicopter, model D, Appendix A. Assume a level flight at a variable advance ratio μ, with a constant tilt angle $\alpha = 5$ degrees. Approximate the rotor to a disk inclined by the same angle on the vertical. Assume that no pitch control is applied to the blades. Calculate and plot the region of reverse flow at a number of selected advance ratios. Discuss how this area changes with the increasing forward flight speed, and how this problem can be solved. Derive the missing data from Table A.15.

4. For the helicopter of reference (Appendix A) consider a gross weight $W = 3,000$ kg, and sea level ISA conditions. Using the method of §13.4, calculate the figure of merit of the rotor, FoM, and the thrust coefficient C_T/σ; plot these quantities versus the flight speed and comment on the results.

5. For the helicopter model in Appendix A evaluate the effects that a change of rotor diameter and rotor speed would have on the rotor profile drag in hover and forward flight.

6. Assuming a constant rotor speed, calculate how the tip Mach number of the model helicopter, aircraft D in Appendix A, changes with altitude in the case that: (a) the aircraft is maintaining a steady hover; and (b) a level forward flight at a maximum speed is maintained.

7. Study methods to reduce the power requirements for the tail rotor, by considering the overall design of the aircraft and its operation.

8. For the helicopter model in Appendix A calculate the L/D ratio as a function of the advance ratio μ at sea level. Consider separately the helicopter as a whole and the main rotor alone. From this calculation find the speed corresponding to maximum L/D. Assume the following gross weights: $W_1 = 2,700$ kg, $W_2 = 3,200$ kg. Follow the method of §13.5. Discuss the results obtained.

9. Consider the reference helicopter, Appendix A, with a take-off gross weight of 3,250 kg at sea level, ISA conditions. Estimate the tail rotor power required, relative to the main rotor's conditions in pure hover. Repeat the calculations,

keeping as a parameter the distance between rotors, x_{tr}, the aircraft weight W, the tail rotor's diameter d_t, and plot the ratio between rotor powers as a function of these parameters. Use Eq. 13.41.

10. Study the impact of rotor diameter on hover performance. Consider the induced and the profile power, OGE hover at sea level, and a gross weight of 2,000 kg. Use a constant value of solidity $\sigma = 0.064$, an average profile drag coefficient $\overline{C}_D = 0.0078$ and a rotational speed of 350 rpm. Find the optimal diameter of the rotor under these conditions. Repeat the analysis in forward flight, $U = 200$ km/h, compare the results and discuss. Does the increase in rotor diameter represent the best option for improved performance?

11. Consider a hypothetical tandem rotor helicopter. The helicopter is to fly forward with front rotor rotating clockwise and rear rotor rotating anticlockwise. Analyze the options of flying in yawed conditions to reduce rotor and airframe vibrations, and decide if port or star-side yaw is preferable.

12. Consider the tandem helicopter model discussed in this chapter, with all the relevant data and performance in Appendix A. Use the computational model explained in §13.9.1 and §13.9.2 to calculate the power split between fore and aft rotors in forward flight. Study how the fore rotor power increases or decreases as a function of the tilt angle $\alpha_{T_{fore}}$, for realistic values of this parameter. Consider an average gross weight of the aircraft at sea level, ISA conditions.

13. Castles and De Leeuw[412] report that for a tandem rotor, whose centers are separated by a distance equal to their diameter, the total induced velocity is

$$\lambda_{h_F} = \frac{v_{i_F}}{\Omega R} = \frac{0.47 C_T}{\mu_F (1 - 3\mu_F^2/2)}, \qquad \lambda_{h_R} = \frac{v_{i_F}}{\Omega R} = \frac{1.25 C_T}{\mu_F (1 - 3\mu_F^2/2)}.$$

Here the subscripts F and R denote the fore and aft rotor, respectively. Under these assumptions, calculate the total induced power of fore and aft rotor in forward flight, and plot the non-dimensional powers P_F/P_h, P_R/P_h versus the advance ratio of the fore rotor, μ_F (P_h is the induced power in hover). Take the missing data from Table A.18.

14. Analyze the reasons why a tandem helicopter flown with a small yaw angle is useful to reduce the power requirements in level forward flight. (*Problem-based learning: additional research is required.*)

15. A general utility helicopter is to be designed with a conventional tail rotor and a vertical fin for improving its longitudinal stability. Study the main parameters of the tail rotor/fin configuration, in order to reduce the tail rotor power, and report briefly on your findings. (*Problem-based learning: additional research is required.*)

Chapter 14

Rotorcraft Maneuver

Ultimately, the jet helicopter may well prove to be the most practical configuration.

Gessow and Myers[352], 1959

The term *maneuver* includes several flight conditions of the helicopter. In this chapter we shall be concerned with those maneuvers that are an essential part of the operation of the aircraft. These are: autorotative performance (for the analysis of flight under engine failure conditions), vortex ring state (when the vortex wake returns on the rotor disk), quartering flight (when the yaw angle of the aircraft is larger) and banked turns. We start from considerations on the flight envelope of the conventional helicopter, to establish the speed and power limits of the aircraft.

14.1 LIMITS ON FLIGHT ENVELOPES

In Chapter 12 we discussed in general terms the flight envelope of the helicopter, with particular reference to the maximum speed. We now discuss some of the technical difficulties arising from the propulsion by a rotary blade at high speeds. The most important problems are the tip Mach number, the stall of the retreating blades, and the cyclic loads on the airfoil sections.

The tip speed and tip Mach number are, respectively,

$$U_{tip} = \Omega R + U_T, \qquad M_{tip} = \frac{U_{tip}}{a}, \tag{14.1}$$

where Ω is the rotor speed, expressed in rad/s, and $U_T = U \cos \alpha_T$ is the in-plane component of the forward speed. Equation 14.1 shows that M_{tip} grows with the forward speed, the rpm, with the rotor radius. M_{tip} increases also with the flight altitude. Eq. 14.1 gives only the free stream Mach number seen by the outer blade sections. In practice, the local air flow acceleration around the blade produces local Mach numbers considerably higher. At a given tip speed, compressibility effects can be reduced by using supercritical blade sections and low pitch angles. We have seen how transonic Mach numbers lead to drag divergence (Figure 13.4), and hence increased power, lower figure of merit, and higher vibrations.

The tip Mach numbers of some production helicopters are summarized in Figure 14.1. In all cases the tip speed is subsonic, with relatively high values in the case of the Boeing-Sikorsky RAH-66. The relatively low tip speed of the tilt rotor V-22 is due to the hybrid nature of this aircraft. The Kaman K-1200 is a helicopter with two intermeshing rotors.

Whereas a high tip speed creates problems of noise, vibration and loss of useful power, a too low tip speed limits the autorotational characteristics of the rotors. Figure 14.2, adapted from Prouty[413], shows the practical limits on the tip speed of the main rotor. The range of acceptable speeds is at the intersection of noise limits,

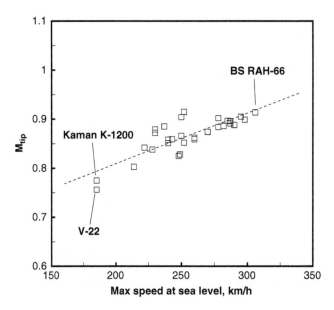

Figure 14.1 *Tip Mach numbers of some helicopters. Note that twin rotors tend to have relatively low M_{tip}.*

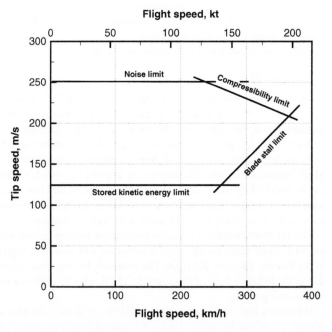

Figure 14.2 *Practical limits on the tip Mach number (the solid lines are approximate, and serve to indicate the speed range of actual helicopters).*

supersonic effects, retreating blade stall and stored kinetic energy. As the aircraft speed increases, the tip speed increases according to Eq. 14.1. To keep the M_{tip} below the divergence Mach number, the condition will have to be

$$M_{tip} = \frac{1}{a}(U_T + \Omega R) < M_{dd}, \tag{14.2}$$

which requires either R or Ω (or both) to decrease as the aircraft moves faster.

The blade tip is important in terms of overall rotor performance, because it is the origin of three-dimensional vortex separation. It also sustains the highest dynamic pressure, and is the source of high drag and noise emission. Considerable research exists in this area, with a large body of work of an experimental nature, and increasing computational research. A brief discussion was given earlier in §12.1.1. A tapered tip with backward sweep is better than a straight tip, although at times it creates an unwanted additional torsional moment on the blade. All modern blade tips converge toward this type of design.

A well-designed blade twist helps with the load distribution on the rotor disk, and tends to improve the figure of merit. For example, the Sikorsky S-58, Sikorsky CH-14E, and Eurocopter EC-365 have a linear twist; the Eurocopter EC-135 has a non-linear twist. The amount of twist is variable, but the maximum is in the range of 15 to 16 degrees.

Dynamic stall is a complex unsteady response of the blade section to changing inflow conditions, such as those created by oscillating, a plunging motion, or varying free stream (dynamic inflow). The aerodynamic forces are generally in a time lag with the inflow, and follow trends that can be considerably different from the steady state airfoil. The dynamic stall results in increased vibrations of the rotor, and increased torsional loads power requirements. If the pitch oscillation occurs at angles of attack below the point of static stall, $\alpha_{C_{Lmax}}$, the aerodynamic coefficients follow a loop that is always in delay with the inflow. If the amplitude of oscillations includes $\alpha_{C_{Lmax}}$, the response is highly non-linear, and the blade section is affected by massive flow separation for about half a cycle.

In summary, the tip speed must be contained within a suitable range to allow the rotorcraft to fly efficiently and safely. This limitation is also critical in the overall speed and productivity characteristics of all rotary-wing aircraft. Fundamental aspects of the physical phenomena are discussed by McCroskey[414,415] and Carr[416], among others.

14.2 KINETIC ENERGY OF THE ROTOR

The kinetic energy for a blade element dm at radial position y is $dE = dm U^2/2$. If m is the mass per unit length of the blade, then $dm = m dy$. The kinetic energy stored by a single blade is found from integration of dE:

$$E = \int_0^R dE = \frac{1}{2}\int_0^R mU^2 dy. \tag{14.3}$$

The speed U in Eq. 14.3 is the effective speed, e.g. the resultant of the rotational and forward speeds, as sketched in Figure 14.3,

$$U(y, \psi) = \left[U_\infty^2 + (\Omega y)^2 + 2U_\infty \Omega y \sin\psi\right]^{1/2}. \tag{14.4}$$

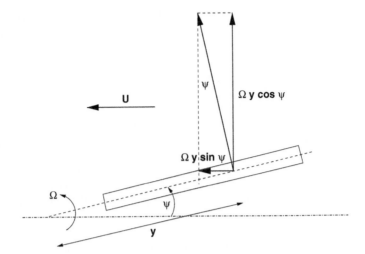

Figure 14.3 *Velocities at the blade's radial position y.*

Integration of Eq. 14.3 with Eq. 14.4 leads to

$$E(\psi) = \frac{1}{2}MU_\infty^2 + \frac{1}{2}\Omega^2 \left(\frac{1}{3}MR^2\right) + U_\infty\Omega \sin\psi \left(\frac{1}{2}MR\right), \tag{14.5}$$

where $M = mR$ is the blade's mass. Introduce the moment of inertia of the blade

$$I_b = \int_0^R my^2 \, dy = \frac{1}{3}mR^3 = \frac{1}{3}MR^2. \tag{14.6}$$

The moment of inertia is calculated from the geometry (including the proper value of the cut-off point), the weight distribution of the blade and the material (reinforced carbon fibers, aluminium, etc.) and the construction of the blade. With substitution of Eq. 14.6, the kinetic energy of the blade becomes

$$E(\psi) = \frac{1}{2}MU_\infty^2 + \frac{1}{2}\Omega^2 I_b + \frac{1}{2}U_\infty\Omega \sin\psi MR. \tag{14.7}$$

In general, the rotor will have N blades with constant azimuthal distance $\Delta\psi = 2\pi/N$. Therefore, the azimuth of the kth blade will be

$$\psi_k = \psi_{k-1} + \frac{2\pi}{N}(k-1). \tag{14.8}$$

For example, if the rotor has two blades, $\Delta\psi = 2\pi/2 = \pi$; hence $\psi_2 = \psi_1 + \pi$. For a four-bladed rotor $\Delta\psi = 2\pi/4 = \pi/2$; hence $\psi_2 = \psi_1 + \pi/2$, etc. Therefore, the total kinetic energy stored by the rotor is

$$E_t(\psi) = \sum_{k=1}^N E(\psi) = \frac{N}{2}MU_\infty^2 + \frac{N}{2}\Omega^2 I_b + \frac{1}{2}U_\infty\Omega MR \sum_{k=1}^N \sin\psi_k. \tag{14.9}$$

It is proved that if the blades are equally spaced the summation is identically zero. For example, for a two-bladed rotor, $N = 2$, then $\sin \psi = -\sin(\psi + \pi)$. Therefore, in this case the kinetic energy of the rotor does not depend on the azimuth angle. Its final expression is

$$E_t(\psi) = \frac{N}{2}MU_\infty^2 + \frac{N}{2}\Omega^2 I_b. \tag{14.10}$$

A final point regarding the derivation of the total kinetic energy of the rotor is that the movement of the blade is considered rigid, and does not take into account the complicated effects of flapping and pitching.

The optimal choice of the number of blades is a problem of aircraft design. Like with any design problem, there are conflicting requirements for cost and performance. Factors to be considered include the weight, mechanical complexity, hub drag, disk loading, and noise. In this context, it will suffice the report on some relevant parametric studies, regarding the hover performance, starting from Knight and Hefner[361], Landgrebe[417] and McVeigh and McHugh[418].

A high rotational speed and moment of inertia are required for increased storage of kinetic energy, to be used when autorotative descent is required. A high mass will lead to weight penalty; a high rotational speed will create transonic aerodynamic problems and increased acoustic emission; at high forward speed there will be the problem of retreating blade stall. Once again, the best solution is a compromise.

14.3 AUTOROTATIVE INDEX

The autorotative index AI is defined as the ratio between the kinetic energy of the main rotor and the helicopter's gross weight, W. Starting from Eq. 14.7, consider the case of pure rotational speed. Divide E by the weight W,

$$\frac{E}{W} = \frac{I_b\Omega^2}{2W}N. \tag{14.11}$$

The autorotative index is the factor

$$AI = \frac{I_b\Omega^2}{2W}. \tag{14.12}$$

According to Eq. 14.12, the AI has the dimensions of a length (meters or ft). Another definition, according to Sikorsky, includes the disk loading

$$AI = \frac{I_b\Omega^2}{WD_L}. \tag{14.13}$$

where $D_L = W/A = W/\pi R^2$. The latter version is preferred in engineering practice; sometimes the definition of AI from Eq. 14.12 is referred to as *energy factor*.

For a given rotor, the larger the autorotation index, the higher the kinetic energy stored in the rotor, relative to the aircraft's weight, and the lower the velocity of autorotational descent. Some values of the autorotation index have been derived by Leishman[324] for some representative helicopters.

Example

We estimate the polar moment of inertia I_b and the autorotative index AI of the blades of the reference helicopter. The airfoil section varies from the OA-212 to the OA-206 (see Appendix A for details). Assume that the average airfoil section is the OA-209. For a unit chord, this airfoil has an area $A_c = 0.06379$ (a summary of other airfoil data is given in Table 14.1). The airfoil chord is $c = 0.40$ m, therefore the scaled airfoil area will be $A_c = A_c \times 0.40^2 = 0.0102064$ m^2.

If ρ_b is the average density of the blade, the element of mass of a blade element is

$$dm = \rho_b dV = \rho_b A_c dy. \tag{14.14}$$

The polar moment of inertia is

$$I_b = \int_{R_i}^{R} y^2 dm = \int_{R_i}^{R} \rho_b A_c y^2 dy = \frac{1}{3}\rho_b A_c (R - R_i)^3 = \frac{1}{3}M(R - R_i)^2, \tag{14.15}$$

where M is the total mass of the blade. The average specific weight of carbon fiber composites is of the order of 2,000 kgf/m^3, corresponding to a specific mass (density) 2,000 kgm/m^3. The inner cut of the blade for the reference helicopter is estimated at $R_i = 0.5$ m.

With this construction, the mass of each blade will be $M = \rho_b A_c (R - R_i) \simeq 71$ kgm. The total mass of the rotor blades is 284 kgm, corresponding to a weight of 284 kgf, or about 13% of the operating weight empty (a reasonable figure). According to Eq. 14.15, the moment of inertia of one blade will be $I_b \simeq 708$ kg m^2. The autorotative index for a gross weight of 3,000 kg is

$$AI = \frac{I_b \Omega^2}{2W} = \frac{708 \times 36.65^2}{2 \times 3,000g} \simeq 16.2 \text{ m} \quad (52.5 \text{ ft}). \tag{14.16}$$

Table 14.1 *Cross-sectional areas and moments of inertia of some rotorcraft airfoils, for unit chord; $I_x =$ pitch inertia moment; $I_z =$ bending inertia moment.*

Airfoil	Area	$I_x \times 10^5$	$I_z \times 10^3$
NACA 0012	0.08126	6.686828	4.4079474
OA-206	0.04251	0.969391	2.2881918
OA-209	0.06381	3.280733	3.4360378
OA-212	0.08497	7.926556	4.5724707
SC1095	0.06568	3.512154	3.6381264
V23010-1.58	0.06741	4.144511	3.5612374
VR-7	0.07756	6.654960	3.9339638
VR-8	0.05183	1.937947	2.5965970
VR-12	0.07676	5.548725	4.3662163

14.4 AUTOROTATIVE PERFORMANCE

One of the fundamental differences between fixed-wing aircraft and rotary-wing aircraft is that loss of engine power on the helicopter is a serious danger to flight safety: the vehicle would rapidly lose altitude and control. Since the rotorcraft gains most of its lift from the operation of the main rotor, a solution must be devised to deal with this event. Most modern helicopters are powered by two gas turbine engines, and failure of one engine leaves room for maneuvering with the remaining engine (*redundancy*). The problem is more serious in the case of tandem rotors and tilt-rotor vehicles, because loss of power on one rotor would inevitably create a catastrophic rolling moment on the vehicle. This is avoided with a construction that delivers power to both shafts. However, consideration must be made for those extreme cases when there is a total engine failure. For Category A rotorcraft, the FAR regulations prescribe that static longitudinal stability must be maintained in autorotation at air speeds from 0.5 times the speed for minimum rate of descent, or 0.5 times the maximum range glide speed.

Autorotation can be considered a three-phase maneuver:

1. The time between loss of engine power and the pilot's reaction. This phase is relatively short, about 3 seconds, but long enough for the rotor to decelerate, and for the aircraft to lose altitude.
2. Stabilized autorotational descent, when the pilot has reacted to the loss of power, and has reduced the collective pitch to bring the aircraft to a steady rate of descent.
3. The landing phase, when the pilot has to reduce the rate of descent to a minimum impact velocity, by increasing the pitch and flaring up the aircraft.

A safe landing speed is about 2.5 m/s; up to 6 m/s there can be considerable damage to the aircraft and consequences to the passengers; a higher landing speed leads to a crash. One way to slow down the descent of the helicopter is to use the main rotor as a windmill (*autorotation*). Rotation itself is induced by the descending vehicle. The idea is to use the rotor to create as much drag as possible in the downward flight, and therefore to minimize the impact velocity.

A number of classical studies exist in this field. Slaymaker and Gray[419,420] developed a semi-empirical method, on the basis of their experiments of free falling rotors, with reference to the effects of the rotor's inertia, disk loading and pitch setting, and rate of variation of the pitch (*flare-up*). This followed the theoretical work of Nikolsky and Seckel[421,422], in addition to the work of Gessow and Myers[423]. Computational work in this field includes that of Houston[424,425], who has solved flight dynamic problems[426]. A review of the aerodynamics of autorotating systems (including Lanchester's propeller and the finned missile) is available in Lugt[427].

14.4.1 Steady Autorotative Performance

Assume in first analysis that the rotor has transitioned from power mode to autorotation mode, and that the aircraft has no forward speed. This implies that the kinetic energy of the rotor is expended to slow down its descent. In other words, the kinetic energy of the rotor is used to overcome the induced power and the blades' profile power. Hence the rotor moves around freely without effective torque. When this balance is guaranteed, then the descent velocity will be minimal.

In Chapter 5 we derived the general expressions for thrust, torque and power coefficients for a propeller. The same expressions hold in this case, when we change the reference system (the speed is vertical, instead of horizontal). The torque and power coefficients are

$$dC_P = dC_Q \simeq \frac{N(dL\phi + dD)y}{\rho A(\Omega R)^2 R} = \frac{N}{2}\rho cdy U^2 \frac{(C_L\phi + C_D)y}{\rho A(\Omega R)^2 R}. \tag{14.17}$$

Now replace the definition of solidity, Eq. 5.23, normalize the radial coordinate by $r = y/R$, and simplify

$$dC_Q \simeq \frac{1}{2}\sigma \frac{U^2}{(\Omega R)^2}(C_L\phi + C_D)r dr. \tag{14.18}$$

The total velocity is the resultant of the rotational velocity of the blade section, the induced velocity v_i and the descent velocity v_s (sinking speed, a *negative velocity*).

$$\frac{U^2}{(\Omega R)^2} = \left(\frac{v_s + v_i}{\Omega R}\right)^2 + \left(\frac{\Omega y}{\Omega R}\right)^2 = \left(\frac{v_s + v_i}{\Omega R}\right)^2 + r^2. \tag{14.19}$$

Define the induced velocity ratio

$$\lambda_s = \frac{v_s + v_i}{\Omega R} = \frac{v_s}{\Omega R} + \frac{v_i}{\Omega R}. \tag{14.20}$$

With the substitution of Eq. 14.20 and Eq. 14.19, the element of torque coefficient becomes

$$dC_Q \simeq \frac{1}{2}\sigma(\lambda_s^2 + r^2)(C_L\phi + C_D)r dr. \tag{14.21}$$

It can be shown that the term λ_s^2 is very small, and can be neglected in all the practical calculations (see Problem 3). The most important approximation is on the inflow angle ϕ,

$$\tan\phi \simeq \phi = \frac{v_s + v_i}{\Omega y} = \frac{v_s + v_i}{\Omega R}\frac{R}{y} = \frac{\lambda_s}{r}. \tag{14.22}$$

With this expression of the inflow angle, the C_Q becomes

$$dC_Q \simeq \frac{1}{2}\sigma r^2 \left(C_L\frac{\lambda_s}{r} + C_D\right)r dr. \tag{14.23}$$

Now consider the element of thrust, and follow the same blade element analysis

$$dT \simeq dL = \frac{N}{2}\rho cdy U^2 C_L. \tag{14.24}$$

The thrust coefficient will be

$$dC_T = \frac{dT}{\rho A (\Omega R)^2} = [\cdots] = \frac{1}{2}\sigma C_L \frac{U^2}{(\Omega R)^2} dr = \frac{1}{2}\sigma C_L r^2 dr. \tag{14.25}$$

Equation 14.25 can be introduced in the torque coefficient, Eq. 14.23

$$dC_Q = dC_T \lambda_s + \frac{1}{2}\sigma r^3 C_D dr. \tag{14.26}$$

The total torque is found from integration of dC_Q along the blade. We need to know the actual inflow conditions and the aerodynamic coefficients at each blade section. However, we make an educated guess regarding the mean values of these coefficients and assume that the inflow velocity is constant. This approximation will suffice to find a first-order estimate of the descent speed v_s. Hence the integral is

$$C_Q = C_T \lambda_s + \frac{1}{2}\sigma \int_0^1 r^3 C_D dr = C_T \lambda_s + \frac{\sigma}{8}\overline{C}_D. \tag{14.27}$$

The condition of steady autorotative descent requires $C_Q = 0$, therefore

$$C_T \lambda_s + \frac{\sigma}{8}\overline{C}_D = 0. \tag{14.28}$$

Equation 14.28 in terms of λ_s yields

$$\lambda_s = -\frac{\sigma \overline{C}_D / 8}{C_T}. \tag{14.29}$$

The corresponding descent speed is

$$\frac{v_s}{\Omega R} = -\left(\frac{v_i}{\Omega R} + \frac{\sigma \overline{C}_D / 8}{C_T}\right). \tag{14.30}$$

The $\overline{C}_D$ in Eq. 14.30 is the average C_D of the blade section; it includes the profile drag and the lift induced drag. If the pilot has managed to convert the helicopter into steady autorotative descent, then $W = T$, and the induced velocity v_i is found as in the case of hover

$$\frac{v_s}{\Omega R} = -\left(\sqrt{\frac{C_T}{2}} + \frac{\sigma \overline{C}_D / 8}{C_T}\right). \tag{14.31}$$

If $\overline{C}_D \simeq 0$, then

$$\frac{v_s}{\Omega R} = -\sqrt{\frac{C_T}{2}} = -\frac{v_i}{\Omega R}. \tag{14.32}$$

Equation 14.32 is equivalent to the condition of *ideal autorotation*, which occurs when the descent speed is equal and opposite to the average induced velocity at the disk,

$$P = T(v_s + v_i) = 0. \tag{14.33}$$

In summary, for the calculation of steady autorotative descent follow algorithm below.

Computational Procedure

1. Read the helicopter's data.
2. Set the thrust coefficient C_T.
3. Calculate the descent speed from Eq. 14.31.

The C_T can be increased to study the effect of disk loading on the autorotational descent. For the calculation of the equivalent drag coefficient of the rotor in steady descent, recall that $W = D_{eq}$, therefore

$$C_{D_{eq}} = \frac{2W}{\rho A v_s^2}. \tag{14.34}$$

An example of calculation is shown in Figure 14.4, where we have plotted the descent speed $v_s/\Omega R$ and the equivalent drag coefficient $C_{D_{eq}}$ in steady autorotative descent for the reference helicopter at sea level with a gross weight of 3,000 kg.

The $C_{D_{eq}}$ varies greatly, depending on the C_T. A flat plate in normal flow at high Reynolds numbers has $C_D \simeq 1.17$. In the present context, an estimate for the induced

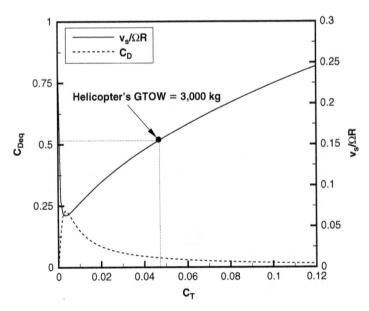

Figure 14.4 *Steady autorotative descent velocity and $C_{D_{eq}}$ for the reference helicopter, sea level, standard conditions.*

velocity in hover is $v_h = \sqrt{W/2\rho A} \simeq 10.35$ m/s. This corresponds to a $C_T \simeq 0.0047$, and hence a $C_{D_{eq}} = 0.52$ with a constant descent speed of 34 m/s.

Clearly, one way to reduce the descent speed is to reduce the disk loading; this can be achieved either by increasing the rotor's diameter, or by decreasing the aircraft's gross weight. For reference, to increase the effective drag coefficient to 1.17, we need a $C_T \simeq 0.017$, and therefore $v_s/v_h \simeq -1.9$. Seddon[391] proposed to estimate the effective drag coefficient from $C_D = 4/(v_s/v_h)$, which seems optimistic.

In Figure 14.5 we have plotted v_s for the actual diameter, and for a larger diameter. An increase of rotor diameter by 25% decreases the descent rate by 20%. Increasing the rotor's diameter is not without complications: a large diameter has poor performance, both in hover and in forward flight. Note that this analysis considers the main rotor alone, not the full helicopter.

A theoretical method for the calculation of the autorotational performance in forward flight consists of developing the forward flight power one step further. Consider the power equation in level flight, Eq. 13.4. Normalize this equation, so that

$$C_Q = \frac{P_{level}}{\rho A (\Omega R)^2 R} = 0. \tag{14.35}$$

This condition also requires that

$$P_i + P_o + P_p + P_{tr} + P_t = 0. \tag{14.36}$$

Therefore, the condition on the induced power is

$$P_i = -(P_o + P_p + P_{tr} + P_t). \tag{14.37}$$

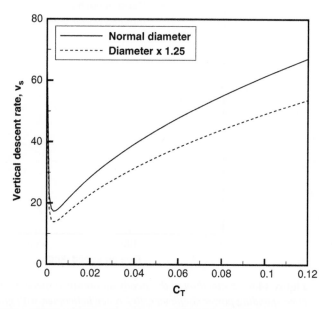

Figure 14.5   *Effect of rotor diameter on autorotative descent.*

The value of P_i from Eq. 14.37 is introduced in the general expression for the induced power in forward flight

$$\frac{P_i}{P_h} = \frac{\lambda_s}{\lambda_h}. \tag{14.38}$$

This equation has to be solved for the unknown λ_s. Starting from the computational procedure in §13.4, we have to add a few lines to the computer program.

Computational Procedure

1. Calculate the induced power from Eq. 14.37.
2. Calculate the induced velocity factor λ_s from Eq. 14.38.
3. Calculate the autorotational descent speed from

$$v_s = \lambda_s \Omega R. \tag{14.39}$$

4. Increase the true air speed and repeat the procedure.

In light of the previous analysis, the result is quite straightforward, see Figure 14.6. The minimum descent speed is estimated at $v_s = 13.5$ m/s (2,630 ft/min) at a forward speed $U \simeq 125$ km/h (67.5 kt). This descent speed is high, though not as severe as the vertical autorotation case. In order to reduce this speed even further, the pilot must attempt to pitch up the aircraft when nearing the ground, so as to increase the overall drag at the expense of the rotor's energy. Note that this analysis has considered the full helicopter.

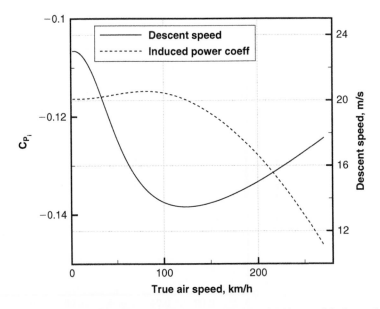

Figure 14.6 *Estimation of the steady autorotative speed in forward flight, and corresponding power coefficient. Reference helicopter, with gross weight $W = 3,000$ kg at sea level.*

14.4.2 Transient Autorotative Performance

In general, the pilot will have a few seconds to take the aircraft into steady autorotative descent. There will be a reaction time to consider. In those initial moments the rotor speed starts decreasing and the aircraft will lose altitude. The rotor speed used in the calculation of steady descent is, in fact, lower than the rotational speed in normal operation. The pilot will have to take the aircraft toward the free autorotational mode by decreasing the pitch, and thus the thrust.

The certification rules (FAR Part 29) require that a range of rotor speeds and pitch limits are established, and that a warning must be available under all conditions (including power-on and power-off flight), when the rotor speed approaches a critical value.

To estimate the initial transient, we write the dynamic equation for the aircraft, and the dynamic equation for the rotor. The dynamic equation for the aircraft is

$$m\frac{\partial^2 h}{\partial t^2} = m\frac{\partial v_s}{\partial t} = T - W, \tag{14.40}$$

$$\frac{\partial v_s}{\partial t} = g\left(\frac{T}{W} - 1\right). \tag{14.41}$$

Use the definition of C_T in the latter equation

$$\frac{\partial v_s}{\partial t} = g\left(\frac{C_T \rho A(\Omega R)^2}{W} - 1\right). \tag{14.42}$$

The rate of variation of Ω is found from the dynamic equation for the rigid rotor around the rotational axis

$$I_b\frac{\partial \Omega}{\partial t} = Q_e - Q_a, \tag{14.43}$$

where Q_e is the engine torque and Q_a is the aerodynamic torque. Under normal operation $Q_e = Q_a$. At the point of engine failure $Q_e = 0$; the rotor will decelerate according to the equation

$$\dot{\Omega} = \frac{\partial \Omega}{\partial t} = -\frac{Q_a}{I_b}. \tag{14.44}$$

Now introduce the definition of torque coefficient and Eq. 14.27 in Eq. 14.44,

$$\dot{\Omega} = -\frac{C_Q}{I_b}\rho A(\Omega R)^2 R = \frac{\rho A(\Omega R)^2 R}{I_b}\left(C_T\lambda_s + \frac{\sigma}{8}\overline{C}_D\right), \tag{14.45}$$

$$\dot{\Omega} = -\frac{\rho\pi R^5}{I_b}\left(C_T\lambda_s + \frac{\sigma}{8}\overline{C}_D\right)\Omega^2. \tag{14.46}$$

A few additional comments on Eq. 14.46 are required. The $\overline{C}_D$ can be considered a constant; C_T can be considered constant before the pilot reacts to the situation, then it is quickly reduced by reduction of the pitch angle; λ_s cannot be a constant. The problem is further complicated by the fact that the rotor is descending in the same direction as its slipstream, and ends in the vortex ring state (see below), where the theory breaks down. There is no easy way to calculate the average induced velocity v_i, and no easy way to calculate λ_s. The only thing we know is that it should slowly decrease to zero as the aircraft descends, e.g. the rotor reaches its point of ideal rotation. Steady state experiments and analysis[324,428,429] indicate that at the point of steady autorotation $v_i/v_h \simeq 1.85$. Here we solve the problem for these initial seconds required for the pilot to react, and therefore C_T and λ_s are constant and equal to the hover condition. A sensitivity analysis can be carried out, with λ_s gradually decreasing as the rotor loses altitude. The problem is reduced to solving the system

$$\frac{\partial v_s}{\partial t} = g(c_1 C_T \Omega^2 - 1), \tag{14.47}$$

$$\dot{\Omega} = c_2 \left(C_T \lambda_s + \frac{\sigma}{8}\overline{C}_D\right)\Omega^2, \tag{14.48}$$

with the constant coefficients

$$c_1 = \frac{\rho A R^2}{W}, \qquad c_2 = -\frac{\rho A R^5}{I_b}. \tag{14.49}$$

The initial conditions are

$$t = 0, \qquad \Omega = \Omega_o, \qquad v_s = 0. \tag{14.50}$$

Computational Procedure

1. Read all the aircraft data.
2. Set flight altitude.
3. Set boundary conditions, Eq. 14.50.
4. Integrate Eq. 14.47 and Eq. 14.46 in the time domain using one of the well-known methods, for example a fourth-order Runge-Kutta method.

The solution of the problem is shown in Figure 14.7. The vertical acceleration starts from 1g (free fall) and it gradually decreases. The descent velocity is nearly parabolic. The corresponding decrease in rotational velocity is shown in Figure 14.8. A loss of over 20% of rotational speed is estimated for the first 6 seconds.

The calculation presented was based on the performance of the rotor only. In practice, the vertical drag of the fuselage helps decrease the descent speed, though not to a value that can be considered safe for landing.

We can study the effects of rotor diameter, aircraft weight, polar moment of inertia and other parameters on the transient autorotative performance; to minimize these effects, see Problem 4.

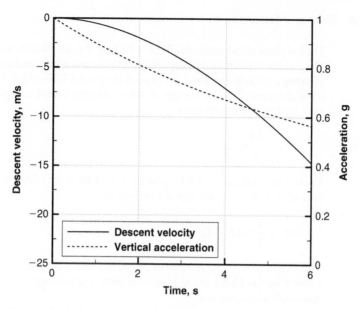

Figure 14.7 *Transient autorotative descent; simulation with constant C_T for the first few seconds; vertical acceleration normalized with the acceleration of gravity, g.*

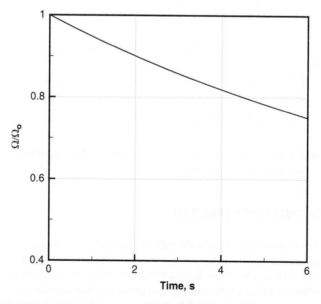

Figure 14.8 *Rotational speed of the rotor for transient autorotative descent.*

14.4.3   Flare and Touchdown

The last phase of autorotational flight is the flare, e.g. the pitching up of the helicopter. The amount of pitch-up will depend on the pilot and operational constraints, such as avoiding a tail strike and loss of vision of the ground. The vehicle is slowed down by using the residual kinetic energy. If ΔE is the total kinetic energy usable, then the estimated time to touchdown is

$$\Delta t \simeq \frac{\Delta E}{P_{OGE}}. \tag{14.51}$$

The kinetic energy of the rotor is given by Eq. 14.10. If the vehicle was in hover before the engine failure, the actual usable energy is

$$\Delta E = \frac{N}{2} I_b (\Omega^2 - \Omega_1^2), \tag{14.52}$$

where Ω_1 is the minimum rotational speed required to produce a thrust $T \sim W$. With a speed Ω_1 the blades are at maximum pitch and produce the maximum C_L. It is reasonable to assume that

$$T = C_T \rho A (\Omega R)^2 = C_{T_1} \rho A (\Omega_1 R)^2, \tag{14.53}$$

$$C_T \Omega^2 \simeq C_{T_1} \Omega_1^2. \tag{14.54}$$

If we insert this equivalence into Eq. 14.52, we have

$$\Delta E = \frac{N}{2} I_b \Omega^2 \left(1 - \frac{C_T}{C_{T_1}} \right), \tag{14.55}$$

where C_T is the nominal thrust coefficient before the failure and C_{T_1} is the maximum thrust coefficient.

If the helicopter was in forward flight before the engine failure, then

$$\Delta E = \frac{N}{2} I_b \Omega^2 \left(1 - \frac{C_T}{C_{T_1}} \right) + \frac{N}{2} M (v_s^2 + V^2). \tag{14.56}$$

In this case, v_s denotes the descent speed before the start of the flare and V is the forward component of the speed.

14.5   HEIGHT/VELOCITY DIAGRAM

The scenario described in the previous section (vertical flight) is unlikely to occur in practice. When the helicopter has a forward flight component the problem is different: there are regions in the flight envelope in which the helicopter can land safely, and others in which a safe landing is not possible. These regions in the altitude/speed diagram are sometimes called *dead man curves*. This term derives from the unfortunate fact that pilots have lost their life trying to determine safe autorotational envelopes by

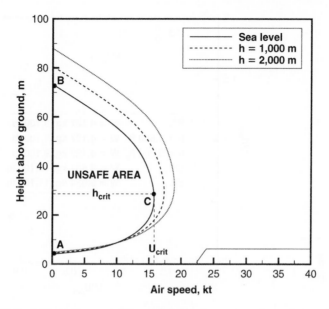

Figure 14.9 *h/V envelopes at different heights above ground; constant weight.*

flight testing. These trials are done by skilled pilots. They start at a fixed altitude and gradually decrease the speed, while simulating a power failure. They attempt to recover control of the aircraft and to make a safe landing. A helicopter in simulated autorotation, which is unable to make a safe landing, must be recovered to powered flight. Therefore, autorotational tests have to be planned carefully, as described by Cooke and Fitzpatrick[4].

One example of an h/V diagram is given in Figure 14.9. The chart is only valid at one gross weight and one atmospheric condition, and changes, also considerably, with the aircraft's weight. The chart illustrates the regions where a total engine failure will not allow a safe landing in autorotation. There are two disjoint areas on unsafe operation. The left side of the chart shows the low-speed and high-altitude performance of the helicopter. Clearly, the pilot would want this area to shrink as much as possible. The boundaries of the h/V diagram can also be interpreted as constraints on the flight path during take-off and landing. By operating the aircraft within the boundaries of the safe envelope, ground operations can continue.

The low-speed envelope is characterized by a number of critical points: (A) the minimum hover altitude, below which a safe autorotational landing is not possible; (B) the maximum hover altitude, above which a safe autorotational landing is not possible; and (C) a maximum air speed at medium altitudes, below which a safe autorotational landing is not possible. The latter point corresponds to a critical speed and a critical altitude. These points have been found to depend linearly from both the weight and the density altitude, therefore some scaling can be done to reduce them to a single curve for each helicopter.

Flight tests on various helicopters (Pegg[430]) have indicated that the critical altitude is nearly constant, and equal to about 29 m (94 ft) from the ground. The critical speed, instead, depends on the speed of minimum power, U_{mp}, and the thrust coefficient

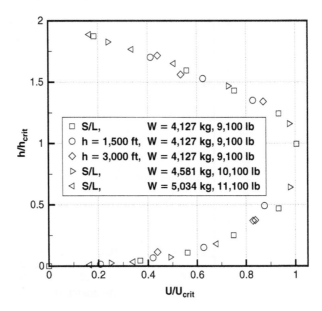

Figure 14.10   *Low-speed h/V envelope of the Sikorsky H-34. Data from Pegg[430].*

C_T/σ. A suitable approximate semi-empirical equation, elaborated from the data presented by Pegg, is

$$U_{crit} = 2.8 U_{mp} + 5\frac{C_T}{\sigma} - 165 \quad (kt), \tag{14.57}$$

where both U_{crit} and U_{mp} are given in knots, and U_{mp} will be calculated in §15.4. If the flight data at different gross weights and flight altitudes are normalized with the critical height and speed, the curves collapse into a single envelope. This is shown in Figure 14.10 for the helicopter Sikorsky H-34 Choctaw, a first-generation utility helicopter of the US Navy.

As discussed earlier, the control of the helicopter in autorotation requires a sharp increase in the collective pitch. A time history of the pitch for the helicopter Sikorsky H-34 is shown in Figure 14.11. The figure shows the response time, which is about 3 seconds.

Gessow and Myers[423] published results of autorotating helicopters over a range of forward speeds, and found that the minimum descent speed was intermediate, and corresponding to the largest L/D. Results of investigations on the height/velocity curve can be found in Pegg[430], who also derived a semi-empirical method to correlate the chart to different helicopter weights and atmospheric conditions. Carlson and Zhao[431] have proposed a calculation method for tilt-rotor aircraft using an optimum control theory.

14.6   THE VORTEX RING STATE

The helicopter operates in the *vortex ring state* when the blades travel in the wake that has been previously released by other blades. The main indicator of vortex ring state

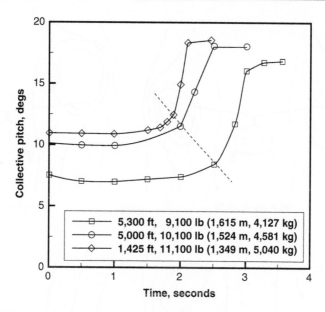

Figure 14.11   *Collective pitch history for the helicopter H-34 in hover at one diameter above the ground. Data from Pegg*[430].

is the increase in mechanical vibration at relatively low frequency, power oscillations, and loss of control effectiveness, accompanied by increased noise. In particular, the vortex ring occurs during autorotational descent, in vertical powered descent and in other flight conditions when the tip path plane moves in the same direction as its slip-stream. A simultaneous phenomenon is the blade/vortex interaction (BVI). Most flight measurements indicate that the helicopter can enter the vortex ring state without compromising the flight recovery. The flow pattern for a vertical descent was visualized by Brotherhood[432] as early as 1949, from flight testing on the Sikorsky R-4B.

A number of operational states can be identified, as sketched in Figure 14.12, adapted from Brotherhood[432]. The descent rates have been estimated from the flight data. Case (a) refers to a low descent rate in vertical flight. Some streamlines are pushed down by the prevailing downwash, but re-enter the rotor disk from the tip, because the rotor is traveling downwards. This course of events is roughly independent of the blade azimuth, and gives rise to the vortex ring, a well-defined vortex structure enveloping the rotor. Case (b) refers to an increased descent rate, in which the vortex ring is mostly confined to a region above the disk. A region of reversed flow and some turbulence coexist with the vortex ring. Case (c) shows a sketch of events for increased descent velocity. The vortex ring disappears, and is replaced by a turbulent flow above the rotor; this is the *turbulent wake state*. The rotor operates in ideal autorotation when the mean mass flow through the disk is zero. This occurs when $v_c = -v_i$.

Finally, at even higher descent rates the slipstream assumes a more structured shape, and the rotor operates like a *windmill brake*. In this case the transfer of power is from the air to the rotor, and the slipstream will expand from below to above the rotor disk.

The momentum theory is valid when there is a unidirectional flow through the rotor disk. In the windmill brake state, uniformity in the flow through the disk is more or less

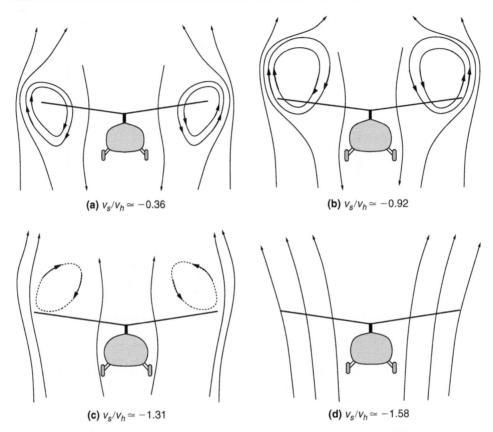

(a) $v_s/v_h \simeq -0.36$ **(b)** $v_s/v_h \simeq -0.92$

(c) $v_s/v_h \simeq -1.31$ **(d)** $v_s/v_h \simeq -1.58$

Figure 14.12 *Vortex ring state for helicopter in vertical descent. Flow patterns elaborated from Brotherhood[432]. Data used: $W = 1,245$ kg; $d = 11.58$ m; $h = 1,000$ m.*

re-established. This allows the axial momentum theory to be used again. Following the analysis of the axial climb, §12.4, if v_s is the rate of descent, the rotor thrust is

$$T = -\dot{m}w = -\rho A(v_s + v_i)w = -2\rho A(v_s + v_i)v_i. \tag{14.58}$$

The negative sign is required because now the mass flow rate is inverted, see Figure 14.12, case (d). Simplification of this equation yields

$$v_i^2 + v_s v_i + \frac{T}{2\rho A} = 0. \tag{14.59}$$

The solution is

$$v_i = -\frac{1}{2}v_s \pm \frac{1}{2}\sqrt{v_s^2 - \frac{2T}{\rho A}} \simeq -\frac{1}{2}v_s \pm \frac{1}{2}\sqrt{v_s^2 - \frac{2W}{\rho A}}. \tag{14.60}$$

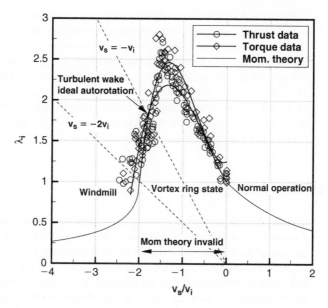

Figure 14.13   *The vortex ring state from rotor measurements, Castles and Gray[433]; constant chord, untwisted blades; $\lambda_i = v_s/v_h$.*

We can introduce the value of v_h from Eq. 12.1 in Eq. 14.60, and divide through by v_h, so that

$$\frac{v_i}{v_h} = -\frac{1}{2}\frac{v_s}{v_h} \pm \sqrt{\left(\frac{v_s}{2v_h}\right)^2 - 1}. \qquad (14.61)$$

This equation makes sense only if $v_s \geq 2v_h$, e.g. when the rate of descent is more than twice the induced velocity in hover. Hence only the positive sign gives a physical solution,

$$\frac{v_i}{v_h} = -\frac{1}{2}\frac{v_s}{v_h} + \sqrt{\left(\frac{v_s}{2v_h}\right)^2 - 1}. \qquad (14.62)$$

A plot of this ratio is shown at the bottom left of Figure 14.13. Analysis of the rotor performance at intermediate rates of descent ($-2 < v_i/v_h < 0$) has been the domain of experimental research for many years. To cite some work of interest in this area, Castles and Gray[433] performed wind tunnel force tests on four model rotors and compared the thrust, induced velocities and vertical descent with the blade element theory; Azuma and Obata[434] and Washizu et al.[428] published results of experimental work on the vortex ring wake; Lee[435] and Bryson et al.[436] solved the autorotational problem as an optimal control problem, with an approximate model for the vortex ring state; and Leishman and his co-workers[437,438,439] and Newman[440] published extensively on numerical simulation methods, based on unsteady wake analysis.

Figure 14.13 shows the summary of operation modes of the helicopter rotor in a single graph. The symbols denote experimental data of Castles and Gray, obtained with a 6-foot (1.85 m) diameter rotor, with constant chord, and untwisted blades. The airfoil section was a NACA 0015, and the rotor's solidity was $\sigma = 0.05$. The induced velocity factor was calculated from both thrust and torque measurements. These experiments were limited to the vortex ring state, when the induced velocity factor $-2 < v_i/v_h < 0$. The point of ideal autorotation is found from Eq. 14.33. The line $v_s = -v_i$ intersecting the experimental data gives the value of the corresponding induced velocity factor; this is estimated at about 1.9.

14.7 TAKE-OFF AND LANDING

The problem of take-off and landing (terminal area operations) may be subject to a number of lateral and vertical constraints, depending on the space available and the size of the helipad. The take-off performance must be determined so that, if one engine fails, the rotorcraft can either return safely to base or continue the take-off and climb in compliance with the minimum climb regulations. For these reasons, a lot of research has been devoted to the simulation of OEI operations. Yoshinori and Kawachi[441] solved the take-off problem for Category-A helicopter operations. Cerbe and Reichert[442] optimized the take-off performance of a general utility helicopter (the Eurocopter BO-105), with all engines operating and with one engine failure, based on the critical decision point of the pilot. Schmitz[443,444] used the optimal control theory to optimize the take-off and landing trajectory of a heavy-lift helicopter. Similar optimal control analysis has been carried out by Zhao *et al.*[445,446] on a UH-60A. In any case, the helicopter is modeled as a point mass.

14.8 TURN PERFORMANCE

In this section we discuss the helicopter performance during a banked turn. With a leap of imagination, we reduce the helicopter to a single point mass. The tail rotor forces and moments are neglected. The main rotor forces are reduced to a thrust at the center of the rotor. Whalley[447] developed a model for determining the maximum maneuvering performance of a helicopter. His model uses variable pitch and roll damping, maximum pitch and roll rate, and maximum load-factor capability. Three maneuvers were investigated: a 180-degree turn, a longitudinal pop-up, and a lateral jink.

Another case of helicopter turn includes the nap-of-the-Earth flight (or terrain flight). This is a flight close to the Earth's surface during which air speed and/or altitude are changed to follow the contours in order to avoid detection. See Kelley *et al.*[448] for a description of this and other flight maneuvers.

First, consider the helicopter performing a turn. All the turn quantities are indicated with the subscript "t". The helicopter is banked by an angle ϕ. As usual, there is the problem of finding the point where all the forces are applied. The rotor forces are applied at the rotor disk; the body forces on the airframe are applied at the center of gravity. Then there are the aerodynamic forces, which are applied to the center of pressure (unknown), and the tail rotor forces, applied at the center of the tail rotor. In

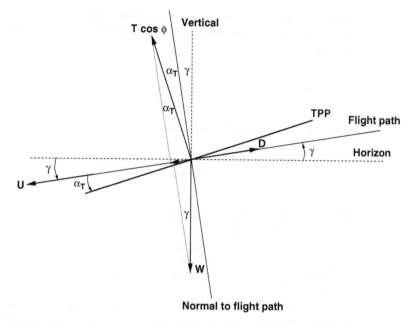

Figure 14.14 *Helicopter's turn radius as a function of normal load factor, sea level, ISA conditions.*

addition, one must consider the moments arising from the banked configuration. The moment created by the center of gravity tends to restore the straight flight condition of the aircraft.

The component of the thrust on the *y–z* plane is $T \sin \phi$, see Figure 14.14. The forces on the aircraft for a steady state (coordinated) turn on a horizontal plane are:

$$T_t \cos \alpha_T \sin \phi - m \frac{U^2}{\chi} = 0, \tag{14.63}$$

$$T_t \cos \alpha_T \cos \phi - W = 0, \tag{14.64}$$

where χ is the radius of curvature of the flight path during the turn. By combining Eq. 14.63 and Eq. 14.64 we find the angle of banking

$$\tan \phi = \frac{U^2}{g \chi}. \tag{14.65}$$

The bank angle is found to be independent on the rotor thrust and on the tilt angle. The moment created by the center of gravity would tend to restore the straight flight condition of the aircraft. If we neglect the effect of the tail rotor, the turn generates a moment around the center of the main rotor

$$M_t = W \sin \phi \, h_{cg}, \tag{14.66}$$

where h_{cg} is the distance between the center of the rotor and the center of gravity. If n is the normal acceleration factor, $n = T_t/W$, then the bank angle ϕ can be expressed in terms of n from Eq. 14.64,

$$\cos\phi = \frac{W}{T_t} = \frac{1}{n}. \tag{14.67}$$

From Eq. 14.65 the radius of curvature of the turn is

$$\chi = \frac{U^2}{g}\frac{1}{\tan\phi} = \frac{U^2}{g}\frac{1}{\sqrt{n^2-1}}, \tag{14.68}$$

where we have used the equivalence $\tan^2\phi + 1 = 1/\cos^2\phi$, along with Eq. 14.67. Modern high-performance helicopters have a limit value of n of about 3.5. The FAA requires that a commercial helicopter withstands a normal acceleration of at least $1.15g$ with a 30 degree bank turn at VNE (never-to-exceed speed). A plot of the turn radius versus normal load factor is shown in Figure 14.14 for two reference air speeds.

The radius of turn found in this analysis may seem a large number, when one thinks of the agility required by a modern helicopter. In fact, helicopters can turn around themselves and dive down at phenomenal speeds. This analysis is based on the assumption of steady state correctly banked turn. From the instantaneous radius of turn we can calculate the turn rate $\dot\psi$

$$\frac{U}{\chi} = \dot\psi = \frac{g}{U}\sqrt{n^2-1}, \tag{14.69}$$

having replaced Eq. 14.68. When the helicopter is turning, its lift capability is increased, due to unloading of the retreating blade, and loading of the advancing blade. As a consequence, stall is alleviated on the retreating blade. This effect appears for turns in both directions, regardless of the sense of rotation of the rotor.

14.9 POWER REQUIRED FOR TURNING

In the general case, the instantaneous flight path of the helicopter during a banked turn is inclined by an angle γ on the horizon. The main rotor is inclined by the angle $\alpha_T + \gamma$ with respect to the vertical plane, as illustrated in Figure 14.15. The essential aspect of this flight is that the thrust component on the vertical plane tangent to the trajectory is $T\cos\phi$, where ϕ is the banking angle. Unless this thrust is increased, the helicopter will descent and decelerate. There are two ways to perform this analysis: with the energy method and with the balance of the forces in the relevant planes, as illustrated in Figure 14.15. The latter method is more elaborate.

The total energy of the helicopter can be written as

$$E = Wh + \frac{1}{2}mU^2 + E_{mr} + E_{tr}, \tag{14.70}$$

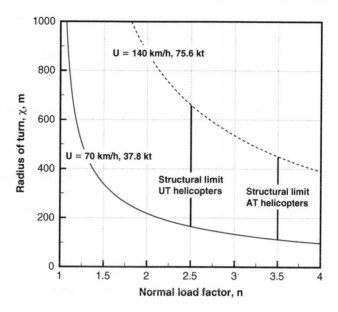

Figure 14.15 *Summary of forces and angles on the vertical plane through the flight path.*

where U is the aircraft's speed; E_{mr} and E_{tr} are the main and tail rotor energy, respectively. The total energy divided by the weight W is the energy height

$$h_E = \frac{E}{W} = h + \frac{1}{2g}U^2 + \frac{E_{mr}}{W} + \frac{E_{tr}}{W}, \tag{14.71}$$

The change of energy height with respect to time is the climb or descent rate, or specific excess power

$$\frac{\partial h_E}{\partial t} = \frac{\partial h}{\partial t} + \frac{U}{2g}\frac{\partial U}{\partial t}. \tag{14.72}$$

Since the rotational speed is maintained constant during the manoeuvre, there are no appreciable changes in E_{mr} and E_{tr}. Hence

$$\frac{\partial h_E}{\partial t} = \frac{\partial h}{\partial t} + \frac{U}{2g}\frac{\partial U}{\partial t} = \frac{T - D}{W}U. \tag{14.73}$$

First, consider a turn at constant altitude ($\gamma = 0$, $h = $ constant). The energy equation in the vertical plane tangent to the flight path will be

$$\frac{U}{2g}\frac{\partial U}{\partial t} = \frac{T_t \cos\phi - D}{W}U, \tag{14.74}$$

$$\frac{\partial U}{\partial t} = \frac{2}{m}(T_t \cos\phi - D_t) = \frac{2}{m}\left(\frac{T_t}{n} - D_t\right), \tag{14.75}$$

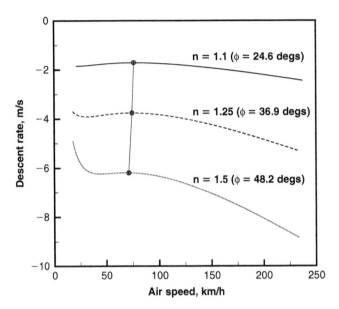

Figure 14.16 *Helicopter turn at constant speed and thrust. Graph shows descent rates versus air speed at selected load factors. Sea level conditions; gross weight W = 3,000 kg.*

If during the turn the thrust is maintained equal to its value before the turn, then $T_t = T$. Therefore, the helicopter cannot overcome the drag, which has either increased or remained constant, and must decelerate as it turns. If the thrust is increased to $T_t = Tn$, then the helicopter can perform a coordinated turn at constant speed and constant altitude, assuming that there is no change in drag.

Next, assume that the speed is maintained constant, and that the helicopter is allowed to climb or descent. The energy equation in the vertical plane tangent to the flight trajectory becomes

$$\frac{\partial h_E}{\partial t} \simeq v_s = \frac{T_t \cos \phi - D_t}{W} < 0 \tag{14.76}$$

Therefore, the aircraft must descend.

We have seen that the average tilt angle α_T is related to the drag. The combination of momentum equations leads to

$$\tan \alpha_T = \frac{D_t}{W} \simeq \frac{D}{L}. \tag{14.77}$$

Therefore, the descent angle with the aircraft's speed follows that of the inverse of the glide ratio.

Figure 14.16 shows the calculated descent speed of the reference helicopter with an $AUW = 3,000$ kg for a given load factor (or bank angle). How can we avoid descending if the speed is to be maintained constant? – The thrust has to increase by a factor n;

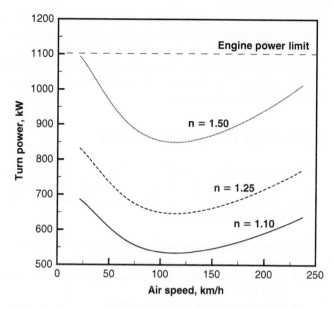

Figure 14.17 *Helicopter turn at constant speed and altitude. Graph shows power required versus air speed at selected load factors. Sea level conditions; gross weight W = 3,000 kg.*

this produces a zero descent speed. Therefore, the power increases by a factor $n^{3/2}$. For a given load factor n, the power required for a coordinated turn at constant speed and altitude may exceed the engine power available. If the drag does not change much during the turn, the turn power follows the trend of the level power. This is shown in Figure 14.17.

The figure shows that at intermediate speeds and load factors there is still some considerable excess power

$$v_s = \frac{\Delta P}{W} = \frac{P_{engine} - P_t}{W} > 0. \tag{14.78}$$

This power can be used for climbing during the coordinated turn.

14.9.1 Unrestricted Turn

We have discussed the case of a coordinated turn, and have neglected the effects of the tail rotor. We have implicitly assumed that the tail rotor thrust is to be used for longitudinal control. However, if the tail rotor thrust is increased while performing a turn, the helicopter can turn around itself at a higher rate. This operation is no longer a coordinated turn. If T_{tr} is the rotor thrust required for straight level flight, an increase in tail rotor thrust ΔT_{tr} will make the aircraft spin around its center of gravity at the rate

$$\dot{\psi} = h_{tr} \Delta T_{tr}, \tag{14.79}$$

where h_{tr} is the distance between the tail rotor shaft and the aircraft's center of gravity. By using a combination of main rotor tilt to perform a coordinated turn, and additional tail rotor thrust, the helicopter will turn inward, and perform a sharper turn than a properly banked turn. The tail rotor thrust can be used to accelerate the aircraft in yaw. For example, a criterion is to obtain a yaw rate of 15 degrees/s within 1.5 seconds with sufficient control margin.

14.10 MORE ON TAIL ROTOR PERFORMANCE

The tail rotor works in a complex aerodynamic environment, in the wake of the main rotor vortices, and in the interference of the tail boom and various control surfaces (horizontal fin, vertical rudder), and the ground proximity. The function of the tail rotor thrust is three-fold:

- To provide the means for yaw/torque balance, in conjunction with the torque on the airframe created by the main rotor.
- To give a good degree of maneuverability of the aircraft under all flight conditions.
- To compensate for gyroscopic effects.

The last point needs clarification. The tail rotor is a gyroscope rotating at high speeds. During yaw the rotational axis is moved, and it creates a precession moment that has to be counterbalanced. Compensation is created by differential lift, and hence from differential pitch. Another aspect of tail rotor operation in hover and low forward flight (say less than 80 km/h) is called *quartering flight*. There are three critical zones, as shown in Figure 14.18. The tail rotor is rotating clockwise, and the main rotor is assumed to be rotating anticlockwise. The zero-azimuth point is when the blade is parallel to the forward speed, in the upwind position.

The shaded zone 1 (right side wind), roughly at azimuth angles $\psi = 45$ to 90 degrees, with relative winds of 35–50 km/h, the thrust required is increased, due to the interaction with tip vortices moving downstream from the main rotor. This situation is also shown in Figure 14.19, with a main rotor in anti-clockwise rotation. It is sometimes called *quartering flight*. These tip vortices also affect the flow around the fin and the vertical tail.

In this particular setting, the main rotor tip vortex is responsible for a tangential velocity component on the blade section of the tail rotor, which reduces the sectional lift and hence the overall thrust. Some experiments suggest that a pusher tail rotor may provide more thrust than a tractor rotor. A pusher tail rotor is designed in such a way as to shield the tail rotor with the fin, and to push the fin against the main rotor tip vortices.

The shaded zone 2 in Figure 14.18 (aft side wind) is critical when hovering in ground effect, because of the interaction between the tail rotor and the ground vortices. Finally, in the shaded zone 3 in Figure 14.18 (left side wind), the tail rotor enters the vortex ring state, which leads to rotor instability, large oscillations in thrust, and eventually to loss of control. The vortex ring state can be limited by bottom-forward rotational movement, with a horizontal stabilizer below the tail rotor. The problem is also limited by high tail rotor disk loading, and small diameter.

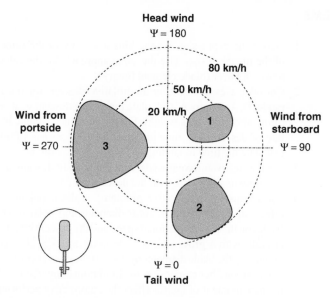

Figure 14.18 *Critical flight conditions in hover. Tail rotor rotating clockwise, and the main rotor rotating anticlockwise; forward speed from above.*

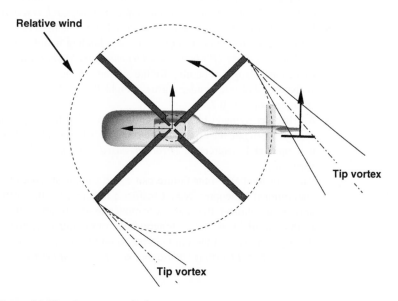

Figure 14.19 *Quartering flight.*

PROBLEMS

1. Derive the expression for the kinetic energy of the rotor, Eq. 14.5, in terms of the tip speed U_{tip}. Use the same approach for the calculation of the profile power in hover (blade element theory).

2. Consider a case of steady autorotative descent for the reference helicopter, model D, Appendix A. Calculate the rate of descent v_s at sea level as a function of the aircraft's gross weight. For the reference weight $W = 3,000$ kg, plot the rate of descent as a function of altitude, from $h = 1,000$ m to sea level, and calculate the time to landing.

3. For the steady autorotative descent theory, §14.4, study and discuss the effects of approximating $\lambda_s^2 \simeq 0$ on the descent speed.

4. For a transient autorotative problem, as discussed in §14.4.2, calculate the effects of aircraft weight, rotor diameter and polar moment of inertia of the blades on the loss of rotational speed and the descent rate. Consider the reference aircraft, with a gross weight $W = 3,000 \pm 50$ kg; also, consider the effect of increasing the blade diameter by 25%. If the construction of the blade is changed from carbon fiber composites to aluminium alloys ($\rho_b = 2,600$ kg/m^3) how will the new moment of inertia affect the autorotative performance?

5. A helicopter flying at low speeds, level flight, with atmospheric side winds of considerable strength, produces a side load on the tail boom. The side force is operating on the tail section opposite to the tail rotor thrust. Investigate the nature of the problem, produce engineering sketches to highlight the local effects and find a technical solution from the available technical literature to overcome the problem. (*Problem-based learning: additional research is required.*)

6. Investigate suitable engineering methods that can be used to produce a height-velocity chart (*dead man's curve*) for safe landing in autorotation, in case of total power failure. (*Problem-based learning: additional research is required.*)

7. Calculate the angle of climb γ for the reference helicopter using the theoretical analysis of §14.9. Consider a gross weight $W = 3,250$ kg at sea level, and an advance ratio $\mu = 0.2$.

8. A helicopter flying at high speed is likely to suffer from a problem called *tail shake*. Investigate the origins of the problem, when and how it occurs, and how it is practically overcome. (*Problem-based learning: additional research is required.*)

9. Partial or total tail rotor failure can cause significant control problems on a conventional helicopter. Now consider a helicopter with a vertical tail and a horizontal stabilizer, flying at low to moderate speeds (up to 100 kt). Analyze the methods available to restore longitudinal trim in flight, without compromising the flight safety, and the limits imposed on the control by the forward speed. (*Problem-based learning: additional research is required.*)

Chapter 15

Rotorcraft Mission Analysis

I always believed that the helicopter would be an outstanding vehicle for the greatest variety of life-saving missions and now, near the close of my life, I have the satisfaction of knowing that this proved to be true.
The last letter written by Igor Sikorsky, October 25, 1972

We devote this chapter to the basic calculations of range and endurance, to the analysis of the payload/range charts of conventional helicopters, and to some essential concepts of mission analysis and fuel estimates. We shall also derive the conditions for speed of minimum power and maximum range. For the helicopter, endurance is more important than range. In Chapter 11 we discussed the hover endurance of the conventional helicopter.

15.1 SPECIFIC AIR RANGE

The specific range is the distance flown in level flight with one unit of fuel (by volume, weight or mass). The fuel flow is proportional to the engine power, therefore the specific air range (*SAR*) is

$$SAR = \frac{\partial R}{\partial m} = \frac{U}{\dot{m}_f} = \frac{1}{SFC} \frac{U}{P}, \tag{15.1}$$

where P is the level flight power, calculated earlier. The specific air range includes the effects of atmospheric winds. In absence of winds, U is equal to the ground speed. The calculation of the fuel flow and the specific range for our reference aircraft is readily done, if the forward flight analysis is already available. The maximum SAR is found at the point of maximum U/P, or minimum P/U.

In first approximation, consider a constant average fuel flow over the speed range (about $6.58 \cdot 10^{-7}$ N/s/watt, or $2.367 \cdot 10^{-3}$ N/h/watt, or 2.367 N/h/kW, or 0.2413 kgf/h/kW, or 0.531 lb/h/kW).

A graphical analysis of the specific range of our reference helicopter is shown in Figure 15.1, for two reference AUW, at sea level. The specific air range reaches a maximum value at intermediate speeds; then it decreases slowly as the aircraft flies at the upper end of its flight envelope.

The calculation method is the following:

Computational Procedure

1. Set the operational conditions: TAS, AUW, density altitude, etc.
2. Use the forward flight power analysis of §13.4 by calling the routine `Assembly Power`, that returns the level flight power P under the specified conditions.

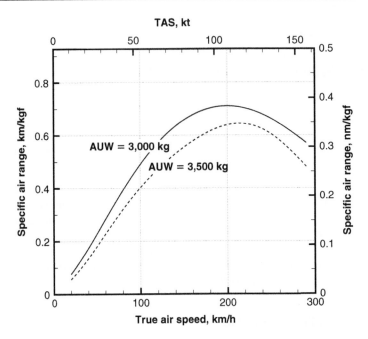

Figure 15.1 *Specific air range versus flight speed at sea level, ISA conditions, at AUW as indicated.*

3. Interpolate the fuel flow from the engine's charts (see also Figure A.20) to find the corresponding value of the fuel flow.
4. Calculate the SFC from $SFC = \dot{m}_f/P$ (in the most useful units).
5. Calculate the SAR from Eq. 15.1 (in the most useful units).
6. Advance the speed and repeat the calculations.

Interpolation of the engine data is done in a variety of ways: for example by fitting a high-order polynomial, or a cubic spline.

The specific range is a function of several factors, the same factors affecting the power requirements for level flight. Therefore, charts of SAR tend to be complicated by the number of parameters involved: aircraft mass, flight altitude, ISA conditions, weather-related atmospheric conditions, rotor speed, flight speed, and disk loading.

Example

To give an idea of the order of magnitude of the specific range, consider the aircraft $AUW = 3,000$ kg, at a reference speed $U = 140$ km/h (75.5 kt) at sea level, ISA conditions. Assume a value of the specific fuel consumption $SFC = 6.58 \cdot 10^{-7}$ N/s/watt. The level flight power is estimated at $P \simeq 570$ kW. Therefore, the fuel flow will be

$$\dot{m}_f = SFC\,P = 6.58 \cdot 10^{-7} \times 570 \cdot 10^3 = 6.58 \times 5.50 \cdot 10^{-2}$$
$$= 37.51 \cdot 10^{-2} = 3.751 \cdot 10^{-1}\,\text{N/s}.$$

Conversion to other popular units yields

$$\dot{m}_f = 3.751 \cdot 10^{-1} \, \text{N/s} = 137.65 \, \text{kgf/h} = 302.83 \, \text{lb/h}.$$

The specific range will be

$$SAR = 103.68 \, \text{m/N} = 1.017 \, \text{km/kgf} = 0.549 \, \text{nm/kgf} = 0.249 \, \text{nm/lb}.$$

If the helicopter of reference were to run at this speed starting with full tanks, then the endurance would be 6.5 hours (not counting fuel reserve, take-off climb, and unusable fuel).

15.2 NON-DIMENSIONAL ANALYSIS OF THE SAR

Due to the number of parameters involved in the SAR, it is convenient to follow a non-dimensional analysis. First, we have seen that the fuel flow is proportional to the engine power, $\dot{m}_f = SFC \, P$. The SFC is dependent on the speed. However, exact curves can only be found in the manufacturers' charts; these charts give the fuel flow versus air speed at different altitudes and atmospheric conditions. In non-dimensional terms,

$$\dot{m}_f = f(P) = f\left(\left[\frac{P}{\sigma_1 \omega^3}\right]\sigma_1 \omega^3\right), \tag{15.2}$$

and by using the result from Eq. 11.9,

$$\dot{m}_f = f\left(\frac{W}{\sigma_1 \omega^2}, \frac{U}{\omega}, h, \frac{\omega}{\sqrt{\theta}}\right). \tag{15.3}$$

By using a similar algebraic trick on the air speed, we find

$$U = \left(\frac{U}{\omega}\right)\omega. \tag{15.4}$$

Finally, combine Eq. 15.4 and Eq. 15.3,

$$SAR = \frac{U}{\dot{m}_f} = f\left(\frac{W}{\sigma_1 \omega^2}, \frac{U}{\omega}, h, \frac{\omega}{\sqrt{\theta}}, \frac{1}{\sigma_1 \omega^2}\right) = f\left(\frac{W}{\sigma_1 \omega^2}, \frac{U}{\omega}, \frac{\omega}{\sqrt{\theta}}, \frac{1}{\delta}\right). \tag{15.5}$$

The last term is found from

$$\frac{1}{\sigma_1 \omega^2} = \frac{\theta}{\theta \sigma_1 \omega^2} = \frac{1}{\delta}\left(\frac{\sqrt{\theta}}{\omega}\right)^2. \tag{15.6}$$

Therefore, the SAR can be plotted in terms of the referred weight, for a given referred air speed and referred rotor speed (or Mach number). The qualitative behavior of the curve is similar to that of Figure 15.1, and is shown in Figure 15.2.

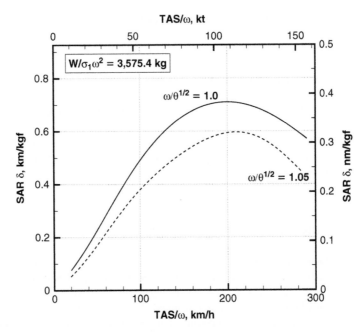

Figure 15.2 *Referred SAR versus referred TAS at constant referred weight.*

Example

One method of calculation is the following. Modify the forward flight routine `AssemblyPower` by calculating with the referred weight, the referred SAR and the referred TAS. Combine this routine with the program on page 413, which calculates the effective fuel flow and the effective SFC from the gas turbine charts.

Start with an $AUW = 3,500$ kg. First, assume a variation of referred rotor speed to a higher value, for example $\omega/\sqrt{\theta} = 1.05$. If we assume a 2% change in the rotor speed ($\omega = 1.02$), the relative temperature is $\theta = 0.9856$, which corresponds to a pressure altitude $\delta = 0.926$, and hence to an altitude $h = 0.644$ km. The referred weight for this case is 3,575.4 kgf. The result is the solid line in Figure 15.2.

Second, return to normal operating conditions at standard atmosphere at sea level (for example, $\omega = 1$, $\sigma_1 = 1$, $\delta = 1$). Hence, the rotorcraft must fly at sea level. If the referred weight is to be equal to the former case, then $W/\sigma_1 \omega^2 = W = 3,575.4$ kg. The result is the dashed line in Figure 15.2.

15.3 ENDURANCE AND SPECIFIC ENDURANCE

The endurance is the flight time. The specific endurance is the inverse of the fuel flow

$$E_s = \frac{1}{\dot{m}_f}. \tag{15.7}$$

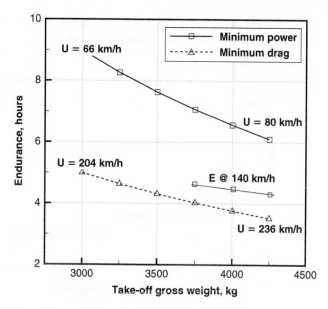

Figure 15.3   *Helicopter endurance at various flight conditions, sea level ISA.*

The endurance is the integral of E_s between the terminal points of the fuel

$$E = \int_i^e E_s \, dm = \int_i^e \frac{dm}{\dot{m}_f} \simeq \frac{\Delta m}{\dot{m}_f}. \tag{15.8}$$

The last equivalence is valid if the specific endurance is constant. The integral Eq. 15.8 is calculated between the start and end of the cruise. The maximum endurance is obtained with the minimum fuel consumption, and hence from flight at minimum power.

Figure 15.3 shows the computed endurance for operation at minimum power and minimum drag, at the corresponding speeds. These data are compared with the performance quoted by the manufacturer. The latter data are given at a constant speed $U = 140$ km/h (75.5 kt), with standard tanks, though it fails to specify the flight altitude, and the cruise fuel as a percentage of the total fuel. The endurance at minimum power is on average 70% higher than the endurance at minimum drag.

15.4   SPEED FOR MINIMUM POWER

From the analysis of forward flight characteristics, we are now in a position to evaluate the helicopter speed for minimum power. Graphically, this speed is found from the horizontal tangent to the power curve in the P/U diagram, as shown by point A in Figure 15.4.

The numerical solution of the power equation allows us to investigate quickly the flight envelope of the aircraft, for example the effects of gross weight, flight altitude, atmospheric conditions, etc.

One such result is shown in Figure 15.5 at a set of altitudes and weights, up to the MTOW. For each operation point we have stored the performance index and produced

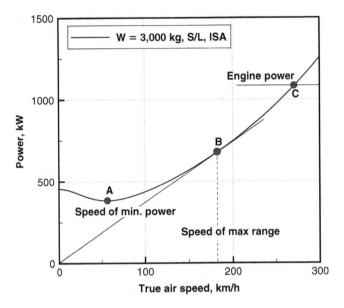

Figure 15.4 *Speed of minimum power, maximum endurance, and maximum absolute speed for the reference helicopter (helicopter model D, Appendix A). AUW = 3,000 kg, sea level, ISA conditions. Calculated performance.*

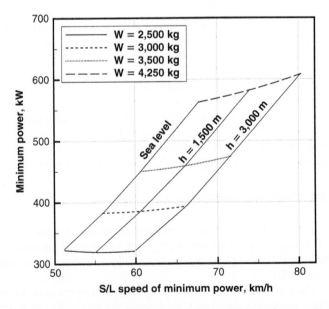

Figure 15.5 *Helicopter speed of minimum power U_{mp} and corresponding power requirements, ISA conditions, at the gross weight indicated (helicopter model D, Appendix A).*

the chart. The speed of minimum power is relatively low, in the range of 50 to 80 km/h, depending on the gross weight. As the weight increases, U_{mp} also increases.

15.5   SPEED FOR MAXIMUM RANGE

Graphically, this speed is found from the tangent to the power curve in the P/U diagram, as shown by point B in Figure 15.4. Computationally, the maximum-range speed is found by simple book-keeping when calculating the power requirements at increasing flight speeds.

For the calculation of the range we can use an expression similar to Breguet's equation

$$R = \int_e^i SAR \, dm. \tag{15.9}$$

If the aircraft flies at a constant speed, altitude and engine power, then the specific range is constant. Hence

$$R \simeq SAR \, \Delta m. \tag{15.10}$$

Example

For our reference helicopter, if we assume that only 85% of the fuel can be used for cruise, we have about 965 liters; this corresponds to about 760 kg (jet A fuel). Therefore, using the previous analysis of the specific range, Eq. 15.2, the range is estimated by

$$R = 103.67 \times 760 \; 9.81 \, \text{m} = 773 \, \text{km} \; (417 \, \text{nm}). \tag{15.11}$$

The maximum range reported for this helicopter with standard tanks is 895 km, therefore the analysis proposed falls 14% short of expectations. However, the data available in Table A.15 do not specify the aircraft's weight and the cruise speed that achieve this cruise range.

From the numerical analysis, the speed of minimum power is estimated between 180 and 220 km/h (97 to 119 kt), data from Figure 15.6, calculated from the forward flight analysis of the preceding sections. The range increases almost linearly with a decrease in GTOW. The maximum range at each altitude is found at the minimum GTOW.

The minimum GTOW for maximum range is found as follows: sum of the operating empty weight (2,250 kg), the maximum fuel (897 kg), the pilot and various operational items. This leads to an estimated weight of 3,250 kg. The maximum range found in this analysis is close to the one quoted by the manufacturer, and it is achieved at a relatively high cruise altitude. More reasonably, the helicopter will cruise at an altitude between sea level and 1,500 m, therefore the range will be sensibly lower than the one reported.

This analysis serves to point out that the maximum range of the aircraft is a performance parameter specified by the aircraft configuration, flight speed and flight altitude.

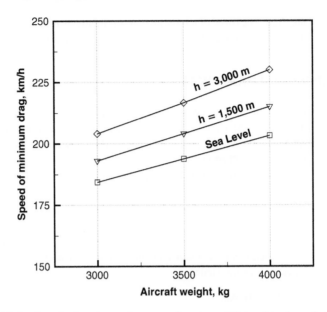

Figure 15.6 *Speed of minimum drag as a function of flight altitude and aircraft weight, ISA conditions.*

15.6 FUEL TO CLIMB

The exact fuel to climb can only be calculated if the flight trajectory is specified in advance; this is not always possible. A first-order estimate is derived from the energy method. The difference in energy between the helicopter on the ground and the helicopter at altitude h and forward speed U is given by

$$\Delta E = Wh + \frac{1}{2}mU^2 + \Delta E_k^{rot}, \tag{15.12}$$

where E_k^{rot} is the difference in kinetic energy of the rotor. Since the main rotor operates at constant rpm, this contribution is essentially due to the forward speed, see Eq. 14.10. It is instructive to verify whether this term can be retained or neglected.

The fuel burn can be estimated with a variety of methods. For example, we can write the increase in total energy of the aircraft as

$$\Delta E \simeq m_f c_p \, \bar{\eta}_1 \bar{\eta}_2, \tag{15.13}$$

where c_p is the specific heat of combustion of the fuel, $\bar{\eta}_1$ is the average thermodynamic efficiency of the gas turbine, and $\bar{\eta}_2$ is the average propulsion efficiency. A reasonable assumption, derived from the statistical analysis of gas turbine operation during climb, is $\bar{\eta}_1 \simeq 0.45$. The overall propulsion efficiency is defined in §13.7 as the ratio

$$\eta_2 = \frac{TU \sin \alpha_T + P_c}{P_e} = \frac{W(U \tan \alpha_T + v_c)}{P_e}. \tag{15.14}$$

The first term denotes the contribution to forward flight; the second term is the climb power. The overall propulsive efficiency accounts for losses in energy that are not used for increasing the total energy of the aircraft: the rotor's profile power, the transmission power and the parasitic power lead to a loss of energy. Even the tail rotor power is a form of a loss, because it is used for flight stability, rather than propulsion. With this approximation, the fuel burn is found from

$$m_f c_p \bar{\eta}_1 \bar{\eta}_2 \simeq Wh + \frac{1}{2} mU^2 + \Delta E_k^{rot},$$ (15.15)

or

$$m_f \simeq \frac{Wh + mU^2/2 + \Delta E_k^{rot}}{c_p \, \bar{\eta}_1 \bar{\eta}_2}.$$ (15.16)

This method is valid only if the aircraft climbs to its cruise altitude without additional maneuvers, such as turning and partial descent. Alternatively, one can integrate the actual fuel flow during the flight. Another approximate method consists of multiplying the fuel flow at the climb power by the climb time – which has to be estimated by other means.

Example of Climb Fuel Calculation

Consider the reference helicopter with a GTOW of 4,000 kg. The helicopter has to climb to a cruise altitude of 700 m from a sea level helipad. The cruise speed is 234 km/h (65 m/s, 126.34 kt). The increase in total energy is

$$\Delta E = Wh + \frac{1}{2} mU^2 + \frac{N}{2} MU^2.$$

where m is the mass of the aircraft, and M is the total mass of one blade. This was calculated earlier (page 388): $I_b \simeq 71$ kg. Therefore, the increase in energy is

$$\Delta E = 27.468 \cdot 10^6 + 8.450 \cdot 10^6 + 0.600 \cdot 10^6 = 36.518 \cdot 10^6 \text{ joule}.$$

The fuel to climb is estimated from Eq. 15.16. The combustion heat of the aviation fuel is about 43.5 MJ/kg.

   We assume that the climb power is equal to the AEO maximum continuous power, e.g. 550 kW for each gas turbine. Therefore, from Eq. 15.14, using $v_c \simeq 8$ m/s, $\overline{U} \simeq 30$ m/s, we estimate a propulsive efficiency $\bar{\eta}_2 \simeq 0.36$. This means that only 36% of the engine power is used for increasing the energy of the aircraft; the rest is lost. Hence the fuel is

$$m_f \simeq \frac{36.518 \cdot 10^6}{43.500 \cdot 10^6 \times 0.45 \times 0.36} \simeq 5.2 \text{ kg}.$$

By way of comparison, if the aircraft were to climb at the maximum climb rate of 8 m/s (Figure A.21), it would take about 88 seconds to reach its cruise altitude; at an average fuel flow of 270 liters/h, this leads to 6.6 liters, or 5.3 kg. To this we need to add the fuel for acceleration to the cruise speed.

Table 15.1 *Selected helicopter weight data. Payload ratio P/R calculated for internal load only. Data for Mil-10 are estimated. P/R refers to total load.*

Helicopter	Ext. Load	Pay (kg)	MTOW (kg)	P/R
CH-47SD	11,340	12,900	24,500	0.526
CH-53E	13,700	11,800	33,300	0.354
CH-54/S-64	9,200		19,700	0.467
Mil-6		12,000	42,500	0.282
Mil-10	11,000	4,000	24,700	0.607
Mil-26	8,000	20,000	56,000	0.357
V-22	6,800	9,100	27,450	0.368
EH-101		5,500	14,600	0.377

A final consideration is that the increase in the kinetic energy of the rotor is about 2% of the potential energy of the aircraft, and the fuel to climb represents only 0.5% of the maximum fuel.

In a mission analysis the fuel to descend and land is taken from the reserve fuel. During descent, the rotor operates at limit pitch, hence at limited power.

15.7 PAYLOAD/RANGE DIAGRAM

The helicopter is economically viable if it can transport a useful load. The weight of the aircraft is the sum of the operating empty weight, the fuel weight and the payload. The useful load is the sum of fuel and payload. When the total weight reaches the maximum take-off weight, the payload has to decrease to accommodate additional fuel. A large fuel weight is required for longer range or endurance. One of the main performance parameters is the payload/range chart. Simple range performance equations, such as the Breguet equation, are not available, but the range can be calculated by using the power, speed, and specific fuel consumption of the engines.

The data in Table 15.1 report the internal payload and the maximum external payload, where available, for selected heavy-lift helicopters. The data on external load have been averaged. It is possible for most of these helicopters to lift heavier weights for shorter times and to transport them over shorter distances. A graphical summary of payload performance is shown in Figure 15.7 for two Eurocopter vehicles. The performance of both Eurocopter EC-120B and EC-365 are calculated at sea level, ISA conditions, with standard tanks, with a pilot mass of 80 kg, at the recommended cruise speed, without reserve fuel. In addition, the range performance of the EC-365 is given with the additional long-range fuel tanks.

Some helicopters can lift amazingly large external loads (called *sling loads*). These loads are hooked to the end of a cable running from the bottom of the airframe. The Boeing CH-47D, as other helicopters, has three hooks for balancing the sling load and providing stability to the aircraft. The Sikorsky CH-53E helicopter is capable of lifting up to 14.5 metric tons, transporting the load 50 nautical miles and returning to base. A typical load would be a 7,260 kg M-198 howitzer* or 11,800 kg light

---

* A cannon that delivers shells at a medium velocity and high trajectory.

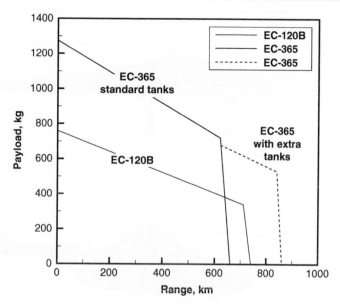

Figure 15.7   *Payload/range of two Eurocopter vehicles, EC-120B and EC-365.*

Figure 15.8   *Sikorsky S-64 Sky Crane.*

armoured vehicle. The aircraft can also retrieve downed aircraft, including another CH-53E.

One of the most impressive heavy-lift helicopters is the Sikorsky S-64 Sky Crane (1962), a helicopter designed to deliver large external bulk loads, see Figure 15.8 and Table 15.1. This helicopter has operated around the world, and moved records amounts of loads across impassable terrain, oceans, mountains, and tall buildings. It is still operated in some countries in fire fighting and for the movement of containers.

There is no consensus on how big and heavy a helicopter can be. The largest helicopter programs go back to the 1960s and 1970s. After that time, the size of heavy helicopters seems to have leveled off, or even decreased, as pointed out by

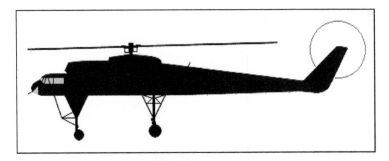

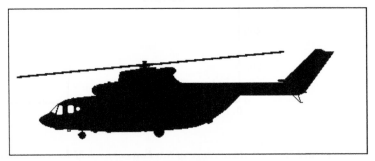

Figure 15.9 *Two large Mil helicopters, Mil-10 (top), crane helicopter, and Mil-26 (bottom), a heavy-lift helicopter; the graphics are drawn at the same scale.*

Stepniewski[449]. By all accounts, the Mil-26 (Figure 15.9) is the largest helicopter ever built and capable of heavy-lift operations. The massive Kellett-Hughes XH-17 Flying Crane (1952) had the largest rotor diameter ever built, 40.48 m. The helicopter was designed to carry large bulk loads over limited distances (20,000 lb/9,072 kg over 50 miles). This project was unsuccessful.

A tilt quad-rotor (e.g. a vehicle with four rotors), based on the Boeing-Sikorsky V-22 tilt rotor concept, would be capable of lifting up to 20,000 kg of internal load. This value is, in fact, similar to the maximum payload of the Mil-26 and the C-130J. For analysis of this concept, see Hirschberg[450]. Meier and Olson[451] analyzed the practicality of the helicopter size in terms of productivity, and concluded that the best size is related to the frequency of use and radius of action. Furthermore, there is the possibility of performing twin-lift operations, e.g. carrying large bulk loads with two helicopters. For problems of helicopter control in twin-lift operations see Menon *et al.*[452]

The Mil Mi-12/V-12 is to date the world's largest prototype helicopter. A second prototype set seven load-carrying records in 1969: a 31,030 kg load was lifted to 2,951 m and the following year 40,204 kg was transported a distance of 2.25 km*. The helicopter was lifted by two Mil-6 rotors mounted at the ends of outrigger wings.

A measure of payload performance is given by the payload ratio, e.g. PAY/MTOW (Figure 15.10). This ratio is quite variable, and depends on the specific aircraft, engine power installed, and mission requirements.

---

* Source: Fédération Aéronautique Internationale (www.fai.org).

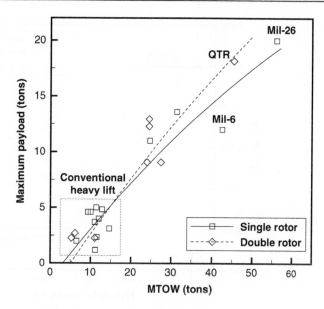

Figure 15.10 *Payload of some heavy-lift helicopters.*

Figure 15.11 shows the effect of the aircraft's GTOW. The excess weight above the 3,250 kg (minimum weight for maximum range) is the payload weight.

We now define the aircraft's envelope in the payload/range diagram, by taking a detailed look at the effect of payload on the range. The gross take-off weight is

$$GTOW = W = OEW + FUEL + PAY. \tag{15.17}$$

By inserting Eq. 15.17 into the range Eq. 15.10, we find

$$R = \frac{U}{P} \frac{W - OEW - PAY}{gSFC}. \tag{15.18}$$

In Eq. 15.18 the factor g accounts for the fact that the SFC is given in units of mass per unit of time per unit of power, in coherence with the earlier discussion. For a given GTOW, the aircraft range decreases linearly with the increasing payload. For maximum range we must use the optimal value of U/P, e.g. the speed of minimum drag. The range is maximum when the payload is zero; this is the *ferry range*. We can define a set of curves from Eq. 15.18 at selected GTOW, *for a given flight altitude*. As the GTOW decreases, the curves shift toward the bottom of the graph. The limit of the payload/range envelope is determined by the ferry range condition at all GTOW,

$$R = \frac{U}{P} \left(\frac{W - OEW}{gSFC} \right). \tag{15.19}$$

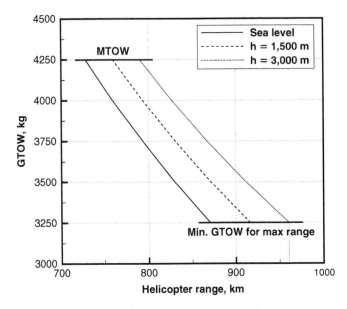

Figure 15.11 *Maximum range versus take-off gross weight (GTOW) at different flight altitudes, ISA conditions, as indicated.*

This line is slightly inclined from the payload axis, as shown in Figure 15.12. One can imagine that a series of payload/range diagrams can be calculated, depending on the cruise speed, the cruise power, and the cruise altitude. For the calculation, proceed as follows:

Computational Procedure

1. Set the payload weight.
2. Calculate the cruise fuel from Eq. 15.17 at the given GTOW.
3. If the cruise fuel is less than the maximum fuel (full tanks), then use the full tanks to calculate the range from Eq. 15.18.
4. Increase the payload weight, and repeat the procedure.

The cruise fuel is set to 90% of the full tanks. More accurate values of the cruise fuel weight fraction will depend on the type of mission. This algorithm can be used for all aircraft weights. The payload/range line corresponding to MTOW will give the upper limit of the diagram. The range limit is the envelope of the points at which the payload is limited by fuel. The results of the present analysis are shown in Figure 15.12.

From the specifications of the aircraft, we see that there is the possibility of installing extra fuel tanks for an additional 180 liters. We now want to study the range performance of the aircraft with these additional tanks. The MTOW does not change, therefore one can argue that the additional range is achieved at a cost to the maximum payload that can be carried. The corrected payload/range diagram is shown in Figure 15.13.

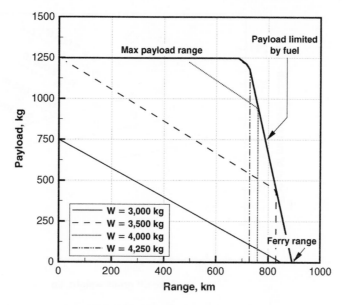

Figure 15.12 *Payload/range diagram at sea level, ISA conditions, some GTOW indicated, helicopter model of Appendix A.*

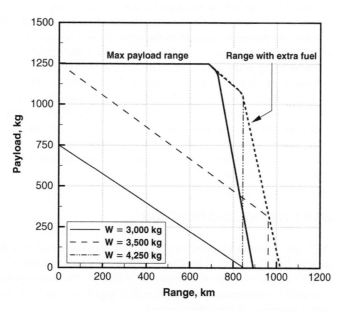

Figure 15.13 *Payload/range diagram with extra fuel tanks, at sea level, ISA conditions, some GTOW indicated, helicopter model of Appendix A.*

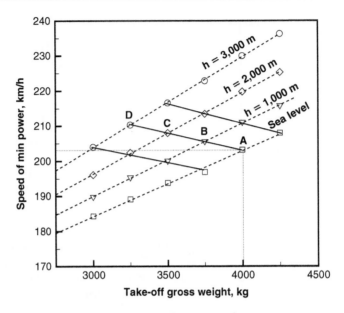

Figure 15.14 *Optimal range versus GTOW, ISA conditions.*

With the additional tanks, there is a gain of about 120 km, but no gain in maximum payload. The shape of the envelope remains the same, and is representative of most conventional helicopters. Over a long cruise, the aircraft flight altitude and speed will have to change to readjust to the optimal conditions created by the change in weight. The optimal operation point V–h can be found in the V/W chart, as shown in Figure 15.14. If the aircraft starts its cruise at sea level with $GTOW = 3,750$ kg, it will have to climb to $h = 1,000$ m (point B), because the weight is reduced to 3,500 kg by the fuel burn. By following the path A-B-C-D, the aircraft will always fly at the speed of minimum power, and therefore it will achieve its maximum absolute range.

In some military operations hover and low-speed loiter are more important than range. Power requirements and fuel consumptions in hover are higher. If loiter is done at the speed of minimum power, the fuel consumption is minimum, hence the endurance is maximum.

15.8 COMPARATIVE PAYLOAD FRACTION

In Chapter 3 we saw examples of payload fractions for some heavy-lift aircraft. Since the viability of any aircraft is ultimately related to carrying payload of some sort from somewhere to somewhere else, economic viability is the most important consideration in aircraft performance. Speed is not everything: a payload has to be delivered in the largest quantities at the minimum cost.

Some data on existing helicopters extracted from various publications are summarized in Figure 15.15. By looking at the data, we note that the helicopter and the V/STOL aircraft are less productive than the fixed-wing aircraft. The payload fractions shown indicate that the gap in productivity increases with the aircraft range.

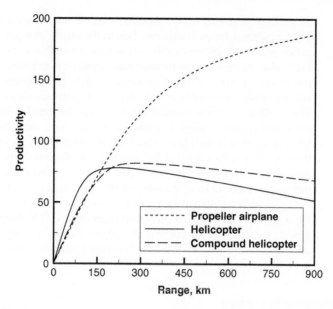

Figure 15.15   *Comparative productivity of some aircraft.*

A comparative analysis with the heavy-lift airplanes shows that the conventional helicopter has a sensibly lower payload fraction. A relevant productivity parameter is

$$E = U_{block} \frac{PAY}{OEW},\tag{15.20}$$

where U_{block} is the so-called *block speed*, e.g. the distance between terminals divided by the time needed to reach the destination. This definition includes the time for taxiing and the time spent in traffic patterns (take-off, loiter, landing). The definition does not include the fact that the helicopter can land in more, seemingly inaccessible, places. Therefore, the helicopter concept has a niche of considerable importance in the air transport market. Values of the productivity factor for the current generation of transport helicopters are in the range of 140 to 180 km, if the economical cruise speed is used; they are below 100 km if the actual block speed is used.

By using the productivity factor, we find that the fixed-wing airplane is more efficient at large payload ratios and large mission ranges. However, at lower ranges both the helicopter and the V/STOL aircraft are more effective. Therefore, the helicopter is the aircraft of choice, not necessarily when VTOL characteristics are required, but also for short mission ranges, in the 100 nautical miles (180 km) limit.

15.9   MISSION ANALYSIS

The final task of the performance analysis is to calculate the gross take-off weight for a specified mission. This analysis requires the estimation of the fuel load, for given payload, required range and type of operation. In other words, we need to identify all the flight segments for the specified mission, and to calculate the fuel, time and flight conditions for each segment. Due to the variety of flight missions that can be prescribed, it will not be possible to be exhaustive on this. Basically, we have to learn to calculate the mission fuel for each flight segment, then have a book-keeping system that will provide the weight of the aircraft and the fuel required. Military operations are clearly more difficult to analyze, because they include various maneuvering capabilities. Only a statistical analysis will provide the detailed knowledge of the fuel consumption during complex operations. Furthermore, the number of parameters to take into account is fairly large. One thing is to provide a first-order estimate, another thing is to calculate a detailed flight path that minimizes the fuel consumption from point to point. One such analysis was done by Slater and Erzberger[453], who showed that fuel savings can be of the order of 10% compared to a non-optimal flight.

The mission fuel calculation will have to be iterative, because the fuel for each flight segment will depend on the initial guess of the GTOW. Stepniewski and Keys[110] provide some details on this type of analysis. Simple military mission calculations are available in Newman[166] for the Westland Lynx helicopter. A suitable method is the following.

Computational Procedure

1. Read the aircraft's data.
2. Guess the GTOW, by using the OEW and the payload, compatible with the certified MTOW.
3. Calculate the fuel requirements for each flight segment.
4. Add all the segment fuels and recalculate the GTOW.
5. If the GTOW < MTOW, use the newly calculated GTOW to repeat the analysis at point 3. If GTOW > MTOW, redo the calculation at point 3 with MTOW.
6. If the last computed GTOW > MTOW, then the required mission is impossible. We need to decrease the payload, or the range, add additional tanks, etc.

The mission fuel calculation at point 3 may have to be calculated three or four times before convergence. The convergence criterion is that the change in calculated GTOW is less than 1% from the previous value.

For such analyses, it is essential to have simple calculation methods that allow for a first-order estimate of the fuel required. The cruise fuel will be calculated with the method of §15.1, assuming a constant speed. The climb fuel can be calculated from Eq. 15.16.

PROBLEMS

1. Calculate the range/payload envelope for the reference helicopter (Appendix A) at a take-off gross weight $GTOW = 3{,}750$ kg, for a cruise flight at constant speed and altitude $h = 1{,}000$ m, ISA conditions. Assume that the engine power decreases with the atmospheric density according to Eq. 12.37. Use the method described in §15.7.
2. Calculate the endurance of the reference helicopter at the speed of minimum power and minimum drag, if the aircraft is to fly straight level at an altitude

$h = 1,000$ m. Use the definition of §15.3 and the data from Figure 15.5 and Figure 15.6.

3. Our reference helicopter is to be used for rescue operations on a maximum range of 200 nautical miles from the coast. A typical mission requirement includes responding to SOS from marine vessels in distress. Make a scenario of possible rescue operations, from warm-up to return-to-base, estimate the aircraft gross take-off weight in the various segments of the mission, and calculate the fuel requirements, the range and endurance. Finally, make an assessment on the feasibility of this mission, and recommend changes in performance and/or aircraft configuration necessary to adapt the helicopter to this type of mission. (Use the data in Appendix A.)

4. The helicopter of reference (model D, Appendix A) is to climb from sea level to a cruise altitude of 700 m and a cruise speed $U = 230$ km/h. The gross take-off weight is $W = 3,500$ kg. Calculate the value of the potential energy, the kinetic energy of the aircraft, and the change in kinetic energy of the rotor, from Eq. 15.12, and discuss whether any of the terms can be neglected. (The calculation of the kinetic energy of the rotor can done from Eq. 14.9.)

5. Calculate the fuel to climb for the reference helicopter with a GTOW $= 3,500$ kg. Assume that the aircraft takes off from sea level, standard conditions, and that the final state is an altitude of 500 m with an air speed 100 km/h (54 kt). Use the energy method, and assume that the average propulsion efficiency is 68% and the thermal efficiency of the gas turbines is 45%.

6. Calculate the specific air range for the helicopter of reference, assuming that the level flight power is 600 kW at a speed of 200 km/h. Provide the result in km/kgf, nm/kgf, nm/lb. Assume that the SFC is $4.7 \cdot 10^{-7}$ N/s/watt.

7. The standard fuel tank of our reference helicopter contains 1,135 liters. Assume that only 85% of this fuel can be used for cruise from point to point. Calculate the endurance of the reference helicopter (model D) at the cruise speed.

Part III
V/STOL and Noise Performance

Chapter 16

V/STOL Performance

An efficient, operational V/STOL aircraft has proven to be an elusive target over the last thirty years.

B. McCormick[454], 1978

V/STOL is an acronym for Vertical/Short Take-off and Landing Aircraft. This category includes aircraft that are capable of performing most of the flight conditions of fixed-wing aircraft. However, take-off, landing and ground performance can be considerably different from CTOL aircraft. Other peculiar problems of V/STOL vehicles are lift augmentation in the terminal phases, transition from vertical to forward flight (and vice versa), operation at high angles of attack, and engine installation. There are, of course, additional performance requirements on aerodynamics, propulsion, structures and flight control systems.

V/STOL aircraft make up the most exotic and unfortunate category of flight vehicle, including several aircraft that never took off, many that flew only experimental tests, and a handful that have entered production. According to McCormick[105] the history of V/STOL aircraft is full of *disillusionment*.

This chapter will only focus on those peculiar performance indices of real-life V/STOL aircraft, and a few essential concepts for the future. It will not be possible to review the various design options in this class of vehicle. Essential concepts in V/STOL performance are available in AGARD R-710[455]. In addition, McCormick[105] published an interesting short history of V/STOL flight research, and a book[456] on the aerodynamics of V/STOL. Hirschberg[457] published another historical account of V/STOL research. Maisel *et al.*[458] reviewed the research on V/STOL leading to the experimental tilt-rotor XV-15 (precursor of the modern Boeing-Sikorsky V-22). The analysis of some modern design concepts is discussed by Talbot *et al.*[342] The effects of small vector/thrust angles and their optimal values on subsonic commercial jets were discussed by Gilyard and Bolonkin[236].

16.1 HOVER CHARACTERISTICS

There are two types of hover used by a V/STOL: by the action of a rotor/propeller, or by the action of a jet. The methods presented for estimating hover performance (thrust, power, end endurance) are essentially the same. Transition to airplane mode is different, and requires unsteady aerodynamic analysis.

The second case, pertinent to some V/STOL aircraft, is hover capability by means of one or more jets (for example, BAe Harrier, Lockheed Joint Strike Fighter (JSF), and a number of experimental airplanes). This capability, and the related stability and control issues, are strongly dependent on the ground effect. The technical means for producing these jets and the aerodynamic problems associated with propulsion are the subject of V/STOL aerodynamics and aircraft design.

For an aircraft out of ground, lift is created by a free jet flow into stationary surroundings. A single jet flow spreads out according to physical mechanisms that depend on the Reynolds number and on the turbulence entrainment. Substantial research exists in this area. The reader can start with Donaldson and Snedeker[459] and Krothapalli *et al.*[460], where some fundamental experiments and correlations are shown. For V/STOL research, Kotansky[461] is a useful resource, as is Bellavia *et al.*[462,463] A review of the OGE hover and transition flight is available in Saddington and Knowles[464]. Various semi-empirical formulas exist for estimating the jet development (diameter, average speed, turbulent development).

From our point of view, the jet flow must generate enough thrust to overcome the aircraft's weight. The jet must produce a thrust equivalent to the weight to maintain a stable hover condition. With a one-dimensional analysis, if U_j is the average speed of the jet at the engine nozzle, and $\dot{m}_j$ is the jet mass flow rate, then the resulting thrust equation must be

$$T = W \simeq (\dot{m}U)_{jet} = 2(\rho A U^2)_{jet}. \tag{16.1}$$

The thrust required is generated by a suitable combination of high mass flow rate and high jet speed. The amounts required can be easily calculated. For a $GTOW = 10,000$ kg, and a jet Mach number limited to $M = 1.5$ at sea level, the mass flow requirements are about 195 kg/s, or 160 m^3/s of air at atmospheric conditions. This mass flow has to be drawn from the surroundings of the aircraft, as shown in Figure 16.1. The corresponding disk loading (defined as the weight of the aircraft divided by the total area of the nozzles) is at least two orders of magnitude higher

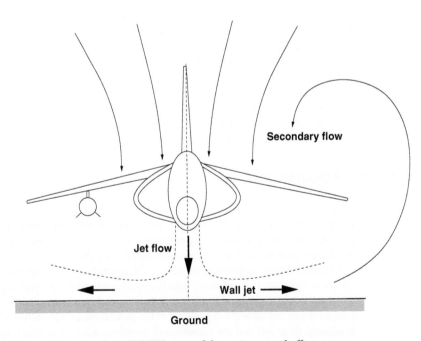

Figure 16.1 *Single-jet V/STOL aircraft hover in ground effect.*

than the conventional helicopters. A maximum disk loading of about $22,000 \, kg/m^2$ is estimated for the BAe Sea Harrier.

Another way to look at direct lift is to consider the thrust requirements on the engine. Clearly, an OGE hover requires $T/W > 1$, albeit not much higher than one. This is called *thrust margin*. By comparison, a CTOL jet aircraft can operate with $T/W = 0.20$.

The jet speed has to be limited for several reasons, including noise, vibrations, and heat fatigue. Equation 16.1 gives only a rough estimate of the thrust generated by a vertical jet. Thrust requirements for hover are in excess of the aircraft gross weight, due to losses in lift attributable to a number of factors. The most important are: (1) jet-induced losses; (2) hot gas reingestion; and (3) margins required for maneuver and acceleration. These losses depend on the relative ground clearance, and are greater in proximity to the ground. The US Military standards require a minimum $0.1g$ vertical acceleration capability with a $T/W = 1.05$. Therefore, the installed thrust must be 5% in excess of the gross weight. In this context, *installed thrust* is the difference between the gross thrust and the jet-induced losses.

16.2   JET-INDUCED LIFT

The effects created by impinging jets on V/STOL vehicle performance are significant, and potentially dangerous if not properly understood. Skifstad[465] reviewed the physics of the jets for V/STOL applications and highlighted several aerodynamic aspects, including the induced aerodynamic field, the free jet, the impinging jet and the multiple jets. Consider first the case of a single-jet V/STOL aircraft. The jet-induced losses are due to a downwash created by the jet, because this draws air from the surroundings. This field of velocities creates a further downward push, particularly around the wings, which are most exposed. Jet nozzles under the wings compound this problem.

The presence of the ground changes the characteristics of the flow, because impingement is followed by lateral spreading of the jet, rising of hot gases, and return of the gases back onto the airplane. One such case is sketched in Figure 16.1. Depending on various geometrical conditions between the aircraft and the ground, three main cases are identified:

1. The jet spreads radially as a thin sheet with rapidly decaying speed; the reduced pressure creates a suck-down effect. Clearly, this is to be avoided. Calculation of the spreading depends on a number of factors, including the aircraft configuration, the geometry of the jet, and the position of the aircraft above the ground.
2. The jet spreads radially and creates a peripheric vortex ring. The vortex ring augments the jet lift. This is the *hovercraft* principle.
3. The jet spreads out, but it creates an upward radial flow bouncing off the ground (fountain effect), which may either be favorable (cushion) or adverse (suckdown). The conditions under which these events occur depend on the geometry of the aircraft. They can be controlled by using strakes (BAe Sea Harrier) at appropriate positions. The cushion effect cannot be used to reduce the engine thrust installed, because the thrust margin must always be based on the worst possible scenario.

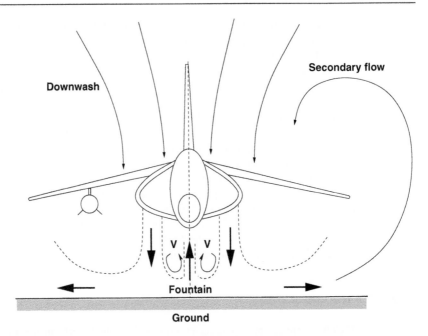

Figure 16.2 *Twin-jet V/STOL aircraft hover in ground effect showing the fountain effect.*

A strong suck-down effect is likely to occur in the single-jet V/STOL, while a twin jet, with closely spaced jets (Figure 16.2), reduces this effect to a few percent of the total thrust. Properly configured twin jets produce favorable fountain effects that almost cancel the suck-down. A detailed investigation on side-by-side jets is available in Louisse and Marshall[466].

Hot-gas reingestion arises from the fact that exhaust gases bouncing off the ground and spreading around the aircraft mix with the surrounding area thanks to the buoyancy effect. They increase the local temperatures (by a few degrees) and may return to the engine's inlets reducing the engine efficiency. This leads to a thrust loss of about 4% on the BAe Sea Harrier. Another phenomenon is the hot-gas fountain, similar to the one described, in which the mixing of hot gases with the surrounding air is more limited; the engines may breathe more polluted air, with temperatures considerably higher than the ambient temperature. Thrust losses in this case can be as much as 8%. A typical example of jet lift losses due to the various causes is shown in Figure 16.3, elaborated and adapted from a number of publications (AGARD CP-242[12], AGARD R-710[455]).

Two cases are shown in Figure 16.3: a case of *clean* aircraft; and an aircraft with strakes. The use of strakes below the aircraft, at appropriate locations, can inhibit some of the processes discussed. The strakes are capable of limiting the overall lift loss to about 2% of the engine thrust. For lower hover positions of the aircraft there are jet flow instabilities that are responsible for the non-linear effects indicated by the lift curves. The maximum losses are limited to 4 or 5%, and are gradually reduced to about 1.5% in OGE operation. Therefore, a thrust margin of 5% is required, as indicated earlier.

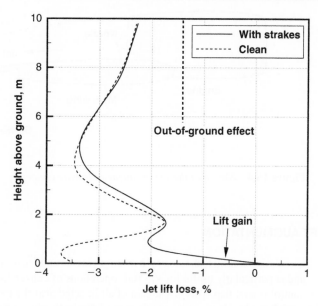

Figure 16.3 *Jet lift loss due to ground interference, for two aircraft configurations.*

16.2.1 Estimation of Jet-induced Fountain Lift and Suck-down

According to Bellavia *et al.*[463], the total loss of lift, relative to the net thrust, can be estimated from an expression like

$$\frac{\Delta L}{T} = \frac{\Delta L_{OGE}}{T} + \frac{\Delta L_f}{T} + \frac{\Delta L_{sf}}{T} + \frac{\Delta L_{sr}}{T}, \qquad (16.2)$$

where ΔL_{OGE} is the loss of lift out-of-ground effect (free air), ΔL_f is the fountain lift, and ΔL_{sf} and ΔL_{sr} are the forward and rear suck-down lift, respectively. The expressions for these losses depend on a number of semi-empirical parameters, in addition to several specific parameters, such as the jet diameters, the distance between jets, and the planform area. Various models exist for the evaluation of the terms in Eq. 16.2, as summarized by Walters and Henderson[467].

For an aircraft with two jets, the loss of OGE lift can be written as

$$\frac{\Delta L}{T} = k\sqrt{\frac{A_p}{A_j}} \left(\frac{2\pi d}{d_e}\right)^{1.58} p_r^{-1/2}, \qquad (16.3)$$

where A_p is the planform area of the aircraft, A_j is the cross-sectional area of the jet at the exit of the nozzle, d is the jet diameter, d_e the equivalent diameter, p_r is the pressure ratio of the nozzle; finally, $k = 10^{-4}$ is a constant. Plotting of this function reveals that the OGE jet-induced losses are in the range of 0.5% for realistic values of the nozzle pressure ratio ($p_r = 2$ to 6).

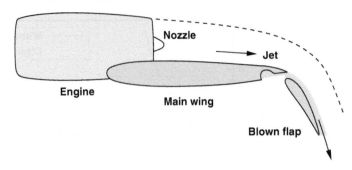

Figure 16.4 *Sketch of the upper-surface-blowing concept.*

16.3 LIFT AUGMENTATION

Circulation control technology is a means of increasing the lift without flaps and slats, and is particularly useful for V/STOL operations of conventional aircraft. One such control is a jet flap. This is a stream of air injected near the trailing edge to energize weak boundary layers prone to separation. Basically, the jet flaps simulate the effect of the flap, without the external mechanisms of the flaps.

The use of a conventional jet diverted downward is an intermediate case between pure jet lift and propeller lift. It is aimed at improving the ground performance of the aircraft in conjunction with aerodynamic lift. The methods giving a lift augmentation by diverting the jet are called *vectored-thrust* systems. In Chapter 4 we briefly discussed a number of high-lift systems, which are used for augmenting the lift during the terminal phases and maneuvers of the aircraft. These systems, and other more complex powered systems, can be an integral part of the aircraft, and provide better performance near the ground.

Lift augmentation by deviation of the engine thrust is a common means of increasing performance of V/STOL vehicles. Upper-Surface Blowing (USB) uses the concept of lift augmentation by reduction of the upper side pressure. This is achieved by diverting the hot gas from the engine nozzle over the wing, see Figure 16.4. One such application is identified on the STOL vehicles Antonov An-72 and An-74, and on some experimental aircraft, such as the Boeing YC-14 (Nichols and Englar[468]) and NASA's QSRA (Shovlin *et al.*[104]). These propulsion systems allow the aircraft to operate at double the C_L of that of CTOL aircraft (see also Figure 4.7).

The Externally Blown Flap (EBF) uses the concept of diverting the jet from the lower side of the wing. In practice, this is done by using the jet in coordination with a multi-element wing system, which serves to divert the jet, see Figure 16.5. One such application is identified in the Lockheed YC-15 and Lockheed C-17 military transport aircraft (Thompson[469]).

Both concepts allow slow and precise landings on short runways and in extreme weather, such as in Arctic conditions. Additional benefits of the vectored thrust can be achieved in flight conditions other than take-off and landing. For example, a vectored nozzle can increase the turn rate and decrease the turn radius, Lee and Lan[470] and Schneider and Watt[471]. Thus, the aircraft is capable of performing unrestricted turns at high speed (see also Problem 3).

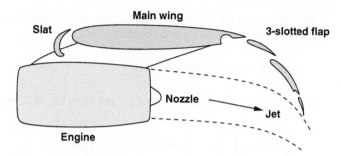

Figure 16.5 *Sketch of the externally-blown-flap concept.*

In air combat situations, vectored nozzles can increase the deceleration rates, while applying extra lift. The additional deceleration is due to the forward component of the gross thrust. V/STOL aircraft such as the BAe Harrier can add up to 0.75 g in deceleration rates using this technique.

16.4   CALCULATION OF SHORT TAKE-OFF

The use of powered lift systems that provide thrust at an angle ϵ with the aircraft longitudinal axis leads to a change in the flight equations. We must note that only the gross thrust can be vectored, while the intake momentum drag operates in the direction of the flight path. With this caution, we use the symbol T for gross thrust, instead of net thrust, as we have done earlier; T_o denotes the thrust that is actually vectored or diverted. The momentum equation along the horizontal runway is

$$m\frac{\partial U}{\partial t} = T\cos\epsilon - D - R. \tag{16.4}$$

The lift is augmented by the vector thrust $T_o \sin\epsilon$, therefore the sum of all forces in the vertical direction is

$$W - L - T_o \sin\epsilon, \tag{16.5}$$

and the rolling resistance is

$$R = \mu(W - L - T_o \sin\epsilon). \tag{16.6}$$

Solution of this equation is carried out in the same way as the CTOL, after noting that there is a constant factor $\cos\epsilon$ on the thrust term. Recall Eq.7.18 and write the lift-off run equation as

$$\frac{1}{2}U_{lo}^2 = \frac{1}{m}\int_o^x [T_o\cos\epsilon - D - \mu(W - T_o\sin\epsilon - L)]\,dx. \tag{16.7}$$

or

$$\frac{1}{2}U_{lo}^2 = \frac{1}{m}\int_o^x [T_o\cos\epsilon - \mu(W - T_o\sin\epsilon) - D - \mu L]\,dx. \tag{16.8}$$

We can consider the thrust as a constant, and equal to the static thrust, therefore we collect all the known terms in the coefficient c_1,

$$c_1 = T_o \cos \epsilon - \mu(W - T_o \sin \epsilon). \tag{16.9}$$

By replacing c_1 and the definition of C_D and C_L in Eq. 16.7, we find

$$\frac{1}{2}mU_{lo}^2 = \int_o^x \left[c_1 - \frac{1}{2}\rho A U^2 (C_D - \mu C_L) \right] dx. \tag{16.10}$$

Now we need a few approximations. The first is that the sum of the aerodynamic terms in the integrand function be a constant. In reality, the C_L is a combination of aerodynamics and propulsion characteristics of the aircraft, and cannot be calculated exactly without engine considerations.

The second approximation is that the acceleration is also constant. We have discussed the implications of this approximation in §7.4. See also Krenkel and Salzman[197] for further analysis. If we follow the same strategy as in §7.4, we find an estimate for the lift-off run of the STOL aircraft

$$\frac{1}{2}mU_{lo}^2 = c_1 x_{lo} - \frac{1}{2}\rho A (C_D - \mu C_L)\frac{1}{2}U_{lo}^2 x_{lo}, \tag{16.11}$$

$$x_{lo} = \frac{mU_{lo}^2/2}{T_o \cos \epsilon - \mu(W - T_o \sin \epsilon) - \rho A (C_D - \mu C_L)U_{lo}^2/4}. \tag{16.12}$$

Equation 16.12 can be manipulated so as to eliminate the lift-off speed, and to show the thrust ratio T/W. The take-off run requires the calculation of the other phases of flight: flare, rotation and climb to clear a screen at 35 or 50 ft. If we neglect in first approximation the flare and rotation, we find the take-off run by summing Eq. 16.12 to the airborne distance

$$x = \frac{h}{\tan \gamma}, \quad \tan \gamma = \frac{T - D}{W}. \tag{16.13}$$

An analysis of interest is the comparison between the lift-off run with and without vectored thrust. This is shown in Figure 16.6, which is relative to the supersonic jet aircraft. We have assumed engine thrust and drag characteristics not dependent on the speed, and equal to those at $M = 0$. The other conditions are sea level, take-off speed $U_{to} = 300$ km/h (162 kt), and dry runway conditions at sea level.

The reduction of take-off field is about 25% with a 25 degree thrust angle. If an analysis is carried out into the weight effects on the take-off run, then we see a considerable difference in x_{lo}. This is shown in Figure 16.7.

In the past, some aircraft were designed for very high lift. Notable in this field is the NASA QSRA, which was capable of achieving lift coefficients as high as 10, as shown in Figure 16.8, thanks to its upper-surface-blowing flaps.

However, it is not true that high lift is always beneficial. Depending on the thrust ratio, the take-off run can in fact increase instead of decreasing. The take-off run

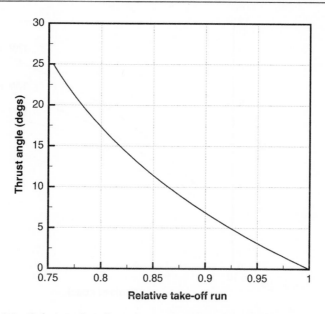

Figure 16.6 *Relative take-off run for jet aircraft model C with vectored thrust.*

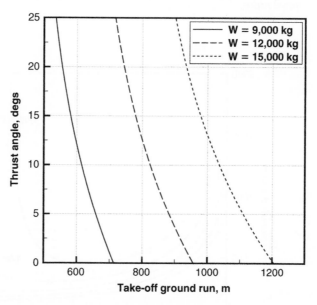

Figure 16.7 *Take-off run for jet aircraft model C with vectored thrust, military thrust, sea level ISA conditions.*

is found from the sum of Eq. 16.12 and Eq. 16.13. The second segment (airborne distance) may offset the benefits of high C_L and vectored thrust. This is shown in Figure 16.9, which resulted from the solution of Eq. 16.12 and Eq. 16.13 with a fixed vector thrust angle $\epsilon = 10$ degrees.

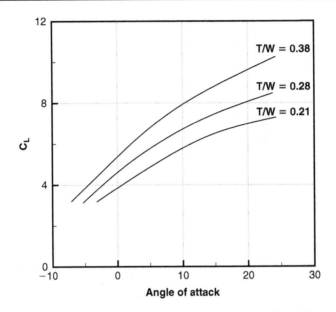

Figure 16.8 *Estimated vectored thrust lift coefficients for the QSRA at three thrust ratios.*

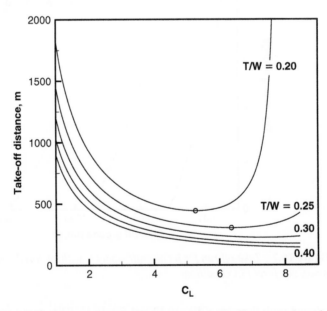

Figure 16.9 *Effects of C_L on STOL aircraft take-off run, at selected thrust ratios; $\epsilon = 10$ degrees.*

16.5 SKI JUMP

For aircraft based on a carrier ship there are several constraint factors on take-off and landing, because the shipboard flight deck is limited. In general, there are additional accelerating and decelerating gear, such as catapults, wires, and parachutes. However, the performance of these aircraft is clearly influenced by the wind conditions, and their performance is usually termed *wind-over-deck*. One type of take-off sometimes used by high-performance aircraft from an aircraft carrier is the *ski-jump*, as sketched in Figure 16.10. This take-off can be a catapult or non-catapult jump from an inclined deck.

Clarke and Walters[472] have simulated the ski jump of the Grumman F-14A and the McDonnell–Douglas F-18A with a three-degree of freedom model, and compared their results with some ramp-assisted take-offs. They included non-linear models for the thrust, for the landing gear loading (spring/strut system), and control systems dynamics. The maneuver relies on the fact that the aircraft is capable of gaining lift as soon as it leaves the deck, so that the maximum height drop from the deck is above sea level in the (statistically) worst ocean waves. The variation of height drop is considerably dependent on the launch speed. For the BAe Sea Harrier the maximum height drop is about 3 m once in 3.6 million take-off operations – by all means a safe operation. However, the safe limits of a ski jump are set by the minimum lift-off speed that provides zero climb rate.

The basic equations of the flight trajectory are

$$\dot{x} = U \cos \gamma, \tag{16.14}$$

$$v_c = \dot{h} = U \sin \gamma, \tag{16.15}$$

$$\dot{U} = \frac{1}{m}(T \cos \epsilon - D) - g \sin \gamma, \tag{16.16}$$

$$U\dot{\gamma} = \frac{1}{m}(T \sin \epsilon + L) - g \cos \gamma, \tag{16.17}$$

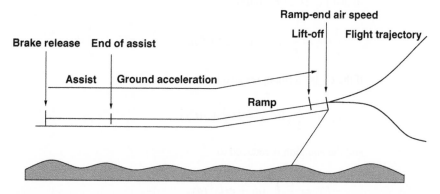

Figure 16.10 *Ski jump from aircraft carrier and height drop.*

where γ is the climb angle, and ϵ is the angle of thrust calculated with respect to the longitudinal axis of the aircraft. These equations are valid from lift-off. The acceleration of the aircraft from brake release to lift-off is given by equations similar to the ones discussed in §16.4. However, there are some changes to be considered: (1) the presence of an inclined ramp; and (2) the acceleration provided by the catapult. The catapult has a nearly impulsive effect, and at the point of release from the aircraft there will be a rapid change of acceleration.

The ski jump simulation can be split in two parts: the ground run till the lift-off point; and the initial airborne phase. Let us consider first the effect of the ramp angle γ_o on the take-off run, assuming that there is no assist to the acceleration. The momentum equation is written along the ramp direction, inclined by γ_o, which for this purpose is assumed to be constant. The ramp adds a constant term in Eq. 16.4,

$$m\frac{\partial U}{\partial t} = T\cos\epsilon - D - R - W\sin\gamma_o, \tag{16.18}$$

where the rolling resistance is

$$R = \mu(W\cos\gamma_o - L - T\sin\epsilon). \tag{16.19}$$

We assume again that the thrust is not dependent on the speed, and is equal to the static thrust. Solution of Eq. 16.19, according to the methods previously described, is the following:

$$\frac{1}{2}U_{lo}^2 = \frac{1}{m}\int_o^x \left[T_o\cos\epsilon - \frac{1}{2}\rho A(C_{D_o} + kC_L^2)U^2 - \right. \tag{16.20}$$
$$\left. \mu(W\cos\gamma_o - T_o\sin\epsilon) - \frac{1}{2}\mu\rho A C_L U^2 - W\sin\gamma_o\right]dx.$$

A number of constant parameters can be identified. We call

$$c_1 = \mu(W\cos\gamma_o - T_o\sin\epsilon) - W\sin\gamma_o, \tag{16.21}$$

so that Eq. 16.21 becomes

$$\frac{1}{2}U_{lo}^2 = \frac{1}{m}\int_o^x \left[c_1 - \frac{1}{2}\rho A(C_{D_o} + kC_L^2 + \mu C_L)U^2\right]dx. \tag{16.22}$$

If the C_L is constant, we also have

$$c_2 = -\frac{1}{2}\rho A(C_{D_o} + kC_L^2 + \mu C_L), \tag{16.23}$$

and the equation is reduced to

$$\frac{1}{2}U_{lo}^2 = \frac{1}{m}\int_o^x (c_1 + c_2 U^2)\,dx, \tag{16.24}$$

which has a well-known form. Unfortunately, the C_L is not constant, and the integral must be solved numerically. We can now take into account the actual variation of thrust, lift coefficient, and ramp angle by a numerical solution of Eq. 16.21:

$$\frac{1}{2}U_{lo}^2 = \frac{1}{m}\sum_{i=1}^{n}\left(c_1 + c_2 U^2\right)_i \Delta x_i, \qquad (16.25)$$

with

$$c_{1_i} = \mu\left(W\cos\gamma_o - T_o\sin\epsilon\right)_i, \quad c_{2_i} = \frac{1}{2}\rho A\left(C_{D_o} + kC_L^2 + \mu C_L\right)_i. \qquad (16.26)$$

If there is no assist, the equation can be solved with a constant acceleration, as described in §7.4. The calculation of the airborne phase requires the integration of the differential equations, Eqs 16.14–16.17, with appropriate initial conditions. These conditions can be found from the calculation of the take-off run, Eq. 16.21, or with specified starting conditions. Figure 16.11 shows the effects of GTOW on the flight path, and indicates that if the aircraft is too heavily loaded there is the risk of crash landing in the ocean. For all the cases shown in Figure 16.11 the starting conditions are: Mach number at take-off $M = 0.22$; lift coefficient $C_L = 0.8$; ramp angle $\gamma_o = 9$ degrees; ramp elevation $h = 14$ m from nominal water level; thrust angle $\epsilon = 4$ degrees. Similar trade-off studies can be done in terms of ramp angle, lift coefficient (use of high-lift devices, vectored thrust), and take-off speed.

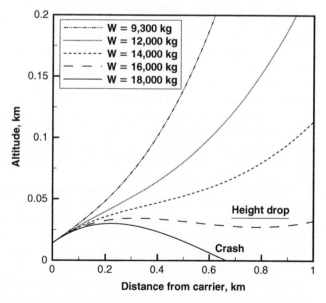

Figure 16.11 *Weight effect on ski jump from aircraft carrier. Summary of calculated flight paths.*

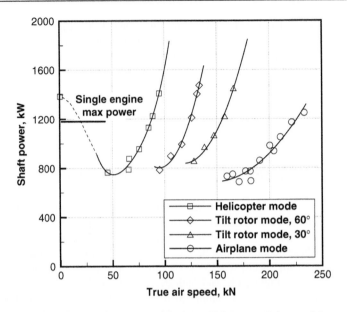

Figure 16.12 *Forward flight power of the XV-15 tilt rotor, elaborated from Roberts and Deckert* [474].

16.6 CONVERTIPLANES OR TILT ROTORS

A convertiplane is an aircraft that can change the direction of the thrust from vertical to horizontal and vice versa. Basically, this can be done by tilting the rotors/propellers. A considerable amount of literature is available on tilt-rotor concepts. Some basic performance analysis is available in Hafner[473].

Since the requirements for hover and cruise are substantially different, the optimal configuration of the aircraft is a compromise. Disk loadings are higher than conventional helicopters, and can exceed $100\,kgf/m^2$ (10^3 Pa). Figure 16.12 shows the power requirements for level flight of the experimental tilt-rotor XV-15 from flight testing. One curve shows the power curve in helicopter mode. At the other end of the chart there is the power required for level flight in airplane mode. Between these two operating modes there is a set of curves obtained with flight at intermediate tilt angles. A proper analysis of the conversion from helicopter to airplane mode requires an unsteady aerodynamics calculation.

Operation in helicopter mode is possible only at low forward speeds; in the airplane mode low speeds are not possible, due to stall problems on the lifting surfaces. Therefore, the aircraft has the possibility of extending its flight envelope, to include the characteristics of the helicopter and the propeller-driven airplane. The Boeing-Sikorsky V-22 and the Agusta-Bell BA609 work on the same principle.

The tilt rotor has a peculiar flight envelope relative to the angle of thrust. In fact, there is a restricted range of thrust angles at any given speed. Safe operation outside these boundaries is not possible. Height/velocity diagrams for the XV-15 in OEI operation near the ground, and optimal OEI take-off flight paths using optimal control theory, have been discussed by Carlson and Zhao[431,475].

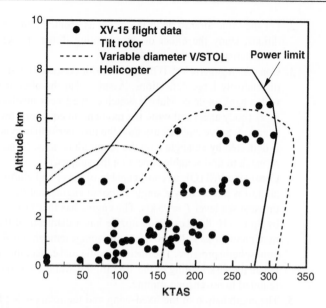

Figure 16.13 *Estimated V/STOL flight envelopes for conceptual tilt-rotor and variable-diameter rotor aircraft. Flight data from Churchill and Dugan[476].*

16.7 V/STOL FLIGHT ENVELOPES

The flight envelopes of V/STOL aircraft are different than the envelopes shown so far. First, V/STOL aircraft can maintain their altitude at zero ground speed by hovering. This is achieved either by jet lift of by rotor propulsion. Second, the maximum speeds obtainable are somewhat lower. Flight envelopes of various V/STOL concepts (tilt wing, variable diameter tilt rotor, shrouded rotor) are discussed by Scott[340].

Figure 16.13 shows the flight envelopes of three V/STOL vehicles. This figure also shows some of the limitations on the flight envelopes, as they have been discussed in Chapter 6. The combination of vertical take-off/landing capability, in addition to conventional cruise performance, extends the flight envelopes and has the potential to increase the aircraft's productivity.

Transition for vertical lift to aerodynamic lift requires a smooth change of the thrust vector, so that ultimately the vectored portion of the thrust is limited or negligible. Thus, the aircraft starts the cruise with aerodynamic lift only. Experience has indicated that this is an extremely complicated flight condition, with flow instabilities difficult to control. For a jet-lift V/STOL aircraft in transition, there must be enough installed thrust to overcome the gross weight and the resistance to forward speed, at least to a point when enough wing lift has been developed.

PROBLEMS

1. Study the effects of runway conditions on the take-off run of a hypothetical vector thrust on the supersonic jet fighter of reference. Consider sea level ISA conditions, and two take-off weights: MTOW and the combat weight.

2. For the same aircraft as in the previous problem, and the same take-off conditions, study the effect of after-burning thrust at take-off. Use the engine characteristics in Table A. 14.

3. Consider a generic V/STOL aircraft with vectored thrust. The nozzle is capable of relatively large deflections. Assume that the aircraft is to perform a turn, with a bank angle ϕ. Make a sketch of the forces involved in the frontal plane y_b, z_b (body axes) and write the momentum equations for the aircraft in such a situation. Solve and discuss how the turn performance is affected if the thrust is deflected by an angle ϵ on the longitudinal axis. (If necessary, do additional research to find suitable papers on the subject.)

4. Consider a V/STOL aircraft, capable of hover and climb with pure jet lift. The aircraft is powered by a single turbofan engine that delivers a maximum static thrust at sea level $T = 95$ kN. The engine has four nozzles that can be vectored by up to 110 degrees. Each nozzle has a diameter of 0.38 m. The air flow at maximum thrust is 78 kgf/s, is at an average temperature $\theta = 900$ K, and roughly atmospheric pressure. The thrust-specific fuel consumption is 0.344 N/h/N. Calculate the disk loading of the aircraft, the average speed of the jet, and the fuel required to hover for 1 minute.

5. The combination of the fixed-wing and the rotary-wing flight envelopes (Figures 6.8 and 16.13, respectively) show a gap in the h/V chart where flight is not possible. In particular, low-speed flight (less than 100 kt) is not achieved by any of the aircraft discussed at the troposphere. Flight in this region could allow detailed measurements of the atmosphere (in particular, humidity, concentration of pollutants, wind shear, etc.). Discuss critically the technical means for extending the flight envelope in this area.

Chapter 17

Noise Performance

The day the Concorde landed at Kennedy airport [New York] the local protesters made so much noise themselves that they were unaware that the aircraft had landed.

M.J.T. Smith[477]

This chapter deals with aircraft noise. Noise is one of many environmental issues that face the aviation industry today. These issues include local air quality, environmental compatibility, and global climate change. These larger problems are beyond the context of specific aircraft performance, and involve the aviation business as a whole. Emission problems are better addressed by textbooks on gas turbines. Archer and Saarlas[158] provide some discussion and data on emissions of several turbojets, turbofans and reciprocating engines. In the present context, we shall be concerned only with the technical aspects of the aircraft. We shall identify the main causes of noise, some methods for noise reduction and the certification procedures required. Excessive noise emission can make an aircraft obsolete. Aircraft certification must meet increasingly tough international regulations (FAR Part 36 and ICAO Annex 16).

The jet engines in the 1950s and early 1960s were extremely noisy. They prompted noise limitations at Heathrow airport, in London, as early as 1959. London Gatwick followed suit in 1968. Nowdays commercial aircraft operations include limits on night flights and taxes levied on the noisiest aircraft at most airports. Furthermore, airports may have to pay for acoustic insulation of properties affected by aircraft noise and loss of value of property near busy airports.

Research into aircraft noise started in earnest in the late 1960s. In 1965, Gebhardt[478] reported on the noise problems of the Boeing B-727, which was introduced into service one year earlier. Crighton[479] estimated that the noise level of a first-generation Boeing B-737 (1965) was the same as the world population shouting at once, while a second-generation aircraft would only produce the same noise as the city of New York shouting at once. However, the energy associated with the noise produced during the take-off of a first-generation Boeing B-737 was enough to cook an egg!

The information available on aircraft noise is rapidly expanding. For the physics fundamentals of sound and acoustics, we refer to Dowling and Ffwocs-Williams[480] and Goldstein[481]. A complementary book, and a fairly good source of information with regards to aircraft applications, is Smith[477]. It includes a detailed discussion on noise sources (power plants, fuselage, propellers, gas turbines), data acquisition and performance prediction, sonic boom, relevant historical notes, and a very extensive bibliography. The CAA and the FAA have staggering amounts of technical publications (continuously updated), which cover just about any aspect of noise. ESDU[482] publishes a technical series on noise, including estimation, propagation and reduction.

17.1 DEFINITIONS OF SOUND AND NOISE

The essential parameters for defining noise are the sound pressure level p or SPL, the acoustic power P, and the acoustic intensity I. The sound pressure level p is calculated with respect to a reference value, $p_o = 2.88 \cdot 10^{-5}$ N/m^2, corresponding to the sound pressure at the threshold of hearing. The sound intensity is the sound pressure times the speed of propagation, or the acoustic power per unit area,

$$I = pa = p\frac{a^2}{a} = \frac{p}{a}\frac{p}{\rho} = \frac{p^2}{\rho a} = \frac{F_p}{A}\frac{r}{t} = \frac{E}{t}\frac{1}{A} = \frac{P}{A}, \tag{17.1}$$

where F_p is the force created by the acoustic pressure p, ρ is the air density, E is the energy, t is the time, r is the distance from the source, and A is the propagation area. For a point source, the propagation area is spherical, so P decays as $1/r^2$; for a line source, the propagation front is cylindrical, and $P \sim 1/r$. Therefore, $P = IA$, or $P \propto p^2$. A measure of the loudness of the sound is given in *Bel*. Since this unit is generally a too large unit, a *decibel* (10^{-1} Bel = 1 dB) is used in practice. The sound loudness in decibels is defined from the ratio between the actual sound intensity I and the sound intensity at the threshold of hearing, I_o,

$$SPL \text{ (dB)} = 10 \log\left(\frac{I}{I_o}\right) = 10 \log\left(\frac{p}{p_o}\right)^2 = 20 \log\left(\frac{p}{p_o}\right), \tag{17.2}$$

where "log" is the symbol for logarithm with base 10. The regulated sound level is indicated as *EPNdB*, or *Effective Perceived Noise in dB*, as measured at appropriate locations. This is the internationally recognized measure of aircraft noise for certification purposes.

The threshold of hearing is set at $20\,\mu$Pa; the corresponding Sound Pressure Level (SPL) is 0 dB, and the relative intensity is 1 dB. From Eq. 17.2, a ten-fold increase/decrease in sound pressure corresponds to ±1 dB change in relative sound intensity; ±1 dB is the minimum perceptible sound difference by a sharp ear (*just noticeable difference*); about 6.9 dB correspond to a doubling of the sound pressure; and $+10$ dB is required before sound is perceptibly louder.

Generally, 120 dB (Concorde at take-off) is the threshold of pain, 110 dB is a very noisy jet aircraft or a discotheque environment, 80–90 dB is a street noise in a busy city, 65 dB is a noisy office in which verbal communication is disrupted; and 50 dB is the average office environment. In a soundproof room the sound pressure level is in the order of 10–20 dB.

The range of frequencies that can be heard by a human ear is between 20 and 20,000 Hz, and the range of intensity is 4 to 120 dB, although this range depends on the person. Human hearing is less sensitive at very low and high frequencies. In order to account for this, some weighting filters can be applied. The most common frequency weighting in current use is "A-weighting", which conforms approximately to the response of the human ear. The A-weighted sound level (dB (A) or dBA) accounts for the fact that the middle to high frequency range (500 to 5,000 Hz) is more annoying to human ears than low frequencies. However, this is partly compensated by the fact that high frequencies show larger decay rates as they propagate through the atmosphere, therefore the most important frequency range from the point of view of annoyance is the 200–2,000 Hz range. The A-weighted overall sound pressure is called OASPL, or OSPL(dBA). Note that OASPL and EPNdB are two different parameters.

For military vehicles, where detection is a critical issue, lower frequencies are also important, because low frequencies have lower attenuation during propagation, and therefore they can be heard over long distances. There is also concern about operations from military airports[483].

If the sound levels from two or more sources have been measured separately, the combined SPL has to be calculated by taking into account that the dB scale is logarithmic. One method is to convert the individual dB values to linear values, add those together, and convert them back to dB with the equation

$$I_{tot} = 10 \log \left(10^{I_1/10} + 10^{I_2/10} + \cdots \right). \tag{17.3}$$

One final parameter is directivity. Directivity provides discrete SPL values at directions from the source, and is generally expressed by a polar diagram, centered on the noise source.

17.1.1 Doppler Effect

The Doppler effect is characterized by frequency changes due to a source of sound approaching or moving away from an observer at a finite speed U, see Figure. 17.1. If the source is stationary, a single frequency f, having a wave length λ, is related to the propagation speed a by $f = a/\lambda$. The period is defined by $T = 1/f$.

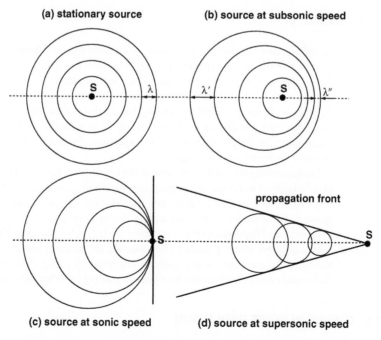

Figure 17.1 *Wave propagation fronts at subsonic and supersonic speeds, and Doppler frequency shift, cases (a) and (b).*

When the source is moving, it has a propagation front with forward and backward traveling waves. The wave traveling backward will have a speed $U + a$ with respect to an observer on the ground, therefore its apparent wavelength will be $\lambda' = (U + a)T$. The wave traveling forward will have an apparent wavelength $\lambda'' = (U - a)T$. Therefore we have the Doppler frequencies

$$f' = \frac{U}{U + a}f, \quad f'' = \frac{U}{U - a}f, \tag{17.4}$$

and in terms of Mach number

$$f' = \frac{M}{M + 1}f, \quad f'' = \frac{M}{M - 1}f. \tag{17.5}$$

At subsonic speed, the volume involved by the sound is $\sim T^3$. At sonic conditions the source will be coincident with the front end of the wave. The volume involved is the infinite space behind the source. At supersonic speeds, the propagation front degenerates into a cone. This propagation front is important in sonic boom analysis.

Sound propagation in the atmosphere is governed by a number of physical parameters. For a spherical propagation front, the noise intensity at a given frequency depends on the temperature, pressure and relative humidity. When propagating from the upper layers of the atmosphere to the ground, the acoustic waves travel through layers of changing atmospheric properties, and often sound like a rumble of varying intensity.

17.1.2 Sources of Noise

Noise emission falls into two categories: tonal noise and broadband noise. Tonal noise is associated to discrete frequencies, e.g. a cyclical motion of some nature, such as the oscillating wake past a bluff body, a cavity, or a blade passing. Tonal noise has distinct pitch at relatively low frequencies. The broadband noise arises from turbulent fluctuations in the flow field and usually lacks a discernible pitch. These fluctuations generally have a wide range of frequencies. A broadband noise has a continuous spectrum in the frequency domain. The spectrum denotes the energy present at all frequencies within a given range.

Understanding the sources of noise is the key to noise reduction technologies. The chief causes of noise are the engines and their subsystems, the airframe, the wing and its control systems (leading- and trailing-edge devices), and the landing gear. Some of these sources are active during the terminal phases of the flight (landing gear, extended flaps, etc.), others are active during all the phases of the flight, though with a different effect. Causes of noise during the terminal phases are important in assessing the environmental impact on communities living close to busy airports. Helicopters and propellers produce noise according to different mechanisms.

17.2 TRENDS IN NOISE REDUCTION

The greatest reduction in aircraft noise in the past has been achieved thanks to the development of high by-pass ratio turbofan engines. For some time, the airframe noise

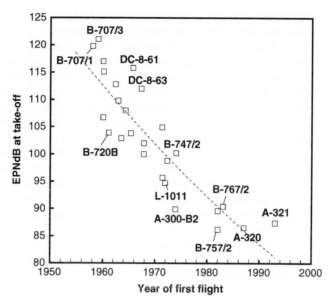

Figure 17.2   *Certified levels of noise emission at take-off for selected aircraft at year of introduction.*

was considered of relatively secondary importance, but this focus has now changed. Crighton[484] defined the airframe noise as the contribution of all non-propulsive sources.

Figure 17.2 shows a trend of the certified noise levels at take-off for selected commercial subsonic jets. The graph shows how, as a combination of more efficient engine and airframe design, the noise levels have been reduced considerably from the first generation of jet-powered aircraft. Further data, and comparison with noise emission from military airplanes, are shown by Waitz *et al.*[485]

The reduction from about 115 dB to 85 dB represents a reduction in SPL by a factor 4. Due to the logarithmic scale, the noise reduction is better than it appears in the chart. Further noise reduction is still possible. By how much? Some specialists believe that technological advances will eventually lead to the development of a *quiet airplane* – one that can be seen but not heard. This is sometimes called the *owl technology*, and was the subject of extensive research in the 1970s (Kroeger *et al.*[486]). This technology has been reviewed by Lilley[487] in an interesting paper, appropriately titled "The Silent Flight of the Owl".

At present, the barred owl (*Strix varia*) is the only bird known to fly *silently* at frequencies above 2 kHz. In some experiments, the owl was made to fly, in a closed chamber, from a 3-m-high perch to a pile of food on the ground. The flight was observed and the noise measured. The average speed was estimated as 8 m/s with a $\overline{C}_L \simeq 1$ and a gross weight 0.6 kg. It was demonstrated that the upper surface of the wing is stalled, or nearly so, and that the flow is nearly turbulent. The removal of the trailing-edge fringe and the leading-edge comb increased the noise to the level produced by other birds. Thus, the evolution of the wings has created efficient physical mechanisms that cut off the perceived noise to 2 kHz, by natural control of the flow separation, its instability and transition to turbulence.

The flight of the owl is of interest to aeronautics, because the bird flies at high angles of attack, like the fixed-wing aircraft in approach configuration. Its noise sources are the turbulent boundary layer on the upper surface of the wings and the exposed legs. However, the owl has acquired the technology (scattering) to silence these instabilities at the frequencies heard by its prey. Although the characteristic length and mass of the owl is several orders of magnitudes below those of the modern large aircraft, it appears that the owl technology is the lower bound of the airframe noise emission. The theoretical *lower bound* of noise emission is that of a clean aircraft flying in level flight with high lift. In other words, the ideal aircraft would be capable of flying at the low speeds required for landing and take-off, while producing high lift; no appreciable noise from the various airframe components would be heard. This limit occurs when the scattering of the boundary layer turbulence starts at the trailing edge.

In light of the knowledge gained from research on the barred owl, and perhaps from many independent ideas, *wing morphing* is considered a promising technology for airframe noise reduction.

17.3 AIRFRAME NOISE OF FIXED-WING AIRCRAFT

Airframe noise is due to the combined effects of the fuselage, the wing system, the high-lift devices and the landing gear. Non-propulsive noise may become the dominant source of noise at take-off and landing. At landing the engines are quietened to meet the noise reduction regulations. Interest in the *ultimate noise barrier* is not recent, see for example Healy[488] and Morgan and Hardin[489].

At cruise condition only the fuselage and the wing system contribute to the Overall Sound Pressure Level (OSPL), with the wing dominating the noise radiation. In many analyses, following the models of Ffowcs-Williams and Hall[490] and Howe[491], the wing is modelled as a semi-infinite flat plate in turbulent flow. For a wing of high aspect-ratio, this is essentially a problem of scattering of turbulent kinetic energy from the trailing edge. The result of one such analysis is briefly summarized below.

The noise intensity just below an aircraft flying at altitude h above the ground, with a speed U, is found from

$$I = \frac{1.7}{2\pi^3} \frac{\rho A U^3 M^2}{h^2} \left[\left(\frac{u_o}{U} \right)^5 \left(\frac{\delta}{\delta^*} \right) \right]_{te} = \frac{1.7}{2\pi^3} \frac{\rho A U^5}{a^2 h^2} \left[\left(\frac{u_o}{U} \right)^5 \left(\frac{\delta}{\delta^*} \right) \right]_{te}, \quad (17.6)$$

where u_o is the characteristic speed for the turbulent flow, and δ and δ^* are the boundary layer thickness and boundary layer displacement thickness at the trailing edge, respectively. As usual, A is the wing area, M is the free stream Mach number, ρ is the free stream air density, and a is the speed of sound. Equation 17.6 ignores the presence of other systems (fuselage, tail plane, engines) and the sweep of the trailing-edge line, which would require a factor $\cos^3 \Lambda_{te}$. Finally, the noise intensity at aircraft positions other than flyover must account for a corrective factor $\sin^2 (\theta/2)$, where θ is the angle between the aircraft and the observer, calculated from the horizon at the downstream end. The corrected sound intensity is

$$I = \frac{1.7}{2\pi^3} \frac{\rho A U^5}{a^2 h^2} \left[\left(\frac{u_o}{U}\right)^5 \left(\frac{\delta}{\delta^*}\right) \right]_{te} \sin^2(\theta/2) \cos^3 \Lambda_{te}. \tag{17.7}$$

The ratio $(\delta/\delta^*)_{te}$ depends on the state of the boundary layer at the trailing edge. For a turbulent boundary layer, this ratio is at least equal to 8. We take the value $(\delta/\delta^*)_{te} \simeq 10$. This value should account for turbulent flow with some separation at relatively high C_L. This ratio, along with the turbulent characteristics u_o/U, can be calculated more precisely with modern computational fluid dynamics codes for the three-dimensional wing. The speed U in Eq. 17.6 is found from the definition of C_L,

$$I = \frac{1.7}{2\pi^3} \frac{\rho A U^3}{a^2 h^2} \frac{2W}{\rho A C_L} \left[\left(\frac{u_o}{U}\right)^5 \left(\frac{\delta}{\delta^*}\right) \right]_{te}, \tag{17.8}$$

$$I \simeq \frac{17}{\pi^3} \frac{U^3}{a^2 h^2} \frac{W}{C_L} \left(\frac{u_o}{U}\right)^5_{te}, \tag{17.9}$$

$$I \simeq c_1 \left(\frac{U^3 W}{h^2 C_L}\right), \tag{17.10}$$

with

$$c_1 = \frac{17}{\pi^3 a^2} \left(\frac{u_o}{U}\right)^5_{te}. \tag{17.11}$$

This coefficient will depend on the local atmospheric conditions (speed of sound, altitude, density), on the Reynolds number, and on the geometry of the aircraft. It is perhaps a fortunate finding that c_1 has a nearly constant value that remarkably fits most experimental data collected in the past 30 years, for systems as diverse as wide-body aircraft and birds – except the barred owl. Typical values of the parameters in Eq. 17.11 are $u_o/U \simeq 0.1$ and $a \simeq 340 \,\text{m/s}$ (near sea level). With these values we find $c_1 \simeq 4 \cdot 10^{-11}$.

Example

Consider an aircraft of the class of the Airbus A-300, with gross weight $W = 150,000 \,\text{kg}$ ($\simeq 1.5 \cdot 10^6 \,\text{N}$), flying over at an altitude of 120 m with an approach speed of 300 km/h (83.3 m/s). The estimated lift coefficient is $C_L \simeq 1.08$ (a relatively high value). Thus, the OSPL is given by

$$OSPL = 10 \log \left(\frac{I}{I_o}\right) = 10 \log \left(\frac{I}{10^{-12}}\right) \tag{17.12}$$

$$\simeq 10 \log \left[0.4 \left(\frac{U^3 W}{h^2 C_L}\right) \right] \simeq 73.5 \,\text{dB}.$$

Therefore, the airframe noise above the observer is not exactly a hush. Remember that this is the noise produced by the wing alone, due to trailing-edge turbulence, and that the other subsystems give an additional contribution.

17.3.1 Airframe Noise at High Lift

What happens if the C_L is increased? High lift is generally associated with large suction peaks on the upper surfaces of the wings, which in turn trigger instability in the boundary layer. This issue is a complex matter for computational aerodynamics. CFD analysis, as discussed by Lockard and Lilley[492], indicates that the average C_L is related to the turbulent quantities by the approximate equation

$$\left(\frac{u_o}{U}\right)^5 \left(\frac{\delta}{y_m}\right)_{te} = \left(1 + \frac{1}{4}C_L^2\right)^4, \tag{17.13}$$

where y_m is the length scale of eddies. Thus, the parameter y_m replaces the displacement thickness δ^* in Eq. 17.8. In areas of adverse pressure gradients $y_m/\delta^* \simeq 0.2$ to 0.3. Operating this substitution in Eq. 17.8, these authors found

$$I = \frac{1.7}{\pi^3} \frac{WUM^2}{C_L h^2} \left(1 + \frac{1}{4}C_L^2\right)^4, \tag{17.14}$$

where C_L is a mean value on the wing. Equation 17.14 does not take into account the presence of partly deployed flaps. Equation 17.6 and Eq. 17.14 show that the sound intensity I depends on the factor

$$F = \frac{WUM^2}{C_L h^2}. \tag{17.15}$$

Equation 17.14 and Eq. 17.15 can be rearranged, so that for a fixed flight altitude above the observer, and for a fixed C_L, they yield an equation in which only the mass and the speed appear explicitly. Equation 17.14 represents the lower bound of aircraft noise, such as it is created by the airframe alone, without engines, undercarriage, tail plane and high-lift devices. Figure 17.3 shows the trend of this lower bound in terms of the factor F. This factor can be obtained for an infinite combination of aircraft weight, speed, lift coefficient and flight altitudes.

The range of the factor $F = WUM^2/C_L h^2$ covers 5 orders of magnitude. A medium size aircraft, such as the Airbus A-300, with a flyover altitude of the order of 100 m will have $F \simeq 6 \cdot 10^3$. A barred owl, instead, flying over with a speed of 8 m/s will have $F \simeq 10^{-1}$, e.g. more than 4 orders of magnitude lower than a transport aircraft.

17.3.2 Noise from Control Surfaces and Landing Gear

The three main components of airframe noise in take-off and landing configurations are leading- and trailing-edge devices and landing gear. Fink[493,494] reported on general studies on the effects of interaction between airframe components, and showed

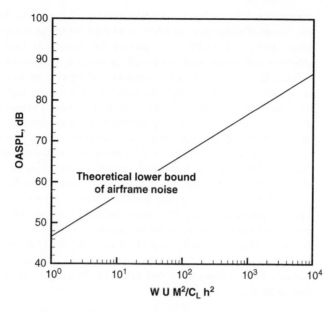

Figure 17.3   *Theoretical lower bound of flyover airframe noise in level flight. Actual noise emission is above the calculated trend line. No consideration for trailing-edge sweep.*

how the component interaction produces a negligible change in noise response, compared with the single components.

The leading-edge noise is located near the trailing edge of the slat. Unsteady flow is identified in the slat cove, at the trailing edge of the slat and on the main wing, where the wing's boundary layer interacts with the slat's turbulent wake. The thickness of the trailing edge promotes further instabilities and acoustic pressures. The dominant frequency of the vortex shedding is comparable with the thickness of the trailing edge, therefore resonance is possible, and can be avoided by repositioning the slat. While suitable aero-acoustic models for the basic slat-wing configuration exist, the actual presence of mechanical brackets that operate the slat complicates matters.

The landing gear consists of a number of cylinders (strut, wheels) that generate noise in all directions. The basic noise emission can be studied by considering the struts as cylinders in normal turbulent flow. The experimental prediction of undercarriage noise in isolation is discussed by Heller and Dobrzynski[495] for a number of bogie configurations (including the Douglas DC-10, the BAC-1-11, the Lockheed TriStar). Kipersztok and Sengupta[496] discussed methods for isolating the different airframe noise components on the Boeing B-747, as did Block[497].

17.3.3   Airframe Noise Reduction

The systems discussed produce noise of similar amplitude, but with different spectral density. In order to reduce the noise emission by a fixed amount of dB (say 4 dB), each component noise will have to be reduced by the same amount. If three noise components have an OSPL $= 80$ dB, from Eq. 17.3, we find that the elimination of one source of noise gives an OSPL $= 84.6$ dB, while the reduction by 4 dB on each source with give an OSPL $= 80.1$ dB (see also Problem 3). Thus, it appears that noise reduction technology must address all the issues at the same time.

Regarding the flap, it may be essential to redesign the mechanical links that operate the systems, and to disrupt the edge vortices. Such systems include fences and microtabs. Noise reduction from the slat could be achieved by proper understanding of the mechanisms of unsteady forcing created by the cove and the slat's trailing edge, in addition to the mechanical links. Undercarriage noise is mostly due to bluff body separation, but it can be improved by removing small exposed parts, by including fairing and other means. Active flow control is a promising method, since it is aimed at increasing lift, and may lead to a removal of some of the flaps and slats.

The reduction of trailing-edge scattering requires the reduction of the flight speed. The fifth-power low in Eq. 17.8 was found for a large number of systems, including gliders and birds – but excluding the barred owl. Therefore, it is now believed that the owl is silent because it is capable of reducing this scattering. It is possible that a few dB may be gained in this area, for example by introducing porous or compliant surfaces, laminar flow control and serrated trailing edges, which gradually reduce the flow to the free stream conditions.

17.4 ENGINE NOISE

Noise produced from the engines is a key element in the overall noise emission. The relationship between engine noise and performance is clarified by a number of parameters, namely: (1) specific thrust T/W; (2) the specific fuel consumption; and (3) the ratio between engine weight and the mass flow rate. It is recognized that for a given thrust level the noise increases with the jet speed; T/W also increases with the jet speed, therefore one solution is to adopt engines with low specific thrust. At a closer inspection, noise is produced by the various components of the engine: the jet, the compressor, the turbine, and the thrust reverser. In this context we will consider only the jet, which is the dominant noise component.

Fundamental studies in this field were done by Lighthill[498,499]. A review of jet noise research up to 1980 by Ribner[500] includes the fundamental physics and some concepts of jet noise suppression. Tam[501] reviewed the problem of supersonic jet noise. Lighthill's analysis shows that the acoustic power of a high-speed jet in stationary surroundings is

$$P \propto \rho_j A_j U_j^3 M^5 = \frac{\rho_j A_j U_j^8}{a_j^5}, \tag{17.16}$$

where A_j is the jet area, $M = U_j/a_j$ the Mach number, and a_j the average speed of sound in the jet. This equation is also known as Lighthill's eighth power law. The mass flow of the jet is $\dot{m} = (\rho A U)$, therefore

$$P \propto \frac{\dot{m} U_j^7}{a_j^5}. \tag{17.17}$$

The far-field noise produced by a high-speed jet behaves as if generated by a monopole; hence, the intensity I decays like the inverse of the distance, $1/r$. In fact,

$$I = \frac{P}{A} \propto \frac{\dot{m}U_j^7}{2\pi r a_j^5} \propto \frac{\dot{m}U_j^7}{a_j^5 r}. \tag{17.18}$$

The corresponding sound pressure level is

$$OSPL\,(\text{dB}) = 10\log\left(\frac{I}{I_o}\right) = 10\log\left(\frac{c_1}{10^{-12}}\frac{\dot{m}U_j^7}{a_j^5 r}\right), \tag{17.19}$$

where c_1 is the constant of proportionality in Eq. 17.18. The importance of the jet speed is evident from U_j^7. It is verified that if we reduce the jet speed by half, Eq. 17.19 gives a reduction in OSPL of 21 dB to an observer at the same distance r. If, on the other hand, the distance of the observer is doubled, but the jet speed is maintained to the original level, then the reduction in OSPL is only 6 dB. This result emphasizes once again the importance of using high by-pass turbofan engines instead of pure jet engines.

If the jet is derived from an engine flying at a speed U, then it is convenient to introduce the engine thrust in Eq. 17.18. A rough approximation for the thrust is $T \simeq \dot{m}(U_j - U)$. Besides, if $U = c\,U_j$, with $c < 1$ a constant factor, then

$$I \sim \frac{T}{(1-c)U_j}\frac{U_j^7}{a_j^5 r}, \quad I \sim \frac{TU_j^6}{a_j^5 r}. \tag{17.20}$$

This equation shows the importance of cutting back the thrust, something that is routinely done in take-off operations. A simple calculation from Eq. 17.20 shows that a reduction of the thrust by 50% leads to a reduction of the OSPL by 3.0 dB – a sensible amount to human hears.

Figure 17.4 shows a projection of noise reduction in dB predicted by Rolls-Royce for one of their series of aero engines. The 2020 target is a reduction of 15 dB, while the current Trent 900, designed to power the huge Airbus A-380, is 7.5 dB quieter than the reference data in the chart (the engine Trent 895). The Quiet Technology Demonstrator on the Trent 800/Boeing-777-200ER showed a jet noise reduction at take-off of 4 dB(A), and inlet fan noise reduction of 13 dB(A).

A method that has been used in the past, and which still has some application in military aircraft, is *noise shielding*. This consists of shielding the engines with the tail plane and the wings, see Broadbent[502] and Jones[503].

17.5   NOISE CERTIFICATION PROCEDURE

All certified aircraft must comply with the most current community noise emission set by international aviation regulatory boards (*ICAO Annex 16, Chapter 3; FAR 36 – Stage 3*[504]). The only exception to this date in commercial aviation was Concorde, which was allowed on discretion to fly from and to selected airports around the world. Concorde was not allowed to fly supersonic over land. The official noise emissions of Concorde were 119.2 dB at take-off, 116.7 dB during approach, and 112.2 dB at the

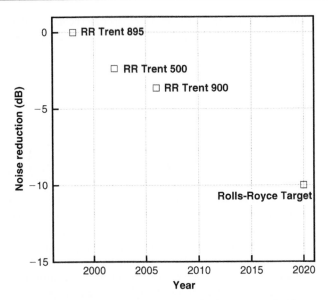

Figure 17.4 *Noise reduction from technological advances on the Trent series of aero engines (data from Rolls-Royce).*

sideline point. In the past several years, major difficulties in designing a sufficiently *quiet* new supersonic air transport aircraft (*SST*), capable of complying with the noise regulations, has left a vacuum in commercial high-speed flight.

The noise regulations extend to nearly every type of powered vehicle. Only military aviation does not comply with any of the noise regulations. Fighter jet exercises are still carried out over protected areas, without any requirement to cut back on noise.

Since the sound loudness decays in space, unique reference points are required for the noise measurement, for the certification of the aircraft, and for comparison of performance between aircraft. Three reference points are considered: the take-off point, the sideline point (or flyover), and the approach (landing) point. The formal testing procedure and current noise limits are described in ICAO Annex 16 and FAR Part 36, and have led to a relationship between noise emission and take-off weight.

The *noise emission at landing* (or approach) is the noise measured by microphones placed 1.0 nautical miles (1,853 m) at the downwind end of the runway. By convention, the aircraft must be 370 ft (121.5 m) above the ground level. The *noise emission at take-off* (community noise) is measured at a point along the center of the runway placed at 21,325 ft (6,500 m) from the brake-release point. The *sideline noise* is measured at a point on the ground placed at 1,474 ft (450 m) from the center of the runway. These reference points are shown in Figure 17.5.

The noise received by a microphone is converted into a digital signal, which must be corrected for variations of atmospheric conditions over the ISA values (temperature, pressure, relative humidity, prevailing winds), for small changes in the flight path (with respect to the requirements), for adjustment in engine power setting, and other items.

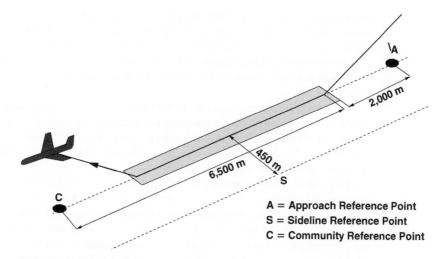

A = Approach Reference Point
S = Sideline Reference Point
C = Community Reference Point

Figure 17.5 *Noise spectra are measured at three points on the ground with aircraft flowing as shown in the graphics.*

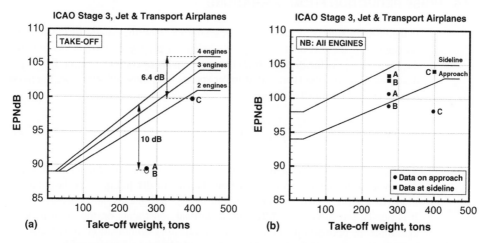

(a)

(b)

Figure 17.6 *Certified noise emissions for the Boeing B-747-400. (A) with engine PW4056, TOW = 272.160 tons; (B) with engine RB-211-52-4G, TOW = 272.160 tons; (C) with engine CF6-80C2B1F, TOW = 396.900 tons. In all cases the flaps are at 10 degree settings.*

The wind has appreciable effects on the measurements. At a down-wind reference point the noise level may increase by a few dB. By measuring upwind or side-wind, the noise level can drop by up to 20 dB. Table B.1 in Appendix B shows some certified noise data for commercial jet aircraft of the current generation. An example of noise compliance for the Boeing B-747-400 is shown in Figure 17.6. The solid lines show the limiting values according to the ICAO Annex 16 Chapter 3 and FAR Stage 3 regulations. The chart shows both the effect of the aircraft's weight and the type of engine. An intermediate weight (as indicated by points A and B in the chart), the compliance at take-off is about 10 EPNdB below the limit. At the MTOW the margin is reduced to 6.4 dB, which corresponds to almost half the limit OSPL. The EPNdB of the Airbus A-380 is estimated at the same level, although at a gross weight about 50% higher.

The new regulations relative to the ICAO Chapter 4 (2006) require that all aircraft have a 10 dB cumulative margin compared to Chapter 3/FAR Stage 3. In other terms, the sum of the take-off, sideline and approach noise should be 10 dB less than the corresponding Chapter 3/Stage 3. In addition, the aircraft must have a 2 dB margin at any two points of the measuring stations.

There is an important problem in the measurement technique, partially evident from the table: twin-engine aircraft are generally less noisy than three-engine aircraft; the latter ones are less noisy than four-engined aircraft. The initial climb angle of these aircraft is different, as shown on Table 7.3. A twin-engine aircraft moves away from the reference point along a flight path with a steeper gradient, therefore it is capable of putting more distance between the source of noise and the microphone.

17.6   NOISE REDUCTION FROM OPERATIONS

If the primary source of noise involves scattering, then the square of the acoustic pressure p^2 is proportional to U^5. For a spherical propagation front this pressure decays as $1/r^2$; r is the distance from the observer. By operating the aircraft at larger distances from the observer, and at lower speeds than those of reference, the noise reduction can be estimated from

$$\Delta dB = 10 \log \left(\frac{U}{U_{ref}} \right)^5 \left(\frac{r_{ref}}{r} \right)^2 . \tag{17.21}$$

Solution of this equation for an aircraft flying at an altitude 10% higher and at speed 10% lower would lead to a noise reduction of about 3.1 dB. In Eq. 17.21 the reduction in speed is more effective than the reduction in flight altitude, as shown in Figure 17.7. It is possible to solve Eq. 17.21 with additional constraints on the flight path, to determine the optimal operation. Studies of trajectory optimization for reduction of noise include those of Zeldin and Speyer[505] and Melton and Jacobson[506].

It has been shown that an optimal solution on landing is to fly the aircraft on a continuous flight path on Instrument Landing System (ILS), with a 3 degree slope. Under these conditions, the aircraft could maintain the speed of minimum power for as long as possible. Flaps and landing gear could be deployed only in the final phase. Noise reduction is estimated at 5 to 10 dB under the flight path at 4 km from the airport. A decrease in height from 150 to 120 m, or an increase in the approach speed from 132 kt to 145 kt leads to an increase in SPL of 2 dB.

At take-off, one idea sometime applied is the *power-down*, as illustrated in the Figure 17.8. This figure shows five different climb profiles. Since the measuring station is placed 6.5 km (3.5 nm) from the point of brake release, it is clear that if the aircraft is capable of increasing its distance from this point and/or decreasing the climb thrust, then the OSPL would decrease.

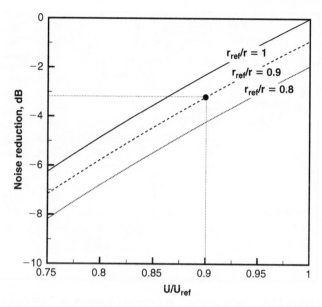

Figure 17.7   *Sound pressure level reduction at approach and landing by reduction in speed and altitude from observer.*

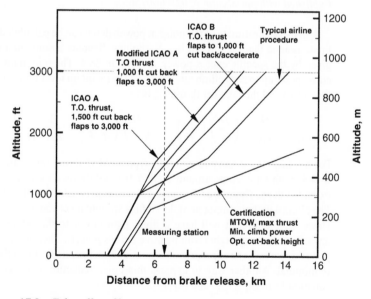

Figure 17.8   *Take-off profiles.*

During the initial climb the aircraft has to reach an altitude at which stable and controlled flight can be maintained even in case of one engine failure. Then it is allowed to cut back the power (by reducing the throttle), and reduce the climb rate, till the aircraft has traveled a distance from the community affected by the noise. In the next climb phase, the engines can be run at full throttle.

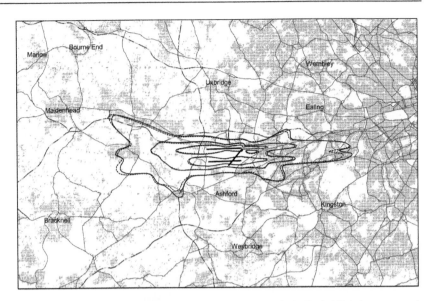

Figure 17.9 *Noise contours at London Heathrow in 2004 (solid lines), compared with contours in 2003 (dotted lines), excluding Concorde operations. Lines denote 57, 64, 69 dBA of noise index. From Monkman et al.[507] © Crown Copyright. Published with permission (C02W0007568).*

A first-order estimate of the noise at power-down can be calculated from Eq. 17.20. First, assume that the aircraft has N engines. Second, assume that the climb rate is given by the specific excess power, as done in §8.4. The minimum thrust is the one corresponding to zero climb rate, or $T = D$ to maintain level flight. The ratio between this thrust and the take-off thrust T_{to} is

$$\frac{T}{T_{to}} = \frac{D}{T_{to}} = \left(\frac{D}{W}\right)\left(\frac{W}{T}\right)_{to} = \left(\frac{D}{L}\right)\left(\frac{W}{T}\right)_{to} = \frac{1}{L/D}\frac{1}{(T/W)_{to}}. \tag{17.22}$$

For a transport-category jet aircraft, typical AEO thrust ratios are in the range 0.2 to 0.4, with an average around $T/W \simeq 0.3$. Glide ratios are about $L/D \simeq 12$ with extended flaps. Therefore, the thrust ratio in Eq. 17.22 is estimated at $T/T_{to} \simeq 0.28$. The aircraft with this amount of thrust flies straight level above the measuring station. However, the aircraft will have a flight path closer to the reference point than in any other climbing flight (with positive excess power). Therefore, it appears that the power-down must be appropriately optimized, so that the OSPL at the reference point is minimal. Another way to express this noise reduction is to write the following equation from Eq. 17.20 and Eq. 17.22

$$\Delta SPL \,(\text{dB}) = 10\log\left(\frac{T_{to}}{T}\right) = -10\log\left(\frac{L}{D}\right) - 10\log\left(\frac{T}{W}\right)_{to}. \tag{17.23}$$

The terminal area operation of aircraft at large airports is closely monitored by the aviation authorities and some environmental organizations. An example is shown in Figure 17.9, which reports noise contours at London Heathrow for 2003 and 2004, reproduced from Monkman et al.[507] The contours denote lines of constant noise index at levels 57, 63 and 69 dBA (the highest dBA is closer to the runways). The data from 2003 are modeled without Concorde operations. An important factor in noise index is the prevalence of flight path direction in and out of the airport. For 2004 the split was 81% westbound and 19% eastbound; for 2003 the data have a split 70% and 30%, respectively. This difference accounts at least partially for the difference in contours. Other factors include an increase in air traffic by 2.5% and the operation of larger and heavier aircraft. The shaded parts in the graph denote built-up areas.

17.7   MINIMUM NOISE TO CLIMB

We have seen that the take-off noise can be reduced by selecting appropriate flight trajectories. The essential parameters of the problem are the aircraft's weight, thrust, speed, jet speed, number of engines, and distance from the reference point. Atmospheric conditions, such as winds and humidity, contribute considerably to the OSPL.

The first case we consider is the climb at constant speed of a subsonic transport aircraft. The problem is to minimize the SPL due to jet noise at the measuring station on the ground, as shown in Figure 17.6. Therefore, we seek the minimum of the sound intensity

$$I \sim \frac{T U_j^6}{a_j^5 r}.$$

(17.24)

The jet speed depends on the engine thrust. The essential relationship is given by the continuity equation

$$T \simeq \dot{m}(U_j - U).$$

(17.25)

If we divide this equation by the corresponding expression at normal take-off values, we find

$$\frac{T}{T_{to}} \sim \frac{U_j(U_j - U)}{[U_j(U_j - U)]_{to}}.$$

(17.26)

The mass flow rate at take-off thrust is known from the engine data. Therefore, we can estimate the jet velocity from

$$U_{j_{to}} \simeq \frac{T_{to}}{\dot{m}_{to}} + U.$$

(17.27)

Solution of Eq. 17.26 with Eq. 17.27 provides the ratio of jet speeds as a function of the thrust ratio T/T_{to}, as shown in Figure 17.10. We estimate that the ratio $U_j/U_{o_{to}}$ is about 0.5 to 1 in the thrust range.

We now modify slightly Eq. 17.24 to take advantage of this result:

$$I \sim \left(\frac{T}{T_{to}}\right) \frac{T_{to} U_j^6}{r} \sim \frac{U_j(U_j - U)}{[U_j(U_j - U)]_{to}} \frac{T_{to} U_j^6}{r}.$$

(17.28)

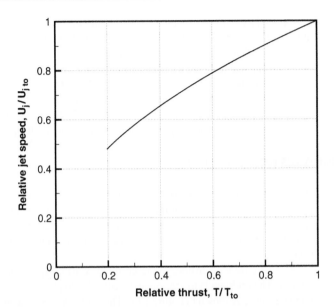

Figure 17.10 *Relative jet speed.*

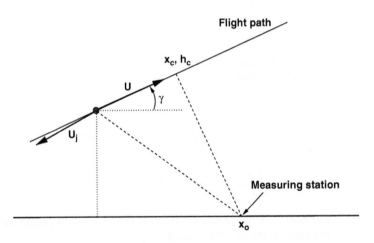

Figure 17.11 *Nomenclature for minimum jet noise climb at constant speed.*

At constant speed, seeking the minimum OSPL is the same as seeking the minimum of Eq. 17.28. The parameters are: the angle γ of the flight path, the thrust and the distance between the aircraft and the measuring station. The initial condition is given by the altitude h_1 (for example, 1,000 ft) and the distance x_1 from brake release. The measuring station is at coordinates $(x_o, 0)$, Figure 17.11.

The climb rate is related to the air speed, the climb angle and the engine thrust by

$$\tan \gamma = \frac{v_c}{U} = \frac{T - D}{W} \simeq \frac{T}{W} - \frac{1}{(L/D)_{to}}. \tag{17.29}$$

The equation of the flight path through (x_1, h_1) is

$$h = x \tan \gamma + h_1 - x_1 \tan \gamma. \tag{17.30}$$

If (x_c, h_c) is the intersection between the flight path (as found with elementary algebra), the normal line through the measuring station, the minimum distance is

$$r = \sqrt{(x_c - x_o)^2 + h_c} = r(\gamma) = r(T). \tag{17.31}$$

Equation 17.31 is a function of γ, or a function of the thrust T. Assigned T, the climb angle γ is calculated from Eq. 17.29. The minimum distance from the flight path is calculated from Eq. 17.31, and the resulting objective function is calculated from Eq. 17.28. The minimum value of the thrust is given by Eq. 17.22; this provides the condition for level unaccelerated flight. As it turns out, when the thrust is cut back, the jet speed decreases faster than the aircraft's distance from the reference point. With a jet speed reduced by 50%, the SPL reduction will be about 21 dB at a given r. Therefore, it appears that a level flight gives the minimum SPL at the measuring station. At any rate, the pilot must follow the take-off flight path set by the relevant regulations.

For noise annoyance, the optimization problem is different, because not only the magnitude of the noise is important, but also the duration, the directivity and the power spectra. Therefore, a suitable optimization problem involves the minimization of the EPNdB and its duration. Another aspect is that the climb might not be along a straight path. Berton[233] solved this problem by assuming that the cumulative noise emission during a time interval is given by

$$\bar{L}_N = \int_o^t L_N d\tau. \tag{17.32}$$

In Eq. 17.32 $\bar{L}_N$ is the accumulated noise level in the flight path from time "o" to time "t" created by a source propagating from the aircraft to the ground. If h_E is the energy height, as defined by Eq. 8.71, the minimum noise trajectory is the trajectory that maximizes the derivative

$$\frac{\partial h_E}{\partial \bar{L}_N}, \tag{17.33}$$

at all points in the trajectory. Equation 17.33 is a necessary condition of maximum increase in energy height for a given increase in noise emission (or minimum noise emission for given increase in energy height).

Minimum climb-out noise problems for VTOL aircraft have been solved by Schmitz and Stepniewski[508] and Henschel *et al.*[509] VTOL vehicles have a higher degree of freedom in the flight envelope, and can keep away from a reference point by climbing through a steep flight path. Antoine and Kroo[510] optimized a conceptual transport aircraft for minimum noise and engine emissions.

17.8 HELICOPTER NOISE

An approaching helicopter can be immediately distinguished from any propeller-driven vehicle from the relatively low frequency of the blade slap. Helicopter noise

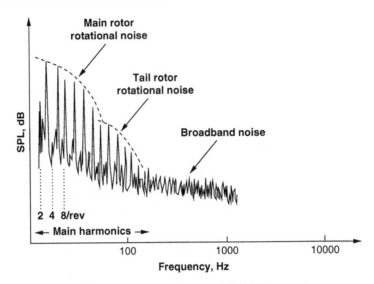

Figure 17.12 *Helicopter far-field noise spectra (qualitative, for approaching aircraft).*

is due to a large number of factors, which fall under two broad categories: (1) the mechanical noise, due to gearbox and vibrations; and (2) aero-acoustics noise, due to aerodynamic and propulsion effects. Mechanical noise is important in close proximity to the aircraft and in the cabin. Aero-acoustics noise is the chief cause of noise at medium to large distances from the aircraft. Since the commercial helicopters often operate around residential areas, helicopter noise has been subject to legal scrutiny for many years.

A presentation of the rotor noise is available in Johnson[354]. Theoretical models and calculation methods are beyond the purposes of this textbook. The literature in this field is vast. Suitable reviews are available in Morfey[511], George[512] and Schmitz[513].

A typical frequency spectrum of noise emission is shown in Figure 17.12. The frequency spectrum contains peaks at the fundamental blade passing frequency of both main and tail rotor. These frequencies and their harmonics are easily identified by the human ear. They contribute to the so-called *rotational noise*, which is caused by steady or cyclically varying loads, volume displacements and non-linear effects on the flow. For a blade rotating at a given rpm, the blade passing frequency is $f = 30/\mathrm{rpm}\,\pi$. If the rotor has N blades, the blade passing frequency is increased by a factor N. When the helicopter is moving away from the observer, the harmonics lose intensity.

Sometimes the blades suffer impulsive loading that results in increased SPL, especially at high frequency. This is the *impulsive noise*. The spectrum also contains non-harmonic signals, especially at high frequencies. This noise component is called *broadband*, and is caused by disturbances not directly associated to the blades' rotation. Broadband noise is ultimately related to the turbulent content of the flow. It can occur under various circumstances, such as operation near the ground, at low speeds, in hover, slow climb, and with atmospheric winds. These can create the conditions for turbulence ingestion noise[514].

17.8.1   Rotational Noise

The essential noise mechanisms of this harmonic noise are the cyclic loads on the rotor blades and the periodic displacement of the air as the blades advance. Higher harmonics are mostly due to the load variations with the azimuthal position of the blades. When the rotor is in forward flight, there can be various causes for variable loading, some of which have been discussed earlier: cyclic inflow, angle of attack, and wake-induced flow. These higher harmonics radiate even at small amplitudes.

Tail rotor rotational noise is important in the 200 to 500 Hz range, which is particularly annoying. Tail rotors produce a larger number of harmonics, because their inflow is non-uniform, due to the ingestion of the main rotor wake. This noise is predominant at intermediate tip Mach numbers (less than 0.7). At higher Mach numbers the noise caused by the thickness of the blades is of the same order as the rotational noise.

17.8.2   Impulsive Noise

Impulsive noise, also called *blade slap*, is a noise created by a blade passing through the slipstream of a preceding blade, through an area of increased turbulence. It is a cracking noise created at the blade passing frequency, and one of the most distinctive helicopter noise radiations. There are two main mechanisms of impulsive noise: blade/vortex interaction (mostly at low speeds), and transonic effects due to the shock at the outer board of the blades, called *high-speed impulsive noise*.

Blade/Vortex Interaction (BVI) occurs under various operating conditions and is the source of non-linear aerodynamics, vibrations and noise emission. BVI occurs not only when a blade intersects the vortex, but also when the vortex passes by at close distance. Another BVI occurs between main and tail rotor, when the tip vortex released by a main rotor blade travels through the disk of the tail rotor. This type of noise is less impulsive than the main rotor BVI. However, this noise proved to be in the 1,000–2,000 Hz frequency range – the most annoying one. The peaks in the acoustic waves result at frequencies at multiple combinations of the main and tail rotor blade passage frequencies (*interaction harmonics*).

Performance studies addressing the influence of BVI on the flight descent speed have focused on the parameters that could be changed to minimize this effect. BVI phenomena proved to be concentrated on the leading-edge region of the blades, with considerable pressure peaks limited to the 10% of the blade chord from the leading edge.

High-speed impulsive noise is due to the shock waves in the transonic/supersonic operation of the outer board sections of the main rotor blades. Tip Mach numbers can be of the order of $M_{tip} \sim 0.9$ with local supersonic flow (see also Figure 14.1).

One example is the Bell UH-1, a helicopter widely used in military operations in the 1960s and 1970s. Its two-bladed rotor, with a relatively large chord ($c = 0.40$ m) rotating at high speed, created a characteristic blade slap noise that could be heard miles away. Flight test acoustic measurements showed that the noise emission of the UH-1 exceeded 100 dB and reached 110 dB in the 50–150 Hz frequency range, at flight speeds as low as 130 km/h (70 kt) and rates of descent of about 2 m/s.

Experimental and flight test investigations proved that as the tip Mach number increases, the far-field acoustic wave generated by the advancing blade evolves toward a sawtooth negative peak. The quick pressure recovery is a sign of a strong shock wave on the blade, and is one of the strongest contributions to rotor noise. At high M_{tip} this type of noise exceeds other noise contributions.

17.9   HELICOPTER NOISE REDUCTION

There are two essential aspects in noise reduction: noise reduction from flight operation procedures, and noise reduction from design and retrofitting the aircraft. There are optimal flight procedures that minimize the acoustic emission. For example, changes in landing speed and descent rate can result in a reduction of the impulsive noise by up to 5 dB. For example, Brentner *et al.*[515,516] have developed a multidisciplinary model of noise emission for a complex maneuver, including take-off, climb with acceleration, banked turn and level flight.

Reduction of helicopter noise not directly related to the engine requires an accurate design, which takes into account the rotational speeds, the M_{tip}, the drag divergence point, and the tip geometry. It was found that the lower vortex noise emission is related to lower blade loading $T/\bar{c}R$.

State-of-the-art methods for the reduction for rotor noise include modulated blade spacing (Sullivan *et al.*[517]), reduced tip speed, blade tip modification, active blade control, variable rpm and others, see Edwards and Cox[518,519]. Tip speed reduction is a primary noise reduction technique, because of the strong relationship between noise, speed and turbulence. However, there are conflicting requirements on the tip speed, due to autorotative performance.

17.10   NOISE CERTIFICATION OF CIVIL HELICOPTERS

Igor Sikorsky's dream of a helicopter in every garage has not become true. Nevertheless, helicopter operations are heavily concentrated within populated areas, and there have been cases where helipads had to be closed due to concerns of community noise.

Like the fixed-wing aircraft, there are a number of flight conditions that must be certified. For the take-off and flyover procedures, the aircraft must operate at its MTOW. For the landing procedure, the aircraft must be at MLW. In all cases the rotor rpm must be the nominal or maximum value allowable. Additional conditions required are that the ambient temperature should not exceed 25°C, and that the atmospheric pressure be equal to the standard value. The ICAO regulations[520], as of 2001, as follows.

For the take-off noise procedure, the helicopter must operate at the certified take-off power along a flight path starting 0.5 km from the measuring station, at an elevation of 20 m (65 feet) above the ground. The reference flight path is a straight line from this point, inclined by an angle defined by the speed of maximum v_c. The aircraft must climb with the speed of maximum climb rate or the minimum take-off speed approved (whichever is greater).

For the flyover noise procedure, the helicopter must be stabilized in level flight at an altitude of 150 m (492 feet), with the lowest of the following ground speeds:

$$V = min(0.9VH, \quad 0.9VNE, \quad 0.45VH + 120\,km/h, 0.45VNE + 120\,km/h), \quad (17.34)$$

where VH is the level flight air speed corresponding to maximum engine power.

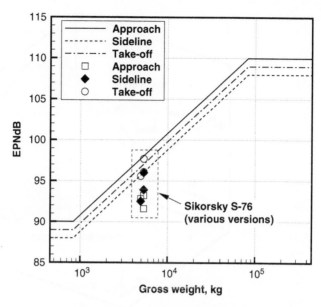

Figure 17.13   *Noise certification requirements for helicopters, according to FAR Stage 2. Symbols denote the Sikorsky S-76A, S-76C and S-76C+, at two reference weights, with the respective engines (Turbomeca Arriel, in all cases).*

For the approach noise procedure, the helicopter must be stabilized to follow a descent path with $\gamma = -6.0$ degrees, at a speed corresponding to the lowest approach speed, or the speed corresponding to maximum climb rate v_c (whichever is lowest), at a stabilized engine power till touchdown. The noise regulations relative to FAR Stage 2 for helicopters are shown in Figure 17.13, along with the certified performance of a typical helicopter, the Sikorsky S-76.

17.11   SONIC BOOM

The term *sonic boom* (or *bang*) is attributed to the noise heard on the ground, and created by an aircraft flying at supersonic speed and high altitude. More precisely, the event consists of a sequence of two shock waves that create a quick pressure rise having the shape of an N-wave. Depending on the atmospheric conditions, the boom can be heard like a sharp bang, a distant rumble beyond the clouds, or not at all, if the shock waves are reflected. Sonic boom can also propagate upward and then reflect back to the lower atmosphere[521].

The *sonic boom carpet* is the footprint on the ground where the boom is heard. The edge of this footprint is a hyperbole (intersection between the Mach cone and the ground), see Figure 17.14. The maximum boom intensity is right below the airplane. For reference, the sonic boom carpet of Concorde was about 70 km.

Various flight tests have indicated that people's tolerance to sonic booms is low, and therefore certification of supersonic aircraft would require the boom to be virtually

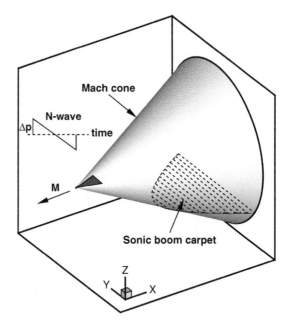

Figure 17.14 *Mach cone and sonic boom carpet.*

silent. Obviously, this annoyance has sparked the interest of many. Among the rational theories, a number of ideas have been proposed to eliminate the sonic boom; some violate the laws of physics. To emphasize the importance of sonic boom on environmental compatibility, we report that studies have been conducted on gray seals and other mammals in the North Atlantic to establish the effects of repeated booms on animal welfare. Extensive behavioral data are now available, see Perry *et al.*[522] Whitehead[523] reported that seals' mating habits showed no effects created by up to three sonic booms a day.

Back to the physics, the strength of the shock at the ground can be characterized by the maximum pressure rise Δp over the average pressure at sea level, p_o. Thus, the problem is to identify the parameters affecting $\Delta p / p_o$. In general, these are Mach number, flight altitude, aircraft weight and a number of geometric parameters, namely: the volume, the length, the reference area of the aircraft, and the shape of the nose. Sonic boom minimization implies minimization of the function Δp, although there are other issues at stake. For example, not only the maximum pressure rise is of interest, but also the duration of the boom. Alternative optimization procedures include the minimization of the integral $\Delta p(t)$, with several constraints on the aircraft (geometry, weight, etc.). The SPL of the sonic boom will be

$$ SPL = 10 \log \left(\frac{\Delta p}{p_o} \right)^2 = 20 \log \left(\frac{\Delta p}{p_o} \right). \tag{17.35} $$

Data from several supersonic aircraft indicate that the initial boom overpressure is in the range of 25 to 150 Pa over a period of 0.1 to 0.2 seconds. The experimental aircraft

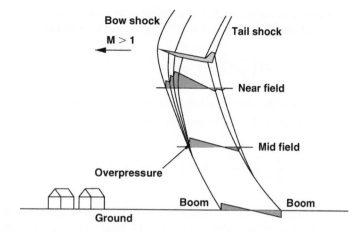

Figure 17.15   *Sonic boom.*

XB-70 of the 1960s considerably exceeded this level, having pressure peaks of over 200 Pa. The maximum overpressure from Concorde was in the range of 100 Pa.

From the SPL equation above, we calculate that the N-wave created by Concorde corresponded to an $SPL \simeq 131$ dB on the ground, while the shock created by the XB-70 was 137 dB, a mere +6 dB, in spite of the doubling of the pressure signature.

The first predictions of sonic boom were based on the models of supersonic flows of projectiles (Witham[524,525]), and lifting wing/body combinations (Walkden[526]), and included provisions for supersonic area rule (Hayes[527]). These initial attempts were limited, among other things, by the consideration of constant thermodynamic properties in the atmosphere. All the theories conclude that the far-field shock pattern is reduced to a front/compression shock and a rear/expansion shock, which give rise to the N-wave. Jones[528] showed that the lower bound of these pressure waves was created by a blunt body. This theory is conceptually important, because it gives the theoretical minimum overpressure on the ground. Hayes pointed out that in the real atmosphere the wave pattern shown in Figure 17.15 coalesces more slowly than in the ideal atmosphere, and may give rise to mid-field pressure waves that propagate undisturbed to the ground. Seebass and George contributed to the theory and to the aerodynamic optimization in a number of publications[529,530], as did Darden[531,532]. For the basic theory see Hayes[533] and Seebass[534]. For more modern advances on the subject, see the collection edited by Darden[535].

Figure 17.15, adapted from Talay[536], shows the effect of shock waves propagating from the aircraft over long distances.

The lower bound of the sonic boom is found below. Assume that the maximum overpressure at the ground is Δp, and that the average atmospheric pressure is p_o. If M is the flight Mach number of the aircraft, and h is its flight altitude, then

$$\frac{(\Delta p/p_o)(h/l)^{3/4}}{k_r \beta^{1/4}} = 1.19 \frac{\gamma}{\sqrt{\gamma+1}} \sqrt{I_F}, \tag{17.36}$$

where k_r is a reflection factor (a typical value would be $k_r \simeq 2$), $\beta = \sqrt{\beta^2 - 1}$, and l is the length of the aircraft. Jones[528] showed that the value of the lower bound of I_F is

$$I_F = 0.25 \left(\frac{1}{2} \beta C_L \frac{A}{l^2} + \frac{A_b}{l^2} \right), \tag{17.37}$$

where A is the wing planform area, and A_b is the cross-sectional area of the aircraft. Combination of the two equations above leads to

$$\frac{(\Delta p/p_o)(h/l)^{3/4}}{k_r \beta^{1/4}} = 0.54 \left(\frac{1}{2} \beta C_L \frac{A}{l^2} + \frac{A_b}{l^2} \right)^{1/2}. \tag{17.38}$$

The term $(\beta/2)C_L A/l^2$ denotes the lift contribution to the sonic boom; the term A_b/l^2 denotes the volume contribution. The left-hand side is a global term that contains the boom overpressure, the flight altitude (relative to the aircraft's length) and the Mach number. Since the overpressure is due to the separate contribution of lift and volume, it is convenient to write

$$\Delta p = \sqrt{\Delta_{p_L}^2 + \Delta_{p_v}^2}. \tag{17.39}$$

The lift-induced overpressure is found from Eq. 17.38

$$\left(\frac{\Delta p}{p_o} \right)_L = 0.54 \, k_r \beta^{1/4} \left(\frac{l}{h} \right)^{3/4} \left(\frac{\beta}{2} C_L \frac{A}{l^2} \right)^{1/2}. \tag{17.40}$$

To simplify the term under the square root, insert the definition of C_L ($C_L = 2W/\rho A U^2$), the equation of state of ideal gases ($p/\rho = \mathcal{R}T$), the definition of speed of sound ($a^2 = \gamma \mathcal{R}T$)

$$\left(\frac{\beta}{2} C_L \frac{A}{l^2} \right)^{1/2} = \frac{\beta^{1/2}}{M} \frac{W^{1/2}}{l} \frac{1}{\rho a^2} = \frac{\beta^{1/2}}{M} \frac{W^{1/2}}{l} \frac{1}{\sqrt{p_a \gamma}}, \tag{17.41}$$

where p_a is the atmospheric pressure at the flight altitude. Finally, if the relative pressure is $p_a/p_o = \delta$, then

$$\Delta p_L = \frac{0.54}{\sqrt{\gamma}} \, k_r \frac{(M^2 - 1)^{1/8}}{M} \left(\frac{l}{h} \right)^{3/4} \frac{W^{1/2}}{l} \sqrt{\frac{p_o^2}{p_a}}, \tag{17.42}$$

$$\Delta p_L = 0.456 \, k_r \frac{(M^2 - 1)^{1/8}}{M} \left(\frac{l}{h} \right)^{3/4} \frac{W^{1/2}}{l} \sqrt{\frac{p_o}{\delta}}. \tag{17.43}$$

In the original formulation of Jones it was assumed that the average pressure on the ground is $\sqrt{p_a p_o}$; this removes the presence of the factor δ in Eq. 17.43. The volume

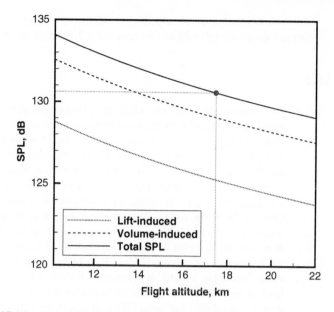

Figure 17.16   *Estimated SPL (dB) on the ground for sonic boom created by supersonic aircraft.*

contribution to the sonic boom is easier to derive. From Eq. 17.36 we find

$$\left(\frac{\Delta p}{p_o}\right)_v = 0.54\, k_r (M^2 - 1)^{1/4} \left(\frac{l}{h}\right)^{3/4} \sqrt{\frac{A_b}{l^2}}. \tag{17.44}$$

If d is the equivalent diameter of the cross-sectional area, so that $A_b = \pi d^2/4$, then

$$\Delta p_v = 0.424\, k_r \beta^{1/4} \left(\frac{l}{h}\right)^{3/4} \left(\frac{d}{l}\right) p_o. \tag{17.45}$$

Equation 17.45 contains a geometrical parameter, the slenderness of the aircraft l/d. The following procedure can be used to calculate the lower bound SPL of a typical supersonic aircraft.

Computational Procedure

1. Read the aircraft data (d/l, W, and other data).
2. Start loop (in altitude, weight or Mach number).
3. Calculate lift-induced overpressure from Eq. 17.43.
4. Calculate volume-induced overpressure from Eq. 17.45.
5. Calculate total overpressure from Eq. 17.39.
6. Calculate SPL of each component from Eq. 17.2.
7. End loop.

A typical solution of the algorithm is shown in Figure 17.16, which has been obtained for an aircraft with slenderness $l/d \simeq 20$ and weight $W = 185,000$ kg.

PROBLEMS

1. The atmospheric conditions affect the propagation of waves and their speed. Recall the fact that the speed of sound is dependent on the absolute temperature. Now consider that noise measurements are being taken at an airport location where the local temperature is $+32°C$, the atmospheric pressure is 1,020 mbar, and the relative humidity is 78%. Find a correction formula for the EPNdB of a certain aircraft, to reduce the measurements to ISA conditions.

2. Calculate the figure of merit of a supersonic aircraft flying at an altitude $h = 30,000$ m at $M = 2.4$. Its gross weight is $W = 200,000$ kg, and its length is $l = 33$ m. Compare this figure with an aircraft, having gross weight $W = 100,000$ kg flying at the same altitude at $M = 1.6$, and with the figure of merit of Concorde, whose data are: $W = 18,5,000$ kg, $l = 63$ m, $M = 2$.

3. For a fixed-wing aircraft on approach, airframe noise is equally produced by landing gear, leading-edge and trailing-edge devices. These components contribute individually the same OSPL of 82 dB, though with a different power spectrum. Calculate the OSPL of the three components in the cases: (a) one component's OSPL is reduced by 5 dB; (b) one component is eliminated; (c) all components' OSPLs are reduced by 5 dB; and (d) no reduction in OSPL is done.

4. A four-engined turbofan aircraft is upgraded with new high by-pass ratio engines that deliver the same thrust with a jet speed 30% lower than the original engines, which was 550 m/s. At the same time, the total mass flow rate is doubled, to 300 kg/s. Calculate the change in OSPL from the jets, assuming that the aircraft's weight is unchanged.

5. For a new supersonic aircraft designed to fly at $M = 2.4$ with an MTOW $= 272,000$ kg (600,000 lb) calculate the sonic boom signature on the ground, including the contributions from the lift and the volume. Assume a slenderness parameter $d/l \simeq 22$, and a flight altitude in the range of 16 to 19 km.

Appendix A

Aircraft Models

The tables and charts presented in the next pages summarize the data used for the calculations shown in this book. They are representative of some real-life aircraft, and give critical insight into aircraft performance data. They are published by several sources, including the manufacturer, the flight manual, the Certificate of Airworthiness, and other documents. The performance characteristics reported by some manufacturers are *statistical data*, based on ISA conditions and no winds. Since the data are determined using statistical methods, they should include the order of approximation. There are a number of sources for additional data, including Jane's [8]. Jenkinson and Marchman[537] provide detailed data of most current commercial airliners.

The data provided include aircraft in various classes: (A) subsonic commercial jet; (B) turboprop transport aircraft; (C) supersonic jet fighter aircraft; (D) general utility helicopter; and (E) tandem helicopter.

A.1 AIRCRAFT A: SUBSONIC COMMERCIAL JET

The aircraft model is the Airbus A-300-600, a modern wide-body passenger aircraft, used for short- to medium-range commercial and passenger flights. The geometrical layout of the aircraft is shown in Figure A.1. The essential data are summarized in

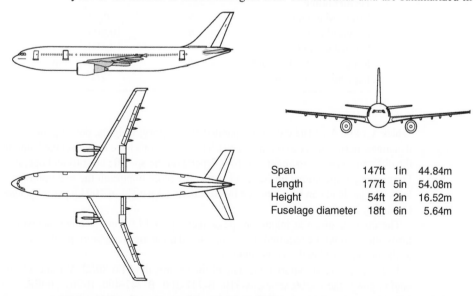

Span	147ft	1in	44.84m
Length	177ft	5in	54.08m
Height	54ft	2in	16.52m
Fuselage diameter	18ft	6in	5.64m

Figure A.1 *Drawings of the Airbus A-300.*

Table A.1 *A-300, basic performance data (weights in kgf).*

Parameter	Symbol	Value	Unit
Weights			
Max take-off weight (±2%)	MTOW	165,000	kg
Maximum ramp weight	MRW	165,900	kg
Operating empty weight (±2%)	OEW	90,100	kg
Max zero-fuel weight	MZFW	130,000	kg
Maximum landing weight	MLW	138,000	kg
Minimum flight weight	MFW	90,000	kg
Maximum payload	PAY	n.a.	kg
Maximum baggage		38,100	kg
Maximum usable fuel		49,600	kg
Max allowable fuel asymmetry		2,000	kg
Power Plant (2 engines)			
CF6-80C2A3	Table A.3		
Thrust angle on FL	ϵ	1.0	degs
Other data	Table A.3		
Aerodynamics			
Zero-lift drag coefficient	C_{D_o}	0.0225	
Induced-lift drag coefficient	k	0.0258	
Max lift coefficient	$C_{L_{max}}$	2.65	
Lift coefficient, landing	C_{L_g}	2.35	
Drag coefficient, landing	C_{D_g}	0.185	
Lift coefficient, take-off	C_{L_g}	1.4	
Drag coefficient, take-off	C_{D_g}	0.085	
Performance			
Cruise Mach number	M	0.82	
Cruise altitude	h	10,800	m
Service ceiling	z	12,500	m
Range at max passengers	R	7,700	km
Passengers (average)		260	

Tables A.1 and A.2. The data are separated by category: weights, power plant, aerodynamics, performance and dimensions. The dimensions have been averaged among the different series. Passenger numbers depend on the specific aircraft and operator. For passenger operations this datum is generally more important the maximum payload. The payload can be estimated by considering a weight of 100 kg per passenger (including baggage).

The range is given at maximum passengers, plus 14.9 tons of other payload. It takes into account the required fuel reserves and other airline operating rules, as well as the presence of head- and tail-winds.

In Table A.3 we summarize the data of the engine, the CF6-80C2. Variants of this engine power the A-300 series, A-310, B-737-200, B-747-400, B-767-300ER, and MD-11(see also Table A.4). Therefore, it represents a useful database for aircraft

Table A.2 *A-300, basic dimensions.*

Parameter	Symbol	Value	Unit
Main wing			
Wing span	b	45.0	m
Wing sweep at LE	Λ_{le}	25.0	degs
Wing area	A	260.0	m^2
Mean aerodynamic chord	MAC	6.61	m
Dihedral angle, wing	φ	7	degs
Tail plane			
Wing span		16.26	m
Wing sweep at QC		34.0	degs
Wing area		69.45	m^2
Aspect-ratio		3.81	
Taper ratio		0.420	
Tail arm		25.60	m
Dihedral angle, wing		5	degs
Other data			
Wheelbase		18.50	m
Track		9.60	m
90 deg avg turning radius		34	m
Distance between engines	b_t	15.7	m
Main wheel diameter		1.245	m

Table A.3 *Engine CF6-80C2A3 basic performance data. Additional data in Figure A.2.*

Performance index	Value	Unit
Maximum static thrust, S/L, ISA	243.6	kN
Take-off thrust, minimum 5 minutes, S/L	262.3	kN
Typical cruise thrust	50.4	kN
Dry weight ($\pm 2\%$)	4,200	kgf
Engine specific thrust, TO, S/L	0.625	kN/kgf
Engine by-pass ratio	5.0	
Pressure ratio	30.4	
Mass flow rate ($\pm 1\%$)	800	kg/s
Average specific fuel consumption, S/L	$9.32 \cdot 10^{-6}$	kgf/N s
Temperatures		
Flat rating temperature at take-off	30	°C
Flat rating temperature at max continuous thrust	25	°C
Turbine exhaust temperature (TO 5 minutes)	920	°C
Turbine exhaust temperature (Max continuous thrust)	925	°C
Turbine exhaust temperature (2 minutes transient)	965	°C

(Continued)

Table A.3 (*Continued*)

Performance index	Value	Unit
Efficiencies		
Intake efficiency	0.98	
Fuel combustion efficiency	0.99	
Mechanical efficiency	0.99	
Intake polytropic efficiency	0.98	
Fan polytropic efficiency	0.93	
Compressor polytropic efficiency	0.91	
Turbine polytropic efficiency	0.93	
Nozzle isentropic efficiency	0.95	
Other data		
High pressure rotor speed	11,055	rpm
Low pressure rotor speed	3,854	rpm
Fan diameter	2.362	m
Maximum diameter	2.691	m
Engine length	4.267	m

Table A.4 *A300, other data.*

Parameter	Symbol	Value	Unit
Speeds			
Min. control speed (air)	VMC	202	km/h
Min. control speed (take-off)	VMC	207	km/h
Max operating Mach number	M_{mo}	0.82	
Max speed, landing gear out		500	km/h
Tire speed limit		362	km/h
Aerodynamic controls			
Max flap deflection		40	degs
Max slat deflection		15	degs
Ground operations			
Max approved airport altitude		2,895	m
Runway visual range		75	m

performance calculations. Additional performance charts are shown in Figure A.2. The CF6-80C2 engine has a lower fuel burn than most engines of its class, and includes a Fully Automatic Digital Engine Control (FADEC), integrated with on-board computers.

Some drag data are shown in Figure A.3 and collected in Tables A.5 and A.6 for reference. The figure shows the C_D versus Mach number at constant C_L. These data can be used for numerical analysis of the range performance.

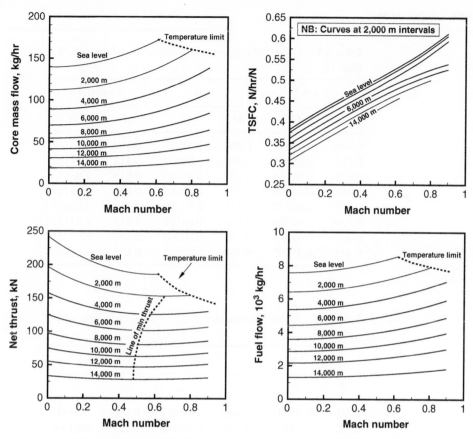

Figure A.2 *Estimated engine performance of the CF6-80: core mass flow, TSFC, thrust and fuel flow.*

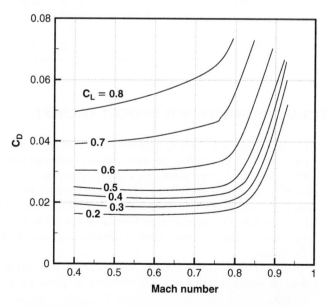

Figure A.3 *Estimated transonic drag rise of model airplane. See also data in Tables A.5 and A.6.*

Table A.5 *Estimated transonic drag rise of model airplane A (part 1).*

$C_L = 0.2$		$C_L = 0.3$		$C_L = 0.4$	
M	C_D	M	C_D	M	C_D
0.400	0.01633	0.400	0.01901	0.400	0.02178
0.556	0.01643	0.550	0.01877	0.513	0.02133
0.677	0.01684	0.672	0.01909	0.615	0.02123
0.755	0.01806	0.755	0.02063	0.698	0.02189
0.807	0.02112	0.811	0.02447	0.764	0.02463
0.850	0.02756	0.853	0.03195	0.821	0.03186
0.891	0.03825	0.892	0.04366	0.874	0.04552
0.931	0.05207	0.929	0.05831	0.927	0.06415

Table A.6 *Estimated transonic drag rise of model airplane A (part 2).*

$C_L = 0.5$		$C_L = 0.6$		$C_L = 0.7$	
M	C_D	M	C_D	M	C_D
0.400	0.02431	0.400	0.02985	0.400	0.03834
0.522	0.02384	0.519	0.03015	0.521	0.04001
0.626	0.02374	0.619	0.03068	0.616	0.04163
0.704	0.02456	0.695	0.03187	0.681	0.04332
0.764	0.02759	0.751	0.03496	0.727	0.04592
0.816	0.03490	0.799	0.04178	0.767	0.05101
0.868	0.04779	0.845	0.05357	0.806	0.05985
0.922	0.06480	0.892	0.06900	0.846	0.07159

A.2   AIRCRAFT B: TURBOPROP TRANSPORT AIRCRAFT

The next model is a heavy turboprop transport aircraft, a four-engine airplane used for logistic, and emergency operations. It is the Lockheed Hercules C-130J, a heavy-lift utility aircraft, used mostly for military operations, but also for the delivery of supplies, rescue and fire fighting. This aircraft has been built in large numbers since the 1960s, and it operates around the world. Some drawings of the aircraft are shown in Figure A.4. The basic data are summarized in Table A.7. The data are grouped according to weights, power plant, aerodynamics, performance and other relevant dimensions.

The aircraft range is strongly dependent on the payload, therefore selected data, as shown in Table A.8, are necessary to assess the cargo capability of the aircraft.

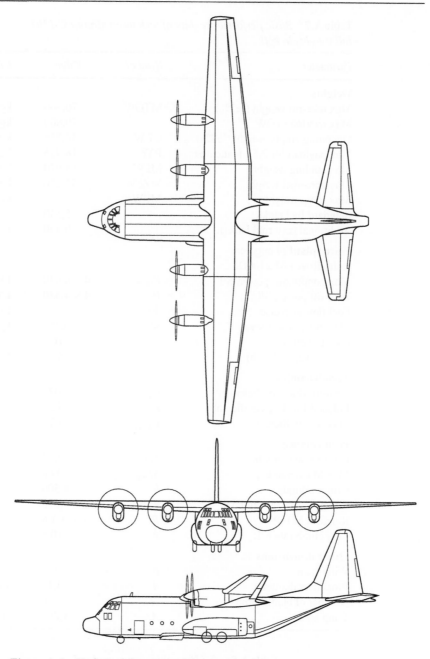

Figure A.4   *Technical drawings of the C-130.*

The power plant is the Allison/Rolls-Royce AE-2100D3, rated at about 3,427 kW at sea level. This is a two-shaft turbofan engine, with a 14-stage compressor, and FADEC. Some estimated engine performance data are given in Figures A.5 and A.6.

Table A.7 *Basic performance data of turboprop aircraft C-130J (all weights in kgf).*

Parameter	Symbol	Value	Unit
Weights			
Max take-off weight (±2%)	MTOW	70,300	kg
Max overload TOW		79,800	kg
Operating empty weight (±2%)	OEW	34,275	kg
Max payload for 2.5g maneuver	PAY	18,955	kg
Max landing weight	MLW	58,970	kg
Max zero-fuel weight	MZFW	53,230	kg
Max ramp weight	MRW		kg
Standard fuel tanks		25,530	l
Additional (external tanks)		10,450	l
Power plant (4 engines)			
Rolls-Royce AE2100D3			
Max continuous power, S/L	P_{max}	$4 \times 4{,}630$	kW
Take-off power, S/L	P_o	$4 \times 4{,}530$	kN
Fuel flow at cruise	$\dot{m}_f$		l/h
SFC, S/L, max power	SFC	0.256	kg/hr/kW
Pressure ratio	p_r	16.6	
Engine weight (each)		870	kg
Aerodynamics			
Zero-lift drag coefficient	C_{D_o}	0.028	
Induced-lift drag coefficient	k	0.035	
Max lift coefficient	$C_{L_{max}}$	2.75	
Performance			
Cruise Mach number	M	0.57	
Max Mach number	M_{max}	0.68	
Cruise altitude	h	8,500	m
Ceiling, w/19,090 kg PAY	z	8,615	m
Range at MTOW	R	Table A.8	
Max climb rate S/L	v_c	10.5	m/s
Other dimensions			
Wing span	b	40.4	m
Propellers diameter	d	4.11	m
Number of blades/prop	N	6	
Wing area	A	162.0	m^2

Table A.8 *C-130J range versus payload.*

Payload (kg)	range (nm)	Notes
15,000	2,049	(estimated)
11,250	2,174	intermediate load
9,000	4,460	with maximum fuel
0	4,522	ferry range

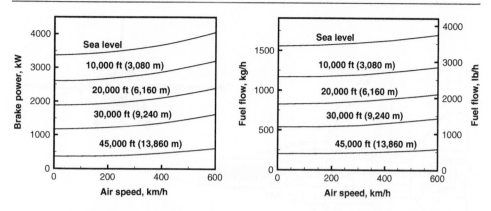

Figure A.5   *AE-2100D3 engine power and fuel flow curves versus aircraft speed at the altitudes indicated, standard conditions.*

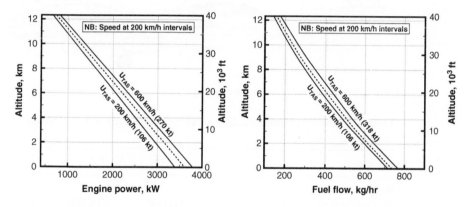

Figure A.6   *AE-2100D3 engine power and fuel flow curves versus flight altitude at the speeds indicated.*

A.3   AIRCRAFT C: SUPERSONIC JET FIGHTER

The next aircraft is a typical high-performance jet fighter. We have taken as a model the Lockheed F-16. This aircraft has been produced in large numbers, and even for the same version different aircraft batches have different instrumentation, weights, weapons systems and performance. Jane's Information Systems[8] reports details of all the derivatives, versions and applications of this aircraft. We have extrapolated some data of interest in the present context. The geometrical details of the aircraft are shown in a three-view format in Figure A.7.

The aircraft has a cropped delta wing blended with fuselage, with vortex control strakes to increase lift at high angles of attack and improve longitudinal stability. The wing section has an average thickness of 4% (NACA 64A-204).

The essential performance data are given in Table A.9. Other data are summarized in Tables A.10 and A.11. The lift and drag characteristics, discussed in the previous

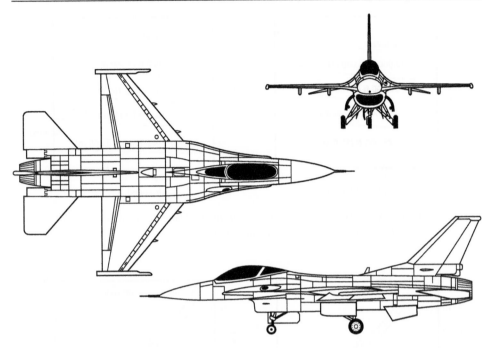

Figure A.7 *Geometry of the fighter aircraft.*

Table A.9 *Basic performance data, supersonic fighter aircraft (estimated) (all weights in kgf).*

Parameter	Symbol	Value	Unit
Weights			
Max take-off weight ($\pm 2\%$)	MTOW	19,100	kg
Operating empty weight ($\pm 2\%$)	OEW	8,900	kg
Maximum payload	PAY	5,700	kg
Typical combat weight		12,000	kg
Standard fuel tanks		3,150	kg
External fuel tanks		1,800	l
Engine's dry weight		1,370	kg
Aerodynamics			
Zero-lift drag coefficient	C_{D_o}	Fig. A.8	
Induced-lift drag coefficient	k	Fig. A.9	
Max lift coefficient	$C_{L_{max}}$	Fig. A.12	
Zero-lift angle	α_o	2.5	degs
Power plant (1 P&W F-100 engine)			
By-pass ratio	BPR	0.63	
Max continuous thrust, S/L	T_{max}	1×86.0	kN
Max thrust w/after-burning		1×110.0	kN
Take-off thrust, S/L	T_o	88.6	kN

(Continued)

Table A.9 (*Continued*)

Parameter	Symbol	Value	Unit
Specific thrust, S/L, MTOW	T/W	0.473	kN/kgf
Fuel flow at cruise	$\dot{m}_f$	Figs A.10–A.11	
Thrust angle on FL	α_T	2	degs
Performance			
Cruise Mach number	M	0.85	
Dash speed, $h = 12,000$ m	M_{max}	2.2	
Service ceiling		15,250	m
Range at MTOW			km
Ferry range		3,100	km
Combat radius		900	km
Limit load factor (turn)	g^+	+9	
Limit load factor ($dU/dt < 0$)	g^-	−3	

Table A.10 *Other dimensions and characteristics, supersonic fighter jet.*

Parameter	Symbol	Value	Unit
Delta wing			
Wing span	b	9.50	m
Wing area	A	28.9	m^2
Wing aspect-ratio	$\mathcal{AR}$	3.1	
Dihedral		0	degs
Wing taper-ratio	λ	0.295	
Leading-edge sweep	Λ_{le}	39	degs
Airfoil thickness (%)	t/c	4.0	
Vertical tail			
Area		0.0226	m^2
Aspect-ratio		1.058	
Airfoil section		biconvex	
Airfoil thickness (%)	t/c	5.3/3.0	
LE sweep		47.5	degs
Horizontal tail			
Area		0.0263	m^2
Aspect-ratio		1.294	
Airfoil section		biconvex	
Airfoil thickness (%)	t/c	6.0/3.5	
LE sweep		40.0	degs
dihedral angle		−10.0	degs
Ailerons			
Authority		−30–+20	degs
Response		30.0	degs/s
Elevons			
Authority		±30	degs
Response		30.0	degs/s

Table A.11 *Other data for supersonic fighter (estimated).*

Parameter	Symbol	Value	Unit
Roll radius of gyration	r_x	0.86	m
Pitch radius of gyration	r_y	2.11	m
Yaw radius of gyration	r_z	2.24	m
Mean aerodynamic chord	MAC	3.45	m
Wing section	64A-204		

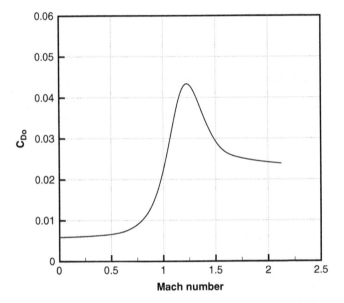

Figure A.8 *Zero-lift drag characteristics as a function of Mach number for supersonic fighter jet, basic configuration.*

chapters, are shown in graphic form in Figures A.8 (zero-lift drag), A.9 (lift), and A.10 ($C_{L_{max}}$). The numerical values of these quantities are given in Table A.12. All the data refer to the configuration A (baseline). Some aerodynamic data have been extrapolated from Fox and Forrest [538], others have been averaged among different versions. Figure A.11 shows the military thrust for F-100 engine; Figure A.12 shows the F-100 engine, after-burning thrust.

Table A.13 and Table A.14 show the engine thrust as a function of Mach number and flight altitude, with and without after-burning.

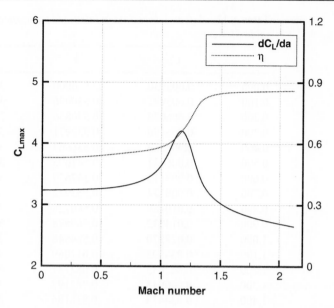

Figure A.9   *Lift characteristics as a function of Mach number for supersonic fighter jet, basic configuration.*

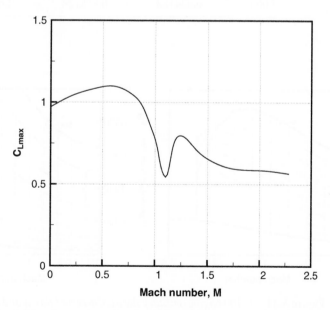

Figure A.10 $C_{L_{max}}$ *versus Mach number, fighter aircraft, baseline configuration.*

Table A.12 *Lift and drag data in tabulated form, speed range up to M = 2.*

M	C_{D_o}	k	C_{L_α}
0.000	0.005930	0.530000	3.236673
0.100	0.005991	0.530096	3.239516
0.200	0.006073	0.530856	3.242833
0.300	0.006190	0.532971	3.247509
0.400	0.006364	0.536604	3.254873
0.500	0.006621	0.541695	3.266857
0.600	0.007115	0.547639	3.286047
0.700	0.008174	0.553972	3.317267
0.800	0.010189	0.560725	3.370291
0.900	0.014232	0.569578	3.463775
1.000	0.022370	0.584548	3.645685
1.100	0.034393	0.617069	4.027599
1.200	0.042833	0.682787	4.144183
1.300	0.040943	0.782197	3.581955
1.400	0.034473	0.832104	3.197555
1.500	0.029339	0.845844	3.024687
1.600	0.026637	0.851330	2.918348
1.700	0.025511	0.853641	2.842454
1.800	0.024931	0.855193	2.784250
1.900	0.024511	0.856645	2.736878
2.000	0.024188	0.858126	2.696331

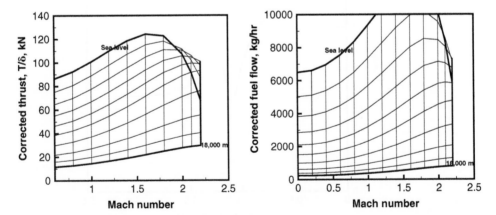

Figure A.11 *F-100 engine, military thrust. Corrected thrust and corrected fuel flow as a function of Mach number at selected altitudes (intervals of 2,000 m from sea level).*

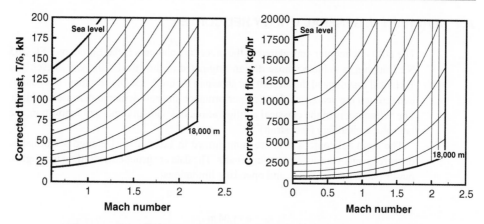

Figure A.12 *F-100 engine, after-burning thrust. Corrected thrust and corrected fuel flow as a function of Mach number at selected altitudes (intervals of 2,000 m from sea level).*

Table A.13 *Engine thrust for supersonic aircraft calculations, after-burning OFF. Thrust in kN, altitudes in km.*

M/h	0	2	4	6	8	10	12	14	16	18
0.0	88.622	72.626	58.847	47.097	37.186	28.924	21.532	15.708	11.459	8.359
0.2	84.082	69.150	56.221	45.144	35.757	27.899	20.801	15.175	11.070	8.075
0.4	83.487	68.952	56.287	45.371	36.071	28.245	21.096	15.390	11.227	8.190
0.6	86.328	71.665	58.785	47.602	38.010	29.887	22.368	16.318	11.904	8.684
0.8	92.227	77.048	63.574	51.765	41.549	32.831	24.630	17.968	13.108	9.562
1.0	100.747	84.849	70.532	57.824	46.707	37.125	27.931	20.376	14.864	10.843
1.2	109.929	93.599	78.572	64.990	52.921	42.376	31.994	23.340	17.027	12.421
1.4	118.569	102.524	87.230	73.012	60.085	48.572	36.836	26.872	19.604	14.301
1.6	124.249	109.957	95.395	81.184	67.778	55.484	42.323	30.875	22.523	16.431
1.8	122.735	112.887	100.973	88.090	75.074	62.536	48.077	35.073	25.586	18.665
2.0	107.332	106.500	100.564	91.380	80.391	68.697	53.404	38.958	28.420	20.733

Table A.14 *Engine thrust for supersonic aircraft calculations, after-burning ON. Thrust in kN, altitudes in km.*

M/h	0	2	4	6	8	10	12	14	16	18
0.0	127.575	103.812	83.542	66.399	52.068	40.227	29.845	21.772	15.883	11.586
0.2	124.242	101.297	81.661	65.026	51.085	39.538	29.361	21.420	15.625	11.399
0.4	127.424	104.110	84.100	67.101	52.816	40.956	30.442	22.208	16.201	11.818
0.6	137.092	112.265	90.886	72.667	57.315	44.532	33.135	24.171	17.633	12.863
0.8	153.732	126.204	102.417	82.074	64.875	50.512	37.622	27.446	20.021	14.606
1.0	178.261	146.762	119.416	95.937	76.015	59.321	44.232	32.268	23.539	17.172
1.2	209.706	173.247	141.415	113.948	90.537	70.839	52.888	38.582	28.146	20.532
1.4	249.331	206.857	169.500	137.059	109.254	85.743	64.108	46.767	34.117	24.889
1.6	297.924	248.492	204.583	166.135	132.946	104.707	78.419	57.208	41.733	30.445
1.8	355.262	298.374	247.134	201.757	162.218	128.302	96.283	70.239	51.240	37.380
2.0	419.632	355.752	296.999	244.117	197.433	156.972	118.082	86.142	62.841	45.843

A.4 AIRCRAFT D: GENERAL UTILITY HELICOPTER

The utility helicopter considered in the preceding chapters is the Eurocopter EC-365 N2 Dauphin, a vehicle whose design originated in the late 1970s (see Roesch[539]). Technical drawings of the aircraft are shown in Figure A.13. Development has continued through the years, and detailed performance and aerodynamic data are available. The helicopter is of the classical configuration, but with an important innovation: the tail rotor technology (fenestron).

The main helicopter's data are summarized in Table A.15. Detailed performance data are given in the accompanying charts. The data are grouped according to weights, power plant, performance, and operating limitations.

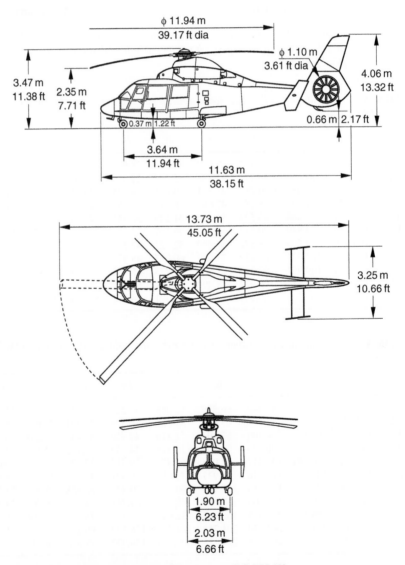

Figure A.13 *Main dimensions of Eurocopter EC-365 N2.*

Table A.15   *EC-365 helicopter performance data (all weights are in kgf).*

Parameter	Symbol	Value	Unit
Weights			
Max take-off weight ($\pm2\%$)	MTOW	4,250	kg
Operating empty weight ($\pm2\%$)	OEW	2,250	kg
Weight with sling load		5,000	kg
Standard fuel tanks		1,135	l
Additional fuel tanks		180	l
Maximum PAY (internal)	PAY	1,600	kg
Power plant			
2 Turbomeca Arriel C2 gas turbines			
OEI power, 30 seconds		680	kW
OEI power, 2 minutes		617	kW
OEI power, unlimited		593	kW
AEO power, take-off (each)		609	kW
AEO power, unlimited (each)		550	kW
Fuel flow at cruise (each)	$\dot{m}_f$	135	l/h
Performance			
Recommended cruise speed	U	260	km/h
Never-to-exceed speed	VNE	300	km/h
Hover ceiling, OGE, TO power, ISA		2,400	m
Hover ceiling, OGE, TO power, ISA $+ 20C$		1,600	m
Service ceiling, ISA	z	4,500	m
Service ceiling, ISA $+ 20$ C		4,100	m
Rate of climb at S/L	v_c	7.5	m/min
Rate of climb, S/L, OEI	v_c	2.7	m/min
Maximum range	R	850	km
Endurance (140 km/h, GTOW $= 3{,}750$ kg)	E	4.5	h
Endurance (140 km/h, GTOW $= 4{,}250$ kg)	E	4.3	h
Passengers		8	
Operating limitations			
Minimum atmospheric temperature		$-40°$C	
Maximum atmospheric temperature		ISA $+ 40°$C	
Overall dimensions			
Distance between shafts	x_{tr}	7.10	m
Fuselage length		12.70	m
Length w/rotor turning		14.43	m
Number of blades, main rotor	N	5	

A.4.1   Main Rotor

Table A.16 is a summary of rotors' geometrical characteristics. The blade sections of the main rotor are ONERA designs, of the family OA-2XX, Figure A.14. A description of these airfoils and the design methodology used is available in Thibert and Philippe[540]. These airfoils have a relatively high divergence Mach number, at least when they are compared with the NACA 0012. Figure A.15 shows the trend in

Table A.16 *EC-365 rotor and fenestron data.*

Parameter	Symbol	Value	Unit
Main Rotor			
Rotor diameter	d	11.94	m
Rotor solidity	σ	0.063	
Rotor height over ground at hub		3.0	m
Airfoil section	Onera	OA-212/209/207	
Pitch at blade tip	θ_o	1	deg
Pitch variation: linear	$d\theta/dR$	−1.5	deg/m
Rotor chord	c	0.40	m
Main rotor speed	rpm	350	
Tail rotor speed	rpm_{tr}	3,650	
Airfoil thickness	t/c	11%	
Tail Rotor (fenestron)			
Tail/main diameter	d_{tr}/d	0.092	
Number of blades	N	11	
Rotor speed	Ω_{tr}	3,665	rpm
Distance between shafts	x_{tr}	7.10	m

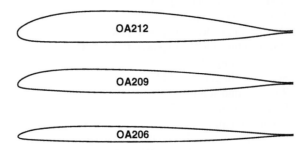

Figure A.14 *Rotorcraft airfoils OA-206, OA-209 and OA-212.*

the M_{dd} for two families of airfoils. For the OA-209 airfoil at zero incidence $\alpha = 0$ we have estimated $C_{L_o} \simeq 0.108$ and $C_{D_o} \simeq 0.007$, $C_{L_\alpha} \simeq 0.130 \; rad^{-1}$ at low subsonic speeds ($M = 0.4$).

The blade tips have a so-called PF design, in which the tips are tapered back at the outer 4% of the radius. Performance of this tip was briefly mentioned in §12.1.3.

A.4.2 Engines

The engines are Turbomeca Arriel C2. These engines are capable of developing a maximum 680 kW emergency power. The power rating is attributed to new engines. The engines tend to evolve and produce more power for the same weight over time. For example, the 2C2 version of the same engine, which may be fitted to other versions of the aircraft has maximum continuous power of 636 kW, and a take-off power of 704 kW. The fuel flow of these engines is shown in Figure A.20 in terms of cruise altitude and gross weight.

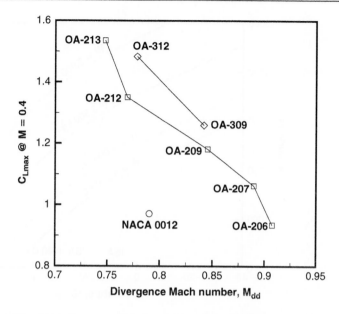

Figure A.15   *Drag divergence Mach number of some ONERA rotor blade profiles.*

The tail rotor is a *fenestron*, a propeller with a large number of blades, housed in a shroud, which reduces noise emission and tail-related accidents[403]. Operational advantages of the fenestron quoted by the manufacturer and in the technical documents include safety in flight and ground operations, excellent yaw maneuverability, and limited noise emission. The main data of the fenestron are given in Table A.16.

A.4.3   Discussion of Data

The reference weights in Figures A.18–A.23 include the MTOW. The engine power of reference includes: the maximum continuous power with two engines running, the take-off power, and the OEI intermediate power. The manufacturer gives the approximation on some of the data, for example weights, service ceiling, rate of climb (depending on atmospheric conditions).

Figures A.16 and A.17 show the hover performance, in- and out-of-ground effect. The aircraft is operated with its two engines at the maximum take-off power. The charts also show the effects of atmospheric temperature on the hover capability.

Figure A.19 is the recommended cruise speed at the weights indicated, as a function of the cruise altitude. The data are valid in ISA conditions, with two engines operating at maximum continuous power. The weights selected range from a light configuration to the MTOW. Figure A.19 shows the recommended cruise speed on a standard day. Note that the recommended speed is sensibly lower than the maximum speed at the same weight and altitude.

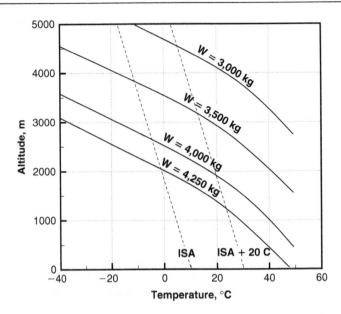

Figure A.16   *EC-365, IGE hover performance, 6 ft from the ground, engines at maximum take-off power.*

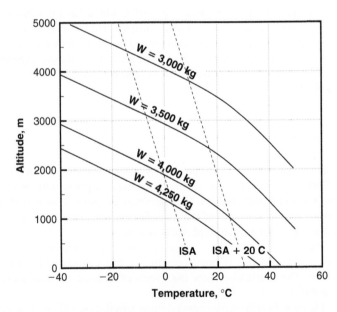

Figure A.17   *EC-365, OGE hover performance, engines at maximum take-off power.*

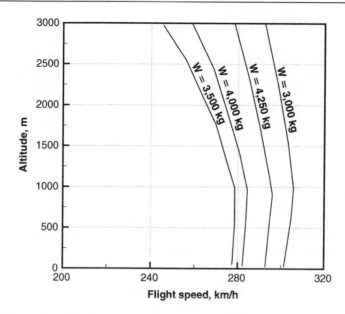

Figure A.18   *EC-365, maximum cruise speed, ISA conditions, two engines at maximum continuous power.*

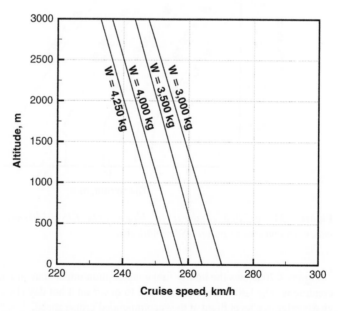

Figure A.19   *EC-365, recommended cruise speed (km/h), ISA conditions, two engines at maximum continuous power.*

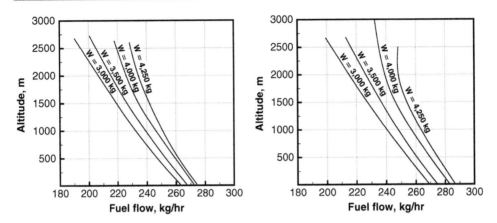

Figure A.20 *EC-365, estimated fuel flow, dual engine operation, standard (left) and non-standard (right) conditions.*

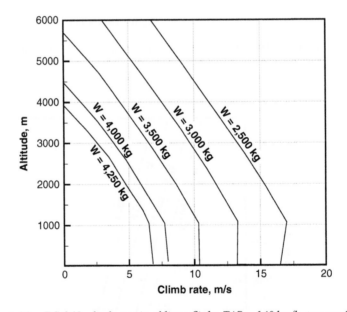

Figure A.21 *EC-365, climb rate in oblique flight, TAS = 140 km/h, two engines at maximum continuous power, ISA conditions.*

Figure A.20 shows the hourly fuel consumption under standard and non-standard conditions. The latter ones are assumed to occur on a hot day (ISA + 20°C). These charts refer to a level flight at the recommended cruise speed.

Figure A.21 shows the climb rate in oblique flight, at a fixed true air speed *TAS* = 140 km/h (75.5 kt), with two engines operating at maximum continuous power, and standard atmospheric conditions. The effects of the atmosphere for a hot day are shown in Figure A.22. Figure A.23 shows the rate of climb at intermediate power

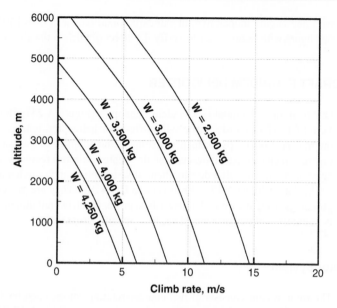

Figure A.22   *EC-365, climb rate in oblique flight, TAS = 140 km/h, two engines at maximum continuous power, ISA + 20° C.*

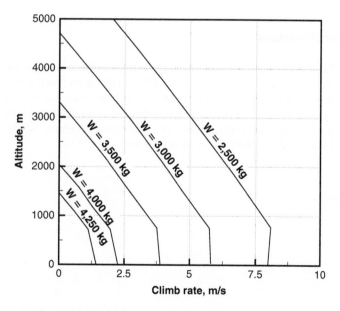

Figure A.23   *EC-365, climb rate in oblique flight, TAS = 140 km/h, OEI, intermediate power, ISA conditions.*

with one engine inoperative, at standard conditions, at $TAS = 140$ km/h. Compare this figure with Figure A.21 to verify the effect of OEI on the aircraft performance.

A.5 AIRCRAFT E: TANDEM HELICOPTER

Our final aircraft model is a tandem helicopter (Figure A.24). This is the Boeing-Vertol CH-47 Chinook, a heavy-lift twin-rotor helicopter, whose missions include the transport of troops, ordnance, artillery, and supplies to the battle zones. Other missions include medical evacuation, fire fighting, aircraft recovery, parachute drops, heavy construction, civil development, disaster relief, and search and rescue.

The data in Table A.17 refer to version CH-47SD. They are presented in different groupings: weights, rotor characteristics, power plant, overall dimensions. Table A.18 is a summary of performance data, for a specific gross weight.

A.5.1 Rotor System

The rotor system consists of two counter-rotating rotors, each having three blades, made of fiber-glass. The main dimensions are given in Table A.17. The blades can be folded manually. The blade section is the Vertol V-23010-1.58. For this airfoil at zero incidence $\alpha = 0$ we have estimated $C_{L_o} \simeq 0.128$ and $C_{D_o} \simeq 0.008$, $C_{L_\alpha} \simeq 0.1175$ rad^{-1} at low subsonic speeds ($M = 0.4$).

A.5.2 Aircraft Versions

Table A.19 shows a summary of performance data relative to four version of the utility helicopter Boeing-Vertol CH-47. The table spans four decades of technological development. It is shown how the empty weight has increased by 26%, and the MTOW by over 50%. The rotor system has undergone a less dramatic change, and the speed has remained substantially constant. In particular, the empty weight has increased by 26%, and the MTOW was increased by more than 51%.

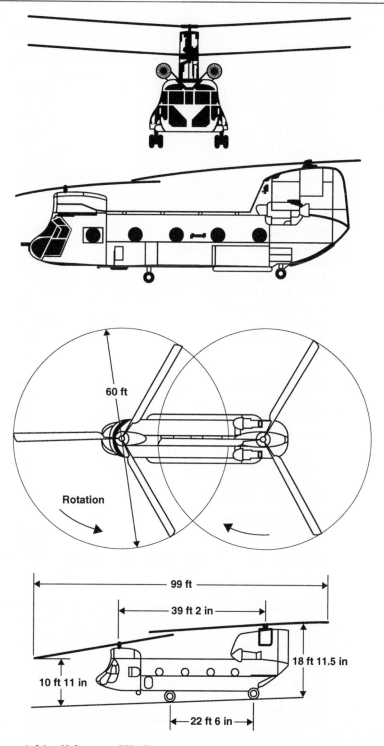

Figure A.24   *Helicopter CH-47.*

Table A.17 *CH-47SD helicopter data (all weights in kgf).*

Parameter	Value	Unit
Weights		
Max take-off weight (±2%)	24,495	kg
Operating empty weight (±2%)	10,185	kg
Max sling load	11,340	kg
Standard fuel tanks	7,828	l
Additional fuel tanks	3 × 3,028	l
Maximum PAY (internal)	12,900	kg
Forward/aft cargo hooks load	2 × 9,072	kg
Central cargo hook load	12,000	kg
Rotor		
Rotor diameter (both)	18.29	m
Number of blades (fiber-glass)	3	
Rotor solidity (each)	0.0423	
Rotor height over ground at hub	5.77	m
Forward rotor mast angle	4.5	degs
Rear rotor mast angle	5.5	degs
Blade section	Vertol V-23010-1.58	
Airfoil thickness	10%	
Blade chord	0.81	m
Pitch at blade tip	3	deg
Rotor twist (linear)	−9.14	degs
Rotor speed	225	rpm
Collective pitch (both)	7.85	degs
Power plant		
2 × Textron Lycoming T55-L712		
AEO power, unlimited (each)	2,237	kW
AEO power, 2 minutes	3,217	kW
Fuel flow at cruise (each)	n.a.	l/h
Overall dimensions		
Distance between shafts	18.85	m
Length w/rotor turning	30.14	m

Table A.18 *CH-47SD performance, gross weight 22,680 kg, ISA.*

Parameter	Value	Unit
Sea level cruise speed	259	km/h
Sea level never-to-exceed speed	287	km/h
Rate of climb	563	m/min
Service ceiling	3,385	m
Hover ceiling, IGE	2,835	m

(Continued)

Table A.18   *(Continued)*

Parameter	Value	Unit
Hover ceiling, IGE, ISA + 20°C	2,185	m
Hover ceiling, OGE	1,675	m
Hover ceiling, OGE, ISA + 20°C	1,005	m
Range with 12,560 kg PAY	1,208	km
Radius of action (1810 kg, $h = 1,200$ m)	935	km
Passengers (troops)	55	
Operating Limitations		
Minimum atmospheric temperature	−40°C	
Maximum atmospheric temperature	ISA + 40°C	

Table A.19   *Boeing Vertol CH-47 helicopter versions.*

	CH-47A	CH-47B	CH-47C	CH-47D
Capacity				
Empty Weight (kg)	8,295	8,925	9,791	10,475
Useful Load (kg)	627	627	635	636
Design Gross Weight (kg)	12,950	14,969	14,969	14,969
Max Gross Weight (kg)	14,969	18,144	20,866	22,680
Rotor Speed (rpm)	230	225/230	235/245	225
Usable Fuel Capacity (kg)	1,663	1,663	3,072	3,037
Performance (14,969 kg/ISA)				
Maximum Cruise Speed at S/L (kt)	110	155	161	158
Maximum v_c at S/L in 30 min (m/min)	628	641	798	955
OGE Hover Ceiling in 10 min (m)	2,710	3,111	4,189	5,298
Service Ceiling (30. min. rating), OEI (m)	616		2,618	3,942
Mission Capability				
External Payload Mission (S/L ISA)				
Takeoff Gross Weight (kg)	14,969	17,894	20,140	22,680
Total Mission Fuel (kg)	1,028	1,154	1,292	1,292
Outbound Cruise Speed (kt)	100	100	100	126
Inbound Cruise Speed (kt)	130	132	137	135
Payload (kg)	5,018	7,188	8,421	10,290
External Payload Mission (616 m/2,000 ft/21 C)				
Takeoff Gross Weight (kg)	14,969	17,327	19,482	22,680
Total Mission Fuel (kg)	989	1,118	1,224	1,242
Outbound Cruise Speed (kt)	100	100	100	112
Inbound Cruise Speed (kt)	120	140	140	137
Payload (kg)	5,090	6,667	7,831	10,327
External Payload Mission (1232 m/4000 ft/35 C)				
Takeoff Gross Weight (kg)	14,106	14,742	18,461	19,459
Total Mission Fuel (kg)	902	1,008	1,153	1,156
Outbound Cruise Speed (kt)	92	100	100	101
Inbound Cruise Speed (kt)	103	144	146	140
Payload (kg)	4,283	4,182	6,882	7,192

Appendix B
Noise data

Table B.1 summarizes the certified noise performance of selected aircraft, in various categories. A full database of certification data is published by the Federal Aviation Administration[541], and is periodically updated. The FAA database provides noise levels of hundreds of aircraft configurations, including the effects of different engine installation. Military aircraft are excluded.

Table B.1 *Certified noise data for selected aircraft.*

	Take-off	*Approach*	*Sideline*
Commercial aircraft			
MD-81	90.4	93.3	94.6
Boeing 767-300	90.4	102	96.6
Boeing 757-200	82.2	95.0	93.3
Airbus A300-600	91.1	99.8	98.6
Airbus A310-300	89.6	98.6	94.3
Airbus A320-300	88.0	96.2	94.4
Airbus A340-300	95.0	97.2	94.7
Business jets			
Bombardier BD-700	80.0	91.0	90.0
Cessna Citation 650 VII	69.3	84.8	
Cessna Citation 750	72.3	90.2	83.0
LearJet 60	78.8	87.7	83.2
LearJet 45	81.9	85.2	93.4
LearJet 31A	79.5	92.8	87.2
Canadair CL-600	80.9	90.3	86.2
Turboprop transports			
DHC-8 Dash 8	80.5	94.8	85.6
SAAB 340B	78.5	91.6	85.6
SAAB 2000	89.0	96.0	94.0
Embraer EMB-120	81.2	92.3	83.5
Canadair CL-600	80.9	90.3	86.2
Fokker 50	81.0	96.7	85.0

(Continued)

Table B.1 (*Continued*)

	Take-off	Approach	Sideline
Utility civil helicopters			
Bell 212	93.3	98.5	
Bell 412	93.2	95.4	
Bell 430	92.4	93.8	91.6
Boeing MD 520N	85.4	89.9	
Boeing MD Explorer	84.1	89.9	83.1
Twin-turbofan commuter aircraft			
Canadair Reg. Jet	78.6	92.1	82.2
Fokker 100	83.4	93.1	89.3
Avro RJ	81.9	97.5	87.2
Amphibians			
Beriev BA40 Albatross	104	102	

Appendix C
Selected Simulation Programs

In this chapter we provide some examples of simulation programs, as they have been developed to provide the numerical simulations shown. We give a few examples of programs in MatLab and Fortran 95. These programs have been stripped of most input/output lines, as well as of most declarations of type.

C.1 ASSEMBLING AIRCRAFT FORCES

A typical `Fortran` program for assembling the aircraft's forces is the following:

```
CDo     = ZeroLiftCoefficient (Mach,alfa)
k       = LiftInducedFactor   (Mach,alfa)
Clalfa  = LiftCurveSlope      (Mach,alfa)
CD      = Cdo + k*CLalfa*alfa**2
CL      = CLalfa*alfa
U       = asound*Mach
Drag    = q*CD*U**2
Lift    = q*CL*U**2
Thrust  = EngineThrust        (Mach,h,throttle)
Weight  = AircraftWeight      ()
```

where the procedure `ZeroLiftCoefficient` interpolates the C_{D_o}, `LiftInducedFactor` interpolates the factor η, and `LiftCurveSlope` interpolates the C_{L_α} from the flight data. A similar procedure, `EngineThrust`, is needed to interpolate the thrust data; `AircraftWeight` updates the weight according to the actual fuel flow. This routine is not essential and can be skipped. It is important to use higher-order interpolation, such as cubic splines. Linear interpolation will not do. In some cases, the flight data can be replaced by interpolating functions, for example polynomials of high order. In the simulations presented, the throttle value was considered constant for all flight conditions. If it changes during the flight (acceleration, turn, etc.), one must take into account the small time lag between throttle position and engine response. This cannot be done without knowing how the engine operates during a transient.

This procedure gives the magnitude of the aerodynamic and propulsion forces. Along with the aircraft weight, they have to be projected along the relevant axes, to provide the momentum equations to solve a variety of flight mechanics problems.

C.2 CALCULATION OF NUMERICAL DERIVATIVES

Most derivatives have to be calculated numerically, either because there is no analytical expression, or because a closed-form solution is too cumbersome. A few lines of code solve the problem. If the function is known to be monotonic, then accuracy on the derivative is not essential, because the procedure converges anyway.

```
subroutine derivatives (f,x,dfdx)

c-- calls procedure ''func'' to calculate the function value at given point

    implicit none
    integer        :: n
    double precision :: x(n),f(n),dfdx(n)
    double precision :: tiny
    parameter (tiny = 1d-4)
    integer :: i
    double precision :: x1,x2

    do i = 1, n
      x1 = (1.d0 - tiny)*x(i)
      x2 = (1.d0 + tiny)*x(i)
      f1 = func()
      f2 = func()
      dfdx(i) = (f2 - f1)/(x2 - x1)
    enddo
```

C.3 OPTIMAL CLIMB OF FIGHTER JET AIRCRAFT

The program below is for MatLab, and solves the problem of minimum fuel to climb for the aircraft model given in Appendix A. Climb can be done with or without afterburning. The program is based on the data in Table A.12 (aerodynamic characteristics) and Table A.13 (engine thrust).

```
m=15181;     % aircraft mass
g=9.81;       % gravitational acceleration
%S=49.24;    % wing area
S=29;
rho1=1.225; % air density
Mac=[0,0.2,0.4,0.6,0.8,1,1.2,1.4,1.6,1.8,2]; %Mach number
x=1:length(Mac);
b=0.025;% step size of calculations
Cla0=[3.236673,3.242833,3.254873,3.286047,3.370291,3.645685,4.144183,3.197555,
    2.918348,2.78425,2.696331]; % Lift incidence coefficients
Cd0o=[0.00593,0.006073,0.006364,0.007115,0.010189,0.02237,0.042833,0.034473,
    0.026637,0.024931,0.024188]; %Drag coefficients
n0=[0.53,0.530856,0.536604,0.547639,0.560725,0.584548,0.682787,0.832104,
  0.85133,0.855193,0.858126]; %n coefficients
h0=[0,2,4,6,8,10,12,14,16,18]*(10^3); %altitude in metres

x1=1:length(Mac);
x2=1:length(h0);
xx=1:b:11;
xt=1:b:10;
```

```
Mach=spline(x1,Mac,xx);
Cla=spline(x1,Cla0,xx);
Cd0=spline(x1,Cd0o,xx);
n=spline(x1,n0,xx);
h=spline(x2,h0,xt);
```

```
T01=[88622     72626   58847   47097   37186   28924   21532   15708   11459   8359;
84082   69150   56221   45144   35757   27899   20801   15175   11070   8075;
83487   68952   56287   45371   36071   28245   21096   15390   11227   8190;
86328   71665   58785   47602   38010   29887   22368   16318   11904   8684;
92227   77048   63574   51765   41549   32831   24630   17968   13108   9562;
100747  84849   70532   57824   46707   37125   27931   20376   14864   10843;
109929  93599   78572   64990   52921   42376   31994   23340   17027   12421;
118569  102524  87230   73012   60085   48572   36836   26872   19604   14301;
124249  109957  95395   81184   67778   55484   42323   30875   22523   16431;
122735  112887  100973  88090   75074   62536   48077   35073   25586   18665;
107332  106500  100564  91380   80391   68697   53404   38958   28420   20733;];
% Thrust values with After-burner off
```

```
FC01=[6506    5419    4459    3621    2899    2284    1711    1248    910     664;
6620    5517    4542    3690    2955    2330    1746    1273    929     677;
6961    5812    4793    3900    3128    2469    1851    1350    985     718;
7534    6311    5220    4259    3424    2709    2033    1483    1082    789;
8336    7017    5830    4776    3854    3060    2300    1678    1224    893;
9344    7923    6625    5459    4429    3533    2662    1942    1417    1033;
10403   8914    7523    6251    5109    4102    3101    2262    1650    1203;
11382   9906    8474    7124    5882    4766    3618    2639    1925    1404;
12018   10719   9358    8002    6705    5502    4200    3064    2235    1630;
11853   11025   9946    8733    7477    6247    4808    3507    2558    1866;
10159   10296   9864    9055    8024    6890    5365    3914    2855    2083;];
% Fuel Flow rate with After-burner off
```

```
T02=[127575    103812  83542   66399   52068   40227   29845   21772   15883   11586;
124242  101297  81661   65026   51085   39538   29361   21420   15625   11399;
127424  104110  84100   67101   52816   40956   30442   22208   16201   11818;
137092  112265  90886   72667   57315   44532   33135   24171   17633   12863;
153732  126204  102417  82074   64875   50512   37622   27446   20021   14606;
178261  146762  119416  95937   76015   59321   44232   32268   23539   17172;
209706  173247  141415  113948  90537   70839   52888   38582   28146   20532;
249331  206857  169500  137059  109254  85743   64108   46767   34117   24889;
297924  248492  204583  166135  132946  104707  78419   57208   41733   30445;
355262  298374  247134  201757  162218  128302  96283   70239   51240   37380;
419632  355752  296999  244117  197433  156972  118082  86142   62841   45843;];
% Thrust values with After-burner On
```

```
FC02=[17748    14346   11471   9064    7068    5432    4020    2932    2139    1560;
18156   14677   11737   9275    7233    5559    4114    3001    2189    1597;
19415   15699   12558   9926    7743    5953    4406    3214    2344    1710;
21630   17499   14004   11074   8642    6647    4920    3589    2618    1910;
24980   20223   16195   12815   10007   7701    5702    4160    3034    2213;
29722   24086   19306   15290   11949   9203    6817    4973    3628    2646;
35886   29117   23366   18526   14495   11174   8282    6041    4407    3215;
43844   35630   28636   22738   17814   13751   10198   7439    5427    3959;
53975   43950   35390   28151   22093   17082   12678   9248    6747    4922;
66618   54380   43892   34992   27519   21320   15839   11554   8429    6149;
82049   67182   54382   43474   34279   26621   19800   14444   10537   7687;];
% Fuel flow rate with After-burner On%
```

```
[x,y] = meshgrid(1:10,1:11);
[xi,yi] = meshgrid(1:b:10,1:b:11);
T1=interp2(x,y,T01,xi,yi,'spline');% Thrust values with After-burner Off
FC1=interp2(x,y,FC01,xi,yi,'spline');% Fuel flow rate with After-burner Off
T2=interp2(x,y,T02,xi,yi,'spline');% Thrust values with After-burner On
FC2=interp2(x,y,FC02,xi,yi,'spline');% Fuel flow rate with After-burner On

CD=zeros([1,length(Cd0)]);% matrix for CD
alpha=zeros([1,length(Cd0)]); %angle of attack
V=zeros([1,length(Mach)]);% air speed
a=340; %speed of sound
D=zeros([length(Mach),length(h)]);%Drag
F=zeros([length(Mach),length(h)]); %excess thrust
Ed=zeros([length(Mach),length(h)]);% rate of change of energy
Edm=zeros([length(Mach),length(h)]); %Energy increase per kg of fuel
rho=zeros([1,length(h)]);
%Dt=[D',D',D',D',D',D',D',D',D',D'];

for i=1:1:length(Mach)
    for p=1:1:length(h)
        rho(p)=rho1*((20*10^3)-h(p))/((20*10^3)+h(p));% air density as function of
        %altitude
        if Mach(i)==0
         alpha(i)=0;
    elseif Mach(i)>=0
        V(i)=a*Mach(i);
    alpha(i)=m*g/(0.5*rho(p)*Cla(i)*(V(i)^2)*S); % angle of attack
    CD(i)=Cd0(i)+n(i)*Cla(i)*(alpha(i)^2); % Drag Coefficient
end
        D(i,p)=CD(i)*0.5*rho(p)*(V(i)^2)*S; %Drag
        F1(i,p)=T1(i,p)-D(i,p); % Excess Thrust for After-burner Off
        F2(i,p)=T2(i,p)-D(i,p); % Excess Thrust for After-burner On
        Ed1(i,p)=F1(i,p)*V(i)'/m; % Rate of change of Energy for After-burner off
        Ed2(i,p)=F2(i,p)*V(i)'/m; % Rate of change of Energy for After-burner On
        Edm1(i,p)=F1(i,p)*V(i)'/FC1(i,p);%Energy increase per kg of fuel
        %for After-burner Off
        Edm2(i,p)=F2(i,p)*V(i)'/FC2(i,p);%Energy increase per kg of fuel
        %for After-burner On
    end
end
Pt=zeros(1,length(Mach));
for k=1:1:length(Mach)
    if max(Edm1(k,:))>max(Edm2(k,:))
 Pt(k)=max(Edm1(k,:));
elseif max(Edm2(k,:))>max(Edm1(k,:))
    Pt(k)=max(Edm2(k,:));
end
end
[i,j]=find(Pt>=0);
Ptx=Pt(j);
%Pt=[max(Edm(1,:));max(Edm(2,:));max(Edm(3,:));
%    max(Edm(4,:));max(Edm(5,:));max(Edm(6,:));max(Edm(7,:));
%max(Edm(8,:));max(Edm(9,:));max(Edm(10,:));max(Edm(11,:))]; % Maximum Energy increase
 per
% kg of fuel with respect to velocity
```

```
i=zeros(1,length(Ptx));
j=zeros(1,length(Ptx));
ht=zeros(1,length(Ptx));
Ptt=zeros(1,length(Ptx));
E=zeros(1,length(Ptx));
dt=zeros(1,length(Ptx));
for c=1:1:length(Ptx)
if c==1
    ht(c)=0;
    j(c)=1;
    i(c)=1;
    elseif c>=1
        if sum(Edm1(:,:)==Ptx(c))==0
 [i(c),j(c)]=find(Edm2(:,:)==Ptx(c));
 Ptt(c)=Ed2(i(c),j(c));% Maximum rate of change of Energy
 ht(c)=h(j(c));
elseif sum(Edm2(:,:)==Ptx(c))==0
    [i(c),j(c)]=find(Edm1(:,:)==Ptx(c));
    Ptt(c)=Ed1(i(c),j(c));% Maximum rate of change of Energy
    ht(c)=h(j(c));
end
  % Altitudes for points of Max Energy change per kg of fuel
 E(c)=0.5*(V(c)^2)+g*ht(c); %Energy levels for points of Max change of Energy
 %with respect to velocity for given altitudes
end
 if c==1
 dt(c)=0;
elseif c>=1
    dt(c)=(E(c)-E(c-1))/Ptt(c); % Times for aircraft to reach
    % different energy levels
end
end
Vt=V(1:length(Ptx));
Time=sum(dt) % Total time for aircraft to reach Max altitude
plot(Vt,ht)
grid on
xlabel('Velocity')
ylabel('Altitude')
title('Minimum Fuel-Energy Glide Path')
```

C.4   OPTIMAL CLIMB RATE OF TURBOPROP

This piece of Fortran code was used for the climb problem described in §8.3.2. In the present case the throttle is fixed to maximum value $\Pi = 1$.

```
c---- go over the number of vertical steps
      do j = 2, nh                       ! number of vertical steps
c----- atmospheric data
      h1 = h*1d-3                        ! conversion from ''m'' to ''km''
      call atmosphere (h1,sigma,delta,theta,asound)
```

```
c----- air density
       rho = rho_sl*sigma

c----- Newton iterations at current altitude to find speed U
       do i = 1, 10                     ! number of Newton iterations

c------- engine power and prop efficiency at current speed
         Power = EnginePower(U,h,throttle)
         eta   = Efficiency (U)

c------- calculate derivatives in residual function at current "U"
         call Derivatives (U,h,detadJ,dPdU,ddldU)

c------- residual
         f   = detadJ*Power*rado + dPdU*eta + c2*dDLdU

c------- derivative of residual
         U1 = U*(1.d0 - tiny)
         U2 = U*(1.d0 + tiny)
         dU = U2 - U1
         call Derivatives (U1,h,detadJ1,dPdU1,ddldU1)
         call Derivatives (U2,h,detadJ2,dPdU2,ddldU2)
         Power1 = EnginePower (U1,h,throttle)
         Power2 = EnginePower (U2,h,throttle)

         f1 = detadJ1*Power1*rado + dPdU1*eta + c2*ddldU1
         f2 = detadJ2*Power2*rado + dPdU2*eta + c2*ddldU2
         dfdU = (f2 - f1)/dU

c------- new aircraft speed with Newton step
         Uold = U
         U    = Uold - f/dfdU

c------- check residual
         Power   = EnginePower(U,h,throttle)
         eta     = Efficiency (U)
         call Derivatives (U,h,detadJ,dPdU,ddldU)
         f   = detadJ*Power*rado + dPdU*eta + c2*ddldU
         if (abs(f) < 1d-3) goto 15           ! exit Newton loop
         if (abs(U - Uold)/Uold < 1d-4) goto 1 ! another exit strategy

caveat: initial residual can be 10^6; Newton iterations converge to about
c       10^{-2} in 6-7 iterations; after that convergence is slow.

       enddo

 1     continue

c----- aerodynamic coefficients
       CL   = 2.d0*Weight/(rho*Area*U**2)
       CD   = CDo + k*CL**2
       drag = 0.5d0*rho*Area*CD*U**2

c----- climb velocity
       vc   = (eta*Power - Drag*U)/Weight

       ......

       enddo          ! go on to the altitude step
```

C.5   CALCULATION OF MISSION FUEL

This routine calculates the mission fuel at constant Mach number and constant C_L. On top of the subroutine we have reported the main free parameters.

```
        subroutine MissionFuel (mf,x,mft1)
c-----------------------------------------------------------------
c--- calculates mission fuel for subsonic jet aircraft.
c-   mf   = mission fuel (output)
c-   x    = range required (input)
c-   mft1 = mission fuel burned in absence of diversions (output)
c-----------------------------------------------------------------

        implicit none
        include 'cruise.inc'    ! all parameters defined here
        integer :: i, itmax
        parameter (itmax = 12)

        g       = 9.807d0    ! acceleration of gravity, m/s^2

c---- ad HOC parameters; some of these may need to be changed
        eta_to = 0.42d0      ! power plant efficiency at take-off
        eta_c  = 0.50        ! power plant efficiency at climb condition
        res    = 0.05d0      ! reserve fuel in percent
        Cp     = 43.5d6      ! specific heat of aviation fuel, Joule
        mu     = 0.025d0     ! rolling resistance (dry runway)
        CLto   = 0.80*CLmax  ! take-off CL

        do i = 1, itmax

c------ store old weight for convergence analysis
          weight_old = weight                  ! N
          mass       = weight/g                ! kg_m

c------ lift coefficient at cruise start
          CL   = 2.d0*weight/(rho*area*U**2)

c------ drag coefficient at cruise start
          CD   = CDo + k*CL**2

c------ glide ratio (const)
          LD   = CL/CD

c------ take-off speed
          Uto = 2.d0*weight/(rho*area*CLto)
          Uto = sqrt(Uto)

c------ manoeuvre fuel (kg_m)
          mfo = 0.0022*mass/eta_to

c------ equivalent all-out range (no extended range), meters
          etam = 0.25d0*Mach + 1.125
          c    = 1.1d0 + 0.5d0*etam
          xdiv = c*(Rdiv + Uhold*thold)*(MLW/weight)
          EOAR = x*(1.d0 + res) + xdiv
```

```
c------ fuel to taxi-out (5 minutes); dominant component is weight
        Utaxi = 5d0                            ! m/s, about 1 mile taxi
        taxitime = 5d0*60d0
        drag  = 0.5d0*rhoo*area*CD*Utaxi**2    ! N
        fflow = fj*(drag + mu*weight)          ! N/s
        fflow = fflow/g                        ! kg_m/s
        mftax = taxitime*fflow                 ! kg_m

c------ fuel to take-off (kg_m)
        mf1 = 0.5d0*mass*Uto**2/(eta_to*Cp)    ! kg_m

c------ fuel to climb (kg_m), weight in Newton
        he  = h1*1d3 + 0.5d0*U**2/g            ! m, energy height
        mf2 = weight*he/(eta_c*Cp)             ! kg_m

c------ cruise fuel (kg_m)
        c1  = (g*fjc/LD)*EOAR/U
        c2  = exp(c1)
        xi  = 1.d0 - 1.d0/c2
        mfc = xi*mass                          ! kg_m

c------ cruise fuel for required range only (kg_m)
        c1  = (g*fjc/LD)*x/U
        c2  = exp(c1)
        xi  = 1.d0 - 1.d0/c2
        mfr = xi*mass                          ! kg_m

c------ reserve fuel
        mres = mfc - mfr                       ! kg_m

c------ total fuel
        mft  = mftax + mfo + mf1 + mf2 + mfc  ! kg_m

c------ total fuel actually USED  (value to return)
        mft1 = mftax + mfo + mf1 + mf2 + mfr  ! kg_m

c------ corrected block fuel
        xi = mft/mass

c------ new weight calculation (N)
        weight = OEW + payload + mft*g         ! N

c------ update aircraft mass
        mass = weight/g                        ! kg_m

c------ write out data
        xkm = x*1d-3                           ! required range  , km
        rkm = EOAR*1d-3                        ! equivalent range, km
        xm  = xkm*0.5399568                    ! required range  , n-miles
        rm  = rkm*0.5399568                    ! equivalent range, n-miles

c------ convergence analysis
        if (abs(weight_old - weight)/weight_old < 0.01) goto 1

        TOW = weight
cx      if (TOW > MTOW ) write(6,*) ' Warning: Calculated TOW > MTOW'

      enddo
```

```
1     continue

        write(6,1003) CL,CD,LD,CLto
        write(6,1000) xkm,xm,rkm,rm
        write(6,1001) mftax,mf1,mf2,mfo,mfc,mfr,mft,mres,mft1,xi,
    &                 mass,weight/g
        write(6,1005) OEW/g,    OEW/weight,
    &                 payload/g,payload/weight,
    &                 mft,      xi,
    &                 weight/g

     if (i < itmax) then
        write(6,902) i
     else
        stop ' ''MissionFuel'' not converged !'
     endif
     dmdx = (mft1/x)    ! rate of fuel burn (kg_m/m)

900  format( ' Starting guess for GTOW          ',f8.1, ' kg_m')
901  format( ' Final    guess for GTOW          ',f8.1, ' kg_m',
    &       / '         after ', i2, ' iterations' )
902  format( '    Converged at iteration # ', i2)
1000 format( ' Segment analysis: ',
    &/ '   Required range                 ',f8.1,' km ',f8.1,' n-miles',
    &/ '   Equivalent all-out range       ',f8.1,' km ',f8.1,' n-miles')

1001 format( ' Fuel breakdown: ',
    & / '    Taxi-out fuel              ',f8.0,'  kg_m ',
    & / '    Take-off fuel              ',f8.0,'  kg_m ',
    & / '    Climb to cruise altitude fuel ',f8.0,'  kg_m ',
    & / '    Manoeuvre fuel             ',f8.0,'  kg_m ',
    & / '    Cruise fuel                ',f8.0,'  kg_m ',
    & / '        for required segment   ',f8.0,'  kg_m ',
    & / '    Mission fuel               ',f8.0,'  kg_m ',
    & / '        reserve fuel           ',f8.0,'  kg_m ',
    & / '        actually used          ',f8.0,'  kg_m ',
    & / '    Block fuel ratio       xi = ',f7.3,
    & / '    New mass estimate          ',f8.0,'  kg_m ',
    & / '        weight                 ',f8.0,'  kg_f ')

1003 format( ' Aerodynamics :',
    & / '              CL  = ', f8.5,
    & / '              CD  = ', f8.5,
    & / '              L/D = ', f8.5,
    & / '              CLto = ', f8.5)

1005 format(' Final weight breakdown (kg_f) :',
    &     /'    OEW  = ', f10.2, 2x, f6.3,
    &     /'    PAY  = ', f10.2, 2x, f6.3,
    &     /'    FUEL = ', f10.2, 2x, f6.3,
    &     /'    GTOW = ', f10.2)
     return
     end
c----------------------------------------------------------------
```

C.6 SUPERSONIC ACCELERATION

This routine calculates the supersonic acceleration at constant altitude or constant angle of attack. Basically, this is a dash from an initial Mach number. Integration of the ODE is done via a routine called rk4 (Runge-Kutta). The second option is unstable.

```
      program SupersonicAcceleration
c------------------------------------------------------------------
c---- Calculates supersonic acceleration of jet aircraft.
c-    Uses tabulated data for engine thrust and drag characteristics
c------------------------------------------------------------------
      implicit none
      include 'flight.inc'

cx    write(6,9000)  ! program version

c---- physical constants
      pi    = acos(-1.d0)
      g     = 9.81d0
      rho_o = 1.225d0

      call  ReadAllFlightData

c---- Aircraft data
      mass   = 1.1d4     ! kg
      Weight = mass*g    ! Newton
      Area   = 28.9      ! m^2

c---- set initial conditions
      Mach   = 0.8d0     ! flight Mach number
      hkm    = 11.d0     ! flight altitude, km
      h1     = hkm*1d3   ! altitude, m
      x1     = 0.d0      ! distance from take off, m

c---- atmospheric conditions
      call atmosphere (hkm,sigma,theta,delta,asound)
      rho = rho_o*sigma

c---- aerodynamic coefficients
      CDo1   = ZeroLiftCoefficient (Mach,alfa)
      k      = LiftInducedFactor   (Mach,alfa)
      Clalfa1 = LiftCurveSlope     (Mach,alfa)

      U2     = (asound*Mach)**2          ! speed^2
      WL     = Weight/Area               ! wing loading
      alfa   = 2.d0*WL/(rho*U2*CLalfa1)  ! angle of attack
      alfa_o = alfa

c---- drag = thrust
      CD   = CDo1 + k*Clalfa1*alfa*alfa
      Drag = 0.5d0*rho*Area*CD*asound**2*Mach**2

c---- excess thrust
      call Engine_Interpol
     &    (fht,fmt,FT,nalt,nspeed,nmax1,nmax1,h1,Mach,Thrust)
      SEP  = (Thrust - Drag)/Weight
      write(6,4000) Weight/g,h1,Mach,Drag*1d-3,alfa*180.d0/pi,CD,SEP
```

```
c---- integration parameters
      dh   = 1.d0        ! meters
      dt   = 5d-2        ! seconds
      time = 0.d0

c---- constant coefficient at constant altitude
      c1 = -0.5d0*rho*asound*Area/mass

c-----integrate ODE
      accmax = 0.d0
      accmin = 1.d5
      Mach_o = Mach

c--- integrate with Runge Kutta
      write(6,*) ' Runge-Kutta integration ...'
      Mach = 0.8d0
      nn   = 1

      do i = 1, nmax

c------ aerodynamic coefficients
      CDo1    = ZeroLiftCoefficient   (Mach,alfa)
      k       = LiftInducedFactor     (Mach,alfa)
      Clalfa1 = LiftCurveSlope        (Mach,alfa)
      CD      = CDo1 + k*Clalfa1*alfa*alfa

c----- interpolate engine thrust
       call Engine_Interpol
     & (fht,fmt,FT,nalt,nspeed,nmax1,nmax1,h1,Mach,Thrust)

c------ current angle of attack
      U2   = (asound*Mach)**2
      alfa = 2.d0*WL/(rho*U2*CLalfa1)

c------ Mach number acceleration
      dMdt = Thrust/(mass*asound) + c1*CD*Mach**2

c------ Runge-Kutta integration
      call rk4 (Mach,dMdt,nn,time,dt,Mach1)

c------ new Mach number
      Mach1 = Mach

c------ air speed acceleration
      dUdt = dMdt*asound

c------ acceleration in "g"
      acc  = dUdt/g
      if (acc > accmax) accmax = acc
      if (acc < accmin) accmin = acc

c------ output
      write(2,1001) time,Mach1,alfa*180.d0/pi,Thrust,acc

      if (Mach > Mach_limit) goto 4

      time = time + dt

      Mach = Mach1
      enddo
4     continue
```

```
c---- FLIGHT AT CONSTANT ANGLE OF ATTACK -------------------------
      write(6,*) ' Flight at constant angle of attack ...'

c---- initial conditions
      Mach    = 0.8d0
      U       = asound*Mach
      gamma   = 1.d-2
      time    = 0.d0
      alfa    = alfa_o

c---- atmospheric data (altitude as given previously)
      call atmosphere (hkm,sigma,theta,delta,asound)

c---- reset accelerations
      accmax = 0.d0
      accmin = 1.d5

c-----integrate ODE by 4th-order Runge-Kutta
      nn      = 2
      do i = 1, 2*nmax

        y(1) = U
        y(2) = gamma

c----- derivatives before integration procedures
         call derivsc (nn,h1,y,dydh)

c------ Runge-Kutta integration
         call rk4c (y,dydh,nn,h1   ,dh,yout)

c------ updated velocities
         U       = yout(1)
         gamma   = yout(2)
         vc      = U*sin(gamma)
         Mach    = U/asound

c------ acceleration in "g"
         dUdt = dydh(1)*vc
         acc  = dUdt/g
         if (acc > accmax) accmax = acc
         if (acc < accmin) accmin = acc

c------ update flight path (Euler integration)
         dt    = dh/vc
         h1    = h1 + dh
         time = time + dt
         x1    = x1 + U*cos(gamma)*dt
         hkm   = h1*1d-3

 15      if (Mach > Mach_limit) goto 5

c------ output
         write(3,1001) time,Mach,vc,gamma*180.d0/pi,x1,hkm

      enddo
 5    continue

 1001 format(6e14.5)
 1002 format(5e11.4,2i6)
 2000 format(2f10.4)
```

```
3000  format('       Time to M = ', f4.2, ' is ', f7.2, ' seconds')
4000  format(/,' Initial Conditions ',
     &      /'       Aircraft Weight   = ', f10.0, ' kg '   ,
     &      /'       Altitude          = ', f10.0, ' metres',
     &      /'       Mach number       = ', f10.2,
     &      /'       Engine Thrust     = ', f10.2, ' kN',
     &      /'       Angle of Attack   = ', f10.2, ' degs',

     &      /'       Drag coeff, CD    = ', f10.4,
     &      /'       Spec Excess Thrust = ', f10.4,/)
4100  format('       Max acceleration  = ', f6.3, ' g',
     &        /'       Min acceleration  = ', f6.3, ' g')

      end
c----------------------------------------------------------------------
```

C.7   ASYMMETRIC THRUST CONTROL

This subroutine in `Fortran` calculates the static control of a subsonic jet aircraft
that has an asymmetric thrust and bank angle, due to an engine failure, as described
in §10.11.

```
      program Asymmetric Thrust Control
c----------------------------------------------------------------------

      implicit none
      integer:: i, nmax
      parameter (nmax = 220)
      character (len = 20) string,filename
      double precision :: pi,g,fac,beta1

c---- atmospheric quantities
      double precision :: h1,sigma1,theta1,delta1,asound1
      double precision :: rho_sl,rho
      parameter (rho_sl = 1.225d0)

c---- speeds
      double precision :: U,VMC,KVMC,Mach,q,Ustall

c---- aircraft data
      double precision :: area,weight,span,drag,thetamax,ximax

c---- matrix
      double precision :: a(3,3),b(3),det
      integer          :: index(3)

c---- aerodynamic coefficients
      double precision :: CLtheta,CNtheta,CYtheta
      double precision :: CLbeta ,CNbeta ,CYbeta
      double precision :: CLxi,CNxi,CYxi
      double precision :: CL,CLmax

c---- control angles
      double precision :: phi,theta,beta,xi
```

```
c---- engines
      double precision :: diam,CDe,bt,T

c---- weights
      double precision :: MTOW,OEW,dw

c---- various factors
      pi = acos(-1.d0)
      g  = 9.807d0
      fac = pi/180d0

c---- read aircraft data
cx    filename = "a300.control"
      filename = "b747-100.control"
      open(1,file = filename, status = 'old')
        read(1,*) MTOW         ! weight, N
        read(1,*) OEW          ! weight, N
        read(1,*) area         ! wing area, m^2
        read(1,*) span         ! wing span, m
        read(1,*) bt           ! moment arm of asymmetric thrust, m
        read(1,*) diam         ! engine diameter, m
        read(1,*) T            ! max asymmetric thrust, kN, sea level
        read(1,*) Mach         ! Mach number before engine failure
        read(1,*) phi          ! bank angle, degs
        read(1,*) h1           ! flight altitude, km
        read(1,*) string
        read(1,*) CLbeta       ! pitch moment deriv wrt sideslip
        read(1,*) CLtheta      ! pitch moment deriv wrt rudder deflection
        read(1,*) CLxi         ! pitch moment deriv wrt aileron deflection
        read(1,*) CNbeta       ! yaw moment deriv wrt sideslip
        read(1,*) CNtheta      ! yaw moment deriv wrt rudder deflection
        read(1,*) CNxi         ! yaw moment deriv wrt aileron deflection
        read(1,*) CYbeta       ! side force coeff deriv wrt sideslip
        read(1,*) CYtheta      ! side force coeff deriv wrt rudder deflect.
        read(1,*) CYxi         ! side force coeff deriv wrt aileron deflect.
        read(1,*) string
        read(1,*) thetamax     ! max rudder deflection (abs value), degs
        read(1,*) ximax        ! max aileron deflection (abs value), degs
        read(1,*) CLmax        ! max lift coefficient
      close(1)


c---- conversion of some factors
      phi = phi*fac            ! rad
      thetamax = thetamax*fac  ! rad
      ximax    = -ximax*fac    ! rad
      T    = T*1d3             ! N, sea level
cx    T    = T*sigma1**0.8     ! N, at altitude


c---- calculate air density at flight altitude
      call atmosphere (h1,sigma1,delta1,theta1,asound1)
      rho = rho_sl*sigma1


c---- run over aircraft weight
      dW     = (MTOW - OEW)/float(nmax)
      weight = MTOW
      do i = 1, nmax
```

```
c----- estimate of engine drag a posteriori (from Torenbeek, G-8)
      Mach = VMC/asound1

c----- calculate CDe with previous estimate of VMC
      CDe =  (0.0785*diam**2 + 2d0/(1.d0 + (0.16*Mach**2))*
     &        (0.25d0*pi*diam**2)*0.0736)/Area

c----- fixed rudder position
      theta  = thetamax

c----- assembly matrix
      a(1,1) = CLbeta
      a(1,2) = CLxi
      a(1,3) = 0d0
      a(2,1) = CNbeta
      a(2,2) = CNxi
      a(2,3) = 2d0*T*bt/(rho*area*span)
      a(3,1) = CYbeta
      a(3,2) = CYxi
      a(3,3) = 2d0*weight*sin(phi)/(rho*area)

c----- assembly RHS
      b(  1) = -CLtheta*theta
      b(  2) = -CNtheta*theta - CDe*bt/span
      b(  3) = -CYtheta*theta

c----- solution
      call LU_Decomposition     (a,3,3,index,det)
      call LU_BackSubstitution (a,3,3,index,b)

      beta = b(1)
      xi   = b(2)
      VMC  = 1d0/sqrt(b(3))

c----- aileron deflection exceeded, skip next step
      if (xi > ximax) goto 2

c----- air speed in knots
      KVMC = VMC*3.6d0/1.853d0

      write(2,2000) weight*1d-3/g,beta/fac,xi/fac,KVMC

c----- fixed aileron position
  2   xi  = ximax

c----- assembly matrix
cx     a(1,1) = CLbeta
cx     a(1,2) = CLtheta
cx     a(1,3) = 0d0                 ! no need to re-calculate this
      a(2,1) = CNbeta
      a(2,2) = CNtheta
      a(2,3) = 2d0*T*bt/(rho*area*span)
cx     a(3,1) = CYbeta              ! no need to re-calculate this
cx     a(3,2) = CYtheta
cx     a(3,3) = 2d0*weight*sin(phi)/(rho*area)

      b(  1) = -CLxi*xi
      b(  2) = -CNxi*xi - CDe*bt/span
      b(  3) = -CYxi*xi
```

```
        call LU_Decomposition    (a,3,3,index,det)
        call LU_BackSubstitution (a,3,3,index,b)
        beta  = b(1)
        theta = b(2)
        VMC   = 1d0/sqrt(b(3))

c----- air speed in knots
        KVMC = VMC*3.6d0/1.853d0

c----- rudder deflection exceeded, skip next step
        if (theta > thetamax) go to 3
        write(3,2000) weight*1d-3/g,beta/fac,theta/fac,KVMC

        Ustall = sqrt(2d0*weight/(rho*area*CLmax))
        write(4,2000) weight*1d-3/g, Ustall*3.6d0/1.853d0

 3      weight = weight - dW

        enddo

 2000 format(4e14.6)

 11     end
c---------------------------------------------------------------------
```

C.8 HOVER POWER WITH BLADE ELEMENT THEORY

The following section of Fortran code has been used for the calculation of the rotor power in hover, with or without drag divergence charts.

```
        thrust = 0d0
        torque = 0d0
        power  = 0d0
        rnb    = float(nb)
        c1     = sigma*CLalfa/16d0
        c2     = 32d0/(sigma*CLalfa)

        call = 0
        Mmax = 0.d0
c----- go over number of spanwise elements
        do i = 1, n

c-------- normalised radius
          r     = y(i)/radius
          theta = thetao + thetaw*r

c-------- induced velocity ratio
          lambda = c1*(sqrt(1.d0 + c2*theta*r) - 1d0)
cx        write(12,1100) r, lambda
          vi(i) = lambda*omega*radius
          Up    = Vc + vi(i)
          Ut    = omega*y(i)
          phi   = lambda/r
```

```
c-------- inflow angle
          alfa   = theta - phi

c-------- lift coefficient
          CL     = CLalfa*alfa

c-------- hover condition
          U      = Ut
          Mach   = U/asound1

c-------- store max Mach number, for reference
          if (Mach > Mmax ) Mmax = Mach

          call   = call + 1
c-------- interpolation of drag divergence curves
          call CoefficientsInterpolation (alfa,Mach,CDo)
          cdall = cdall + cd
          CD     = CDo

c-------- above section can be skipped; use instead CDo = 0.01

c-------- element of lift and drag
          dLift = 0.5d0*rho*chord*dy*CL*U**2
          dDrag = 0.5d0*rho*chord*dy*CD*U**2

c-------- exact equations
cx        dT     = rnb*(dLift*cos(phi) - dDrag*sin(phi))
cx        dQ     = rnb*(dLift*sin(phi) + dDrag*cos(phi))*y(i)
cx        dP     = rnb*(dLift*sin(phi) + dDrag*cos(phi))*omega*y(i)

c-------- approximated equations
          dT     = rnb*(dLift)
          dQ     = rnb*(dLift*phi + dDrag)*y(i)
          dP     = rnb*(dLift*phi + dDrag)*omega*y(i)

c-------- Prantl's tip loss correction
          phi    = lambda/r
          f      = 0.5d0*rnb*(1d0 - r)/(r*phi)
          ff     = (2.d0/pi)*acos(exp(-f))
          dT     = dT*ff
          thrust = thrust + dT
          power  = power  + dP
          torque = torque + dQ

          alfa1  = alfa*180.0/pi
          write(2,1100)  r,alfa1,phi,U,Mach,CL,CD,dT,dP
       enddo

c----- average CD in calculation
       CD = cdall/float(call)
       write(6,1200) CD

c----- thrust and power coefficients
       CT = thrust/(0.5d0*rho*area*Utip**2)
       CP = power /(0.5d0*rho*area*Utip**3)

c----- write out results
       write(6,1300) Mmax
       write(6,2000) thrust*1d-3,torque*1d-3,power*1d-3,CT,CP,CT/sigma
```

C.9 FORWARD FLIGHT POWER OF HELICOPTER

This subroutine in `Fortran` is a typical example of analysis of the forward flight power calculation, described in §13.4. It does not include the trim calculation presented in §13.3.

```fortran
      subroutine AssemblyPower (Plevel,Pe,Pff,Pfftail,Pv,Pa,Ph)
c----------------------------------------------------------------
c-----           Calculates helicopter's level power
c-
c-    on OUTPUT
c-    Pff    = main rotor's induced power in forward flight
c-    Pv     = main rotor's profile power
c-    Pa     = airframe power (due to drag)
c-    Ph     = main rotor's induced power in hover
c-    Plevel = level flight power required
c-    Pe     = available engine power
c-
c-    Calculation based on known tilt angle and forward speed, and given
c-    aircraft AUW, and flight altitude, ISA conditions or otherwise.
c-    Corrections are possible for changes in atmospheric temperature
c-    above/below ISA conditions.
c-
c-    Parameters Required:
c-    1.) equivalent flat plate drag of airframe (for calculation of Pa)
c-    2.) average CD of rotating blades (for calculation of Pv)
c-    3.) drag divergence Mach number (for calculation of compressibility)
c-    4.) rotor position wrt ground (for calculation of ground effect)
c----------------------------------------------------------------
      implicit none

c----- all parameters declared externally in ''copter.inc''
      include 'copter.inc'
      integer :: newton
      parameter (newton = 12)

      double precision :: dMdd
      parameter (dMdd = 1d-3)
      pi = acos(-1d0)

c----- air density at current flight altitude
      call atmosphere(h1,sigma1,theta1,delta1,asound1)

c----- tail rotor's advance ratio
      alfaTR = alfa
      mut = U*cos(alfaTR)/(omegat*radiust)

      do i = 1, 1

      mu  = U*cos(alfa)/(omega*radius)

c----- tip Mach number
      Utip = U*cos(alfa) + omega*radius
      Mtip = Utip/asound1

c----- correction of density for non ISA effect
      if (isa_effect) then
         sigma1 = (theta1/thetao)*(1.d0 - 0.60396*h1/theta1)
      endif
```

```
c----- corrected air density
      rho = rhoo*sigma1                          ! kgm/m^3

c----- engine power at current altitude (not dependent on speed)
      Pe  = PeSL*delta1/sqrt(theta1)             ! kW at altitude
      Pe  = PeSL*1d3                             ! Watt

c----- ideal power in hover (from momentum theory) - not dependent on U
      Pid = Weight*sqrt(Weight)/sqrt(2.d0*rho*Area) ! Watt

c----- power in hover, corrected for non uniform inflow - not dependent on U
      Ph  = kind*Pid                             ! Watt

c----- height above ground, in meters
      hag = (h1 - h1g)*1d3 + zrotor              ! metres

c----- ground effect factor, from Hayden (1976)
cx    kg = 1.d0/(0.9926 + 0.3794*( diameter/hag)**2)

c----- ground effect factor, from Cheeseman-Bennett (1955)
cx    kg = 1.d0/(1.d0 - (0.25d0*radius/hag)**2)

c----- induced velocity factor in hover
      vi      = sqrt( Weight/(2d0*rho*area) )    ! m/s
      lambdah = vi/(omega*radius)

c----- Newton iterations to find solution for inflow ratio
      lambda = lambdah
      if (U .eq. 0.d0) goto 2
      do j = 1, newton
        f = lambda - mu*tan(alfa) - lambdah**2/sqrt(mu**2 + lambda**2)
        df = 1.d0 + lambdah**2*lambda/sqrt( (mu**2 + lambda**2)**3 )
        lambda = lambda - f/df
        f  = lambda - mu*tan(alfa) - lambdah**2/sqrt(mu**2 + lambda**2)
        if (abs(f) < xacc) goto 1                ! exit strategy
      enddo
1     continue

c----- check if something went wrong
      if (abs(f) > xacc ) write(6,*) ' Newton not converged', h1

c----- induced power required in forward flight
2     Pff = Ph*lambda/lambdah                    ! Watt

c----- main rotor thrust
      thrust = 2d0*rho*area*lambda**2 *(omega*radius)**2

c----- correction for ground effect (usually not necessary in f-flight)
      Pff = Pff*kg

c----- airframe power
      q  = 0.5d0*rho*U**2
      Pa = q*efpa*U                              ! Watt

c----- profile power, main rotor (sigma = rotor solidity)
      c1 = (omega**3)*(radius**5)
      c2 = 1.d0 + 3.d0*mu**2 + 3d0*mu**4/8d0
      Pv = 0.125d0*pi*rho*sigma*CDo*c1*c2         ! Watt
```

```
c----- profile power, main rotor, hover
      c2 = 1d0
      Pvo = 0.125d0*pi*rho*sigma*CDo*c1*c2

c----- tail rotor power in forward flight
      Thrust_TR = (Ph + Pvo)/(omega*xtr)
      vi_TR     = sqrt(Thrust_TR/(2d0*rho*areat))
      Pi_TR     = 1.2d0*Thrust_TR*vi_TR
      lambdah   = vi_TR/(omegat*radiust)

c----- tail rotor induced power
      lambda = lambdah
      alfaTR = 4d0*pi/180d0
      ta     = tan(alfaTR)
      do j = 1, 20
        f  = lambda - mut*ta - lambdah**2/sqrt(mut**2 + lambda**2)
        df = 1.d0 + lambdah**2*lambda/sqrt( (mut**2 + lambda**2)**3 )
        lambda = lambda - f/df
        f  = lambda - mut*ta - lambdah**2/sqrt(mut**2 + lambda**2)
        if (abs(f) < xacc) goto 3                 ! exit strategy
      enddo
   3  continue
      Pff_TR = Pi_TR*lambda/lambdah

      c1     = (omegat**3)*(radiust**5)
      c2     = 1.d0 + 3.d0*mut**2
      sigmat = 4d0*0.25/(pi*radiust)
      Pvt    = 0.25d0*pi*rho*0.5*sigmat*CDo*c1*c2      ! Watt

      Pfftail = Pff_TR + Pvt
      FoM     = Pff_TR/(Pff_TR + Pvt)

c----- fenestron analysis, cr = contraction ratio; set cr = sqrt(2)
      muf    = U/(omegaf*radiusf)
      Pfen   = Pff_TR/sqrt(2d0*cr)

      c1     = (omegaf**3)*(radiusf**5)
      c2     = 1.d0 + 3.d0*muf**2
      Pvfen  = 0.25d0*pi*rho*0.5*sigmaf*CDo*c1*c2       ! Watt
      Fom_fen = Pfen/(Pfen + Pvfen)

c----- compressibility effects on main rotor (Gessow-Crim, 1958)
      if (Mtip > Mdd) then
        c1    = rho*area*(omega*radius)**3
        Pcomp = (0.007*dMdd + 0.052*dMdd**2)*sigma*c1
      endif

c----- total power, level flight
      Plevel = Pff + Pv + Pa + Pfftail + Pcomp ! Watt

c------include transmission losses
      Plevel = Plevel*ktran                    ! Watt

c----- drag
      dPv  = 0.125d0*rho*sigma*CDo*(omega*radius)**3*(3d0*mu**2)
      drag = weight*tan(alfa) + (Pa + dPv)/U
      alfa1 = atan(drag/weight)
      alfa = alfa1
      er1  = (thrust*cos(alfa) - weight)/weight
```

```
        er2   = (thrust*sin(alfa) - drag)/drag
        rest  = er1 + er2
        write(80,1001) i,alfa,rest,er1,er2
        if ( abs(rest) < 1d-3 ) goto 9

        enddo

c----- check if calculated power is larger than available engine power
 9      continue
        if (abs(da) > 1d-3) write(61,*) U,da
        write(80,*)
        if (Plevel > Pe) then
           write(6,900) Plevel*1d-3,U*3.6d0
           stop   ! unable to fly at this speed
        endif

 900    format('  ERROR: Calculated Power > Engine Power'
     &        /'                Power = ', f9.3, ' kW',
     &        /'                Speed = ', f9.3, ' km/h')

        return
        end
c-------------------------------------------------------------------
```

Bibliography

[1] The Boeing Corporation, 2004.

[2] Kimberlin R.D. *Flight Testing of Fixed-Wing Aircraft*. AIAA Educational Series, 2003.

[3] Olson W.M. *Aircraft Performance Flight Testing*, AFFTC-TIH-99-01. Air Force Flight Center, Edwards Air Force Base, Edwards, CA, 2002.

[4] Cooke A.K. and Fitzpatrick E.W.H. *Helicopter Test and Evaluation*. AIAA Educational Series (also Blackwell Science), 2002.

[5] Lan C.T.E. and Roskam J. *Airplane Aerodynamics and Performance*. Roskam Aviation and Engineering Corp, Ottawa, Canada, 1981.

[6] Anderson J.D. *Introduction to Flight*. McGraw-Hill, 1985.

[7] IPCC. *Aviation and the Global Atmosphere*. Cambridge Univ. Press, 1999.

[8] Jane's. *Jane's All the World's Aircraft*. Jane's Information Systems (updated every year).

[9] Taylor M.J., editor. *World Aircraft & Systems Directory*. Flight International, 2002.

[10] Gurton B. *The Encyclopedia of Russian Aircraft 1875–1995*. Motorbooks International Publishers & Wholesalers, Osceola, WI 54020, USA, 1995.

[11] Loftin L.K. Quest for performance: The evolution of modern aircraft. Report SP-468, NASA, 1981.

[12] *Performance Prediction Methods*, AGARD CP-242, 1978.

[13] Filippone A. Data and performances of selected aircraft and rotorcraft. *Progr. Aero. Sci.*, 36(8):629–654, 2000.

[14] ESDU. *Airfield Performance (general). Landing*, Performance, Vol. 6 of *Data items 85029, -030, 99015, -016, -017*. ESDU International, London, 1987.

[15] ESDU. *Calculation of Ground Performance in Take-off and Landing*, Performance, Vol. 4. ESDU International, London, 1985.

[16] ESDU. *Representation of Drag in Aircraft Performance Calculations*, Performance, Vol. 2 of *Data Item 81026*. ESDU International, London, 1987.

[17] ESDU. *Approximate Methods for Estimation of Cruise Range and Endurance: Aeroplanes with Turbojet and Turbofan Engines*, Performance, Vol. 5 of *Data Item 73019*. ESDU International, London, 1982.

[18] Bairstow L. *Applied Aerodynamics*. Longman & Greens, London, 1951.

[19] Etkin B. *Dynamics of Flight*. John Wiley & Sons, New York, 1959.

[20] Ashley H. On making things the best – Aeronautical uses of optimization. *J. Aircraft*, 19(1):5–28, 1982.

[21] Bryson A.E. and Ho Y.C. *Applied Optimal Control*. Blaisdell, New York, 1969.

[22] Ashley H. *Engineering Analysis of Flight Vehicles*. Addison-Wesley, 1974.

[23] Hall D. Technical preparation of the airplane Spirit of St Louis. Report TN 257, NACA, July 1927.

[24] Sutter J.F. and Anderson C.H. The Boeing model 747. *J. Aircraft*, 4(5):452–456, 1967.

[25] Lynn-Olason M. Performance and economic design of the 747 family of airplanes. *J. Aircraft*, 6(6), 1969.

[26] Danby T., Garrand W.C., Ryle D.M., and Sullivan L.J., V/STOL development of the C-130 Hercules. *J. Aircraft*, 1(5):242–252, 1964.

[27] Gang Yu., editor. *Operations Research in the Airline Industry*. Kluwer Academic Publishers, 1998.

[28] Von Kármán T. *The Wind and Beyond: Theodore von Kármán Pioneer in Aviation and Pathfinder in Space*. Little Brown & Co., Boston, MA, 1967.

[29] Stevens J.H. *The Shape of the Aeroplane*. Hutchinson & Co. Ltd, London, 1953.

[30] Weyl A.R. *Fokker: The Creative Years*. Putnam & Sons, London, 1965 (reprint 1987).

[31] Gallagher G.L., Higgins L.B., Khinoo L.A., and Pierce P.W. *Naval Test Pilot School Flight Test Manual – Fixed Wing Performance*, USNTPS-FTM-No. 108. US Navy Pilot School, Sept. 1992.

[32] Isaacs R. *Differential Games: A Mathematical Theory with Applications to Warfare, Pursuit, Control and Optimisation*. John Wiley & Sons, 1965.

[33] Perkins C.D. and Hage R.E. *Airplane Performance Stability and Control*. John Wiley, 1949.

[34] von Mises R. *Theory of Flight*. McGraw-Hill, 1945.

[35] AIAA. *Aerospace Design Engineering Guide*. 4th edition, 1998.

[36] Jones R.T. Reduction of wave drag by antisymmetric arrangement of wings and bodies. *AIAA J.*, 10(1):171–176, 1972.

[37] Yechout T.R., Morris S.L., Bossert D.E., and Hallgren W.F. *Introduction to Aircraft Flight Mechanics: Performance, Static Stability, Dynamic Stability and Classical Feedback Control*. AIAA, 2003.

[38] Miele A. *Flight Mechanics. Vol. I: Theory of Flight Paths*. Addison-Wesley, 1962.

[39] NACA. Standard atmosphere – Tables and data for altitudes to 65,800 feet. Report 1235, NACA, 1955.

[40] Minzer R.A., Champion S.W., and Pond H.L. The ARDC Model Atmosphere. Technical Report 115, Air Force Surveys in Geophysics, 1959.

[41] Anon. U.S. Standard Atmosphere. Technical report, US Government Printing Office, Washington, DC, 1962.

[42] Anon. Manual of the ICAO Standard Atmosphere, extended to 80 kilometres (262,500 feet). Technical report, ICAO, 1993, 3rd edition.

[43] Barry R.G. and Chorley R.J. *Atmosphere, Weather and Climate*. Routledge, London, 8th edition, 2003.

[44] Asselin M. *Introduction to Aircraft Performance*. AIAA Educational Series, 1997.

[45] Lynch F.T. and Khodadoust A. Effects of ice accretions on aircraft aerodynamics. *Progr. Aero. Sci.*, 37(8):669–767, 2001.

[46] Kind R.J., Potapczuk M.G., Feo A., Golia C., and Shah A.D. Experimental and computational simulation of in-flight icing phenomena. *Progr. Aero. Sci.*, 34(3):257–345, 1998.

[47] Zhu S. and Etkin B. Model of the wind field in a downburst. *J. Aircraft*, 22(7), July 1985.

[48] Ivan M. A ring-vortex downburst model for flight simulations. *J. Aircraft*, 23(3): 232–236, March 1986.

[49] Hahn K.U. Takeoff and landing in a downburst. *J. Aircraft*, 24(8):552–558, Aug. 1987.

[50] Zhao Y. and Bryson A.E. Approach guidance in a downburst. *J. Guidance, Control and Dynamics*, 15(4):893–900, 1992.

[51] Frost W., Chang H.P., McCarthy J., and Elmore K. Aircraft performance in a JAWS microburst. *J. Aircraft*, 22(7):561–567, July 1985.

[52] Etkin B. Turbulent wind and its effect on flight. *J. Aircraft*, 18(5):327–345, May 1981.

[53] Houbolt J.C. Atmospheric turbulence. *AIAA J.*, 11(4):421–437, April 1973.

[54] Hale F.J. *Aircraft Performances, Selection and Design*. John Wiley, 1984.

[55] Cleveland F.A. Size effects in conventional aircraft design. *J. Aircraft*, 7(6):483–512, Nov. 1970.

[56] Baumann A. Progress made in the construction of giant airplanes in Germany during the war. Report TN 29, NACA, 1920.

[57] Handley Page F. The case for the large aeroplane. *The Aero J.*, 28–49, Jan. 1917.

[58] Kingsford P.W. *Lanchester F.W.: A Life of an Engineer*. Arnold Publications Ltd, London, 1960 (Chapt 9).

[59] Lange R.H. Design concepts for future cargo aircraft. *J. Aircraft*, 13(6):385–392, June 1976.

[60] Whitener P.C. Distributed load aircraft concepts. *J. Aircraft*, 16(2):72–75, Feb. 1979.

[61] Allen J.E. Quest for a novel force: A possible revolution in aerospace. *Progr. Aero. Sci.*, 39(1):1–60, Jan. 2003.

[62] McMasters J.H. and Kroo I.M. Advanced configurations for very large transport airplanes. *Aircraft Design*, 1(4):217–242, 1998.

[63] Anonymous. 747-400 airplane characteristics for airport planning. Report D6-58326-1, The Boeing Corporation, Dec. 2002.

[64] Pennycuick C.J. *Bird Flight Performance: A Practical Calculation Manual.* Oxford Univ. Press, 1989.

[65] Pennycuick C.J. Flight of auks (Alcidae) and other northern seabirds compared with southern procellariiformes: Ornithodolite observations. *J. Exp. Biology*, 128:335–347, 1987.

[66] Pennycuick C.J. Predicting wingbeat frequency and wavelength of birds. *J. Exp. Biology*, 171–185, 1990.

[67] Greenewalt C.H. Dimensional relationships for flying animals. Smithsonian Miscellaneous Collections, 1962. No. 144.

[68] Haldane J.B.S. *A Treasury of Science*, Chapter: On Being the Right Size. Harper, New York, 1958.

[69] von Kármán T. *Aerodynamics – Selected Topics in the Light of their Historical Development.* Cornell University Press, 1954 (Chapter 1).

[70] Pennicuick C.J. *Bird Flight Performance – A Practical Calculation Manual.* Oxford Univ. Press, 1989.

[71] Templin R.J. The spectrum of animal flight: Insects to pterosaurs. *Progr. Aero. Sci.*, 36(5–6), 2000.

[72] Tennekes H. *The Simple Science of Flight.* The MIT Press, Cambridge, MA, 1993.

[73] Staton R.N., editor. *Introduction to Aircraft Weight Engineering.* SAWE Inc., Los Angeles, 2003.

[74] Torenbeek E. *Synthesis of Subsonic Airplane Design.* Kluwer Academic Publ., 1985.

[75] Anon. Aerodrome design manual. Part 3. Pavements (doc 9157p3), 2nd edition, 1983 (reprinted 2003).

[76] Raymer D. *Aircraft Design: A Conceptual Approach.* AIAA Educational Series, 3rd edition, 1999.

[77] Udin S.V. and Anderson W.J. Wing mass formula for subsonic aircraft. *J. Aircraft*, 29(4):725–732, July 1992.

[78] Beltramo M.N., Trapp D.L., Kimoto B.W., and Marsh D.P. Parametric study of transport aircraft systems cost and weight. Report CR 151970, NASA, April 1977.

[79] Kershner M.E. Laminar flow: Challenge and potential. In *Research in Natural Laminar Flow and Laminar-Flow Control*, NASA CP-2487. NASA Langley, March 1987.

[80] Isikveren A.T. Identifying economically optimal flight techniques of transport aircraft. *J. Aircraft*, 39(4), July 2002.

[81] Windhorst R., Ardema M., and Kinney D. Fixed-range optimal trajectories of supersonic aircraft by first-order expansions. *J. Guidance, Control, Dynamics*, 24(4), July 2001.

[82] Leyman C.S. A review of the technical development of Concorde. *Progr. Aero. Sci.*, 23(3):188–238, 1986.

[83] Pitts W., Nielsen J., and Kaattar G. Lift and center of pressure of wing-body-tail combinations at subsonic, transonic, and supersonic speeds. Report R-1307, NACA, Jan. 1957.

[84] Davenport E.E. and Huffman J.K. Aerodynamic characteristics of three slender sharp-edge 74 degrees swept wings at subsonic, transonic, and supersonic Mach numbers. Report TN-D-7631, NASA, July 1974.

[85] Levin D. and Katz J. Dynamic load measurements with delta wings undergoing self-induced roll oscillations. *J. Aircraft*, 21(1):30–36, 1984.

[86] Pekham D.H. Low-speed wind tunnel tests on a series of uncambered slender pointed wings with sharp edges. Report 2613, RAE, 1958.

[87] Polhamus E. A concept of the vortex lift of sharp-edge delta wings based on a leading-edge-suction analogy. Report TN-D-3767, NASA, Dec. 1966.

[88] Polhamus E. Prediction of vortex-lift characteristics by leading-edge suction analogy. *J. Aircraft*, 8(4):193–199, April 1971.

[89] Peake D.J. and Tobak M. Three-dimensional interactions and vertical flows with emphasis on high speeds. AGARDograph AG-252. AGARD, July 1980.

[90] Delery J.M. Aspects of vortex breakdown. *Progr. Aero. Sci.*, 30(1):1–59, Jan. 1994.

[91] Ashley H., Katz J., Jarrah M., and Vaneck T. Survey of research on unsteady aerodynamic loading of delta wings. *J. Fluids & Structures*, 5(4):363–390, July 1991.

[92] Gursul I. Review of unsteady vortex flows over delta wings. *J. Aircraft*, 42(2), March 2005.

[93] Wood R., Wilcox F., Bauer S., and Allen J. Vortex flows at supersonic speeds. Report TP-2003-211950, NASA, March 2003.

[94] Hill W.A. Experimental lift of low aspect-ratio triangular wings at large angles of attack and supersonic speeds. Report RM-A57I17, NACA, Nov. 1957.

[95] Erickson G.E. High angle-of-attack aerodynamics. *Ann. Rev. Fluid Mech.*, 27: 45–88, 1995.

[96] Lamar J.E. and Frink N.T. Aerodynamic features of designed strake-wing configurations. *J. Aircraft*, 19(8):639–646, Aug. 1982.

[97] Lamar J.E. and Frink N.T. Experimental and analytical study of the longitudinal aerodynamic characteristics of analytically and empirically designed strake-wing configurations at subcritical speeds. Report TP-1803, NASA, June 1981.

[98] Erickson G.E. and Murri D.G. Wind tunnel investigations of forebody strakes for yaw control on F/A-18 model at subsonic and transonic speeds. Report TP-3360, NASA, Sept. 1993.

[99] Erickson G.E. and Rogers W.R. Effects of forebody strakes and Mach number on overall aerodynamic characteristics of configuration with 55 degrees cropped delta wing. Report TP-3253, NASA, Nov. 1992.

[100] Brune B.W. and McMasters J.H. *Progress in Aeronautics and Astronautics*, Vol. 125, chapter Computational Aerodynamics Applied to High-Lift Systems. AIAA, 1991.

[101] Callagan J.G. Aerodynamic prediction methods for aircraft at low speeds with mechanical high lift devices. In *Prediction Methods for Aircraft Aerodynamic Characteristics*, AGARD LS-67, pages 2.1–2.52, May 1974.

[102] Obert E. Forty years of high-lift R & D – An aircraft manufacturer's experience. In *High-Lift Systems Aerodynamics*, AGARD, CP-515, Sept. 1993.

[103] Shovlin M.D., Skavdahl H., and Harkonen D.L. Design and performance of the propulsion system for the quiet short-haul research aircraft. *J. Aircraft*, 17(12):843–850, Dec. 1980.

[104] McCormick B.W. *Aerodynamics, Aeronautics and Flight Mechanics*. John Wiley, 2nd edition, 1995.

[105] Smith A.M.O. Aerodynamics of high-lift aircraft system. In *Fluid Dynamics of Aircraft Stalling*, AGARD CP 102, 1972.

[106] *High Lift Systems Aerodynamics*, AGARD CP-515, Sept. 1993.

[107] Gratzer K.B. Analysis of transport applications for high-lift schemes. In *Assessment of Lift Augmentation Devices*, AGARD LS-43, 1971.

[108] Rudolph P.K. High-lift systems on commercial subsonic airliners. Report CR-4746, NASA, Sept. 1996.

[109] Stepniewski W.Z. and Keys C.N. *Rotary Wing Aerodynamics*. Dover Publ., 1988.

[110] Webb T.S., Kent D.R., and Webb J.B. Correlation of F-16 aerodynamics performances predictions with early flight test results. In *Performance Prediction Methods*, AGARD CP-242, May 1978.

[111] White F. *Viscous Fluid Flow*. McGraw-Hill, 1974.

[112] Anderson J.D. *Aircraft Performance and Design*. McGraw-Hill, 1998.

[113] Ackroyd J.D. The United Kingdom's contributions to the development of aeronautics. Part 2: The development of the practical airplane (1900–1920). *Aero. J.*, 104(1042), Dec. 2000.

[114] Küchemann D. *The Aerodynamic Design of Aircraft*. Pergamon Press, 1978.

[115] Bertin J.J. *Aerodynamics for Engineering Students*. Prentice Hall, 4th edition, 2000.

[116] Kuethe A.M. and Chow C.Y. *Foundations of Aerodynamics*. John Wiley, 5th edition, 1997.

[117] Hayes W.D. Linearized supersonic flow. Technical report, North American Aviation, June 1950.

[118] Harmon S.M. and Jeffreys I. Theoretical lift and damping in roll of thin wings with arbitrary sweep and taper at supersonic speeds – supersonic leading and trailing edges. Report TN-2114, NACA, May 1950.

[119] Poisson-Quinton Ph. Parasitic and interference drag prediction and reduction. In *Aircraft Drag Prediction and Reduction*, AGARD R-723, 1985.

[120] Ashley H. and Landhal M. *Aerodynamics of Wings and Bodies*. Addison-Wesley, 1965.

[121] Abbott A. and von Doenhoff I.H. *Theory of Wing Sections*. Dover, 1959 (and later editions).

[122] Allen H.J. and Eggers A.J. A study of the motion and aerodynamic heating of missiles entering the Earth's atmosphere. Report 1381, NACA, 1958.

[123] Vinh N.X., Busemann A., and Culp R.D. *Hypersonic and Planetary Entry Flight Mechanics*. Univ. of Michigan Press, Ann Arbor, MI, 1980.

[124] Larrabee E.E. Aerodynamic penetration and radius as unifying concepts in flight mechanics. *J. Aircraft*, 4(1):28–35, Jan. 1967.

[125] Rossow V.J. and James K.D. Overview of wake-vortex hazards during cruise. *J. Aircraft*, 37(6):960–975, 2000.

[126] Spalart P.R. Airplane trailing vortices. *Ann. Rev. Fluid Mech.*, 30:107–138, 1998.

[127] Rossow V.J. Lift-generated vortex wakes of subsonic transport aircraft. *Progr. Aero. Sci.*, 35(6):507–660, 1999.

[128] Rossow V.E. and Tinling T.B. Research on aircraft/vortex-wake interactions to determine acceptable level of wake intensity. *J. Aircraft*, 25(6):481–492, Dec. 1988.

[129] Rossow V.E. Wake-vortex separation distances when flight-path corridors are constrained. *J. Aircraft*, 33(3):539–546, May 1996.

[130] Rossow V.E. Use of individual flight corridors to avoid vortex wakes. *J. Aircraft*, 40(2):225–231, March 2003.

[131] Jensen E., Toon O., Kinne S., Sachse G., Anderson B., Roland C., Twohy C., Gandrud B., Heymsfield A., and Miake-Lye R. Environmental conditions required for contrail formation and persistence. *J. Geophysical Research*, 103(D4):3929–3936, Feb. 1998.

[132] Schumann U. On conditions for contrail formation from aircraft exhausts. *Meteorologische Zeitschrift*, 5(1):4–23, Jan. 1996.

[133] Detwiler A.G. and Jackson A. Contrail formation and propulsion efficiency. *J. Aircraft*, 39(4):638–645, July 2002.

[134] Sears W.R., editor. *High Speed Aerodynamics and Jet Propulsion*. Princeton Univ. Press, Princeton, NJ, 1957.

[135] Jones R.T. *Wing Theory*. Princeton Univ. Press, Princeton, NJ, 1990.

[136] Katz J. and Plotkin A. *Low Speed Aerodynamics*. McGraw-Hill, 1992.

[137] Küchemann D. and Weber J. An analysis of some performance aspects of various types of aircraft designed to fly over different ranges and different speeds. *Progr. Aero. Sci.*, 9:329–456, 1968.

[138] Schlichting H. and Truckenbrodt E. *Aerodynamics of the Airplane*. McGraw-Hill, 1979.

[139] Nickel K. and Wohlfahrt M. *Tailless Aircraft in Theory and Practice*. Edward Arnold, London, 1994.

[140] Minkkinen G., Wilson T., and Beaver G. An experiment on the lift of an accelerated airfoil. *J. Aircraft*, 13(5):687–689, May 1976.

[141] Nelson R.C. and Pelletier A. The unsteady aerodynamics of slender wings and aircraft undergoing large amplitude maneuvers. *Progr. Aero. Sci.*, 39(2–3):185–248, 2003.

[142] Rom J. *High Angle of Attack Aerodynamics*. Springer-Verlag, 1992.

[143] Althaus D. *Stuttgarter Profilkatalog*. Stuttgart Univ., 1981 (in German, available from Stuttgart University).

[144] Eppler R. *Airfoil Design and Data*. Springer Verlag, 1990.

[145] Selig M.S., Lyon C., Giguere P., Ninham C., and Guglielmo J. *Low Speed Airfoil Data, Vol. 2*. Soar Tech Publ., 1996.

[146] Klein V. Estimation of aircraft aerodynamic parameters from flight data. *Progr. Aero. Sci.*, 26(1):1–77, 1989.

[147] Drela M. Elements of airfoil design methodology. In Henne P.A., editor, *Applied Computational Aerodynamics*, Vol. 125 of *Prog. Astronautics & Aeronautics*. AIAA, 1990.

[148] Selig M.S., Gopalarathnam A., Giguere P., and Lyon C.A. Systematic airfoil design studies at low Reynolds numbers. In *Fixed and Flapping Wing Aerodynamics for Micro Air Vehicle Applications*, Vol. 195 of *Prog. Astronautics & Aeronautics*, chapter 8. AIAA, 1992.

[149] Olason M.L. and Norton D.A. Aerodynamic design philosophy of the Boeing 737. *J. Aircraft*, 3(6), Nov. 1966.

[150] AGARD, *Aerodynamic Drag Prediction and Reduction*, AGARD R-723, 1985.

[151] Seckel E. *Stability and Control of Aircraft and Helicopters*. Addison-Wesley, 1964. (Appendix I contains aircraft performance and stability data.)

[152] Brandt S.A., Stiles R.J., and Bertin J.J. *Introduction to Aeronautics: A Design Perspective*. AIAA, 2004.

[153] Heffley R.K. and Jewell W.F. Aircraft handling qualities data. Report CR-2144, NASA, 1972.

[154] Gunston B., editor. *Jane's Aero Engines*. Jane's Information System, 2002 (updated every year).

[155] Mattingly J.D. *Elements of Gas Turbine Propulsion*. McGraw-Hill, 1996.

[156] Oates G.C. *Aerothermodynamics of Gas Turbine and Rocket Propulsion*. AIAA Educational Series, 1988.

[157] Archer R.D. and Saarlas M. *An Introduction to Aerospace Propulsion*. Prentice Hall, 1996.

[158] Kerrebrock J.L. *Aircraft Engines and Gas Turbines*. The MIT Press, 2nd edition, 1992.

[159] Heywood J.B. *Internal Combustion Engine Fundamentals*. McGraw-Hill, 1989.

[160] Babikian R., Lukachko S.P., and Waitz I. The historical fuel efficiency characteristics of regional aircraft from technological, operational and cost perspectives. *J. Air Transport Management*, 8(6):389–400, 2002.

[161] Goodger E. and Vere E. *Aviation Fuels Technology*. MacMillan, London, 1985.

[162] Good R.E. and Clevell H.J. Drop formation and evaporation of JP-4 fuel jettisoned from aircraft. *J. Aircraft*, 17(7):450–456, July 1980.

[163] Pfeiffer K., Quinn D., and Dungey C. Numerical model to predict the fate of jettisoned aviation fuel. *J. Aircraft*, 33(2):353–362, March 1996.

[164] Clevell H.J. Ground contamination by fuel jettisoned from aircraft in flight. *J. Aircraft*, 20(4):382–384, 1983.

[165] Newman S. *The Foundations of Helicopter Flight*. Arnold Publ., 1994.

[166] Glauert H. *Airplane Propellers*, Vol. 4 of *Aerodynamic Theory*. Dover ed., 1943.

[167] Theodorsen T. *Theory of Propellers*. McGraw-Hill, 1948.

[168]   Biermann D. and Hartman E.P. Wind-tunnel tests of four- and six-blade single- and dual-rotating tractor propellers. Report 747, NACA, 1942.

[169]   Biermann D. and Hartman E.P. The aerodynamic characteristics of full-scale propellers having 2, 3, and 4 blades of Clark Y and R.A.F. 6 airfoil sections. Report 640, NACA, 1938.

[170]   Theodorsen T., Stickle G.W., and Brevoort M.J. Characteristics of six propellers including the high-speed range. Report R-594, NACA, 1937.

[171]   *Aerodynamics and Acoustics of Propellers.* AGARD-CP-366, Feb. 1985.

[172]   Anonymous. *Generalized Method of Propeller Performance Estimation.* Hamilton Standard Report PDB6101, Revision A, Stratford, CT, 1963.

[173]   Conway J.T. Exact actuator disk solutions for non-uniform heavy loading and slipstream contraction. *J. Fluid Mech.*, 365:235–267, 1998.

[174]   Wieselberger. Contribution to the mutual interference between wing and propeller. Report TM-754, NACA, 1934.

[175]   McHugh J. and Eldridge H. The effect of nacelle-propeller diameter ratio on body interference and on propeller and cooling characteristics. Technical report R-680, NACA, 1939.

[176]   Delano J.B. Investigation of the NACA 4-(5)(08)-03 and NACA 4-(10)(08)-03 two-blade propellers at forward Mach numbers to 0.725 to determine the effects of camber and compressibility on performance. Report 1012, NACA, 1951.

[177]   Stack J., Delano E., and Feldman J. Investigation of the NACA 4-(3)(8)-045 two-blade propellers at forward Mach numbers to 0.725 to determine the effects of compressibility and solidity on performance. Report R-999, NACA, 1950.

[178]   Lincoln F.C. Migration of birds – Circular 16. US Fish and Wildlife, 1935 (revised by Zimmermann J.L., 1998).

[179]   Martin B.P. *World Birds.* Guinness Books, 1987.

[180]   Gabrielli G. and von Kármán T. What price speed? *Mechanical Engineering*, 72(10):775–781, 1950.

[181]   Lorentz R. Flight power scaling of airplanes, airships, helicopters: Application to planetary exploration. *J. Aircraft*, 38(2):208–214, March 2001.

[182]   Pinsker W.J.G. Zero rate of climb speed as a low speed limitation for the stall-free aircraft. Report ARC CP-931, Aeronautical Research Council, 1967.

[183]   Press W.H., Teukolsky S.A., Vettevling W.T., and Flannery P.P. *Numerical Recipes.* Cambridge Univ. Press, 2nd edition, 1992.

[184]   Seddon J. and Goldsmith E.L. *Intake Aerodynamics.* Blackwell Science, 1999.

[185]   Bilimoria K.D. and Cliff E.M. Singular trajectories in airplane cruise-dash optimization. *J. Guidance, Control and Dynamics*, 12(3):303–310, May 1989.

[186]   Moes T.R. and Iliff K. Stability and control estimation flight test results for the Sr-71 aircraft with externally mounted experiments. Report TP-2002-210718, NASA, June 2002.

[187]   Palumbo N., Moes T.R., and Vachon M.J. Initial flight tests of the NASA F-15B propulsion flight test fixture. Report TP-2002-210736, NASA, 2002.

[188]   Kermode A.C. *Mechanics of Flight.* Isaac Pitman and Sons, 1956.

[189]   Lancaster J.W. Feasibility study of modern airships. Report CR-137692, NASA, Aug. 1975.

[190]   Eshelby M. *Aircraft Performance: Theory and Practice.* Arnold Publ., London, 2000 (also available by the AIAA).

[191]   Wetmore J.W. The rolling friction of several airplane wheels and tires and the effect of rolling friction on take-off. Technical report R-583, NACA, 1937.

[192]   Harrin E.N. Low tire friction and cornering forces on a wet surface. Report TN 4406, NACA, Sept. 1958.

[193] Yager T.S. Factors influencing aircraft ground handling performance. Report TM-85652, NASA, June 1983.

[194] Agrawal S.K. Braking performance of aircraft tires. *Progr. Aero. Sci.*, 23(2):105–150, 1986.

[195] Kuchinka A.J. Prediction of off-runway takeoff and landing performance. *J. Aircraft*, 3(3):213–218, 1966.

[196] Haftmann B., Debbeler F.J., and Gielen H. Take-off drag prediction for Airbus A-300-600 and A-310 compared with flight test results. *J. Aircraft*, 25(12):1088–1096, Nov. 1988.

[197] Krenkel A.R. and Salzman A. Take-off performances of jet-propelled conventional and vectored-thrust STOL aircraft. *J. Aircraft*, 5(5):429–436, Sept. 1968.

[198] Powers S.A. Critical field length calculations for preliminary design. *J. Aircraft*, 18(2):103–107, 1981.

[199] Vinh N.X. *Flight Mechanics of High-Performance Aircraft*. Cambridge Univ. Press, 1993.

[200] Pinsker W.J.G. The dynamics of aircraft rotation and liftoff and its implication for tail clearance especially with large aircraft. Report ARC R&M 3560, Aeronautical Research Council, 1967.

[201] Gasich W.E. Experimental verification of two methods for computing the take-off ground run of propeller-driven aircraft. Report TN-1258, NACA, June 1947.

[202] Van Hengst J. Aerodynamic effects of ground de/anti-icing fluids on Fokker 50 and Fokker 100. *J. Aircraft*, 30(1):35–40, Jan. 1993.

[203] Perelmuter A. On the determination of the take-off characteristics of a seaplane. Report TM-863, NACA, May 1938.

[204] Parkinson J., Olson R., and House R. Hydrodynamic and aerodynamic tests of a family of models of seaplane floats with varying angles of dead rise – NACA models 57-A, 57-B, and 57-C. Report TN-716, NACA, 1939.

[205] De Remer D. Seaplane takeoff performance – using delta ratio as a method of correlation. *J. Aircraft*, 25(8):765–766, Aug. 1987.

[206] Clifton R.G. and Leonard J.L. Aircraft tyres – An analysis of performance and development criteria for the 70s. *Aeronautical J.*, 76(736):195–216, April 1972.

[207] Currey N.S. *Aircraft Landing Gear Design: Principles and Practices*. AIAA, 1998.

[208] Pinsker W.J.G. The landing flare of large transport aircraft. Report ARC R&M 3602, Aeronautical Research Council, 1967.

[209] Padovan J. and Kim Y.H. Aircraft landing-induced tire spinup. *J. Aircraft*, 28(12): 849–854, Dec 1991.

[210] Frost W., Crosby B., and Camp D.W. Flight through thunderstorm outflows. *J. Aircraft*, 16(11):749–755, Nov. 1979.

[211] Hsin C.C. Arrested landing studies for STOL aircraft. *J. Aircraft*, 11(3):159–165, March 1974.

[212] Wahi M.K. Airplane brake-energy analysis and stopping performance simulation. *J. Aircraft*, 16(10):682–694, Oct. 1979.

[213] Milwitzky B. Generalized theory for seaplane impact. Report R-1103, NACA, 1952.

[214] Smiley R.F. The application of planing characteristics to the calculation of the water-landing loads and motions of seaplanes of arbitrary constant cross-section. Report TN-2814, NACA, 1952.

[215] Bell J.W. Tank tests of two floats for high-speed seaplanes. Report TN-473, NACA, 1933.

[216] Bryson A.E. and Denham W.F. A steepest-ascent method for solving optimum programming problems. *J. Applied Mechanics*, 29(2):247–257, 1962.

[217] Rutowski E.S. Energy approach to the general aircraft performance problem. *J. Aero. Sci.*, 21(3):187–195, March 1954.

[218] Merritt S.R., Cliff E.M., and Kelley H.J. Energy-modelled climb and climb-dash – The Kaiser technique. *Automatica*, 21(3):319–321, May 1985.

[219] Kelley H.J. An investigation of optimal zoom climb techniques. *J. Aero Sci*, 26:794–803, 1959.

[220] Kelley H.J., Cliff E.M., and Weston A.R. Energy state revisited for minimum-time aircraft climbs. AIAA Paper 83-2138. Aug. 1983.

[221] Lanchester F.W. *Aerodonetics*. Constable, 1908.

[222] Etkin B. and Reid L.C. *Dynamics of Flight: Stability and Control*. John Wiley & Sons, 1996.

[223] Nelson. R.C. *Flight Stability and Automatic Control*. McGraw-Hill, 1998.

[224] Campos A., Fonseca L., and Azinheira R. Some elementary aspects of non-linear airplane speed stability in constrained flight. *Progress. Aero. Sci.*, 31(2):137–169, 1995.

[225] Bulirsch R. and Stoer J. Numerical treatment of ordinary differential equations by extrapolation methods. *Numer. Math.*, 8(1):1–13, Jan. 1966.

[226] Bulirsh R. and Stoer J. *Introduction to Numerical Analysis*. Springer-Verlag, 2nd edition, 1993.

[227] Brüning G. and Hahn P. The on-board calculation of optimal climbing paths. In *Performance Prediction Methods*, AGARD CP-242, pages 5.1–5.15, May 1978.

[228] Miele A. Optimum flight paths of a turbojet aircraft. Report TM-1389, NACA, Sept. 1955.

[229] Miele A. General solutions of optimum problems of non-stationary flight. Report TM-1388, NACA, Oct. 1955.

[230] Kelley H.J. and Edelbaum T.N. Energy climbs, energy turns, and asymptotic expansions. *J. Aircraft*, 7(1):93–95, Jan. 1970.

[231] Schultz R.L. and Zagalsky N.R. Aircraft performance optimization. *J. Aircraft*, 9(2):108–114, Feb. 1972.

[232] Calise A.J. Extended energy management method for flight performance optimization. *AIAA J.*, 15(3):314–321, March 1977.

[233] Berton J.J. Optimum climb to cruise noise trajectories for the high speed civil transport. Report TM-2003-212704, NASA, Nov. 2003.

[234] Ardema M.D., Windhorst R., and Phillips J. Development of advanced methods of structural and trajectory analysis for transport aircraft. Report CR-1998-207770, NASA, March 1998.

[235] Ojha S.K. Fastest climb of a piston-prop aircraft. *J. Aircraft*, 30(1):146–148, 1993.

[236] Gilyard G.B. and Bolonkin A. Optimal pitch thrust-vector angle and benefits for all flight regimes. Report TM-2000-209021, NASA, March 2000.

[237] Neuman F. and Kreindler E. Optimal turning climb-out and descent of commercial jet aircraft. SAE Paper 821468, Oct. 1982.

[238] Neuman F. and Kreindler E. Minimum-fuel turning climbout and descent guidance of transport jets. Report TM-84289, NASA, Jan. 1983.

[239] Hale. Effects of wind on aircraft cruise performance. *J. Aircraft*, 16(6):382–387, 1979.

[240] Houghton R.C.C. Aircraft fuel savings in jet streams by maximising features of flight mechanics and navigation. *J. of Navigation*, 51(3):360–367, Sept. 1998.

[241] Visser H.G. Terminal area traffic management. *Progr. Aero. Sci.*, 28:323–368, 1991.

[242] ESDU. *Introduction to the Estimation of Range and Endurance*, Performance, Vol. 5 of *Data Item 73019*. ESDU International, London, 1980.

[243] ESDU. *Estimation of Cruise Range: Propeller-driven Aircraft*, Performance, Vol. 5 of *Data Item 75018*. ESDU, London, 1975.

[244] Bennington M.A. and Visser K.D. Aerial refueling implications for commercial aviation. *J. Aircraft*, 42(2):366–375, March 2005.

[245] Smith R.K. Seventy-five years of inflight refueling. Report R25-GPO-070-00746-1, Air Force History and Museum Program, Washington, DC, Nov. 1998. (Available on the US Government online store; document out of print as of 2005.)

[246] Torenbeek E. and Wittenberg H. Generalized maximum specific range performance. *J. Aircraft*, 20(7):617–622, July 1983.

[247] Abramovitz M. and Stegun I. *Handbook of Mathematical Functions*. Dover, 1972.

[248] Torenbeek E. Cruise performance and range prediction reconsidered. *Progr. Aero. Sci.*, 33(5-6):285–321, May–June 1997.

[249] Shevell R.S. *Fundamentals of Flight*. Prentice Hall, 1983.

[250] Martinez-Val R. and Pérez L. Extended range operations of two and three turbofan engined airplanes. *J. Aircraft*, 30(3):382–386, 1993.

[251] Green J.E. Greener by design – The technology challenge. *Aero J.*, 106(1056):57–113, Feb. 2002.

[252] Cavcar M. and Cavcar A. Optimum range and endurance of a piston propeller aircraft with cambered wing. *J. Aircraft*, 42:212–217, Jan. 2005.

[253] Sachs G. Optimization of endurance performance. *Progr. Aero. Sci.*, 29(2):165–191, 1992.

[254] Page R.K. Performance calculation of jet-propelled aircraft. *J. Royal Aero. Soc.*, 1947.

[255] Ashkenas I.L. Range performance of turbojet aircraft. *J. Aero. Sci*, 15(1):97–101, 1948.

[256] Edwards A.D. Performance estimation of civil jet aircraft. *Aircraft Engineering*, 22(254):94–99, 1950.

[257] Speyer J.L. Non optimality in the steady-state cruise for aircraft. *AIAA J.*, 14(11): 1604–1610, Nov. 1976.

[258] Gilbert E.G. and Parsons M.G. Periodic control and the optimality of aircraft cruise. *J. Aircraft*, 13(10):828–830, Oct. 1976.

[259] Menon P.K.A. Study of aircraft cruise. *J. Guidance, Control and Dynamics*, 12(5): 631–639, Sept. 1989.

[260] Sachs G. and Christodoulou T. Reducing fuel consumption of subsonic aircraft by optimal cyclic cruise. *J. Aircraft*, 24(5):616–622, 1987.

[261] *Recent Advances in Long Range and Long Endurance Operation Aircraft*, AGARD-CP-547, Nov. 1993.

[262] Lissaman P.B.S. and Schollenberger C.E. Formation flight of birds. *Science*, 168: 1003–1005, May 1970.

[263] King R.M. and Gopalarathnam A. Ideal aerodynamics of ground effect and formation flight. *J. Aircraft*, 42(1), 2005.

[264] Lissaman P. Simplified analytical methods for formation flight and ground effect. In *AIAA Aerospace Meeting*, AIAA Paper 2005-0851, Reno, NV, Jan. 2005.

[265] Milne-Thompson L.M. *Theoretical Aerodynamics*. MacMillan, London, 1966.

[266] Munk M.M. The minimum induced drag of aerofoils. Report 121, NACA, 1921.

[267] Hummel D. The use of aircraft wakes to achieve power reduction in formation flight. In *The Characterisation and Modification of Wakes from Lifting Vehicles in Fluids*, AGARD CP-584, pages 36–1, 36–13, May 1996.

[268] Ray R., Cobleigh B., Vachon M., and St John C. Flight test techniques used to evaluate performance benefits during formation flight. Report TP-2002-210730, NASA, Sept. 2002.

[269] Vachon M., Ray R., Walsh K., and Ennix K. F/A-18 performance benefits measured during the autonomous formation flight project. Report TM-2003-210734, NASA, Sept. 2003.

[270] Phillips W.H. Propulsive effects due to flight through turbulence. *J. Aircraft*, 12(7): 624–627, 1975.

[271] AGARD WG 19. *Operational Agility*, AGARD AR-314, April 1994.

[272]   Bryson A.E. and Hedrick J.K. Minimum time turns for a supersonic airplane at constant altitude. *J. Aircraft*, 8:182–187, 1971.

[273]   Hedrick J.K. and Bryson A.E. Three-dimensional minimum-time turns for supersonic aircraft. *J. Aircraft*, 9:115–121, 1972.

[274]   Kelley H.J. Differential-turning optimality criteria. *J. Aircraft*, 12(1):41–44, 1975.

[275]   Kelley H.J. Differential-turning tactics. *J. Aircraft*, 12(2):930–935, Feb. 1975.

[276]   LeBlaye P. *Human Consequences of Agile Aircraft*, NATO RTO-TR-015, chapter Agility: Definitions, Basic Concepts and History. Jan. 2001.

[277]   Jaslow H. Spatial disorientation during a coordinated turn. *J. Aircraft*, 39(4):572–576, 2002.

[278]   Webb P.M.D. *Bioastronautics Data Book*, NASA SP-3006, 1964.

[279]   AGARD. *Current Concepts on G-Protection Research and Development*, AGARD LS-202, 1995.

[280]   Smith A.H. Physiological changes associated with long-term increases in acceleration. In PHA Sneath, editor, *Life Sciences and Space Research XIV*. Akademie-Verlag, 1976.

[281]   Short S. Birth of American soaring flight: A new technology. *AIAA J.*, 43(1):17–28, Jan. 2005.

[282]   Alexander D.E. *Nature's Flyers: Birds, Insects and the Biomechanics of Flight*. The Johns Hopkins University Press, Baltimore, MD, 2002.

[283]   Irving F. *The Paths of Soaring Flight*. Imperial College University Press, London, 1999.

[284]   Prandtl L. Some remarks concerning soaring flight. Report TM-47, NACA, Oct. 1921.

[285]   Klemperer W. Soaring flight. *J. of the Franklin Institute*, 204(3):221–241, 1927.

[286]   Metzger D.E. and Hedrick J.C. Optimal flight paths for soaring flight. *J. Aircraft*, 12(11):867–871, Nov. 1975.

[287]   Bannasch R. *Fixed and Flapping Wing Aerodynamics for Micro Air Vehicle Applications*, Vol. 195 of *Progress in Astronautics and Aeronautics*, chapter Fixed and Flapping Wing Aerodynamics for Micro Air Vehicle Applications, pages 453–472. AIAA, 2001.

[288]   Cone C.D. The soaring flight of birds. *American Scientist*, 50:130–140, 1962.

[289]   Hedenstroem A. Migration by soaring or flapping flight in birds: The relative importance of energy cost and speed. *Phil. Transactions – Royal Society, Series B, Biological Sciences*, 342(1302), 1993.

[290]   Wolfram S. *The Mathematica Book*. Wolfram Research Inc., Champaign, IL, 5th edition, 2003.

[291]   Phillips W.H. Effect of steady rolling on longitudinal and directional stability. Report TN-1627, NACA, June 1948.

[292]   Pinsker W.J.G. Critical flight conditions and loads resulting from inertia cross-coupling and aerodynamic stability deficiencies. Report ARC CP-404, Aeronautical Research Council, 1958.

[293]   ESDU. Rolling moment derivative L_ξ for plain aileron at subsonic speeds. ESDU International, London, Oct. 1992. Data Item 88013.

[294]   ESDU. Stability derivative L_p, rolling moment due to rolling for swept and tapered wings. ESDU International, London, March 1981. Data Item No. Aero A.06.01.01.

[295]   Sandhal C.A. Free-flight investigation of the rolling effectiveness of several delta wings – Aileron configurations at transonic and supersonic speeds. Report RM-L8D16, NACA, Aug. 1948.

[296]   Myers B.C. and Kuhn R.E. High subsonic damping-in-roll characteristics of a wing with quarter-chord line swept 35° and with aspect-ratio 3 and taper ratio 0.6. Report RM-L9C23, NACA, May 1949.

[297]   Myers B.C. and Kuhn R.E. Effects of Mach number and sweep on the damping-in-roll characteristics of wings of aspect-ratio 4. Report RM-L9E10, NACA, June 1949.

[298] Anderson S.B., Ernst E.A., and van Dyke R.D. Flight measurements of the wing-dropping tendency of a straight-wing jet airplane at high subsonic Mach numbers. Report RM A51B28, NACA, April 1951.

[299] Stone D.G. A collection of data for zero-lift damping in roll of wing-body combinations as determined with rocket-powered models equipped with roll-torque nozzles. Report TN 3955, NACA, April 1957.

[300] Sanders C.E. Damping in roll of models with 45°, 60°, 70° delta wings determined at high subsonic, transonic and supersonic speeds with rocket-powered models. Report RM-L52d22a, NACA, June 1952.

[301] Malvestuto F., Margolis K., and Ribner H.S. Theoretical lift and damping in roll at supersonic speeds of thin sweptback tapered wings with streamwise tips, subsonic leading edges, and supersonic trailing edges. Report 970, NACA, 1950.

[302] Jones A.L. and Alksne A. A summary of lateral-stability derivatives calculated for wing plan forms in supersonic flow. Report 1052, NACA, 1951.

[303] Stachowiak S.J. and Bosworth J.T. Flight test results for the F-16XL with a digital control system. Report TP-2004-212046, NASA, March 2004.

[304] Toll T.A. Summary of lateral-control research. Report 868, NACA, 1947.

[305] Skow A.M. Agility as a contributor to design balance. *J. Aircraft*, 29(1):34–46, Jan. 1992.

[306] Azbug M.J. and Larrabee E.E. *Airplane Stability and Control*. Cambridge Univ. Press, 1997.

[307] Rathert G., Rolls L., Winograd L., and Cooper G. Preliminary flight investigation of the wing dropping tendency and lateral control characteristics of a 35° swept-back airplane at transonic Mach numbers. Report RM-A50H03, NACA, Sept. 1950.

[308] Chambers J.R. and Hall R.M. Historical review of uncommanded lateral-directional motions at transonic conditions. *J. Aircraft*, 41(3):436–447, May 2004.

[309] Hall R.M., Woodson S.H., and Chambers J.R. Accomplishments of the abrupt-wing-stall program. *J. Aircraft*, 42(3):653–660, May 2005.

[310] Owens D., Capone F., Hall R., Brandon J., and Chambers J. Transonic free-to-roll analysis of abrupt wing stall on military aircraft. *J. Aircraft*, 41(3):474–484, May 2004.

[311] Williams J.E. and Vukelich S.P. The USAF stability and control digital DATCOM. Report AFFDL-TR-79-3032, Vol. I, Air Force Flight Directorate Laboratory, April 1979.

[312] Torenbeek E. *Synthesis of Subsonic Airplane Design*. Kluwer Academic Publ., 1985. Appendix G-8.

[313] Uehara S., Stewart H.J., and Wood L.J. Minimum-time loop maneuvers of jet aircraft. *J. Aircraft*, 15(8):449–455, Aug. 1978.

[314] Shinar J., Merari D., and Medinah E. Analysis of optimal turning maneuvers in the vertical plane. *J. Guidance, Control and Dynamics*, 3(1):69–77, 1980.

[315] Smith G.A. and Workman G.L. KC-135 acceleration levels during low- and high-gravity trajectories. In P Curreri, editor, *Materials Science in Parabolic Flight*, NASA TM-4456, 1993.

[316] Bahr D.E. and Schulz R.D. Acceleration forces aboard NASA KC-135 aircraft during microgravity maneuvers. *J. Aircraft*, 26(7):687–688, July 1989.

[317] Menon P., Walker R., and Duke E. Flight test maneuver modeling and control. *J. Guidance, Control and Dynamics*, 12(2):195–200, March 1989.

[318] Lee H. Optimal aircraft trajectories for specified range. Report ARC-11282, NASA, 1994.

[319] Barman J.F. and Erzberger H. Constrained optimum trajectories for short-haul aircraft. *J. Aircraft*, 13(10):748–754, Oct. 1976.

[320] Lichtsinder A., Kreindler E., and Gal-Or B. Minimum-time maneuvers of thrust-vectored aircraft. *J. Guidance, Control & Dynamics*, 21(2):244–250, 1998.

[321] Betts J.T. Survey of numerical methods for trajectory optimization. *J. Guidance, Control & Dynamics*, 21(2):193–207, March 1998.

[322] Virtanen K., Ehtamo H., Raivio T., and Hämäläinen R.P. VIATO – Visual interactive aircraft trajectory optimization. *IEEE Trans on Syst., Man. & Cybernetics – Part C: Applications & Systems*, 29(3):409–421, Aug. 1999.

[323] Warner E.P. The prospect of the helicopter. Report TN-107, NACA, 1922.

[324] Leishman G. *Principles of Helicopter Aerodynamics*. Cambridge Univ. Press, 2000.

[325] Pirie P. and Lambermont A. *Helicopter and Autogyros of the World*. Cassell, London, 1958.

[326] Liberatore E.K. *Helicopters before Helicopters*. Krieger Publications, 1998.

[327] Everett-Heath J. *Soviet Helicopters*. Jane's Information Systems, London, 1988.

[328] Boulet J.A. *A History of the Helicopter as Told by its Pioneers, 1907–1956*. Editions France-Empire, Paris, 1984 (limited edition).

[329] Jane's. *Jane's Helicopter Markets and Systems*. Jane's Information Systems (updated every year).

[330] Jane's. *Jane's Aircraft Upgrades*. Jane's Information Systems (updated every year).

[331] Deal P.L. and Jenkins J.L. Investigation of level-flight and maneuvering characteristics of a hingeless-rotor compound helicopter. Report TN-D-5602, NASA, Jan. 1970.

[332] Erickson R., Kufeld R., Cross J., Hodge R., Ericson W., and Carter R. NASA rotor system research aircraft flight-test data report: Helicopter and compound configuration. Report TM-85843, NASA, 1984.

[333] Bergquist R.R. Some problems of design and operation of a 250-knot compound helicopter rotor. *J. Aircraft*, 1(5):252–259, 1964.

[334] de la Cierva J. The autogyro. *J. Royal Aero. Soc.*, 39(239):902–921, 1930.

[335] Leishman J.G. Development of the autogyro: A technical perspective. *J. Aircraft*, 41(4):765–781, July 2004.

[336] Houston S.S. Validation of a rotorcraft mathematical model for autogyro simulation. *J. Aircraft*, 37(3):203–209, May 2000.

[337] Fay J. *The Helicopter in Civil Operations*. Granada, London, 1981.

[338] Wiseman C.H., editor. *The International Countermeasures Handbook*. EW Communications, 1986.

[339] Drees J.M. Prepare for the 21st century. In *7th Alexander Nikolsky Honorary Lecture*, St Louis, MO, 1987.

[340] Scott M.W. Technology needs for high speed rotorcraft. Report CR-177590, NASA, Aug. 1991.

[341] Rutherford J., O'Rourke M., Lovenguth M., and Mitchell C. Conceptual assessment of two high-speed rotorcraft. *J. Aircraft*, 30(2):241–247, 1993.

[342] Talbot P.D., Phillips J.D., and Totah J.J. Selected design issues of some high-speed rotorcraft concepts. *J. Aircraft*, 30(6):864–871, Nov. 1993.

[343] Lynn R.R. Wing-rotor interactions. *J. Aircraft*, 3(4):326–332, 1966.

[344] Perry F.J. The aerodynamics of the world speed record. In *43rd AHS Annual Forum*, St Louis, MO, May 1987.

[345] Bramwell A.R.S. *Helicopter Dynamics*. Butterworth-Heinemann, Oxford, 2nd edition, 2001.

[346] Prouty R.W. Should we consider variable rotor speeds? *Vertiflite*, 50(4):24–27, 2004.

[347] Knowles P.A. The application of non-dimensional methods to the planning of helicopter performance flight trials and the analysis of results. Report ARC, CP-927, Aeronautical Research Council, 1967.

[348] Heyson H.H. A momentum analysis of helicopters and autogyros in inclined descent, with comments on operational restrictions. Report TN-D-7917, NASA, Oct. 1975.

[349] Bailey F.J. A simplified theoretical method of determining the characteristics of a lifting rotor in forward flight. Report 716, NACA, 1941.

[350] Gessow A. and Tapscott R.J. Charts for estimating performance of high-performance helicopters. Report 1266, NACA, 1956.

[351] Tanner W.H. Charts for estimating rotary wing performance at high forward speed. Report CR 115, NASA, 1964.

[352] Gessow A. and Myers G.C. *Aerodynamics of the Helicopter*. The College Park Press, College Park, MD, 2000 (reprint from 1954 edition).

[353] Prouty R.W. *Helicopter Performance, Stability and Control*. Krieger Publications, 1988 (reprint).

[354] Johnson W. *Helicopter Theory*. Princeton Univ. Press, 1980.

[355] Johnson W. CAMRAD II – Comprehensive analytical model of rotor aerodynamics and dynamics. Johnson Aeronautics, Palo Alto, CA, 2000.

[356] Ballin M.G. Validation of a real-time engineering simulation of the UH-60A helicopter. Report TM-88360, NASA, Feb. 1987.

[357] Howlett J. UH-60A Black Hawk engineering simulation program. Vol. 1: Mathematical model. Report CR-177542, NASA, Sept. 1989.

[358] ESDU. *Non-dimensional methods for the measurement of hover performance of turbine-engined helicopters*, Performance, Vol. 14 of *Data Item 73027*. ESDU International, London, 1977.

[359] ESDU. *Non-dimensional methods for the measurement of hover performance of turbine-engined helicopters*, Performance, Vol. 14 of *Data Item 74042*. ESDU International, London, 1974.

[360] Coleman C.P. A survey of theoretical and experimental coaxial rotor aerodynamic research. Report TP-3675, NASA, March 1997.

[361] Knight M. and Hefner R.A. Static thrust of the lifting airscrew. Report TN-626, NACA, 1937.

[362] Knight M. and Hefner R.A. Analysis of ground effect on the lifting airscrew. Report TN-835, NACA, Dec. 1941.

[363] Philippe J.J., Roesch P., Dequin A.M., and Cler A. A survey of recent developments in helicopter aerodynamics. In *Aerodynamics of Advanced Rotor Systems*, AGARD LS-139, 1985.

[364] Scott M.T., Sigl D., and Strawn R.C. Computational and experimental evaluation of rotor tips for high speed flight. *J. Aircraft*, 28(6):403–409, June 1991.

[365] Vuillet A. Rotor and blade aerodynamic design. In *Aerodynamics of Rotorcraft*, AGARD R-781, chapter 3, Nov. 1990.

[366] McKillip R.M. Experimental studies in helicopter vertical climb performance. Report NASA-CR-203585, NASA, Feb. 1996.

[367] Felker F.F. and McKillip R.M. Comparisons of predicted and measured rotor performance in vertical climb and descent. In *50th American Hel. Soc. Meeting*, Washington, DC, May 1994.

[368] Curtiss H.C., Erdman W., and Sun M. Ground effect aerodynamics. *Vertica*, 11(1/2):29–42, Jan. 1987.

[369] Curtiss H.C., Sun M., Putnam W.F., and Hanker E.J. Rotor aerodynamics in ground effect at low advance ratios. *J. Am. Hel. Soc.*, 29(1):48–55, Jan. 1984.

[370] Whitehouse G.R. and Brown R.E. Helicopter rotor response to wake encounters in ground effect. Phoenix, AZ, May 2003.

[371] Heyson H.H. Theoretical study of the effect of ground proximity on the induced efficiency of helicopter rotors. Report TM-X-71951, NASA, 1977.

[372] Betz A. The ground effect on lifting propellers. Report TM 836, NACA, Aug. 1927.

[373] Cheeseman I.C. and Bennett W.E. The effect of the ground on a helicopter rotor. R&M 3021, ARC, 1955.

[374] Hayden J.S. The effect of the ground on helicopter hover power required. In *32nd AHS Annual Forum*, Washington, DC, May 1976.

[375] Fradenburgh E.A. The helicopter and the ground effect machine. *J. Am. Hel.*, 5(4):26–28, Oct. 1960.

[376] Makofski R.A. Charts for estimating the hovering endurance of a helicopter. Report TN-3810, NACA, Oct. 1956.

[377] Weis-Fogh T. Energetics of hovering flight in hummingbirds and in drosophila. *J. Exp. Biology*, 56(1):79–104, 1972.

[378] Greenwalt C.H. *Hummingbirds*. Doubleday & Co., New York, 1960.

[379] Altshuler D.L. and Dudley R. Kinematics of hovering hummingbird flight along simulated and natural elevational gradients. *J. Exp. Biology*, 206(18):3139–3147, 2003.

[380] Brotherhood P. and Stewart W. An experimental investigation of the flow through a helicopter rotor in forward flight. Report ARC R&M 2734, Aeronautical Research Council, 1949.

[381] Nitzer G.E. and Crandall S. A study of flow changes associated with airfoil section drag rise at supercritical speeds. Report TN-1813, NACA, Feb. 1949.

[382] Dadone L.E. US Army helicopter design datcom. Vol. 1: Airfoils. Technical report CR-153247 (USAAMRDL CR-76-2), NASA, May 1976.

[383] Bingham G.J. and Noonan K.W. Two-dimensional aerodynamic characteristics of three rotorcraft airfoils at Mach numbers from 0.35 to 0.90. Report TP-2000 (AVRADCOM-TR-82-B-2), NASA, 1982.

[384] Bousman W.G. Aerodynamic characteristics of SC1095 and SC1094 R8 airfoils. Report TP-2003-212265, NASA, Dec. 2003.

[385] Gessow A. and Crim A.D. A theoretical estimate of the effects of compressibility on the performance of a helicopter rotor in various flight conditions. Report TN-3798, NACA, Oct. 1958.

[386] Powell R.D. Compressibility effects on a hovering helicopter rotor having an NACA 0018 root airfoil tapering to an NACA 0012 tip airfoil. Report RM-L57F26, NACA, Sept. 1957.

[387] Powell R.D. and Carpenter P.J. Low tip Mach number stall characteristics and high tip Mach number compressibility effects on a helicopter rotor having an NACA 0009 tip airfoil section. Technical report TN-4355, NACA, 1958.

[388] Shivers J. and Carpenter P.J. Experimental investigation on the Langley helicopter test tower of compressibility effects on a rotor having NACA 63(sub 2)-015 airfoil sections. Technical report RM-L57F26, NACA, Dec. 1956.

[389] Jewel J.W. Compressibility effects on the hovering performance of a two-blade 10-foot-diameter helicopter rotor operating at tip Mach numbers up to 0.98. Technical report TN-D-245, NACA, 1960.

[390] Wilson F.T. and Ahmed S.R. Fuselage aerodynamic design issues and rotor/fuselage interactional aerodynamics. In *Aerodynamics of Rotorcraft*, AGARD R-781, 1990.

[391] Seddon J. *Basic Helicopter Aerodynamics*. BSP Professional Books, 1990.

[392] Hoerner S.F. *Fluid Dynamic Drag*. published by the author, 1965.

[393] Ahmed S.R. and Amtsberg J. An experimental study of the aerodynamic characteristics of three model fuselages. In 13th European Rotorcraft Forum, Paper No. 20, Arles, France, Sept. 1987.

[394] Filippone A. and Michelsen J.A. Aerodynamic drag prediction on helicopter fuselage. *J. Aircraft*, 38(2):326–333, March 2001.

[395] Balch D.T., Saccullo A., and Sheehy T.W. Experimental study of main rotor/tail rotor/airframe interactions in hover. Vol. 1. Report CR-166485, NASA, 1983.

[396] Berry J.D. and Bettschart N. Rotor-fuselage interaction: Analysis and validation with experiment. Report TM-112859, NASA, 1997.

[397] Gorton S.A., Berry J., Hodges W., and Reis D.G. Rotor wake study near the horizontal tail of a T-tail configuration. *J. Aircraft*, 39(4):645–653, 2002.

[398] RAeS. *Royal Aeronautical Society Conference on Yaw Control Concepts* London, March 1990.

[399] Amer K.B. and Gessow A. Charts for estimating tail-rotor contribution to helicopter directional stability and control in low-speed flight. Report R-1216, NACA, 1955.

[400] Lynn R.R., Robinson F.D., Batra N.N., and Duhon J.M. Tail rotor design. Part I – Aerodynamics. *J. Am. Hel. Soc.*, 15, Oct. 1970.

[401] Cook C.V. Tail rotor design and performance. *Vertica*, 2:163–181, 1978.

[402] Mouille E. and D'Ambra F. The fenestron, a shrouded tail rotor concept for helicopters. In *42nd AHS Annual Forum*, May 1986.

[403] Vuillet A. and Morelli F. New aerodynamic design of the fenestron for improved performance. In *12th European Rotorcraft Forum*, Sept. 1986.

[404] Yeo H., Bousman W.G., and Johnson W. Performance analysis of a utility helicopter with standard and advanced rotors. In *58th AHS Forum*, San Francisco, CA, Jan. 2002.

[405] Gustafson F.B. and Myers G.C. Stalling of helicopter blades. Report R-840, NACA, 1946.

[406] Zilliac G., Long K., and Nixon D. Modeling of H-46 tip strikes. In *AHS 55th Annual Forum*. Montreal, May 1999.

[407] Halliday A.S. and Cox D.H. Wind tunnel experiments on a model of a tandem rotor helicopter. Technical report CP-517, Aeronautical Research Council, 1961.

[408] Yates J.E. Flight measurements of the vibration experienced by a tandem helicopter in transition, vortex-ring state, landing, approach, and yawed flight. Report TN 4409, NACA, Sept. 1958.

[409] Heyson H.H. Preliminary results from flow field measurements around single and tandem rotors in the Langley full-scale wind tunnel. Technical report TN-3242, NACA, Nov. 1954.

[410] Payne P.R. *Helicopter Dynamics and Aerodynamics*. The MacMillan Co., New York, 1959.

[411] Harris F.D. Twin rotor hover performance. *J. Am. Hel. Soc.*, 44(1):34–37, Jan. 1999.

[412] Castles W. and De Leeuw J.H. The normal component of the induced velocity in the vicinity of a lifting rotor and some examples of its application. Report R-1184, NACA, 1954.

[413] Prouty R.W. *Military Helicopter Design Technology*. Krieger Publ. Co., 1988 (reprint).

[414] McCroskey J. Unsteady airfoils. *Ann. Rev. Fluid Mech.*, 15:285–311, 1982.

[415] McCroskey J., Carr L., and McAlister K. Dynamic stall experiments on oscillating airfoils. *J. Aircraft*, 14(1):57–63, Jan. 1975.

[416] Carr L.W. Progress in analysis and prediction of dynamic stall. *J. Aircraft*, 25(1):6–17, 1988.

[417] Landgrebe E. Wake geometry of a hovering rotor and its influence on rotor performance. *J. Am. Hel. Soc.*, 17(4):3–15, 1972.

[418] McVeigh M.A. and McHugh F.J. Influence of tip shape, chord, blade number and airfoil on advanced rotor performance. In *38th AHS Forum*, Anaheim, CA, May 1982.

[419] Slaymaker S.E. and Gray R.B. Power-off flare-up tests of a model helicopter rotor in vertical autorotation. Report TN 2870, NACA, Jan. 1953.

[420] Slaymaker S.E., Lynn R.R., and Gray R.B. Experimental investigation of transition of a model helicopter rotor from hovering to vertical autorotation. Report TN 2648, NACA, March 1952.

[421] Nikosky A.A. and Seckel E. An analytical study of the steady vertical descent in autorotation of single-rotor helicopter. Report TN 1906, NACA, June 1949.

[422] Nikosky A.A. and Seckel E. An analysis of the transition of a helicopter from hovering to steady autorotative vertical descent. Report TN 1907, NACA, 1949.

[423] Gessow A. and Myers M.G. Flight tests of a helicopter in autorotation, including a comparison with theory. Report TN-1267, NACA, April 1947.

[424] Houston S.S. Modeling and analysis of helicopter flight mechanics in autorotation. *J. Aircraft*, 40(4):675–682, 2003.

[425] Houston S.S. Rotor-wake modeling for simulation of helicopter flight mechanics in autorotation. *J. Aircraft*, 40(5):938–945, 2003.

[426] Houston S.S. Analysis of rotorcraft flight dynamics in autorotation. *J. Guid. Control Dyn.*, 25(1):33–39, Jan. 2002.

[427] Lugt H.J. Autorotation. *Ann. Rev. Fluid Mech*, 15:123–147, 1983.

[428] Washizu K., Azuma A., Koo J., and Oka J. Experiments on a model helicopter rotor operating in the vortex ring state. *J. Aircraft*, 3(3):225–230, March 1966.

[429] Gessow A. Review of information of induced flow of a lifting rotor. Report TN-3238, NACA, Aug. 1954.

[430] Pegg R.J. An investigation of the helicopter height-velocity diagram showing effects of density altitude and gross weight. Report TN D-4536, NASA, May 1968.

[431] Carlson E.B. and Zhao Y.J. Prediction of tilt-rotor height-velocity diagrams using optimal control theory. *J. Aircraft*, 40(5):896–905, Sept. 2003.

[432] Brotherhood P. Flow through a helicopter rotor in vertical descent. Report R&M 2735, Aeronautical Research Council, 1949.

[433] Castles W. and Gray R.B. Empirical relation between induced velocity, thrust, and rate of descent of a helicopter rotor as determined by wind-tunnel tests on four model rotors. Report TN-2474, NACA, Oct. 1951.

[434] Azuma A. and Obata A. Flow induced variation of the helicopter rotor operating in the vortex ring state. *J. Aircraft*, 5(4):381–386, 1968.

[435] Lee A. Optimal landing of helicopter in autorotation. Report CR-177082, NASA, July 1985.

[436] Lee A., Bryson A., and Hindson W. Optimal landing of a helicopter autorotation. *J. Guid., Control Dyn.*, 11(1):7–12, Feb. 1988.

[437] Leishman J.G., Baghwat M., and Anatham A. Free-vortex wake predictions of the vortex ring state for single-rotor and multi-rotor configurations. In *58th AHS Forum*, Montreal, June 2002.

[438] Leishman J.G., Baghwat M., and Anatham A. The vortex ring state as a spatially and temporally developing wake instability. *J. Am. Hel. Soc.*, 49(2):160–175, 1004.

[439] Brown R.E., Leishman J.G., Newman S.J., and Perry F.J. Blade twist effects on rotor behaviour in the vortex ring state. In *European Rotorcraft Forum*, Sept. 2002.

[440] Newman S., Brown R., Perry F., Lewis S., Orchard M., and Modha A. Predicting the onset of wake breakdown for rotors in descending flight. *J. Am. Hel. Soc.*, 48(1):28–39, 2003.

[441] Yoshinori O. and Kawachi K. Optimal take-off of a helicopter for category A V/STOL operations. *J. Aircraft*, 30(2):235–240, March 1993.

[442] Cerbe T. and Reichert G. Optimisation of helicopter take-off and landing. *J. Aircraft*, 26(10):925–931, Oct. 1989.

[443] Schmitz F.H. Optimal takeoff trajectories of a heavily loaded helicopter. *J. Aircraft*, 8(9):717–723, Sept. 1971.

[444] Schmitz F.H. and Rande Vause C. Near-optimal takeoff policy for heavily loaded helicopters exiting from confined areas. *J. Aircraft*, 13(5):343–348, May 1976.

[445] Zhao Y. and Chen R.T. Critical considerations for helicopters during runway takeoffs. *J. Aircraft*, 32(4):773–781, 1995.

[446] Zhao Y., Jhemi A., and Robert T. Optimal vertical takeoff and landing helicopter operation in one engine failure. *J. Aircraft*, 33(2):337–346, 1996.

[447] Whalley M.S. Development and evaluation of an inverse solution technique for studying helicopter maneuverability and agility. Report TM-102889, NASA, July 1991.

[448] Kelley H.L., Pegg R.J., and Champine R.A. Flying quality factors currently limiting helicopter nap-of-the-earth maneuverability as identified by flight investigation. Report TN D-4931, NASA, Dec. 1968.

[449] Stepniewski W.Z. Factors shaping conceptual design of rotary-wing aircraft. *1st Alexander Nikolsky Honorary Lecture, 37th AHS Forum*, May 1981.

[450] Hirschberg M. From the past to the future of heavy lift – Part One: Quad tilt concepts. *J. Am. Hel. Soc.*, 62–64(2), 2001.

[451] Meier W.H. and Olson J.R. Efficient sizing of a cargo rotorcraft. *J. Aircraft*, 25(6):538–543, 1988.

[452] Menon P., Prasad J., and Schrage D. Nonlinear control of a twin-lift helicopter configuration. *J. Guid., Control Dyn.*, 14(6):1287–1293, 1991.

[453] Slater G.L. and Erzberger H. Optimal short-range trajectories for helicopters. *J. Guid., Control Dyn.*, 7(4):393–400, 1984.

[454] McCormick B. A critical look at V/STOL technology. AIAA Paper 1978-0152, Aug. 1978.

[455] *Special Course on V/STOL Aerodynamics*, AGARD R-710, 1984.

[456] McCormick B. *Aerodynamics of Vertical/Short Take-off and Landing Flight*. Academic Press, 1967.

[457] Hirschberg. V/STOL: The first half-century. *Vertica*, 43(2):34–54, 1997.

[458] Maisel M.D., Giulianetti D.J., and Dugan D.C. The history of the XV-15 tilt rotor research aircraft from concept to flight. Report SP-2000-4517, NASA, 2000. Monographs in Aerospace History, No. 17.

[459] Donaldson C.P. and Snedeker R.S. Investigation of free jet impingement. Part 1: Free and impinging jets. *J. Fluid Mech.*, 45(2):281–320, Jan. 1971.

[460] Krothapalli A., Rajkuperan E., Alvi F., and Lourenco L. Flow field and noise characteristics of a supersonic impinging jet. *J. Fluid Mech.*, 392:155–181, Aug. 2000.

[461] Kotansky D.R. Jet flowfields. In *Special Course on V/STOL Aerodynamics*, AGARD R-710, April 1984.

[462] Bellavia D.C., Wardwell D.A., Corsiglia V.R., and Kuhn R.E. Suckdown, fountain lift, and pressures induced on several tandem jet V/STOL configurations. Report TM-102817, NASA, March 1991.

[463] Bellavia D.C., Corsiglia V.R., Kuhn R.E., and Wardwell D.A. On the estimation of jet-induced fountain lift and additional suckdown in hover for two-jet configurations. Report TM-102268, NASA, Aug. 1991.

[464] Saddington A.J. and Knowles K. A review of out-of-ground-effect propulsion-induced interference on STOVL aircraft. *Progr. Aero. Sci.*, 41(3–4):175–191, 2005.

[465] Skifstad J.G. Aerodynamics of jets pertinent to VTOL aircraft. *J. Aircraft*, 7(3):193–205, May 1970.

[466] Louisse J. and Marshall F.L. Prediction of ground effects for VTOL aircraft with twin lifting jets. *J. Aircraft*, 13(2):123–127, 1976.

[467] Walters M. and Henderson C. Development and validation of the V/STOL aerodynamics, stability and control manual. *J. Aircraft*, 20(5):450–455, 1983.

[468] Nichols J.H. and Englar R.J. Advanced circulation control wing system for Navy STOL aircraft. *J. Aircraft*, 18(12):1044–1050, Dec. 1981.

[469] Thompson J.D. YC-15 powerplant system design and development. *J. Aircraft*, 12(12):954–959, Dec. 1975.

[470] Lee P. and Lan C.E. Effect of thrust vectoring on level turn performance. *J. Aircraft*, 29:509–511, May 1992.

[471] Schneider G.R. and Watt G.W. Minimum time turns using vectored thrust. *J. Guid. Control Dyn.*, 12(6):777–782, Nov. 1989.

[472] Clarke J.W. and Walters M.M. CTOL ski jump: Analysis, simulation and flight test. *J. Aircraft*, 23(5):382–389, May 1986.

[473] Hafner R. Domain of the convertible rotor. *J. Aircraft*, 1(6):350–359, 1964.

[474] Roberts L. and Deckert W.L. Recent progress in V/STOL technology. Report TM-84238, NASA, Aug. 1982.

[475] Carlson E.B. and Zhao Y.J. Optimal short take-off of tilt-rotor aircraft in one engine failure. *J. Aircraft*, 39(2):280–289, 2003.

[476] Churchill G.B. and Dugan D.C. Simulation of the XV-15 tilt-rotor research aircraft. Report TM-84222, NASA, March 1982.

[477] Smith M.T.J. *Aircraft Noise*. Cambridge Univ. Press, 1989.

[478] Gebhardt G.T. Acoustical design features of Boeing Model 727. *J. Aircraft*, 2(4): 272–277, 1965.

[479] Crighton D.G. Model equations of nonlinear acoustics. *Ann. Rev. Fluid Mech.*, 11: 11–33, 1979.

[480] Dowling A. and Ffwocs-Williams J. *Sound and Sources of Sound*. Ellis Horwood, 1983.

[481] Goldstein H. *Classical Mechanics*. Addison-Wesley, 2nd edition, 1980.

[482] ESDU. Aircraft noise series. ESDU International, London, 2004. 7 Volumes with 32 Data Items.

[483] Shahadi P.A. Military aircraft noise. *J. Aircraft*, 12(8):653–657, 1975.

[484] Crighton D.G. Airframe Noise. In: Hubbard H., editor. *Aeroacoustics of Flight Vehicles: Theory and Practice*, NASA Ref. Publication 1258, WRDC TR-90-3052, 1991.

[485] Waitz I., Kukachko S., and Lee J. Military aviation and the environment: Historical trends and comparison to civil aviation. *J. Aircraft*, 42(2):329–339, 2005.

[486] Kroeger R.A., Gruska H.D., and Helvey T.C. Low speed aerodynamics for ultra quiet flight. Report TR-971-75, US Air Force Flight Directorate Laboratory, 1971.

[487] Lilley G.M. A study of the silent flight of the owl. AIAA 98-2340, 1998.

[488] Healy G.J. Measurement and analysis of aircraft far-field aerodynamic noise. Report CR-2377, NASA, 1974.

[489] Morgan H.G. and Hardin J.C. Aircraft noise – The next aircraft noise barrier. *J. Aircraft*, 12(7):622–624, July 1975.

[490] Ffowcs-Williams J.E. and Hall L.H. Aerodynamic sound generation by turbulent flow in the vicinity of a scattered half plane. *J. Fluid Mech.*, 40(657–670), 1970.

[491] Howe M.S. A review of the theory of trailing-edge noise. *J. Sound & Vibration*, 61(3):437–466, 1978.

[492] Lockard D.P. and Lilley G.M. The airframe noise reduction challenge. Report TM-213013, NASA, 2004.

[493] Fink M.R. Noise component method for airframe noise. *J. Aircraft*, 16(10):659–665, 1979.

[494] Fink M.R. and Schlinke R.H. Airframe noise component interaction studies. *J. Aircraft*, 17(2):99–105, 1980.

[495] Heller H.H. and Dobrzynski W.M. Sound radiation from aircraft wheel-well/landing gear configurations. *J. Aircraft*, 14(8):768–774, 1977.

[496] Kipersztok O. and Sengupta G. Flight test of the 747-JT9D for airframe noise. *J. Aircraft*, 19(12):1061–1069, 1982.

[497] Block P.J.W. Assessment of airframe noise. *J. Aircraft*, 16(12):834–841, Dec. 1979.

[498] Lighthill M.J. On sound generated aerodynamically — Part I. *Proc. Royal Soc. London, Sect. A*, 211:564–587, 1952.

[499] Lighthill M.J. On sound generated aerodynamically — Part II. *Proc. Royal Soc. London, Sect. A*, 222:1–32, 1954.

[500] Ribner H.S. Perspectives on jet noise. *AIAA J.*, 19(12):1513–1526, Dec. 1981.

[501] Tam C.K.W. Supersonic jet noise. *Ann. Rev. Fluid Mech.*, 27:17–43, 1995.

[502] Broadbent E.G. Noise shielding for aircraft. *Prog. Aero. Sci.*, 17:231–268, 1976.

[503] Jones D.S. The mathematical theory of noise shielding. *Prog. Aero. Sci.*, 17:149–229, 1976.

[504] FAA. FAR Part 36: Noise Standards: Aircraft type and airworthiness certification. (Title 14: Aeronautics and Space, National Archives of Regulations, US Government.)

[505] Zeldin S. and Speyer J.L. Maximum noise abatement trajectories. *J. Aircraft*, 11(2): 119–121, 1974.

[506] Melton R.G. and Jacobson I.D. Aircraft trajectories for reduced noise impact. *J. Aircraft*, 20(9):798–804, Sept. 1983.

[507] Monkman D., Rhodes D., and Deeley J. Noise exposure contours for heathrow airport 2004. Technical report ERDC-0501, ERDC, Directorate of Airspace Policy, Civil Aviation Authority, 2004.

[508] Schmitz F.H. and Stepniewski W.Z. Reduction of VTOL operational noise through flight trajectory management. *J. Aircraft*, 10(7):385–394, July 1973.

[509] Henschel H., Plaetschke E., and Schutze K. Minimum noise climb-out trajectories of VTOL aircraft. *J. Aircraft*, 11(7):429–433, 1974.

[510] Antoine N.E. and Kroo I.M. Aircraft optimization for minimal environmental impact. *J. Aircraft*, 41(4):790–797, July 2004.

[511] Morfey C.L. Rotating blades and aerodynamic sound. *J. Sound & Vibrations*, 28:587–617, 1973.

[512] George A.R. Helicopter noise: State-of-the-art. *J. Aircraft*, 15(11):707–715, Nov. 1978.

[513] Schmitz F.H. Rotor noise, in *Aeroacoustics of Flight Vehicles: Theory and Practice*. Vol. 1, chapter 2, NASA Reference Publication 1258, 1991.

[514] Simonich J., Amiet R., Schlinker R., and Greizer E. Helicopter rotor noise due to ingestion of atmospheric turbulence. Report CR-3973, NASA, May 1986.

[515] Brentner K., Lopes L., Chen H., and Horn J. Near real-time simulation of rotorcraft acoustics and flight dynamics. *J. Aircraft*, 42(2):347–358, March 2005.

[516] Brés G., Brentner K., Perez G., and Jones H. Maneuvering rotorcraft noise prediction. *J. Sound & Vibration*, 275(3–5):719–738, 2004.

[517] Sullivan B., Edwards B., Brentner K., and Booth E. A subjective test of modulated blade spacing for helicopter main rotors. *J. Am. Hel. Soc.*, 50(1):26–32, 2005.

[518] Edwards B. and Cox C. Revolutionary concepts for helicopter noise reduction – S.I.L.E.N.T. Program. Report CR-2002-211650, NASA, 2002.

[519] Edwards B. Psychoacoustic testing of modulated blade spacing for main rotors. Report CR-211651, NASA, May 2002.

[520] ICAO. Environmental Protection, Annex 16 to the Convention on International Civil Aviation. Vol. 1: Aircraft Noise. Appendix 1, paragraph 8.6, 2001.

[521] George A.R. and Kim Y.N. High-altitude long-range sonic boom propagation. *J. Aircraft*, 16(9):637–639, Sept. 1979.

[522] Perry E.A., Boness D.J., and Insley S.J. Effects of sonic booms on breeding gray seals and harbor seals on Sable Island, Canada. *J. Acoust. Soc. Am.*, 111(1):599–609, 2002.

[523] Whitehead A.H. Impact of environmental issues on the high-speed civil aircraft. In *Fluid Dynamic Research on Supersonic Aircraft*, NATO RTO-EN-4, pages 12–1, 12–25. 1998.

[524] Witham G.B. The behaviour of supersonic flows past a body of revolution far from the axis. *Proc. Royal Soc. London – Part A*, 201(1064):301–348, March 1950.

[525] Witham G.B. On the propagation of weak shock waves. *J. Fluid Mech.*, 5(3):290–318, Sept. 1956.

[526] Walkden F. The shock pattern of a wing-body combination far from the flight path. *Aero. Quarterly*, 9(2):164–194, May 1958.

[527] Hayes W.D. Brief review of the basic theory. In R Seebass, editor, *Sonic Boom Theory*, NASA SP-147. Jan. 1967.

[528] Jones L.B. Lower bounds for sonic bangs. *Aero. J.*, 65:433–437, 1962.

[529] Seebass R. and George A.R. Sonic boom minimization including both front and rear shock waves. *AIAA J.*, 10(10):2091–2093, 1969.

[530] Seebass R. and George A.R. Design and operation of aircraft to minimize their sonic boom. *J. Aircraft*, 11(9):509–517, 1974.

[531] Darden C.M. Sonic-boom minimization with nose-bluntness relaxation. Report TP-1348, NASA, Jan. 1979.

[532] Darden C.M. Sonic boom theory: Its status in prediction and minimization. *J. Aircraft*, 14(6):569–576, 1977.

[533] Hayes W.D. Sonic boom. *Ann. Rev. Fluid Mech.*, 3:269–290, 1971.

[534] Seebass R. Sonic boom theory. *J. Aircraft*, 6(3):177–184, 1969.

[535] Darden C.M., editor. *High-Speed Research: Sonic Boom*, number NASA CP-3172, Hampton, VA, Oct. 1992 (2 volumes).

[536] Talay T.A. Introduction to the aerodynamics of flight. Report SP 367, NASA, 1975.

[537] Jenkinson L.R. and Marchman J.F. *Aircraft Design Projects for Engineering Students*. AIAA and Butterworth-Heinemann, 2003.

[538] Fox M.C. and Forrest D.K. Supersonic aerodynamic characteristics of an advanced F-16 derivative aircraft configuration. Report TP-3355, NASA, July 1993.

[539] Roesch P. Aerodynamic design of the Aerospatiale S 365N 2 helicopter. In *6th European Rotorcraft Forum*, Sept. 1980.

[540] Thibert J.J. and Philippe J.J. Studies of airfoils and blade tips for helicopters. In *Prediction of Aerodynamic Loads on Rotorcraft*, AGARD CP-334, May 1982.

[541] FAA. Noise levels for US certified and foreign aircraft. Advisory Circular AC-36, Nov. 2001 (periodically updated).

Index

Printed and bound by CPI Group (UK) Ltd, Croydon, CR0 4YY

03/10/2024

01040335-0013